TEXTBOOK IN APPLIED MECHANICS

M.M. MALHOTRA
R. SUBRAMANIAN
P.S. GAHLOT
B.S. RATHORE

TECHNICAL TEACHERS' TRAINING INSTITUTE
CHANDIGARH

NEW AGE

NEW AGE INTERNATIONAL (P) LIMITED, PUBLISHERS

New Delhi • Bangalore • Chennai • Cochin • Guwahati • Hyderabad
Jalandhar • Kolkata • Lucknow • Mumbai • Ranchi

First Edition : 1994
Reprint : 2005

NEW AGE INTERNATIONAL (P) LIMITED, PUBLISHERS
4835/24, Ansari Road, Daryaganj,
New Delhi - 110 002
Visit us at : www.newagepublishers.com

Offices at :
Bangalore, Chennai, Cochin, Guwahati, Hyderabad, Jalandhar,
Kolkata, Lucknow, Mumbai and Ranchi

Rs. 200.00

ISBN : 81-224-0645-9

5 6 7 8 9 10

Published by New Age International (P) Ltd.,
4835/24, Ansari Road, Daryaganj, New Delhi-110 002 and
printed in India at Chaman Offset, New Delhi-110 002.

FOREWORD

Technical Teachers' Training Institute, Chandigarh is one of the four regional institutes established by the Ministry of Human Resource Development, Government of India, for the development and quality improvement of polytechnic education in the northern region. An important activity undertaken by the institute towards this end is the development of instructional materials, both print materials like textbooks, laboratory manuals etc. and non-print materials like transparencies, charts, video films etc. This book on Applied Mechanics written by the faculty members of this institute is the outcome of our efforts in this direction.

The objective of this book is to provide an elementary textbook on the subject which is suitable for the student for self-study. The authors have made special efforts to explain the concepts and principles in a lucid manner and to illustrate them with examples and sketches wherever necessary. I would like to congratulate the authors for their commendable job of bringing out such a book. It is hoped that the book will be useful to the student community as much as to the teachers and that it will contribute towards better learning by students.

S. R. GOWARIKER
Chairman
Board of Governors
T.T.T.I., Chandigarh

PREFACE

Knowing it well that a number of excellent books are available in the market on the subject 'Applied Mechanics', the authors have tried to add one more to this list. The major purpose of undertaking this venture has been to make available a text book in Applied Mechanics to the students entering the first year of Engineering Programmes, with a view to develop in them a sound understanding of the basic concepts and principles of the subject. This is considered important because Applied Mechanics is a core subject and includes all the basics which need to be learnt by an engineering student to facilitate learning of other applied engineering subjects in his programme of study as well as for solving practical problems in the field of work.

While writing the book, the authors have taken into consideration the entry level of the students and sequenced the content in a manner that facilitate easy grasp of the subject.

Each chapter of the book starts with the expected learning outcomes followed by inputs of the related concepts, principles, illustractive examples, review questions and problems for practice. The purpose of including the expected learning outcomes at the beginning of the chapter is to raise expectations in the learner of what he would achieve at the end of reading a chapter. Each basic concept and principle has been explained at length by the use of examples, non-examples and applications to real life problems drawn from daily experiences of the learner.

Major emphasis, while presenting the subject matter, had been laid on developing conceptualisation and understanding by students. This is evident from the way the solutions of illustrative examples have been presented. The solution includes both the diagramatic representation of the problem as well its freebody diagram. Such an approach, it is considered, will develop the necessary understanding of the principles of mechanics and their use by the students.

Exercises given at the end of each chapter include questions of different types. It has been especialy constructed to provide necessary practice and feedback to the students for achieving the intended learning outcomes.

Systeme Internationale (SI) units of measurement have been used throughout the book. Units of measurements following this system are given in Appendix I of the book.

Teachers of polytechnics have provided information on the general difficulty experienced by students in the learning of this subject. This has helped the authors to orient the writing of the book in a manner that would help overcome such difficulties.

The authors wish to express their sincere thanks to these teachers and to Mr. Radhakishna for going through the manuscript and giving valuable suggestions.

Miss Nandini Kapoor has helped in editing and checking the manuscript of the book for which we are thankful to her. The authors thank Mr. Shadi Ram, Mr. Ramesh Kumar & Mr. Yogender Kaushal for excellent assistance provided in typing the manuscript.

Thanks are also due to Mr. Gurdev Singh and Mr. B.K. Sharma for drawing the sketches. We have all praise for M/s Wiley Eastern Ltd. for the care and speed they have displayed in bringing out this volume.

AUTHORS

CONTENTS

1. INTRODUCTION

After having mastered the content of this chapter, you will be able to:

- Define the term 'Mechanics'
- Explain with examples the types of problems in Mechanics.
- State and explain the basic principles of Mechanics
- State the basic physical quantities used in mechanics and the corresponding units of measurement
- Explain the procedure for formulating and solving problems in mechanics.

CHAPTER 1

Introduction

During your school days, you have studied Mechanics as one of the topics in Physics. You will recall that Mechanics deals with problems related to the motion of bodies as also the conditions under which a body can assume a particular state of motion.

In Engineering, be it civil, mechanical, electrical or electronics, there are innumberable problems that are related to the determination of the state of motion of bodies under the action of a set of forces e.g.

- Is a dam stable or will it tilt due to the water pressure acting on it?
- What is the power of a motor that should be used for raising a particular quantity of water through a certain height in a certain amount of time?

For solving all such problems, an engineer needs to be well versed with the application of the principles of mechanics.

You will in this book therefore, study the various principles of Mechanics and understand them thoroughly by making use of them in solving a number of problems.

A thorough understanding of these principles will be of utmost importance to you in understanding the subjects such as Hydraulics, Strength of Materials, Theory of Machines and Structural Analysis and Design that you would study in your coming semesters.

In this chapter, you will understand the meaning of the term Mechanics, and other terms associated with it. You will also know the various principles of Mechanics that are used in solving problems related to the motion of bodies.

Before it is possible to understand the meaning of '*Mechanics*', the *problems* covered under it and the *principles* upon which this branch of science is based, it is necessary to understand certain basic concepts which are explained as under:

1.1 Basic Concepts

1.1.1 *Space*

Space is a region which extends in all directions and contains everything in it e.g. sun and its planets, stars, etc.

The position of a body in space is determined with respect to a reference system. For example, the position of an aircraft in space is determined with respect to the earth. Similarly the position of an object on the earth's surface is determined with reference to a fixed or immovable station on the earth by taking linear and angular measurements.

1.1.2 *Time*

Time is a measure of duration between successive events. The unit of time is a second which is a fraction $\left(\frac{1}{86,400}\right)$ of an average solar day. According to SI units second is taken to be a duration equal to 9 192 631 770 periods of vibrations of a calcium—133 atom.

1.1.3 *Motion*

You must have seen a number of bodies moving on the earth's surface. The earth in turn is rotating about its own axis and travelling around the sun. Again the sun and its planets are moving with respect to stars, which in turn are moving in space. Everything in this universe is, therefore, in motion. Heat, light, sound, electricity, and even life in all its manifestations is a form of motion. It is something which is not absolute but relative. A body is thus said to be in motion, when it changes its position with respect to other bodies. The relative change in position is called motion. Motion involves both space and time. For example, when a body moves from one position to another position, there has to be a time lapse between these two positions of the body. It cannot happen instantly. Also the body covers certain space when it moves from one position to another. Therefore, one can say that the motion of a body involves both time and space.

We often come across a term known as the *state of rest*. A body is said to be in a state of rest, if it does not undergo any change in position with respect to a reference system. It is thus a particular state of motion of a body when it does not change its position with respect to a reference system. For example, a building constructed on the earth's surface will be in a state of rest only with reference to the earth. Actually the building is not at rest, as it is also rotating alongwith the earth about the latter's axis as well as travelling around the sun. Therefore, there is no such thing as an absolute state of rest. Whenever we say that a body is at rest or immobile, it is always with respect to another body, usually the earth.

1.1.4 *Trajectory*

The motion of any body takes place along a definite path with respect to a reference system. The path followed by a body during its motion is called its trajectory, which may be a straight line or a curve. If the path is a straight line, the motion is said to be *rectlinear*. In case it is a curve, the motion is called *curvilinear*.

1.1.5 *Matter*

Any substance, which occupies space is called matter.

1.1.6 *Body*

Any matter that is bounded by a closed surface is called a body.

1.1.7 *Force*

Force is that which changes, or tends to change, the state of motion of a body. It is induced by the action of one body on another, or of a field (e.g. magnetic) on a body. The change in motion due to a force takes place in the direction of the force.

1.1.8 *Equilibrium*

A body acted upon by a system of forces is said to be in equilibrium, if it either remains in a state of rest or continues to move in a straight line with uniform velocity.

1.1.9 *Inertia*

The resistance offered by matter to any change of its state of motion is called inertia.

1.1.10 *Mass*

It is a quantitative measure of inertia or the resistance offered by any substance to the change of its position. For example, let us take two identical blocks (of the same size and shape), one made of wood and the other of steel. Push these blocks on identical surfaces by applying the same force (equal in magnitude and direction). We find that the wooden block moves much faster than the steel block. This shows that the steel block possesses greater inertia, and thus a greater mass than the wooden block.

1.1.11 *Particle*

A particle occupies no space i.e. it has no size, but has a definite mass concentrated at a point. In the solution of engineering problems, many a time, when the size of a body is very small as compared to the path it travels, the body is assumed as a particle.

1.1.12 *Rigid body*

A rigid body is that which does not undergo deformation on the application of forces. In other words, the relative position between its parts does not change. It is impossible to find a body which does not deform under the effect of forces. However, if the deformations are relatively small, the body is considered as rigid. In this book we shall be assuming bodies as rigid.

1.2 Mechanics

Mechanics is a branch of Physical science which deals with the motion of bodies under the action of forces. In other words, it determines the state of motion of bodies under the effect of forces, or conversely, it determines the forces imparting a certain state of motion to bodies. Mechanics, therefore, forms the basis for understanding the behaviour, as well as the design and analysis of all engineering structures and machines. All modern developments and research in the field of strength and stability of soils, structures and machines, fluid flow, vibrations, molecular, atomic and sub-atomic behaviour are dependent upon the principles of mechanics. Study of mechanics is, therefore, of vital importance for all those engaged in engineering profession.

1.3 Branches of Mechanics

The types of problems dealt within mechanics are as follows:

1 Determining the trajectory of a body and its displacement, velocity and acceleration at any instant of time without regard to the forces acting on it.

2 Determining the nature of motion of a body under the action of the forces applied, and

3 Determining the conditions under which a body is in equilibrium when acted upon by a system of forces.

Depending upon the problems that it deals with mechanics is logically divided into three parts

1 Kinematics] Dynamics
2 Kinetics]
3 Statics

1.3.1 *Kinematics*

Kinematics deals with the problems related to the determination of the nature of motion of bodies i.e. their displacement, velocity and acceleration without any regard to the forces acting on them, or the masses involved.

1.3.2 *Kinetics*

Kinetics deals with the problems which require the determination of the effect of forces on the motion of a body, or conversely the forces causing a certain motion.

Another term, which is most commonly used is *dynamics*. This includes both the problems of kinematics and kinetics. Dynamics therefore, deals with the motion of bodies under the effect of forces.

1.3.3 *Statics*

Statics deals with the problems concerning the equilibrium of bodies under the effect of forces.

Mechanics is also usually divided into Mechanics of solid rigid bodies, Mechanics of solid deformable bodies (usually known as strength of materials) and Mechanics of fluids, depending upon whether the above problems in mechanics relate to solid rigid bodies, solid deformable bodies or fluids. In this book, we shall be confined only to the study of mechanics of solid rigid bodies.

1.4 Quantities

A physical quantity is one which can be measured. All physical quantities can be classified as Scalar and Vector quantities.

1.4.1 *Scalar Quantity*

Scalar quantities are those which can be completely defined by their magnitude or numerical value only. Speed of a vehicle, say 40 km/hour; the mass of a body, say 30 kg; the temperature of a body, say 50°C are all scalar quantities.

1.4.2 *Vector Quantities*

Vectors are the quantities which require not only magnitude, but also direction (line of action) and sense for completely defining them. For example, if we say that displacement of a body is 10 m, and do not state the direction and the sense in which the displacement has occurred, it does not define the displacement of the body completely. On the other hand, if we say that the displacement of a body is 10 m in a direction N 45° E, (from the north direction 45° towards east) the displacement is completely defined. Similarly, force, velocity, acceleration etc. are vector quantities or vectors. A vector can be represented by a straight line. The length of the line indicates to a certain chosen scale, the magnitude of the vector quantity, and an arrow marked on the line shows the sense of the vector as shown in Fig. 1.1.

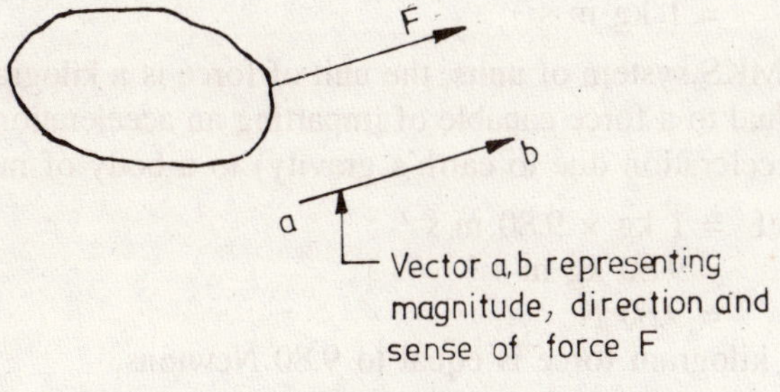

Fig. 1.1 Representation of a Vector quantity

1.4.3 *Units of Measurement*

All the quantities mentioned above have to be measured in certain units for the sake of easy communication and comparison of results. Measurement can be best defined as a comparison of the magnitude of a quantity to be measured with the magnitude of a similar quantity which is considered as a unit. A unit is therefore, a value, quantity or magnitude in terms of which other values or quantities are expressed. A unit is a fixed quantity.

There are a number of systems of units, but we will be discussing here only the following two systems of units which are used in this book.

SI Units (Systems International d' Units)

MKS Units (Metre, Kilogram, Second System of Units)

In both the systems, there are a number of basic quantities i.e. the quantities whose units are independently defined and magnitudes standardised. The basic quantities that have an importance in mechanics are—length, mass and time. The units of all other quantities are derived from those of the basic quantities. Force is one of the most important derived quantities and will be discussed here along with the above three basic quantities.

In both the MKS and SI units, the units for length, mass and time are the same and are given in Table 1.1.

Table 1.1 *Units of measurement for basic quantities in SI and MKS Units*

Physical Quantity	*S.I. Units (Units)*	*M.K.S. Units (Units)*	*Symbol*
Length	Metre	Metre	m
Mass	Kilogram	Kilogram	kg
Time	Second	Second	s

The units of force are, however, different in the MKS and SI units.

In the SI Units, the unit of force is a Newton (N) which is equal to a force that can impart an acceleration of $1 ms^{-2}$ to a body of mass 1 kg.

From Newton's second law, we know that

$$\text{Force (F)} = \text{mass} \times \text{acceleration}$$
$$= \text{kg} \times \text{ms}^{-2}$$
$$\therefore 1\text{ N} = 1\text{ kg m s}^{-2}$$

In the MKS system of units, the unit of force is a kilogram force (kgf), which is equal to a force capable of imparting an acceleration of 9.80 m s^{-2} (average acceleration due to earth's gravity) to a body of mass 1 kg.

$$\therefore 1\text{ kgf} = 1\text{ kg} \times 9.80\text{ m s}^{-2}$$
$$= 9.80\text{ kg m s}^{-2}$$
$$= 9.80\text{ N}$$

$\therefore$ one kilogram force is equal to 9.80 Newtons.

In this book, S.I. Units are used primarily, though a few problems in each topic are solved by using the MKS system of units.

1.5 Dimensional Homogeneity

It is extremely important to ensure that both the sides in any mathematical relation have the same dimensions. This is called dimensional homogeneity.

Dimensional homogeneity is a necessary condition for any mathematical relation to be correct, but it is not a sufficient condition. This is so because the dimensionless coefficients also effect the relation. The three basic quantities used in mechanics are length, mass and time, and are denoted by L, M and T respectively. Using these symbols, the dimensional representation of displacement, velocity and acceleration are as under:

Quantity	Dimensional representation
Displacement	L
Velocity	$\dfrac{\text{Displacement}}{\text{Time}}$ $\dfrac{L}{T} = LT^{-1}$
Acceleration	$\dfrac{\text{Velocity}}{\text{Time}}$ $\dfrac{LT^{-1}}{T} = LT^{-2}$

The following example illustrates the procedure of checking dimensional homogeneity of relations.

Let us analyse dimensionally the equation,

"Force = rate of change of momentum"

$$\text{Force} = \text{mass} \times \text{acceleration}$$
$$= M \times LT^{-2}$$
$$= MLT^{-2} \qquad \text{... (i)}$$

Rate of Change of Momentum

$$= \frac{\text{mass} \times \text{velocity}}{\text{Time}}$$
$$= \frac{M \times (LT^{-1})}{T}$$
$$= MLT^{-2} \qquad \text{... (ii)}$$

Comparing (i) and (ii), we find that the dimensions of force are the same as that of the rate of change of momentum. Thus it can be concluded that the relationship

"Force = rate of change of momentum"

is dimensionally homogeneous.

Consider another example of work-energy principle,

"Work done = Change in the kinetic energy"

Force × Displacement

$$= \frac{1}{2} \text{ mass} \times (\text{velocity})^2$$

Expressing dimensionally, we have

Force × Displacement

$$= F.L$$

$$= MLT^{-2} \,.\, L$$

$$= ML^2T^{-2} \qquad \text{... (i)}$$

Change in Kinetic energy

$$= \frac{1}{2} \text{ mass} \times (\text{velocity})^2$$

$$= \frac{1}{2} (M)\,(LT^{-1})^2$$

$$= \frac{1}{2} ML^2T^{-2} \qquad \text{... (ii)}$$

Thus the dimensions of (i) and (ii) are the same, and hence the equation is dimensionally homogeneous. The dimensionless coefficient $\frac{1}{2}$ in (ii) represents certain constant of integration and does not affect the dimensional homogeneity.

1.6 Reference Frame of Axes

We know that space is a region extending in all directions. Position of a body in space can be defined only by referring measurements to a fixed frame of axes. The measurements may be linear, angular or both. We also know that there is nothing stationary in the universe, and therefore, it is not possible to have a fixed frame of reference.

Most of the problems in engineering deal with motion of bodies on or near the earth's surface. Therefore, for such problems the reference frame of axes is taken on the earth's surface which is assumed fixed for all practical purposes.

For determining the motion of bodies projected in space like rockets and satellites, certain stars are taken as the reference system. With such a reference frame the movement of the earth can also be taken into account. Such problems will not be dealt with in this book.

Let us now consider methods to define the position of a particle with respect to a reference frame. To define the position of a particle in space, three measurements are necessary to be made; e.g. the position 'P' of a particle with respect to the reference frame XYZ is defined by three co-ordinates x, y, z as shown in Fig. 1.2

In case the motion of a particle is taking place in a plane, its position can be defined by only two measurements. This can be done in the following two ways.

1.6.1 *Rectangular Coordinate System*

In this system the position of any particle 'P' is determined by the rectangular coordinates (x, y), where x is the linear distance of the particle 'P' from the

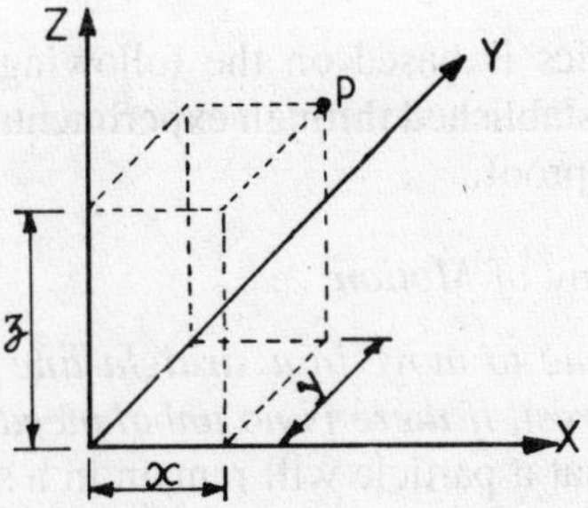

Fig. 1.2 Defining Position of a point 'P' in space.

origin O measured parallel to the X-axis, and *y* is the linear distance of the particle P from the origin O measured parallel to the Y-axis. This is shown in Fig. 1.3.

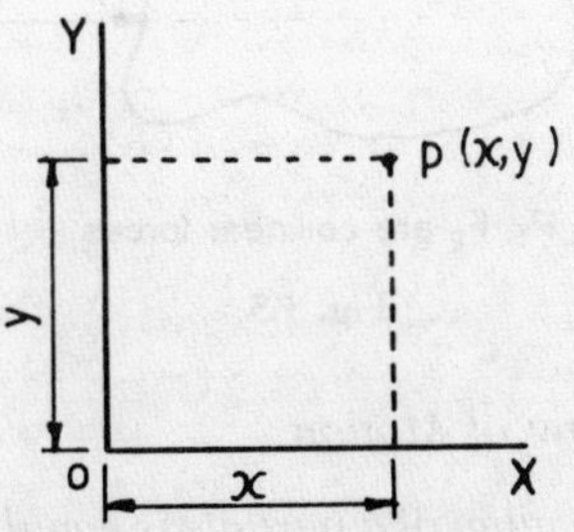

Fig. 1.3 Defining the position of a Point 'P' in a plane through rectangular co-ordinates.

1.6.2 *Polar Coordinates system*

In *Polar Coordinate system* the position of a particle 'P' is defined (r, θ), where r is the radial distance of the particle 'P' from the origin O, and θ is the angle made by the radial line OP with the base line OX.

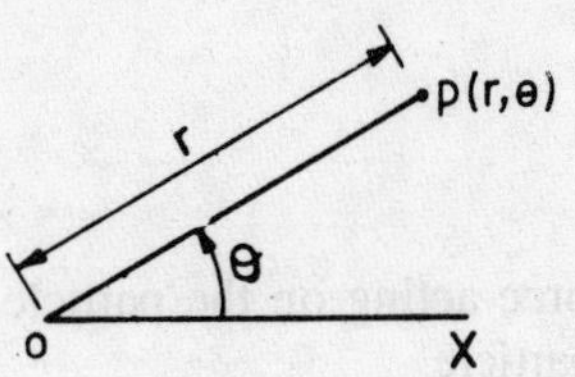

Fig. 1.4 Defining the position of a point 'P' in a plane through Polar Coordinate System.

1.7 Principles of Mechanics

The science of mechanics is based on the following principles which are empirical (i.e. they are established through experimentation and observation), and have no analytical proof.

1.7.1 *Newton's First Law of Motion*

Any particle will continue to move in a straight line with uniform velocity or remain in a state of rest, if there is no unbalanced force acting on it. In other words, it means that a particle will remain in a state of equilibrium, if there is no unbalanced force acting on it. For example, let there be two collinear forces F_1 and F_2 acting on a particle, as shown in Fig. 1.5. If F_1 is equal and opposite to F_2, the particle will be in equilibrium and the unbalanced force will be zero. In case they are not equal, the difference between the two will be the unbalanced force, and the particle will not remain in equilibrium and will move in the direction of the greater force.

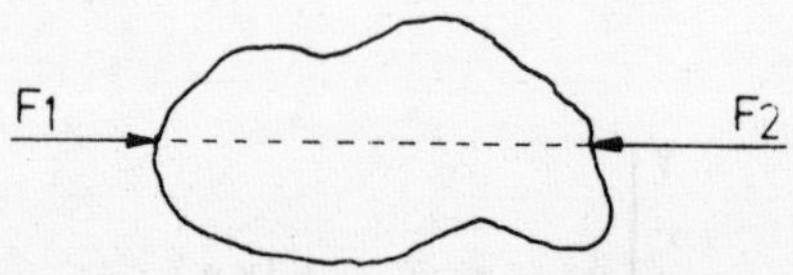

F_1, F_2 are collinear forces

Fig. 1.5

1.7.2 *Newton's Second Law of Motion*

The rate of change of momentum of a particle is equal to the unbalanced force acting on it, and takes place in the direction of the unbalanced force.

The momentum of a particle is the product of its mass and velocity. With the mass remaining constant, the rate of change of momentum will be the product of mass and the rate of change of velocity, which in turn will mean the product of mass and acceleration of the particle.

$$F = \frac{d}{dt}(mv)$$

$$= m\frac{dv}{dt}$$

$$= m.\ a$$

$$\therefore F \propto a$$

where,

F = unbalanced force acting on the particle
m = mass of the particle
v = velocity of the particle
a = acceleration of the particle.

Thus, the acceleration of a particle (with constant mass) is proportional to the unbalanced force acting on it.

1.7.3 *Newton's Third Law of Motion*

To every action, there is an equal and opposite reaction. For example, the action of a body *A* on body *B* (Fig. 1.6) is followed by an equal and opposite reaction of body *B* on body *A*. This means that *action and reaction between any two bodies in contact are equal, opposite in direction and collinear.*

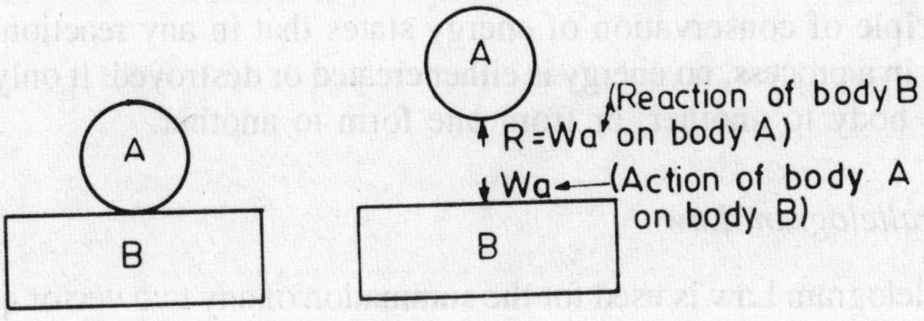

Fig. 1.6

Taking another example of an inkpot resting on a table, the vertical downward force due to the weight of the inkpot (action) is accompanied by an equal and opposite force (reaction) exerted by the table on the inkpot (Fig. 1.7).

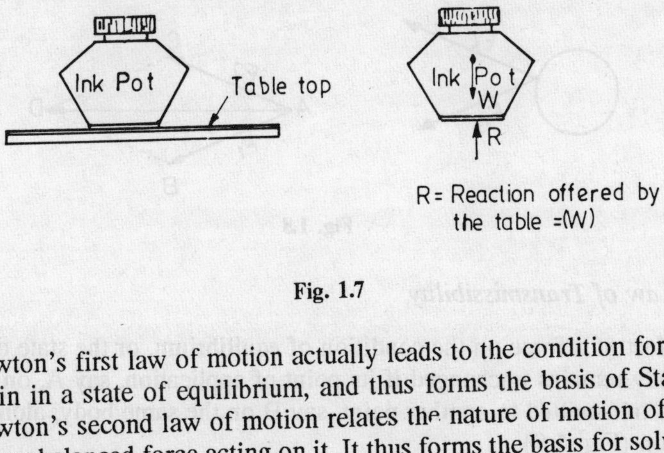

Fig. 1.7

Newton's first law of motion actually leads to the condition for a body to remain in a state of equilibrium, and thus forms the basis of Statics.

Newton's second law of motion relates the nature of motion of a body with the unbalanced force acting on it. It thus forms the basis for solving the problems in Dynamics. Actually, Newton's first law of motion is a particular case of Newton's second law of motion. This is so, because when the unbalanced force acting on a particle is zero, the particle will not accelerate, and will either continue to move with a uniform velocity in a straight line, or remain in a state of rest.

Newton's third law of motion helps in understanding the nature of force. According to this law, a force is **never singular**; it always occurs in pairs of equal and opposite forces.

1.7.4 *Principle of Work and Energy*

Work done by a force, or a set of forces, on a particle in moving it through a distance is equal to the change in the kinetic energy of the particle.

1.7.5 *Principle of Conservation of Energy*

The principle of conservation of energy states that in any reaction within a system or in a process, no energy is either created or destroyed. It only changes from one body to another, or from one form to another.

1.7.6 *Parallelogram Law*

The Parallelogram Law is used for the summation of any two vector quantities. For example, if there are two forces F_1 and F_2 acting at a point O on a body, then the effect of those forces on the body is equal to that of a single force acting at that point. This single force 'R', known as the *resultant* of the forces F_1 and F_2, is given by the diagonal AD of the parallelogram ABDC (Fig. 1.8) which is constructed with its sides AB and AC equal in magnitude (to a certain defined scale) and direction to the forces F_1 and F_2 respectively. This is known as the *parallelogram law of forces.*

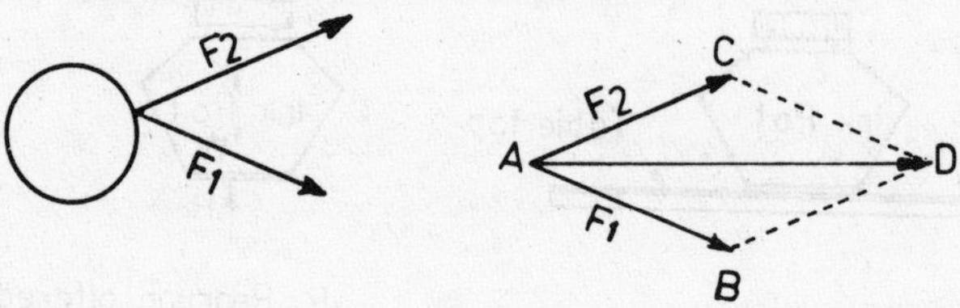

Fig. 1.8

1.7.7 *Law of Transmissibility*

The effect of a force on the condition of equilibrium, or the state of motion of a body remains unchanged if its point of application, say A, on the rigid body is transmitted to another point, say B on the same body, along its line of action (Fig. 1.9).

1.7.8 *Law of Superposition*

According to the law of superposition of forces, the effect of additional forces is simply additive. In other words, the motion imparted to a body by

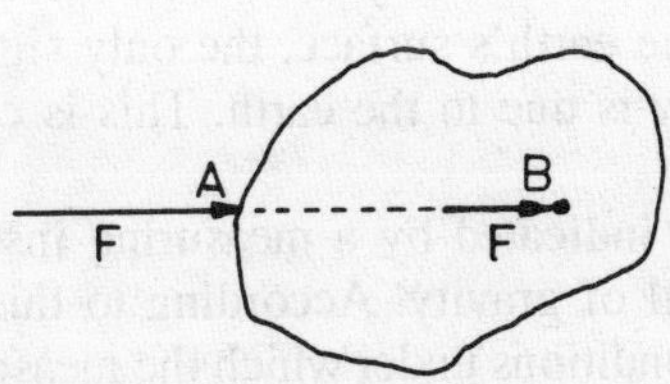

Fig. 1.9

a set of forces F_1, F_2, F_3, ..., F_n will be equal to the sum of the motions imparted by each of the forces taken at a time.

As a corollary to this, the effect of a system of forces on a rigid body is not altered or changed if another system of forces which is in equilibrium is added to, or subtracted from, this system of forces.

1.7.9 *Newton's Law of Gravitational Attraction*

According to this law, any two particles have a mutual force of attraction between them, called the force of gravitation. This mutual force of attraction 'F' is given by the expression

$$F = G\frac{m_1 m_2}{r^2}$$

where,

G = constant of gravitation (6.67×10^{-11} $m^3/(kg\ s^2)$)

m_1, m_2 = Masses of the particles 1 and 2 in kgs

r = Centre to centre distance of the particles in metres

If we take any two bodies on earth's surface, and try to work out the force F between them by using the above expression, we will find that it is very very small. For example, if there are two bodies of mass 50 kg each, and their centre to centre distance is 1.0 m, the gravitational force between them will be

$$F = \frac{6.67 \times 10^{-11} \times 50 \times 50}{1 \times 1}$$

$$= .000000166 \text{ N}$$

The only significant force of gravitation on earth's surface is between the body and the earth itself, called the weight of the body.

1.7.10 *Weight*

The weight of a body in space is defined as the resultant force of gravity acting on it due to all the other bodies in the space. Accordingly there are points in space where the weight of the body is zero, because at these points, the resultant force of gravity due to all the other bodies in space is zero. However, near the earth's surface, the only significant force of gravitation acting on a body is due to the earth. This is called the weight of the body.

Weight of a body indicated by a measuring instrument is the reaction of that body to the pull of gravity. According to this the weight of a body will depend upon the conditions under which the measurement has been taken. For example, the weight of a body recorded by a measuring instrument will be zero, when the body *alongwith the instrument* is falling freely, because the acceleration of the body is the same as that of the measuring instrument. Similarly an astronaut in a rocket which takes off from the earth at an acceleration of 10 g will weigh 11 times his weight as per the measuring instrument.

1.8 Description of Problems by Mathematical Models

A problem exists in the form of a physical phenomenon. Let us say that it is required to determine the motion (i.e. displacement, velocity, acceleration at any instant) of a steel ball under the effect of a force. The elements of the problem are: the ball, the force acting on the ball and the surface on which it is moving. All these constitute a physical phenomenon. The ball is made of a number of molecules separated by small distances, and is thus not a continuum. Similarly, the ball deforms under the effect of the force acting on it. The force acting on the ball may not exactly pass through its centre of gravity. The surface on which the ball is moving may be rough, and the roughness may be changing along its length. Air friction may resist the motion of the ball. There are, therefore, a number of factors that are involved in the solution of this problem. It is almost impossible to take all these factors into account, because by doing so, the problem may become very complicated and impossible to solve analytically. Hence in the solution of any physical problem, certain, simplifying assumptions are made and a physical phenomenon is represented by a physical model. In the above example, the ball may be assumed to be a rigid body with the force acting through its C.G. The contact surfaces may be assumed to be smooth. The physical model is shown in Fig. 1.10. To make the problem more simplified, the ball may be assumed as a particle.

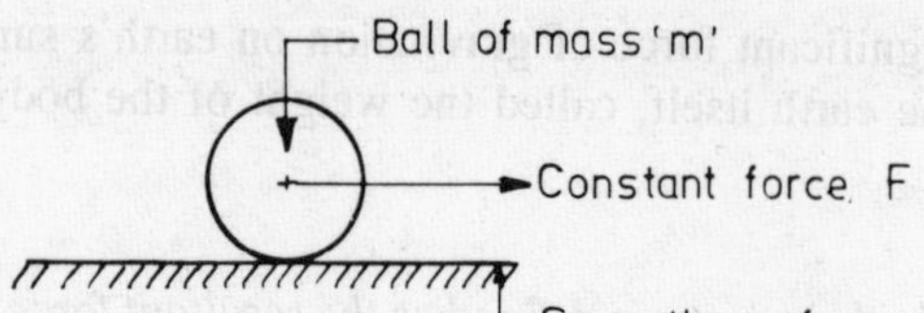

Fig. 1.10

Further, the physical model is converted into mathematical model which is a set of equations describing the relationship between the known and unknown quantities. For example, in the present case, the set of equations describing the mathematical model is as under:

$$F = ma$$

where,

F = the unbalanced force acting on the ball of mass m

a = acceleration of the ball.

By substituting the known quantities in these equations, it is possible to find the unknown quantities. In the solution of most of engineering problems, it becomes necessary to use the above method.

1.9 Procedure of Solving a Problem

In solving a problem in mechanics, it is always better to organise and present the work in a neat and systematic manner. This will not only help the students in eliminating the usual personal errors, but will also facilitate easy checking of the solution. Following is the suggested procedure for solving a problem:

1 Understand the problem thoroughly and represent it by drawing a neat freehand sketch. Indicate on the sketch the relevant data given in the problem.
2 List all the information or the data provided in the question.
3 Write the information desired, or the results asked for in the problem. This is very important in order to plan the approach for a solution.
4 Draw a free body diagram of a body or its parts whenever required. The concept of free body diagram is explained in Chapter 3 of this book.
5 Think of the principles, or equations, relating to the given quantities and the results desired. Write them clearly, before substituting in them the given quantities. At the time of substituting the quantities, ensure that the units are consistent.
6 It must be ensured that in the solution of any mathematical relation, the units of any quantity, say length, mass etc. are kept the same. For example in the relation, $F = \frac{mv}{t}$, if the units of force (F) is in Newtons, then the unit of mass (m) must be in kg, that of velocity (v) in ms^{-1}, and time (t) in seconds.
7 The limit of accuracy in finding the results should be consistent with that of the given data. If the data is given upto two places of decimal, the result may also be worked out upto this limit. The accuracy obtained from logarithmic tables or slide rule is usually acceptable.
8 Never forget to write the units for each quantity worked out. Also be sure to check the units written for each quantity.

Review Questions

1.1 'State of rest is not something absolute, but is relative'. Explain.
1.2 Differentiate between mass and weight.
1.3 List the types of problems in mechanics.
1.4 When is the motion of a particle called rectilinear?
1.5 Newton's first law of motion is included in Newton's second law of motion. Explain.
1.6 A body is moving in a straight line with uniform velocity. State whether it is in equilibrium. If so, explain why.
1.7 State Newton's law which explains that a force always occurs in pairs.
1.8 State the principle of transmissibility.
1.9 What are the units of force in SI and MKS systems of units? How many Newtons are equal to 1 kilogramme force?
1.10 Classify the following under scalar and vector quantities: Displacement, Speed, Area, Force, Volume, Weight, Velocity, Time.
1.11 A spring balance records the weight of a body as 2 N. What weight of this body will the spring balance record if both are allowed to fall freely? Explain the results.
1.12 A man is holding a bucket of water in his hands. Show by drawing a diagram the action and reaction of the pull applied by the man to hold the bucket.
1.13 Show that the following equation has dimensional homogeinity.

$$V_2^2 - V_1^2 = 2\,a\,S$$

where,

V_2, V_1 = Velocities of a particle at positions 2 and 1.
a = acceleration of the particle
S = distance travelled by the particle.

2. FORCES

After having mastered the content of this chapter, you will be able to:

- Explain the characteristics of a force
- Classify and explain force systems with illustrative sketches
- Explain the terms 'Resultant' & 'Equilibrant'
- State the parallelogram law of forces
- Apply the parallelogram law to determine the resulotant of two (or more) forces meeting at a point.
- Derive the cosine law from the parallelogram of forces.
- Explain the concept of component of a force
- Resolve a force into components along given directions
- Explain the term force polygon
- Draw the force polygon to determine the resultant of a concurrent force system
- Determine the resultant of a concurrent force system by resolution & composition of forces
- Explain the term moment of a force
- State and prove the 'Varignon's Theorem of Moments'
- State the general principle of moments
- Determine the resultant of non-concurrent force systems by resolution & composition and using the principle of moments.
- Explain the concept of funicular polygon
- Determine graphically the resultant of non-concurrent force systems using force polygon and funicular polygon
- Compare and contrast graphical and analytical methods for the determination of resultant of coplanar force systems.

CHAPTER 2

Forces

2.1 Introduction

You already know that Applied Mechanics is the science that deals with force, and the effect of force on bodies. Or, in short, we may say that Applied Mechanics deals with the motion of bodies, state of rest being considered as a special case of motion.

In order to understand the principles of mechanics, it is, therefore, essential that force should be studied in detail and its effects understood fully. As an example, if a body is acted upon by two forces, say 10 N and 20 N, the net force acting on the body need not be 20 + 10 = 30 N or 20 − 10 = 10 N. This is so because force is a vector quantity and cannot be added or subtracted using the same principles as for scalar quantities. This chapter deals with the basic principles involved in the study of force vectors as applied to the problems in mechanics. Also we are dealing with only the external effects of forces on bodies, i.e., we are treating the bodies as rigid.

We start the discussion with the distinguishing features of force as a vector quantity.

2.2 Concept of Force and Force Systems

2.2.1 *Force*

Force, as has been defined earlier, is that which changes or tends to change the state of rest or of uniform motion of a body. This is the fundamental effect of a force.

Force is a vector quantity and, therefore, it requires both magnitude and direction to be specified.

Force normally is exerted through physical contact between bodies. Some forces, like magnetic and gravitational forces, do exist without physical contact between bodies. A magnet attracts iron filings which are some distance away and the earth applies a gravitational force on all bodies near the surface of the earth without physical contact.

Any physical quantity is measured to determine its magnitude by using certain units. The unit of force is the Newton (N) as per the S.I. (System International) and Kilogram force (Kgf) in M.K.S. System

Table 2.1 gives the units and sub-units of force as per both S.I. and MKS systems.

Table 2.1 *Units of Force*

SI System	*MKS System*
Unit—Newton (N)	Unit-Kilogram force (kgf).
1000 N = 1 kilo Newton (KN)	1000 gf = 1 Kilogram force (kgf)
1000 KN = 1 Mega Newton (MN)	1000 kgf = 1 tonne (tf)
1 N = 0.012 kgf.	1 kgf = 9.81 N

2.2.2 *Graphical Representation of Force*

In the solution of problems, forces are shown diagramatically for a better visualisation of the physical phenomenon. In such cases only the point of application, the line of action and the sense of the force are shown. This diagram is called the space diagram. It does not represent the magnitude of the force to any scale. In order to solve the problem graphically, forces are represented by vectors.

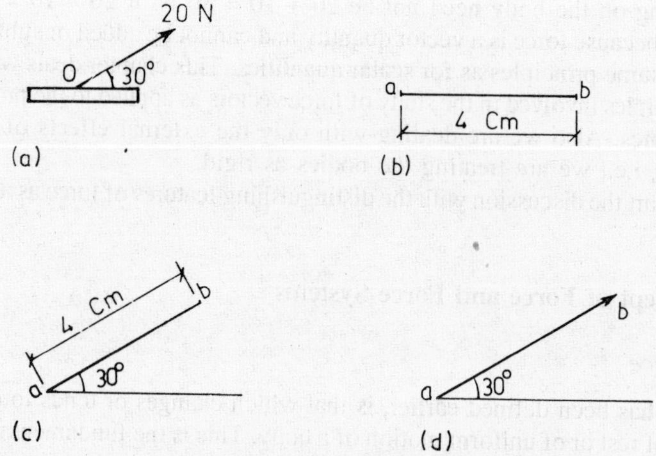

Fig. 2.1 Graphical Representation of force vector

The representation of a force vector is shown in figure 2.1. A force vector has 4 characteristics, namely, (i) magnitude of the force vector, (ii) direction or line of action, (iii) sense and (iv) the point of application. All these four characteristics have to be represented. This is done by the method given below.

Suppose we have to represent graphically a force of 20 N acting on a plate as shown in Fig. 2.1 (a). First we choose a scale to represent the magnitude, say 1 cm = 5 N. A line *ab* 4 cm long (Fig. 2.1 b) will represent the magnitude of the force. This line is drawn inclined at 30° to the horizontal (Fig. 2.1 c) and shows the direction or line of action of the force. An arrow is placed at the end of the vector to show the sense of the vector.

2.2.3 *Characteristics of a Force*

It was mentioned earlier that force is a vector quantity and always has a magnitude and a direction associated with it. In order to specify a force completely, not only magnitude, direction and sense are important but the point of application may also become important.

To illustrate this we take the case of a wheel on which a force of 100 N is acting, differently in each case, as shown in Fig. 2.2 (a), (b), (c) and (d). In cases (a) and (b), the force has the same magnitude, direction and sense. But in case (a) the wheel will rotate anti-clockwise, while in case (b) the wheel will rotate clockwise. In cases (b) and (d), the magnitude, direction and point of application are the same and only the sense is different. In (b) the rotation will be clockwise, and in (d) anti-clockwise. Same is the case with (a) and (c).

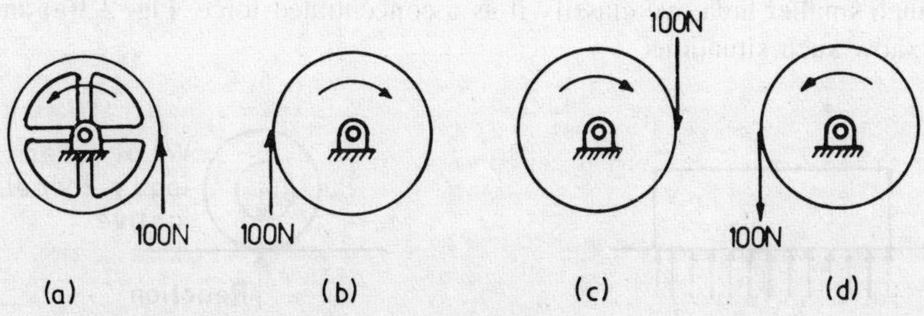

Fig. 2.2. Characteristics of force.

2.2.4 *Effect of Force Acting on a Rigid Body*

A force acting on a rigid body produces one or both of the following two effects: (i) Linear displacement, and (ii) Turning or rotating effect. Both the above effects essentially result in a change in the state of rest or motion of a body.

2.2.5 *Types of Forces*

Force can be classified in different types, such as active and reactive, or body and surface forces.

Active and Reactive Forces

It has already been mentioned that force is exerted through physical contact between bodies. Force therefore, is the action of one body on another. The bodies, in fact, exert force on each other, thus giving a pair of forces. Forces, therefore, always *exist in pairs and an isolated force does not exist.*

The force exerted by one body on another is called an active force (action), and the equal and opposite force exerted by the second body on the first is called the reactive force (reaction). Action and reaction are always equal and opposite. In any situation, it is very necessary and important to distinguish between active and reactive forces.

Body and Surface Forces

Another way to classify forces is as body and surface forces. The force due to gravity, i.e., the force due to the attraction of the earth, acts on each and every particle of a body and is known as the body force. It is to be remembered that the body force is exerted by an external agency, in this case the earth.

Surface forces are those which act on the surface of bodies. Kicking a football is an example of a surface force applied on the football.

Surface forces can further be classified as distributed or concentrated forces. When two bodies are in contact, a force is always acting on a certain area. When a force is acting over a large area on the surface of a body, we may classify it as a distributed force. When a relatively large force is acting on a much smaller area, we classify it as a concentrated force. Fig. 2.3(a) and (b) show such situations.

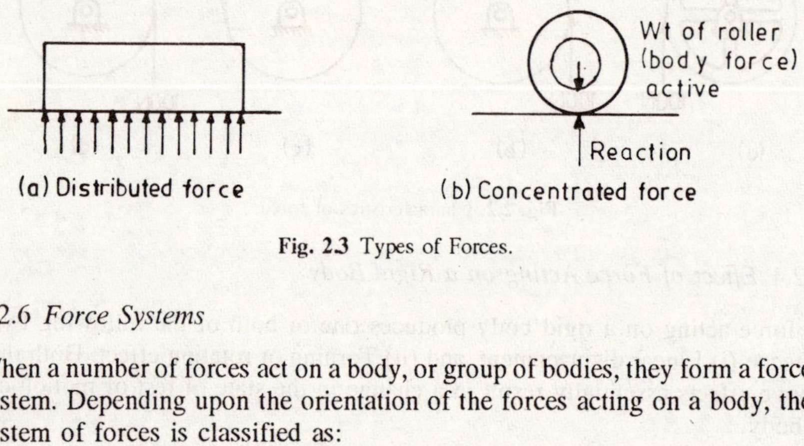

Fig. 2.3 Types of Forces.

2.2.6 *Force Systems*

When a number of forces act on a body, or group of bodies, they form a force system. Depending upon the orientation of the forces acting on a body, the system of forces is classified as:

(i) Coplanar Force System
(ii) Non-coplanar Force System

Coplanar force system consists of a set of forces with their lines of action lying in the same plane.

Non-coplanar force system consists of a set of forces whose lines of action do not lie in the same plane.

Both coplanar and non-coplanar force systems can further be classified into (i) concurrent force system and (ii) non-concurrent force system.

In general, therefore, we can have 4 types of force systems.

(i) Coplanar, Concurrent Force System

Figure 2.4 shows a coplanar concurrent force system. In this system, all the forces are in the same plane and their lines of action meet at a point point 'O' in Fig. 2.4), which is known as the point of concurrency of the force system.

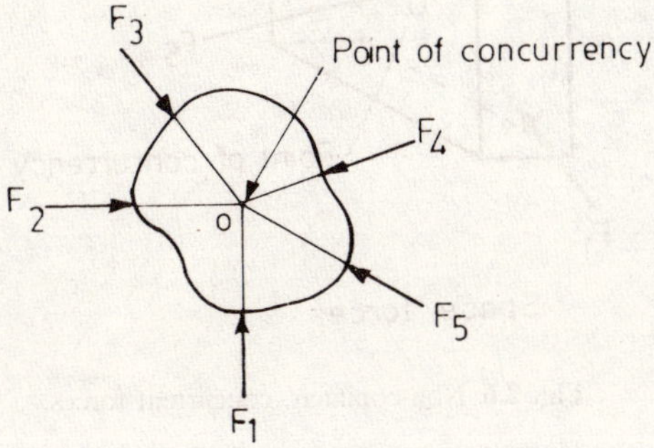

Fig. 2.4 Concurrent forces.

(ii) Coplanar, Non-concurrent Force System

In this system, the forces acting on a body are in the same plane, but their lines of action do not meet at the same point. This system of forces can further be classified as (a) Coplanar parallel force system (Fig. 2.5a), in which the lines of action of all the forces are parallel and (b) Coplanar, non-parallel force system (Fig. 2.5b), in which the lines of action of the forces are neither parallel nor concurrent.

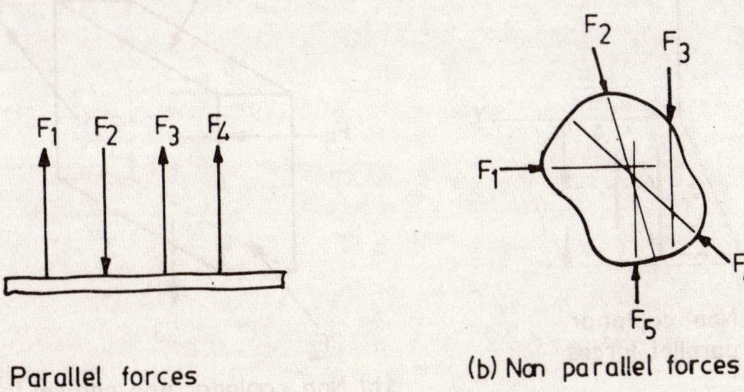

Fig. 2.5 Coplanar, non-concurrent forces.

(iii) Non-coplanar, Concurrent Force System

In this system, the forces acting on a body do not lie in the same plane, but their lines of action do meet at a point in space. Point 'O' in Fig. 2.6 is the point of concurrency.

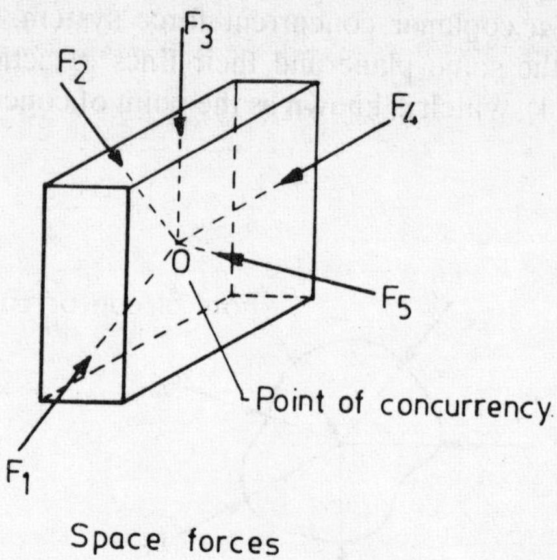

Fig. 2.6 Non-coplanar, concurrent forces.

(iv) **Non-coplanar, Non-concurrent Force System**
In this system, the forces acting on a body neither lie in the same plane, nor their lines of action meet at a point. This system of forces is further classified as non-coplanar, parallel force system and non-coplanar, non-parallel force system. Fig. 2.7(a) and (b) illustrate these systems of forces.

In this book we shall be dealing with only coplanar force systems.

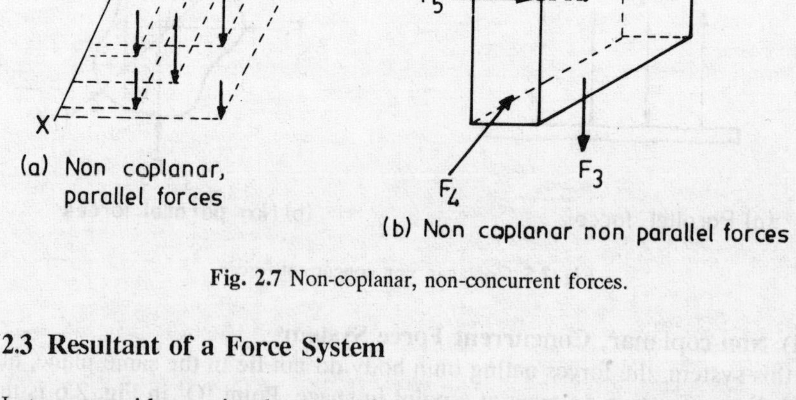

Fig. 2.7 Non-coplanar, non-concurrent forces.

2.3 Resultant of a Force System

Let us consider a simple case of two forces acting on a body. From the definition of force, we can say that these two forces will tend to change the state of rest or motion of the body. If we replace the given two forces by a single force which has the same effect on the body as the two forces, then such a force is called the resultant of the given two forces. The concept of the resultant of forces is very important, as this helps in reducing the given force system into a single force.

The resultant of a force system, therefore, is the single force which has the same effect on the body as the given force system.

To illustrate, the displacement caused by a force is always in the line of action of the force in its sense. Instead of calculating the displacements produced by individual forces and summing them up, the forces system itself can be reduced to a single resultant force and the effect of the resultant force will then give the total effect of the forces of the system. Fig. 2.8 diagramatically represents the resultant of a force system.

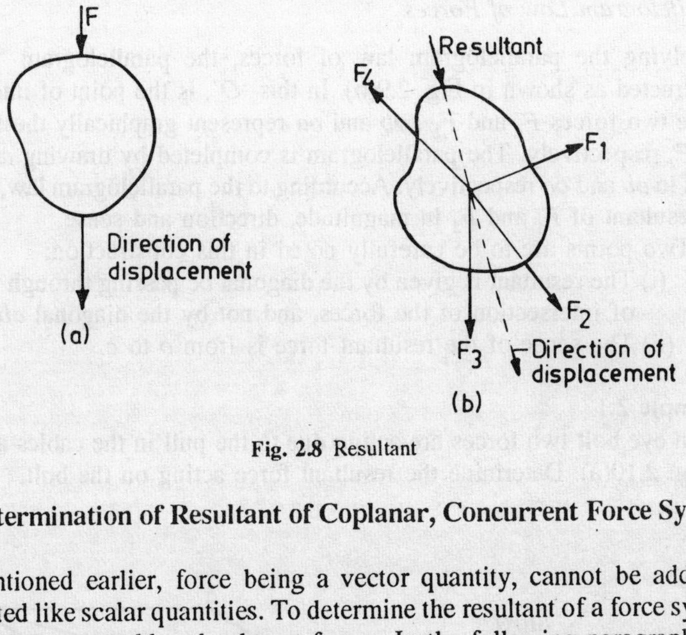

Fig. 2.8 Resultant

2.4 Determination of Resultant of Coplanar, Concurrent Force Systems

As mentioned earlier, force being a vector quantity, cannot be added or subtracted like scalar quantities. To determine the resultant of a force system, it is necessary to add and subtract forces. In the following paragraphs, we will discuss the determination of the resultant of coplanar concurrent force systems.

2.4.1 *Two Concurrent Forces*

Let F_1 and F_2 be two concurrent forces acting on a body. F_1 is horizontal and F_2 makes an angle with the horizontal as shown in Fig. 2.9(a).

According to the principle of transmissibility, F_1 and F_2 can be shifted alongwith their lines of action to the point '*O*', the point of concurrency of these forces as shown in Fig. 2.9(a). The problem is to determine the resultant

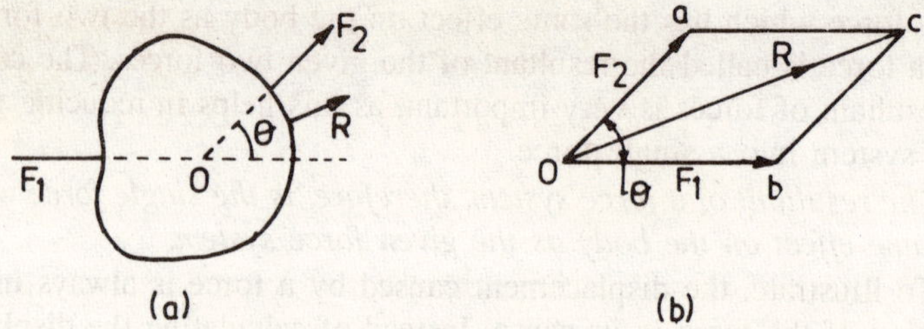

Fig. 2.9 Two Concurrent forces

of the forces F_1 and F_2. This can be done by using the Parallelogram Law of Forces or the Triangle Law of Forces.

Parallelogram Law of Forces

Applying the parallelogram law of forces, the parallelogram '*oacb*' is constructed as shown in Fig. 2.9(b). In this '*O*', is the point of intersection of the two forces F_1 and F_2 . *ob* and *oa* represent graphically the forces F_1 and F_2 respectively. The parallelogram is completed by drawing *ac* and *bc* equal to *ob* and *oa* respectively. According to the parallelogram law, *oc* gives the resultant of F_1 and F_2 in magnitude, direction and sense.

Two points are to be carefully noted in this construction:

(i) The resultant is given by the diagonal *oc* passing through the point of intersection of the forces, and not by the diagonal *ab*, and

(ii) The sense of the resultant force is from *o* to *c*.

Example 2.1

On an eye bolt two forces are acting due to the pull in the cables as shown in Fig. 2.10(a). Determine the resultant force acting on the bolt.

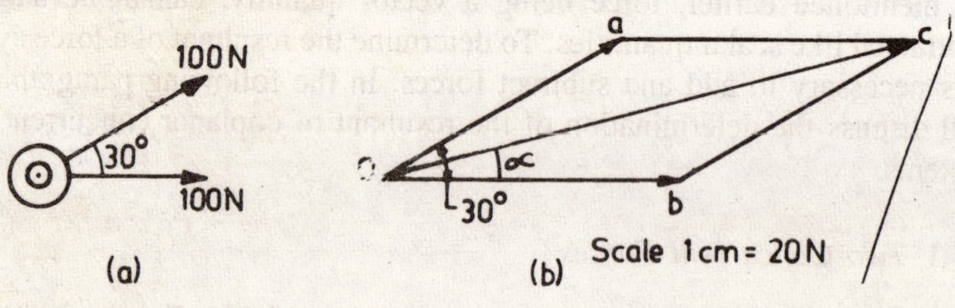

Fig. 2.10

Solution

From the figure we see that two concurrent forces of 100 N each are acting at an angle of 30° on the eye bolt. To determine the resultant, we construct the parallelogram shown in Fig. 2.10(b). The resultant is completely represented by *oc* (from *o* to *c*), in magnitude, direction and sense. By measurement from Fig.2.10(b), we get

$$\text{Length of } oc = 9.65 \text{ cms}$$

$$\therefore \text{ Magnitude of the resultant} = 9.65 \times 20 = \mathbf{193.0\ N.}$$

$$\text{Angle } `\alpha' = 15°$$

The resultant is, therefore, a force of magnitude 193.0 N, and acts at an angle of 15° with the horizontal force of 100 N.

Analytical Solution of the Parallelogram Law

The solution given above is a graphical solution. It is also possible to determine the resultant analytically.

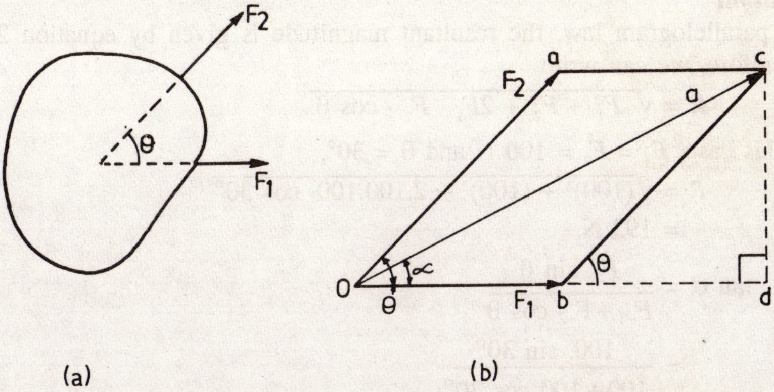

Fig. 2.11

Fig. 2.11(a) shows two concurrent forces F_1 and F_2 acting on a body. Fig. 2.11(b) shows the parallelogram (diagramatically) drawn with the two forces F_1 and F_2. In Fig. 2.11(b), $oa = bc = F_2$, and $ob = ac = F_1$. *oc* gives the resultant 'R'. Draw *cd* perpendicular to *ob*, or *ob* produced.

Now, *odc* is a right angled triangle.

$$oc^2 = od^2 + dc^2$$

$$= (ob + bd)^2 + dc^2 \quad (\because od = ob + bd)$$

From triangle *bdc*,

$$bd = bc \cos\theta$$

$$dc = bc \sin\theta$$

Substituting these values,

$$oc^2 = (ob + bc\cos\theta)^2 + (bc\sin\theta)^2$$

$$= ob^2 + bc^2\cos^2\theta + 2.\,ob.\,bc.\cos\theta + bc^2\sin^2\theta$$

$$= ob^2 + bc^2 + 2\ ob.\ oc.\ \cos\theta$$

Or $R^2 = F_1^{\ 2} + F_2^{\ 2} + 2F_1 \cdot F_2 .\ \cos\theta$

$$R = \sqrt{F_1^{\ 2} + F_2^{\ 2} + 2F_1 \cdot F_2 .\ \cos\theta} \quad \text{... (2.1)}$$

The direction of the resultant R can also be calculated from Fig. 2.11(b). Let the resultant make an angle, 'α', with the force F_1.

$$\tan\alpha = \frac{cd}{od} = \frac{cd}{ob + bd} = \frac{F_2 \sin\theta}{F_1 + F_2 \cos\theta} \quad \text{...(2.2)}$$

Knowing the forces F_1 and F_2 and the angle 'θ' between them, eq. 2.1 and eq. 2.2 can be used to determine the magnitude and direction of resultant of forces F_1 and F_2.

Example 2.2
Solve example 2.1 analytically (Fig. 2.10).

Solution
By parallelogram law, the resultant magnitude is given by equation 2.1. Therefore, we can write

$$R = \sqrt{F_1^2 + F_2^2 + 2F_1 \cdot F_2 \cdot \cos\theta}$$

In this case, $F_1 = F_2 = 100$ N and $\theta = 30°$,

$$\therefore \quad R = \sqrt{(100)^2 + (100)^2 + 2.100.100.\ \cos 30°}$$

$$= 193 \text{ N}$$

$$\therefore \quad \tan\alpha = \frac{F_2 .\ \sin\theta}{F_{-1} + F_{-2} \cos\theta}$$

$$= \frac{100.\ \sin 30°}{100 + 100 \cos 30°}$$

$$= 0.268$$

$$\alpha = 14.96°$$

These results agree closely with the graphical solution obtained in Example 2.1 within the limits of graphical accuracy.

Law of Triangle of Forces
The construction of a triangle of forces is shown in Fig. 2.12. F_1 and F_2 are the two forces acting on the body. From a point 'O', representing the point of concurrency, draw $ob = F_1$ and $bc = F_2$ to a chosen scale as in Fig. 2.12(b). oc now gives the resultant in magnitude, direction and sense.

It is easy to see that the law of triangle of forces is just a corollary of the Parallelogram Law. In Fig. 2.12(b) and (c) are shown two possible constructions of the triangle of forces. The two differ only in the order of representing the forces. If F_1 and F_2 are drawn in the same order, triangle *obc* is obtained and if the order is F_2, F_1 we get triangle *oac*. In Fig. 2.12(d), the two triangles are shown together. When these two triangles are joined, it results in a parallelogram. Thus the law of triangle of forces is a corollary to the parallelogram law of forces.

The construction of the triangle is to be carefully noted. In Fig. 2.12(b), *ob* is drawn equal to F_1 in magnitude, direction and sense, *bc* is then drawn from point *b* and equal to F_2 in the same order. *O* is now joined with *c*. *oc* then represents the resultant in magnitude direction and sense (*o* to *c*). Fig. 2.12(c) is drawn in the same way, taking the force F_2 first.

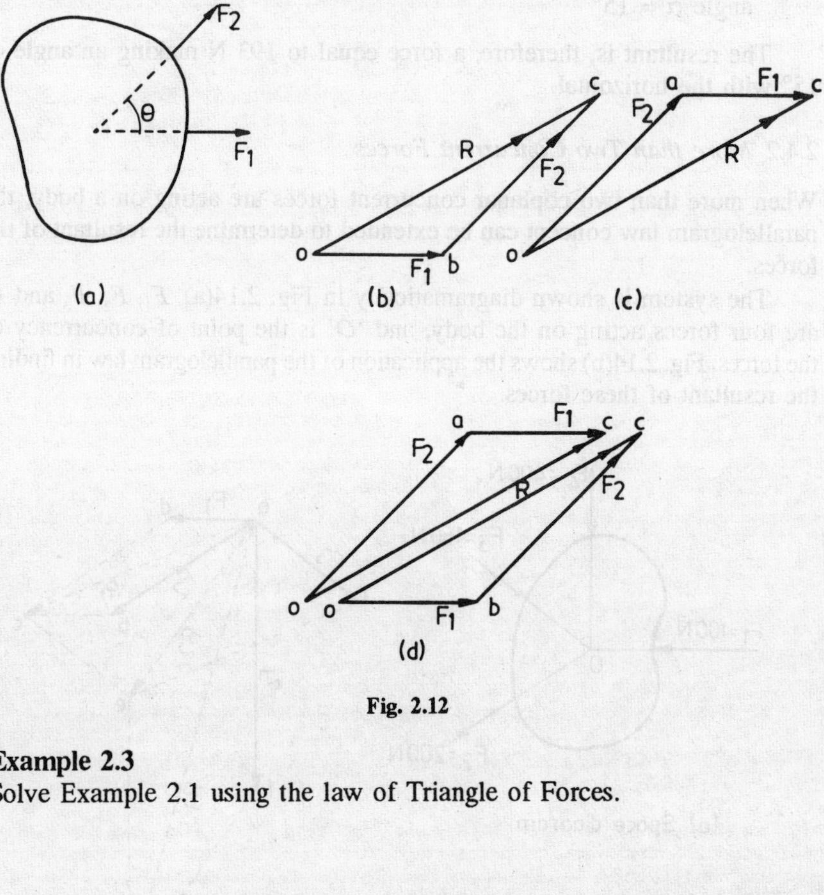

Fig. 2.12

Example 2.3
Solve Example 2.1 using the law of Triangle of Forces.

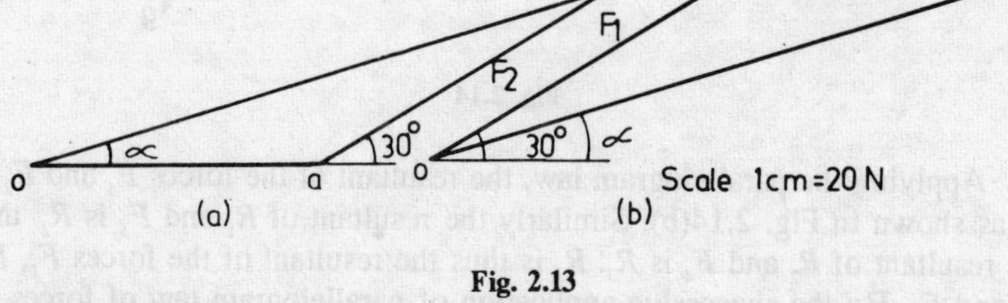

Fig. 2.13

Solution
Fig. 2.13 illustrates the solution. We draw the triangle starting with the horizontal 100 N force as in Fig. 2.13(a) or the inclined 100 N force as in Fig. 2.13(b).

The resultant is given by *ob* or *oa*.
By measuring from Fig. 2.13(a) or Fig. 2.13(b).

$oa = 9.65$ cm

Resultant = 193 N

angle $\alpha = 15°$

The resultant is, therefore, a force equal to 193 N making an angle of 15° with the horizontal.

2.4.2 *More than Two Concurrent Forces*

When more than two coplanar concurrent forces are acting on a body, the parallelogram law concept can be extended to determine the resultant of the forces.

The system is shown diagramatically in Fig. 2.14(a). F_1, F_2, F_3 and F_4 are four forces acting on the body, and '*O*' is the point of concurrency of the forces. Fig. 2.14(b) shows the application of the parallelogram law in finding the resultant of these forces.

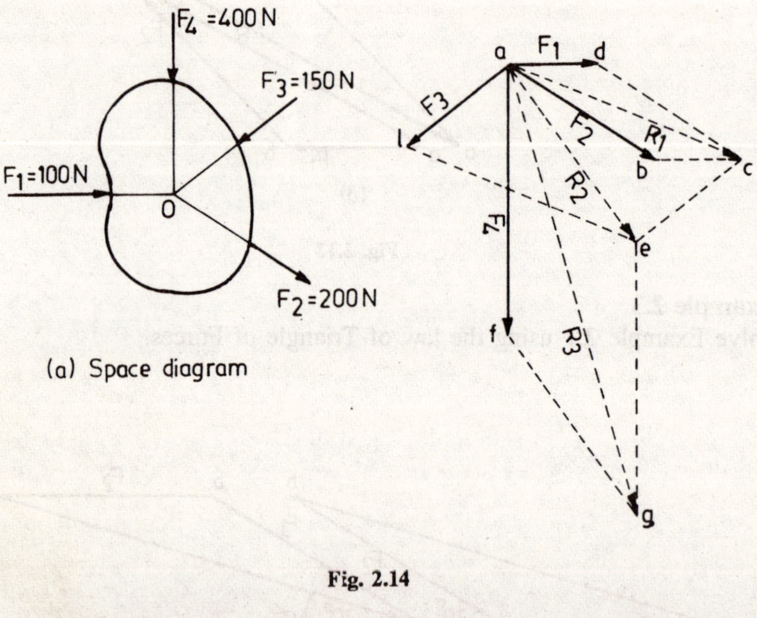

Fig. 2.14

Applying the parallelogram law, the resultant of the forces F_1 and F_2 is R_1 as shown in Fig. 2.14(b). Similarly the resultant of R_1 and F_3 is R_2, and the resultant of R_2 and F_4 is R_3. R_3 is thus the resultant of the forces F_1, F_2, F_3 and F_4. By the successive application of parallelogram law of forces, it is, therefore, possible to determine the resultant of more than two concurrent forces.

One important point to be noticed from the above construction is that the resultant of such a force system always passes through the point of concurrency of the forces system.

Fig. 2.15(a) shows another way to determine the resultant of the force system using the triangle law of forces. In this construction, *oa* and *ab* are

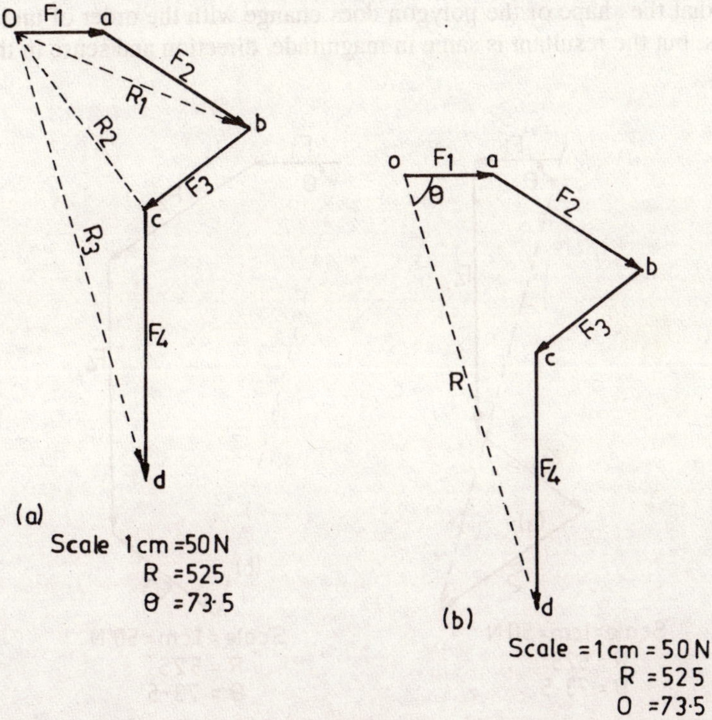

Fig. 2.15

drawn first, representing F_1 and F_2, and the resultant is *ob* from triangle *oab*. Then *bc* is drawn representing F_3 and *oc* is the resultant of *ob* and F_3 or F_1, F_2 and F_3. *cd* is now drawn representing force F_4 and *od* is the resultant of forces *oc* and F_4, or F_1, F_2, F_3 and F_4.

Force Polygon

From Fig. 2.15(a), we can deduce an important result regarding the resultant of F_1, F_2, F_3 and F_4. If we omit the intermediate resultants *ob* and *oc* in Fig. 2.15(a), we get what is known as a force polygon shown in Fig. 2.15(b). Each side of this polygon represents vectorially a force of the force system, i.e. forces F_1, F_2, F_3 and F_4 are represented graphically by the sides *oa*, *ab*, *bc*, and *cd* of the polygon respectively. The closing line *od* of this polygon represents the resultant '*R*' in magnitude, direction and sense. It can be said that if there are a number of concurrent forces acting on a body, their resultant is represented in magnitude, direction and sense by the closing line, joining the first point to the last point, of the force polygon. The remaining sides of the polygon represent the forces on the body in magnitude, direction and sense.

An important point to be noticed about the force polygon is that the order of taking the forces of the system for drawing the polygon does not affect the resultant in any way. In Fig. 2.16(a) and (b), two polygons are drawn for the same force system given in Fig. 2.14(a). These force polygons differ from those drawn in Fig. 2.15(b) only in the order of taking the forces. It can be seen that the shape of the polygon does change with the order of taking the forces; but the resultant is same in magnitude, direction and sense in the two cases.

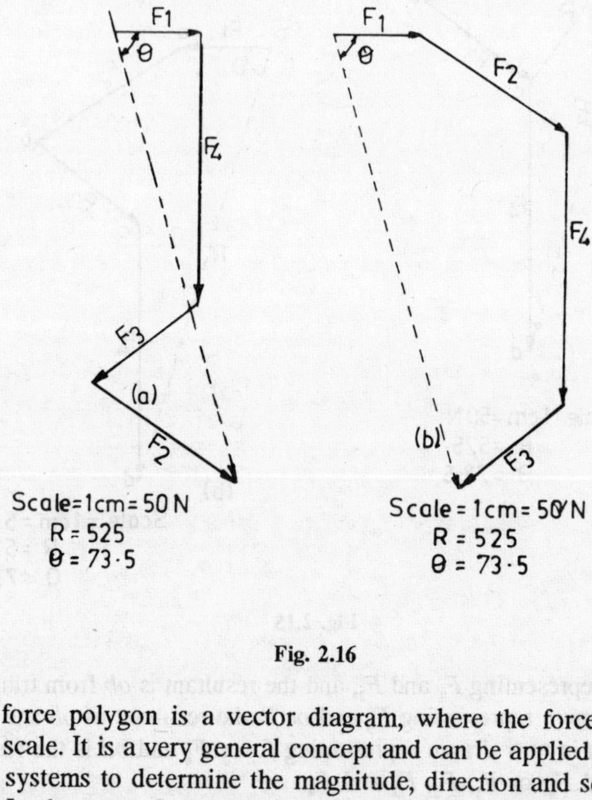

Fig. 2.16

The force polygon is a vector diagram, where the forces are represented to scale. It is a very general concept and can be applied to all coplanar force systems to determine the magnitude, direction and sense of their resultant. In the case of concurrent, coplanar force systems, we can easily determine the point of application of the resultant of the system which passes through the point of concurrency of the force system. In the case of non-concurrent, coplanar force systems, the magnitude, direction and sense can be obtained by the force polygon, but the location of the resultant in the space diagram has to be obtained by other methods, which will be explained later.

Example 2.4

Four concurrent coplanar forces are acting on body as shown in Fig. 2.17. Determine the resultant of the force system in magnitude, direction and sense.

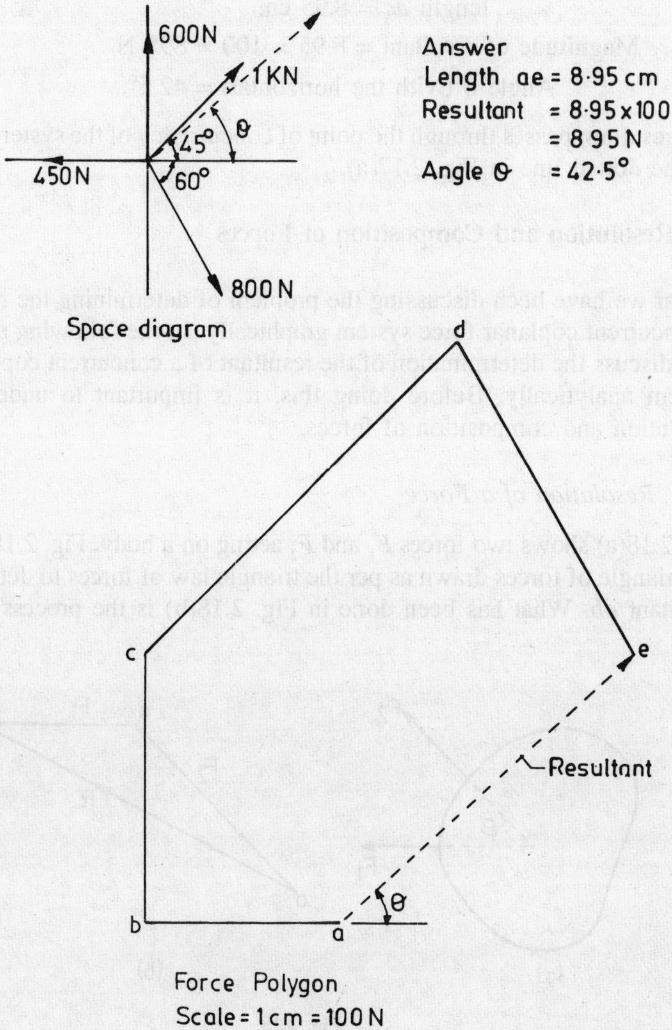

Fig. 2.17

Solution

A suitable scale is first chosen to draw the force polygon. The scale chosen is 1 cm = 100 N. The force polygon is shown in Fig. 2.17(b).

The polygon is started from the point '*a*' by drawing *ab* = 450 N in magnitude, direction and sense. The construction of the polygon is continued by drawing '*bc*' representing the 600 N force, '*cd*' representing the 1000 N force, and finally '*de*' representing the 800 N force. Now the line joining the first point '*a*' to the last point '*e*' is drawn to give the resultant in magnitude, direction and sense (*a* to *e*). From Fig. 2.17(b) by measurement,

$$\text{length } ac = 8.95 \text{ cm}$$

$$\therefore \text{ Magnitude of resultant} = 8.95 \times 100 = 895 \text{ N}$$

$$\text{Angle } \theta \text{ (with the horizontal)} = 42.5°.$$

The resultant passes through the point of concurrency of the system as shown by the dotted line in Fig. 2.17(a).

2.5 Resolution and Composition of Forces

So far we have been discussing the problem of determining the resultant of a concurrent coplanar force system graphically. In the following articles, we will discuss the determination of the resultant of a concurrent coplanar force system analytically. Before doing this, it is important to understand the resolution and composition of forces.

2.5.1 *Resolution of a Force*

Fig. 2.18(a) shows two forces F_1 and F_2 acting on a body. Fig. 2.18(b) shows the triangle of forces drawn as per the triangle law of forces to determine the resultant *ob*. What has been done in Fig. 2.18(b) is the process known as

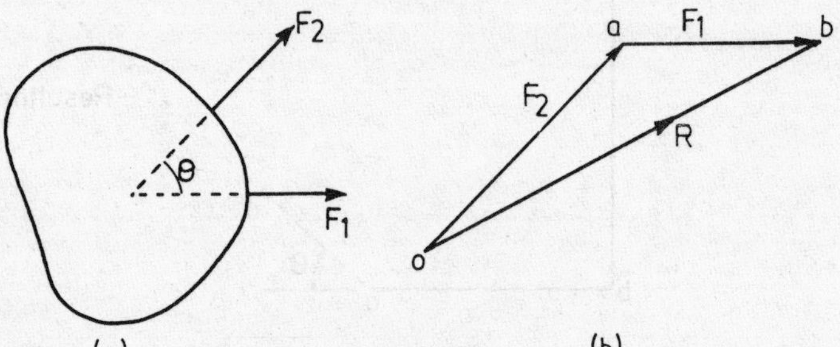

Fig. 2.18

the composition of forces. In the composition of forces, two or more forces are combined into a single force. In this case *oa* and *ab* are combined into a single force *ob*. Forces *oa* and *ab* are called the **components** of the force *ob*. For a given force *ob*, if we determine the components *oa* and *ab*, the process is known as **resolution**. The force *ob* is resolved into 2 components *oa* and *ab*. It is clear that the two components must intersect at a common point on the line of action of the force *ob*.

To illustrate the concept of components further, let us take the case of a triangle *abc* shown in Fig. 2.19(a). If this is a triangle of forces, then forces *ac* and *cb* are the components of the force *ab*. Note carefully the arrows placed on the sides of the triangle.

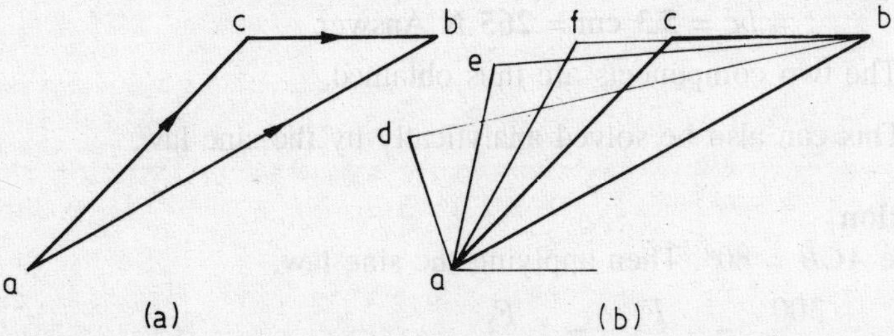

Fig. 2.19

Now if *ab* is the given force, and we have to determine its components, we see at once from Fig. 2.19(b) that there are many possible solutions. Any triangle formed on the side *ab* will give us the two components of the force *ab*. Thus *ad* and *db*, *ae* and *eb* and *af* and *fb* all could be the components of *ab*. The components become unique only when we specify the triangle by stipulating the values of the components, or the directions of the components.

Example 2.5

A force F of 300 N is acting in the direction AB as shown in Fig. 2.20. Find the components of this force along directions AC and BC shown by dotted lines.

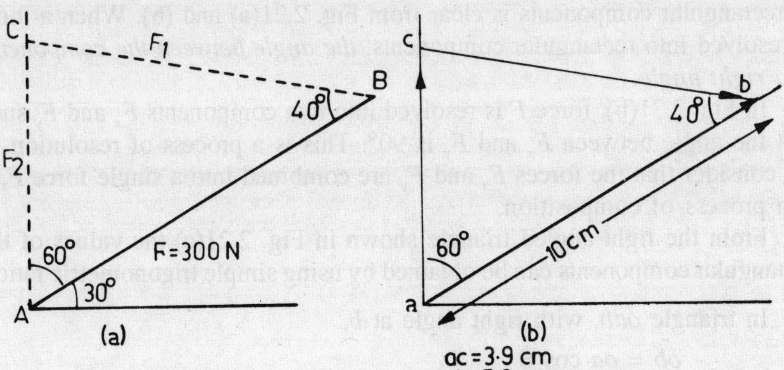

Fig. 2.20

(i) The problem can be solved graphically as shown in Fig. 2. 20(b).

We start by drawing a line *ab* at 30° to the horizontal and equal to 10 cm in length, choosing a scale of 1 cm = 30 N for the force diagram. We set off a line at 60° to *ab* at *a*, and a line 40° to *ab* at *b*. The two intersect at C. We measure *ac* and *bc*.

$$ac = 3.9 \text{ cm} = \mathbf{195 \text{ N}} \text{ Answer}$$

$$bc = 5.3 \text{ cm} = \mathbf{265 \text{ N}} \text{ Answer}$$

The two components are thus obtained.

(ii) This can also be solved analytically by the sine law.

Solution

Angle ACB = 80°. Then applying the sine law,

$$\frac{300}{\sin 80} = \frac{F_1}{\sin 60} = \frac{F_2}{\sin 40}$$

where F_1 and F_2 are components along directions AC and BC respectively. Solving,

$$F_1 = \frac{300 \sin 60}{\sin 80} = \mathbf{263.8 \text{ N}} \text{ from } A \text{ to } C$$

$$F_2 = \frac{300 \sin 40}{\sin 80} = \mathbf{195.8 \text{ N}} \text{ from } C \text{ to } B$$

2.5.2 *Resolution of a Force into Rectangular Components*

In computing the resultant of a given force system analytically, a combined process of resolution and composition, is adopted. In doing this, what is known as rectangular components are very convenient in determining the resultant. In Fig. 2.21 is shown the rectangular components of a force *F*. The concept of rectangular components is clear from Fig. 2.21(a) and (b). When a force is resolved into rectangular components, *the angle between the components is a right angle.*

In Fig. 2.21(b), force *F* is resolved into two components F_x and F_y such that the angle between F_x and F_y is 90°. This is a process of resolution. If we consider that the forces F_x and F_y are combined into a single force *F*, it is a process of composition.

From the right angled triangle shown in Fig. 2.21(c) the values of the rectangular components can be obtained by using simple trigonometric ratios.

In triangle *oab*, with right angle at *b*,

$$ob = oa \cos \theta$$

$$F_x = F \cos \theta$$

and $$ba = oa \sin \theta$$

$$F_y = F \sin \theta$$

The above results are very important in resolving a force into its rectangular components.

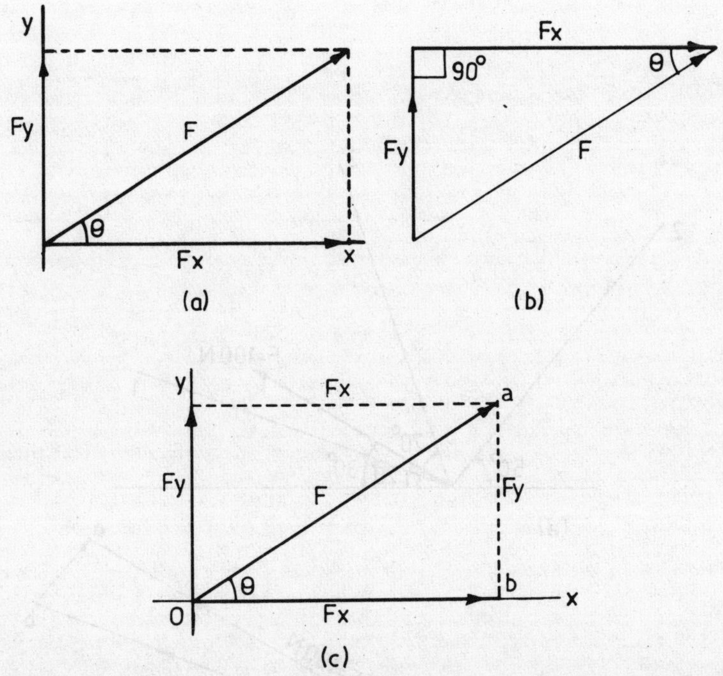

Fig. 2.21

Example 2.6

A force of 100 N acts at point 'o' (Fig. 2.22). Its line of action makes an angle of 30° with the horizontal. Determine

(i) the two components of this force along the directions (1) and (2), and

(ii) the two rectangular components of the force, one of the forces *oc* being along direction (3).

Solution

(i) We draw the line '*oa*' to represent the 100 N force making an angle of 30° with the horizontal. We draw the lines *ob* and *ab* at 20° and 50° with the horizontal. The components are *ob* and *ba*. By measurement [Fig. 2.22(b)]

$$ob = 8.4 \text{ cm} = 84 \text{ N}$$

$$ba = 1.5 \text{ cm} = 15 \text{ N}$$

(ii) We draw the line *oa* representing the force *F*. The line '*oc*' is drawn at 40° to *oa*, and *ac* is drawn perpendicular to *oc*. *oc* and *ca* are the components. By measurement from Fig. 2.22(c), we get

$$oc = 5.7 \text{ cm} = 76 \text{ N}$$

$$ca = 4.8 \text{ cm} = 64 \text{ N}$$

Analytically (i) from a diagramatic sketch of triangle *oba* as in Fig. 2.22(d),

$$oba = 70°, \text{ using the sine law,}$$

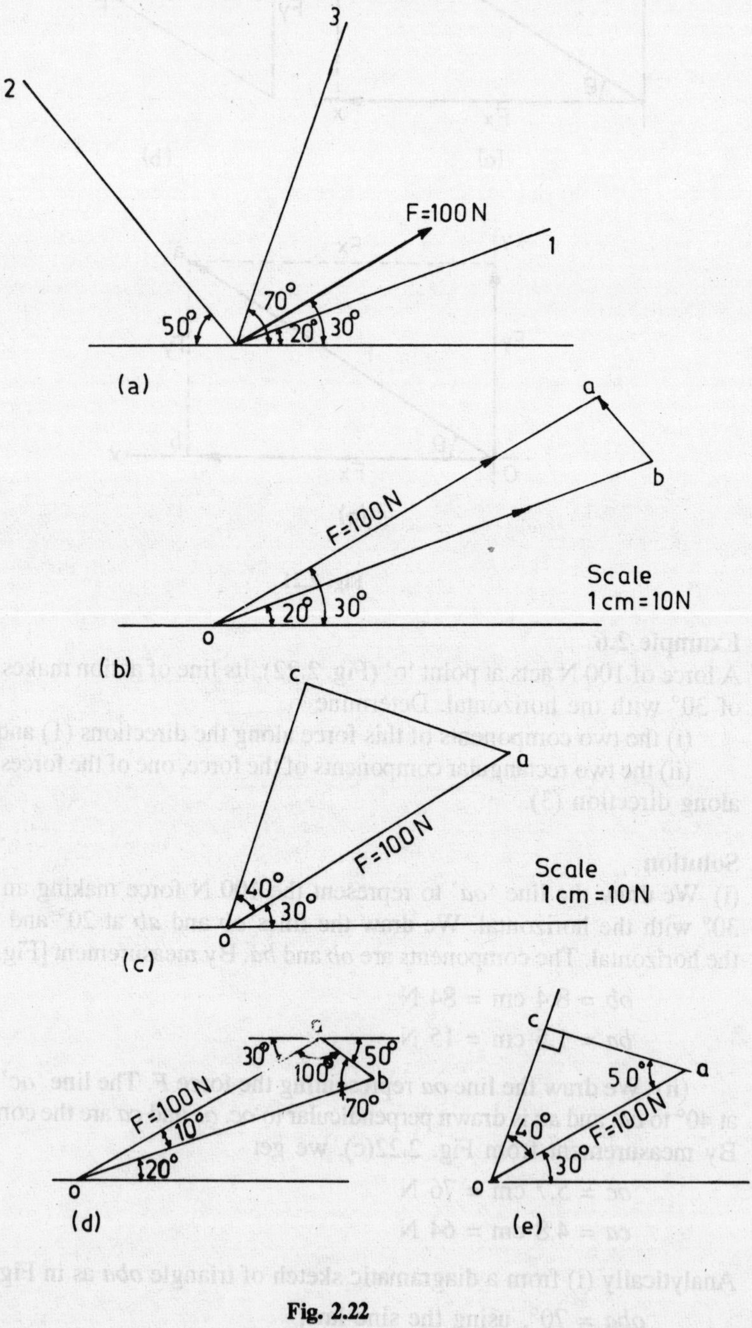

3
2
F=100N
1
70°
50°
20°
30°
(a)
a
b
F=100N
Scale
1cm=10N
20°
30°
o
(b)
c
a
F=100N
Scale
1 cm=10 N
40°
30°
o
(c)
a
30°
100°
50°
F=100N
b
70°
10°
20°
o
(d)
c
a
50°
40°
F=100N
30°
o
(e)

Fig. 2.22

$$\frac{oa}{\sin 70°} = \frac{ob}{\sin 100°} = \frac{ba}{\sin 10°}$$

$$oa = 100 \text{ N}; \; ba = 18.5 \text{ N} \; ob = 105 \text{ N}$$

(ii) from a diagramatics sketch of Fig. 2.22(e), angle $oac = 50°$

$$oc = oa \cos 40° = 100 \times 0.766 = 76.6 \text{ N}$$

$$ca = oa \cos 50° = 100 \times 0.642 = 64.2 \text{ N}$$

2.5.3 *Analytical Determination of the Resultant of a Concurrent Coplanar Force System*

The process of resolution and composition can be very conveniently used in determining the resultant of a concurrent, coplanar force system.

Fig. 2.23(a) shows a concurrent, coplanar force system consisting of four forces F_1, F_2, F_3 and F_4 acting on a particle.

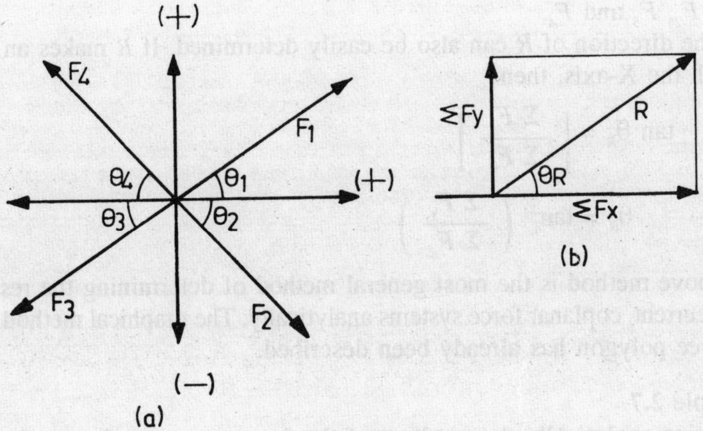

Fig. 2.23

The point of concurrence of the force system is taken as the origin, and rectangular co-ordinate axes X- and Y- are selected as shown. The forces F_1, F_2, F_3 and F_4 can now be resolved into rectangular components, which have directions along the X and Y axes. The components of force F_1 for example, are $F_1 \cos \theta$ along the X-axis, and $F_1 \sin \theta$ along the Y-axis (Fig. 2.23(b). The components are taken positive when they act along the positive directions of X and Y axes. Proper signs are assigned to the components as per sign convention already established.

The calculations are conveniently done in a tabular form as shown in Table below.

Force	*Angle with X-axis*	*X-component* $(\rightarrow +)$	*Y-component* $(\uparrow +)$
F_1	θ_1	$+F_1 \cos \theta_1$	$+F_1 \sin \theta_1$
F_2	θ_2	$+F_2 \cos \theta_2$	$-F_2 \sin \theta_2$
F_3	θ_3	$-F_3 \cos \theta_3$	$-F_3 \sin \theta_3$
F_4	θ_4	$-F_4 \cos \theta_4$	$+F_4 \sin \theta_4$

$$\Sigma F_x = F_1 \cos \theta_1 + F_2 \cos \theta_2 - F_3 \cos \theta_3 - F_4 \cos \theta_4$$
$$\Sigma F_y = F_1 \sin \theta_1 - F_2 \sin \theta_2 - F_3 \sin \theta_3 + F_4 \sin \theta_4$$

You should carefully note the sense of the components and satisfy yourself about the signs placed for the X- and Y- components.

The given force system is now reduced to only two forces, one acting in the direction of the X-axis (ΣF_x) and another acting in the direction of the Y-axis (ΣF_y) Fig. 2.23c. Since the angle between ΣF_x and ΣF_y is 90°,

$$R = \sqrt{(\Sigma F_x)^2 + (\Sigma F_y)^2}$$

Since R is the resultant of ΣF_x and ΣF_y, consequently it is also the resultant of F_1, F_2, F_3 and F_4.

The direction of R can also be easily determined. If R makes an angle θ_R with the X-axis, then

$$\tan \theta_R = \left[\frac{\Sigma F_y}{\Sigma F_x} \right]$$

or
$$\theta_R = \tan^{-1} \left(\frac{\Sigma F_y}{\Sigma F_x} \right)$$

The above method is the most general method of determining the resultant of concurrent, coplanar force systems analytically. The graphical method using the force polygon has already been described.

Example 2.7

Determine analytically the resultant of the four concurrent forces shown in Fig. 2.24.

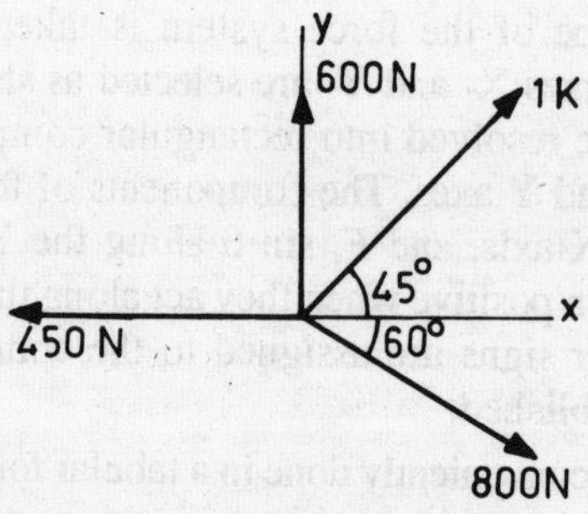

Fig. 2.24

Solution

The procedure is to determine the X- and Y- components of all the forces, and then determine the resultant as

$$R^2 = (\Sigma F_x)^2 + (\Sigma F_y)^2$$

and $$\tan \theta_R = \frac{\Sigma F_y}{\Sigma F_x}$$

where R is the resultant making an angle θ_R with the X-axis. The calculations are shown in the Table below:

Force	*Angle with X-axis*	*X component (F cos θ)*	*Y component (F sin θ)*	*Remarks*
1000 N	45°	+ 1000 cos 45° = 707 N	+ 1000 sin 45° = 707 N	
800 N	60°	+ 800 cos 60° = 400 N	− 800 sin 60° = −692.8 N	
450 N	Zero	− 450 cos 0° = −450 N	Zero	
600 N	90°	Zero	+ 600 sin 90° = 600 N	
Total		ΣF_x = 657 N	ΣF_y = 614.2 N	

Resultant $R = \sqrt{(657)^2 + (614.2)^2} = 900$ N

angle $\theta_R = \tan^{-1}\left(\frac{614.2}{657}\right)$

$= \tan^{-1}(0.935)$

$= 43.10°$

Example 2.8

Determine the resultant of the concurrent, coplanar force system shown in Fig. 2.25.

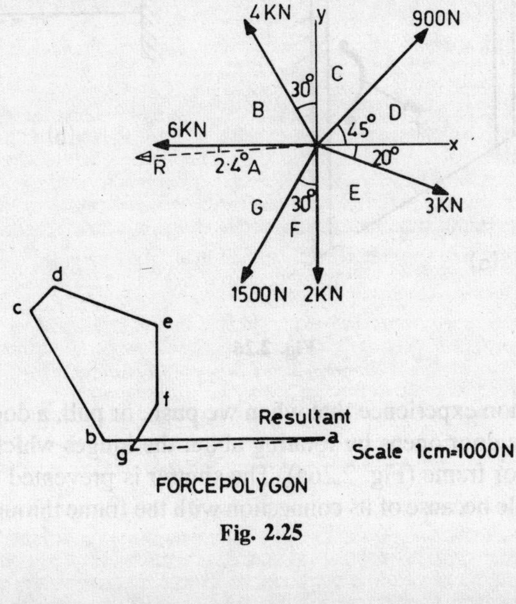

Fig. 2.25

Solution

The graphical solution to the problem is shown in Fig. 2.25. The forces are denoted by letters eg. *AB* represents 6 KN force *BC* represents 4 KN force etc. The force polygon is laid out as shown in Fig. 2.25(b) to a scale of 1 cm = 1000 KN. The force polygon representing the given forces is a-b-c-d-e-f-g. The line joining the first point (a) to the last point gives the resultant. Thus, *ag* in the sense *a* to *g* is the resultant. The magnitude and direction of the resultant can be measured from the figure. ag = 5.3 cm.

Resultant magnitude = 5300 N

$\angle bag$ is 2.4°.

2.6 Turning Effect of Forces

So far we have dealt with concurrent coplanar force systems, where the turning effect of a force does not exist. Turning effect of a force becomes very important when we deal with non-concurrent force systems. The concept of the turning (or rotating) effect of a force will be examined in detail before taking up non-concurrent force systems.

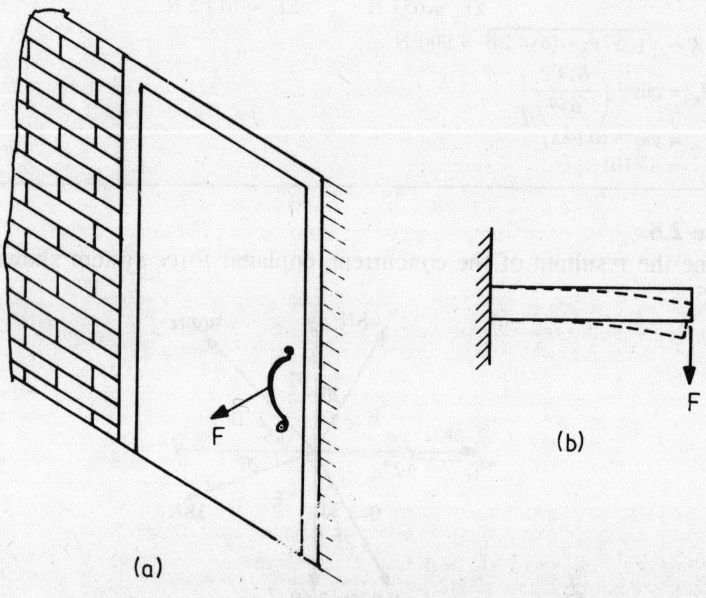

Fig. 2.26

It is a common experience that when we push, or pull, a door by holding at the handle, the door opens by rotating about the hinges which connect the shutter to the door frame (Fig. 2.26a). The shutter is prevented from moving forward as a whole because of its connection with the frame through the hinges.

You might have also noticed that the handle is placed as far away from the hinges as possible. You can easily find out for yourself that if you have to open a door, it requires a larger force when applied near the hinges than when the force is applied away from the hinges.

Now think of a bar, firmly embedded in a wall and projecting out, as shown in Fig. 2.26(b). This is commonly known as a cantilever. If you hold the bar at its end and hang yourself free, you can see the bar taking up the shape shown by the dotted lines. This curving or rotation of the bar is more if it is thin. Your weight essentially provides the turning effect on the bar because of which it bends. The turning effect of a force is measured by a quantity known as the 'moment'.

2.6.1 *Moment*

Moment of a force is a measure of the turning effect of the force about an axis.

Moment 'M' of a force F about an axis is defined as the product of the force 'F' and the perpendicular distance 'd' from the axis of rotation to the line of action of the force.

$$M = F \times d$$

Fig. 2.27(a) shows all the qauntities in the above definition. The moment of the force 'F' about the Z-axis is $F \times d$. The point o where the perpendicular drawn from the line of action of the force meets the Z-axis is known as the moment centre, distance 'd' is known as the moment arm, and Z-axis as the moment axis (axis of rotation).

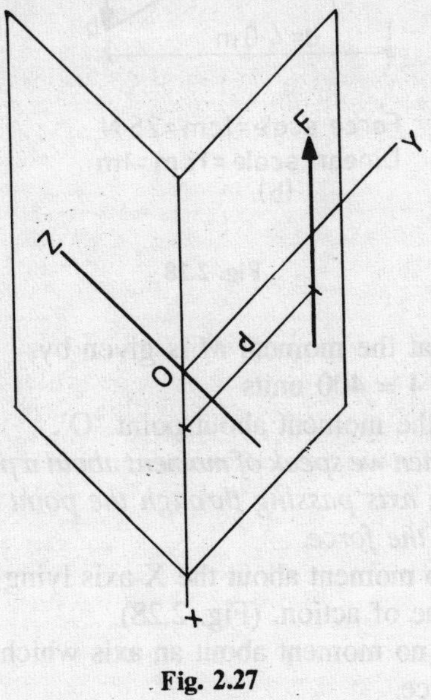

Fig. 2.27

The following points must be carefully noted from Fig. 2.27:

(i) The moment arm is at right angles to both the line of action of the force and the moment axis.

(ii) The force lies in a plane perpendicular to the moment axis.

(iii) The moment of a force is dependent upon the magnitude of the force, as well as the moment arm.

Let us now consider a force of 100 N and determine its moment about an axis through 'O' at a perpendicular distance of 4 m from the line of action of the force (Fig. 2.28).

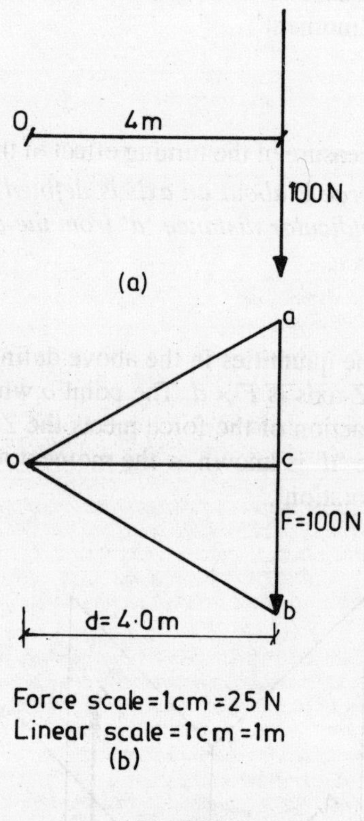

Fig. 2.28

We can easily see that the moment M is given by,

$$M = 100 \times 4 = 400 \text{ units}$$

M is also known as the moment about point 'O'.

You must note that *when we speak of moment about a point we actually mean the moment about an axis passing through the point and perpendicular to the plane containing the force.*

(iv) The force has no moment about the X-axis lying in the same plane, or parallel to its line of action. (Fig. 2.28)

(v) The force F has no moment about an axis which intersects the line of action of the force.

(vi) We havė seen that moment is the product of the force and the moment arm. If the force is measured in Newtons and moment arm in metres, then, the unit of moment can be derived from the equation for moment.

$$\text{Moment} = \text{Force} \times \text{distance}$$
$$= \text{Force (N)} \times \text{distance (m)}$$
$$= \text{Force} \times \text{distance (Nm)}$$

The unit of moment is thus Newton meter or a multiple of this unit, like Kilonewton-metre, Newton millimetre, etc.

Therefore, we can conclude that a force has no moment about an axis which is parallel to it and an axis which intersects the line of action of the force; for example, in Fig. 2.27 force F has no moment about X-axis.

In Fig. 2.28(b) a geometrical interpretation of the moment of force is given. *ab* represents the force of 100 N to the scale shown. Distance '*d*' is equal to 4 m represented by *oc* to a linear scale. The moment of a force about point '*o*' = $F \times d$.

You will note that $F \times d$ is twice the area of the triangle *abo*. Therefore, *the moment of a force F about the moment centre 'o' is equal to twice the area of the triangle formed by joining the ends of the force represented to scale with the moment centre.*

In Fig. 2.28(b) we find that the area of the triangle is given by,

$$\text{Area} = \frac{4 \text{ cm} \times 4 \text{ cm}}{2} = 8 \text{ Sq. cm}$$

$$\text{Moment} = \frac{4 \times 25 \times 4 \times 1}{2} = 200 \text{ Nm}$$

Therefore, twice the area of the triangle gives the moment of the force in Nm.

2.6.2 *Nature of Moment*

In Fig. 2.29 is shown a force system consisting of two forces F_1 and F_2. O is the moment centre and the moment axis is perpendicular to the plane of the paper at 'O'. The moment arm of forces F_1 and F_2 are d_1 and d_2 respectively.

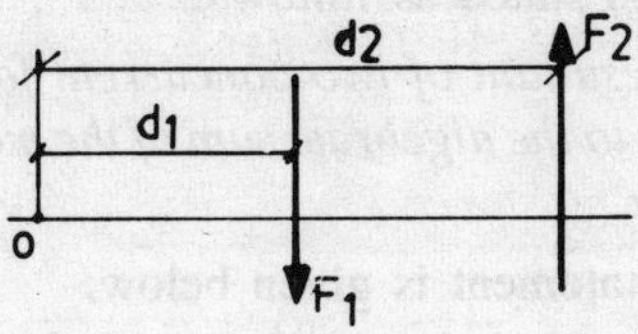

Fig. 2.29

Moment of Force F_1 about 'o' = $F_1 \times d_1$
Moment of Force F_2 about 'o' = $F_2 \times d_2$.

If you assume that *OA* (Fig. 2.29) represents the plane of a door shutter with hinges at '*o*', you can easily understand the difference between these moments. While force F_1 tends to rotate the door in a clockwise direction, the force F_2 rotates the door in an anti-clockwise direction. If $F_1 \times d_1 = F_2 \times d_2$, the door will not rotate at all. Therefore, the moment also has to be assigned a sign depending upon its nature, whether it is clockwise or anti-clockwise.

Let us take another case of turning effect of a force. In Fig. 2.30 is shown the plane of a revolving type of door which is commonly used in compound walls to limit the entry to only one person at a time. The forces F_1 and F_2 have different sense, but you can easily see that they rotate the door in the same direction, in this case anti-clockwise. The nature of the moment of a force, therefore, depends upon the sense of the force as well as the position of the axis of rotation. Therefore, while taking the moments of a large number of forces, we have to assign proper signs to the moments, depending upon the nature of the moment due to each force.

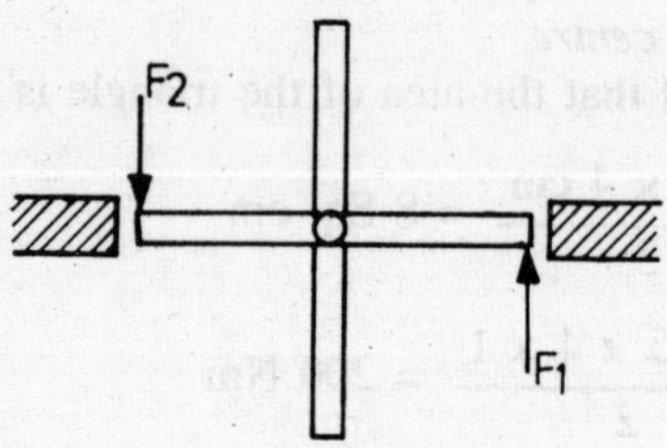

Fig. 2.30

Moment is also a vector quantity like force. In order to specify the moment, it is not enough to state the magnitude of the moment. We have also to mention the nature of the moment and the axis about which moment is taken.

2.6.3 *Principle of Moments—Varignon's Theorem*

A very important principle with regard to moments is known as 'Varignon's Theorem. This principle is stated as follows:

> *The moment of the resultant of two concurrent forces about any point in their plane is equal to the algebraic sum of the moments of these forces about the same point.*

The proof of the above statement is given below:

Let F_1 and F_2 be two forces acting at an angle θ, and let R be their resultant as shown in Fig. 2.31.

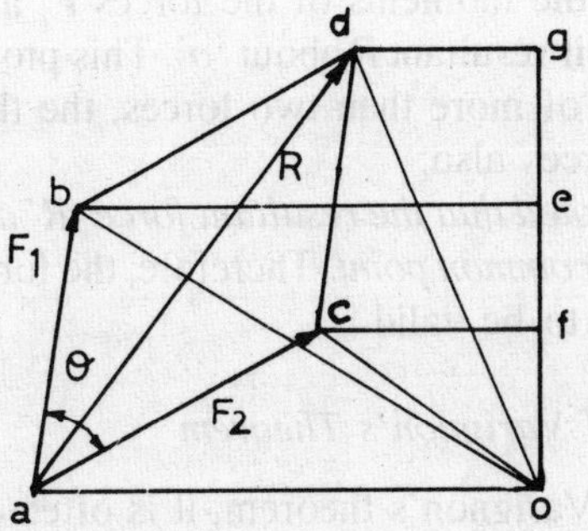

Fig. 2.31

abdc is the parallelogram of forces, with its sides *ab* and *ac* representing the forces F_1 and F_2 respectively. Diagonal *ad* represents the resultant R of the forces F_1 and F_2.
Select any point '*o*' as the moment centre, join *oa*, *ob*, *oc* and *od*. Draw *og* perpendicular to *oa*. Draw *cf*, *be* and *dg* perpendicular to *og*. Then,

Moment of force F_1 about '*o*',

$$M_1 = F_1 \times \text{perpendicular distance from } o \text{ to } ab.$$
$$= 2 \times \text{area of triangle } abo.$$

Similarly,

Moment of F_2 about '*o*', $M_2 = 2 \times$ area of triangle *aco*
and Moment of R about '*o*', $M = 2 \times$ area of triangle *ado*.

For triangles *abo*, *aco* and *ado*, *oa* is the common base and the altitude of these triangles are *oe*, *of* and *og*.

$$\therefore \quad M_1 = 2 \times oa \times oe \times \frac{1}{2}$$

$$M_2 = 2 \times oa \times of \times \frac{1}{2}$$

$$M = 2 \times oa \times og \times \frac{1}{2}$$

Since $oe = gf$, being the projection of $ab = cd$ on og,

$$M_1 + M_2 = oa \times oe + oa \times of$$
$$= oa\,(of + oe)$$
$$= oa\,(of + gf)$$
$$= oa \times og$$
$$= M$$

$$\therefore \quad M_1 + M_2 = M$$

That is, the sum of the moments of the forces F_1 and F_2 about '*o*' is equal to the moment of their resultant R about '*o*'. This proves Varignon's theorem. If R is the resultant of more than two forces, the theorem can be extended to the remaining forces also.
It is to be carefully noted that the resultant force 'R' and the component forces should intersect at a common point. Therefore, the forces should be concurrent for the above proof to be valid.

2.6.4 *Application of Varignon's Theorem*

In accordance with Varignon's theorem, it is often convenient to determine the moment of a force about a point '*o*' (i.e. about an axis through '*o*' perpendicular to the plane containing the force) as the sum of the moments of its components. The following problems will illustrate the point.

Example 2.9
Determine the moment of the force of 100 N about point 'O' (Fig. 2.32).

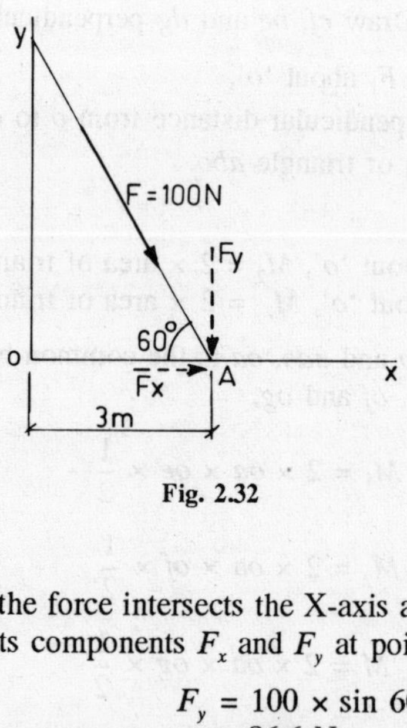

Fig. 2.32

Solution
The line of action of the force intersects the X-axis at point A. Resolve the force of 100 N into its components F_x and F_y at point A.

$$F_y = 100 \times \sin 60$$
$$= 86.6 \text{ N}$$
$$F_x = 100 \times \cos 60$$
$$= 50 \text{ N}$$

F_x has no moment about point o, since its line of action passes through point '*o*'. Applying the principle of moments.

Moment of force of 100 N about '*o*' = Moment of F_y + Moment of F_x
= $F_y \times 3$
= 86.6×3
= **259.8 Nm** Clockwise

Example 2.10

In Fig. 2.33, calculate the moment of the 1000 N force about point *A*.

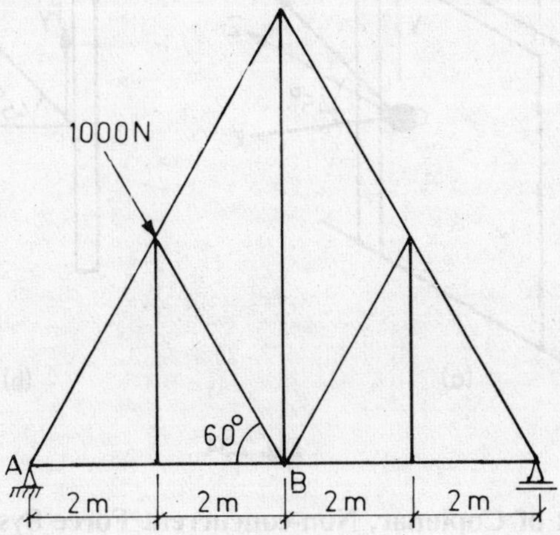

Fig. 2.33

Solution

If we know the perpendicular distance from *A* to the line of action of the force, we can directly calculate the moment. An easier method is to apply Varignon's theorem.

Moment of the 1000 N force about A = Moment of the vertical components of 1000 N force at (B) about A

= 1000 sin 60 × 4.

= **3464 Nm** clockwise.

Example 2.11

On a door knob, a force of 100 N is applied at an angle of 45° with the horizontal, and in a plane perpendicular to the plane of the door as shown in Fig. 2.34. Determine the turning effect of the force about the hinges.

Solution

The 100 N force is in a plane perpendicular to the plane of the door. If we resolve this force into vertical and horizontal components as shown in Fig. 2.34(b), the vertical component F_y will have no turning effect about the hinges as its line of action is parallel to the hinges. Therefore, the turning effect of the 100 N force is equal to the moment of the horizontal component F_x about the line of hinges.

Turning effect = F_x × 0.90

= 100 × cos 45 × 0.90

= **63.63 Nm**

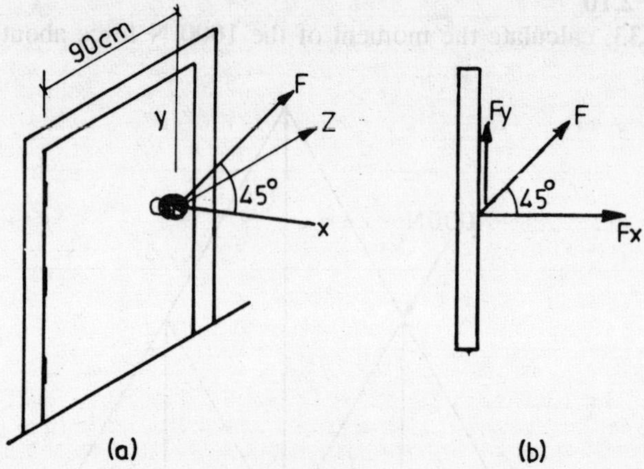

Fig. 2.34

2.7 Resultant of Coplanar, Non-concurrent Force Systems

In art. 2.4 and 2.5, we described the graphical and analytical methods for the determination of the resultant of coplanar concurrent force systems. In this article, we will discuss methods to determine the resultant of coplanar, non-concurrent force systems.

Coplanar, non-concurrent, non-parallel force systems are the most general coplanar force systems and the concurrent and parallel force systems are only particular cases of these. We will, therefore, first describe the general approach to the solution of coplanar, non-concurrent, non-parallel force systems, and thereafter the special case of parallel force systems.

2.7.1 *Graphical Method*

Fig. 2.35 shows a general (non-concurrent, non-parallel) coplanar force system consisting of forces F_1, F_2, F_3 and F_4. We can find the resultant of these forces by successive application of parallelogram law.

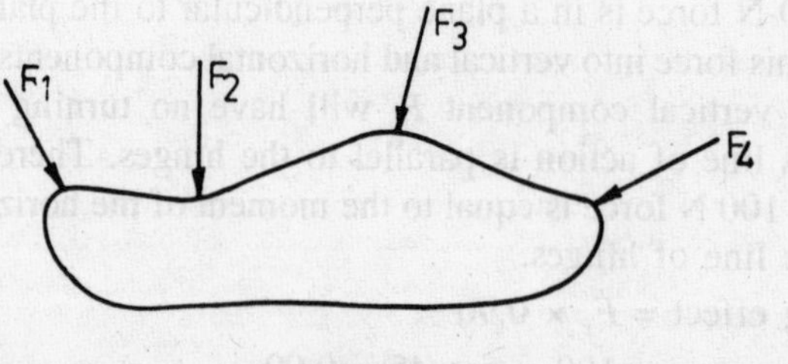

Fig. 2.35

Fig. 2.36 shows the application of parallelogram law. The lines of action of F_1 and F_2 intersect at point (1) and their resultant is determined by parallelogram law as R_1. The lines of action of R_2 and F_3 intersect at point (2), and their resultant is again determined by drawing a parallelogram, as R_2. R_2 is the resultant of F_1, F_2 and F_3. This process can be repeated and the final resultant R can be determined as shown in Fig. 2.36.

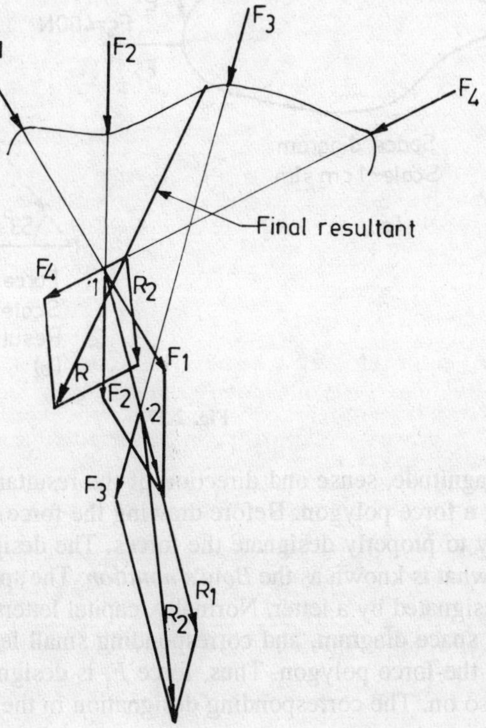

Fig. 2.36

The above process is rather tedious, though only the simple law of parallelogram of forces is involved. It has two disadvantages, namely: (i) the lines of action of the forces may not intersect within the boundaries of the paper, and (ii) it may happen at some stage that the forces turn out to be parallel to each other. In such a situation, the method cannot be applied. We describe below a more elegant and versatile graphical method to determine the resultant of a general coplanar force system.

2.7.2 *Funicular Polygon*

Fig. 2.37(a) shows a general coplanar force system. Note carefully that forces are not drawn to scale in this diagram. Only a linear scale is employed to locate the forces in their correct geometrical positions. This diagram is known as a *space diagram*, as it should be, because only a space (linear) scale is employed. The sense of any force is shown by an arrow as usual.

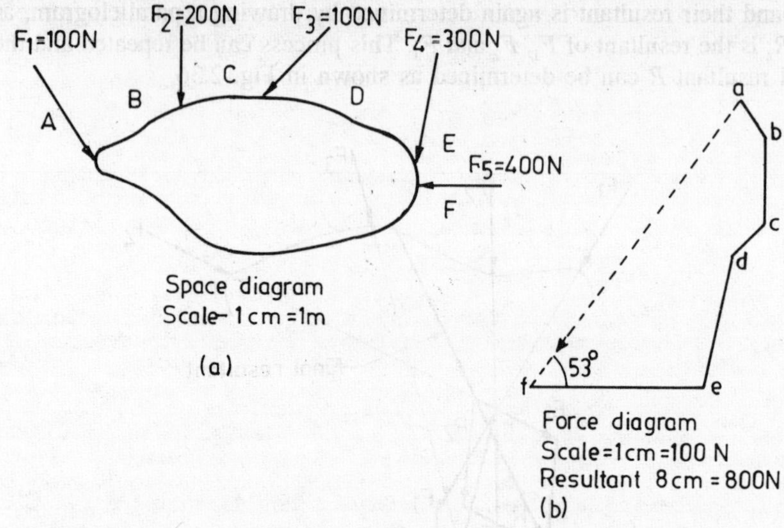

Fig. 2.37

The magnitude, sense and direction of the resultant can be determined by drawing a force polygon. Before drawing the force polygon, however, it is necessary to properly designate the forces. The designation of the forces is done by what is known as the *Bow's notation.* The space between any two forces is designated by a letter. Normally, capital letters $A, B, C, D \ldots$ are used in the space diagram, and corresponding small letters, $a, b, c, d \ldots$ are used in the force polygon. Thus, force F_1 is designated as AB, force F_2 as BC and so on. The corresponding designation in the force polygon is ab, bc and so on.

Having designated the forces by Bow's notation, the force polygon is drawn as shown in Fig. 2.37(b). The scale chosen for the force polygon is 1 cm = 100 N. We start by drawing a line parallel to the line of action of F_1 and of length 1 cm (100 N). We designate this as '*ab*'. We then draw the line *bc* parallel to the line of action of force F_2 and of length 2 cm (200 N). We designate the end point of this line as '*c*'. We continue this process until all the forces are represented. In Fig. 2.37(b), *f* is the last point so obtained. We join *a* to *f* and *af* (from *a* to *f*) represents the resultant in magnitude, direction and sense. The magnitude of the resultant can be scaled off from this diagram.

A basic difference between the resultant of a concurrent force system and that of a non-concurrent force system is evident at once. The force polygon determines the magnitude, sense and direction of the resultant in both cases. In the case of a concurrent force system, this information is sufficient to locate the resultant in space because it is known that it passes through the point of concurrency of the system. The resultant is, therefore, completely determined once the force polygon is drawn. It is not possible by this construction alone to locate the resultant of a non-concurrent force system in the space diagram.

A simple and ingeneous way of locating the resultant is by what is known as the *funicular polygon.* The method of constructing the funicular polygon is explained below and is shown in Fig. 2.38. The space diagram and force diagram are redrawn from Fig. 2.37.

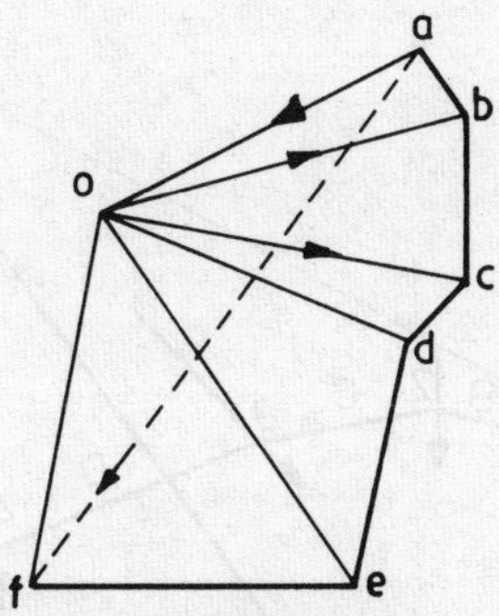

Fig. 2.38

The steps in the construction of the funicular polygon are given below:

(i) A point '*o*' is selected at a convenient distance from the force polygon. Join point '*o*' to points *a, b, c, d, e* and *f* which are the vertices of the force polygon. Point '*o*' is known as the 'pole'.

(ii) A line *O1* is drawn in space *A* of the space diagram parallel to the line *oa* in the force diagram. This line may be drawn anywhere in space *A* such that it conveniently intersects the line of action (or its extension) of the 100 N force. We designate this intersection point as '1'.

(iii) From point 1, we draw the line 1–2 parallel to *ob* in the force diagram. The point of inter-section of this line and the line of action of the 200 N force (or its extension) is point '2'. Note that the line 1–2 is in space *B* and is parallel to line *ob.*

(iv) We continue the above process and draw the lines 2–3 parallel to *oc* in space *C*, 3–4 parallel to *od* in space *D*, 4–5 parallel to *oe* in space *E*, and 5–6 parallel to *of* in space *F*. We thus get a continuous chain of lines 0-1-2-3-4-5-6 known as the *funicular polygon.* Note that the funicular polygon is drawn in the space diagram.

(v) We produce the lines 0–1 and 5–6 (the first and last lines) in either direction such that they intersect at point '*p*'.

(vi) Point '*p*' is a point in the line of action of the resultant. We, therefore, draw a line parallel to *a-f* through point '*p*', and mark the sense of the resultant on this line.

The resultant is thus completely determined. *The force polygon gives the magnitude, direction and sense of the resultant and the funicular polygon locates the resultant in space.*

We mentioned that point '*p*' is on the line of action of the resultant. This can be proved as follows:

In Fig. 2.39, the space diagram, force diagram and funicular polygon are redrawn from Fig. 2.38.

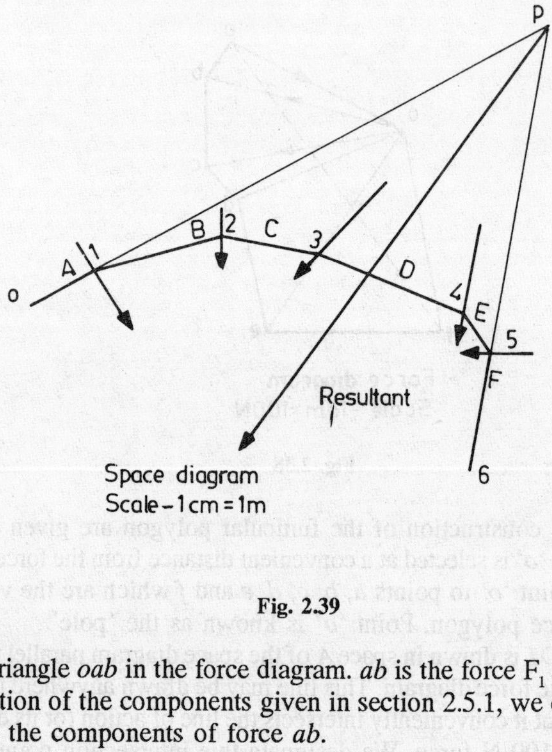

Fig. 2.39

Take triangle *oab* in the force diagram. *ab* is the force F_1 = 100 N. From the description of the components given in section 2.5.1, we can say that *ao* and *ob* are the components of force *ab*.

Similarly, the components of other forces will be as given in the table below:

Force	*Component 1*	*Component 2*
ab	*ao*	*ob*
bc	*bo*	*oc*
cd	*co*	*od*
de	*do*	*oe*
ef	*eo*	*of*

Notice the way the components are written. Here *bo* means the components is in the sense *b* to *o*, and *ob* means the component acting from *o* to *b*.

Since in the force diagram *bo* and *ob* are equal in magnitude, collinear and opposite; they cancel each other. The only components left, then, are *ao* and *of*, which are represented by lines 1–0 and 5-6 respectively, in the space diagram. The given force system by such resolution into components is equal to the forces represented by 1-0(*ao*) and 5-6(*ob*) which intersect at point '*p*', through which the resultant of *ao* and *of* must pass. Since *ao* and *of* are the two forces representing the force system F_1, F_2, F_3, F_4 and F_5, point '*p*' is a point in the line of action of the resultant of this force system.

The above procedure can further be explained clearly in the funicular polygon shown in Fig. 2.40. Here the resolution into components is done in the space diagram itself. Thus, force F_1 is resolved into two components along 0-1 and 1-2, force F_2 along 2-1 and 2-3 and so on. It is evident that in this process all the intermediate components vanish and the two components left unbalanced act along 1-0 and 5-6. Their intersection, therefore, locates the resultant in space.

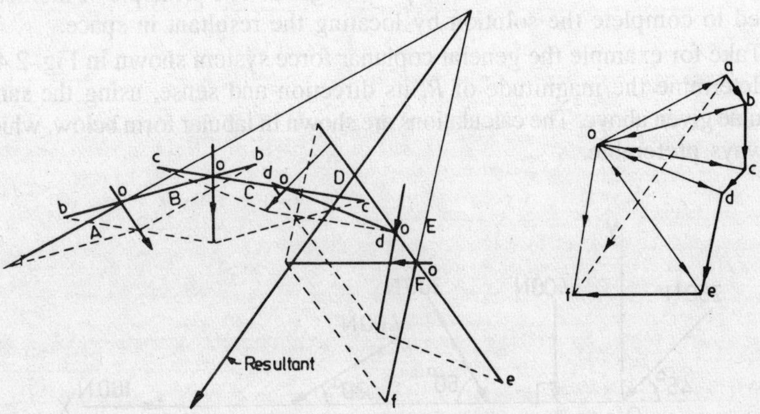

Fig. 2.40

The funicular polygon, therefore, is an ingeneous way of resolving the given force system into components such that these components reduce automatically to two concurrent components from which the line of action of the resultant may be obtained.

The concept of funicular polygon is very important in the graphical analysis of force systems. The reader is advised to go through its construction and proof given above carefully.

2.7.3 *Analytical Method*

The analytical method for the determination of resultant of a non-concurrent coplanar force system uses the principle of rectangular components (by resolution and composition) as given in section 2.5.3, and the principle of moments.

Here again, the basic difference in the approach to the determination of the resultant of a concurrent, coplanar force system and that of non-concurrent force system must be carefully understood.

In the case of a concurrent force system, the resultant always passes through the point of concurrence. Thus in art 2.5.3, we used the formulae,

$$F_x = F \cos \theta_x$$

and $$F_y = F \sin \theta_x$$

and $$R = \sqrt{(\Sigma F \cos \theta_x)^2 + (\Sigma F \sin \theta_x)^2}$$

Thus the magnitude of the resultant is known from R, and the direction and sense of the resultant is known from the values and signs of ΣF_x and ΣF_y. The problem can be seen to have been solved at this point as there is no further computation necessary regarding the location of the resultant.

In the case of a non-concurrent force system, the same approach is used in determining the resultant. But the solution is not complete, since the location of the resultant is not known in the space diagram. The principle of moments is used to complete the solution by locating the resultant in space.

Take for example the general coplanar force system shown in Fig. 2.41. We determine the magnitude of R, its direction and sense, using the same formulae given above. The calculations are shown in tabular form below, which is always preferable.

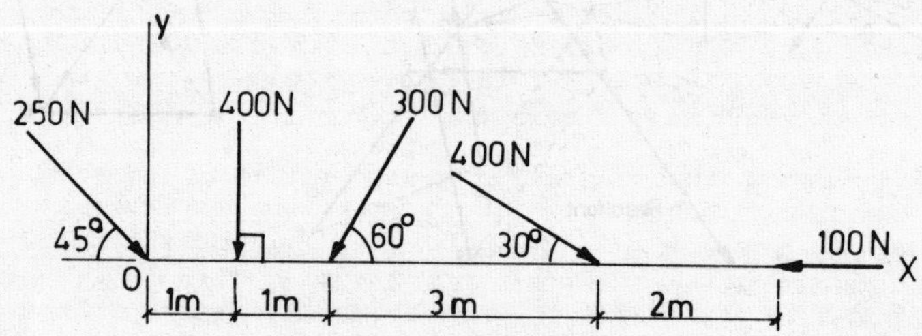

Fig. 2.41

Force	*Angle with X-axis* θ_x	*X-component*	*Y-component*
250 N	45°	176.8	–176.8
400 N	90°	–	–400.0
300 N	60°	– 150.0	–259.8
400 N	30°	+346.4	–200.0
100 N	0°	– 100.0	–

$$\Sigma F_x = 273.2\text{N}$$

$$\Sigma F_y = -1036.6\text{N}$$

$$R = \sqrt{(273.2)^2 + (-1036.6)^2}$$

$$= \mathbf{1072\ N}$$

$$\tan \theta_x = \frac{-1036.6}{273.2}$$
$$= 3.794$$
$$\theta_x = 75.2°$$

For the X- and Y- axes shown in Fig. 2.41, the reader should carefully verify the calculations shown in table above.

Now the question arises—where is the resultant located in the space diagram? We take the help of the principle of moments to answer this question. In section 2.6.3, we proved the Varigon's theorem of moments for two concurrent forces. It is easy to see that the same proof is valid for any number of concurrent forces. We state below a generalized principle of moments, without proof, for any coplanar force system:

The algebraic sum of the moments of the forces of any Coplanar force system about any moment centre in their plane is equal to the moment of their resultant about the same moment centre.

Accepting the above principle to be true, we apply this principle to the force system shown in Fig. 2.41. We take moments about the point 'o' (i.e. about an axis through 'o' perpendicular to the plane of the paper). We compute the moments for clarity in tabular form shown below:

Force (N)	*Components* (N)	*Perpendicular distance from 'o' (m)*	*Moment about 'o' (Nm) (Clockwise +)*
250 N	X – 176.8	0	0
	Y – 176.8	0	0
400 N	X – 0	—	0
	Y – 400	1	+ 400
300 N	X – 150	0	0
	Y – 259.8	2	+ 519.6
400 N	X – 346.4	0	0
	Y – 200	5	+ 1000.0
100 N	X – 100	0	0
	Y – 0	0	0
Total:			1919.6

The algebraic sum of the moments of the force system is, therefore, equal to 1919.6 Nm. We calculated the moment of each force as the algebraic sum of the moments of its components which is true according to Varignon's theorem. The resultant of the force system has been calculated as 1072 N. If this resultant is at a perpendicular distance of '*d*' from point '*o*', then the moment of the resultant about '*o*' is equal to 1072 d Nm. Therefore,

$$1072 \text{ d} = 1919.6$$
$$\text{d} = \frac{1919.6}{1072}$$
$$= \mathbf{1.79 \text{ m.}}$$

The position of the resultant is, therefore, as shown in Fig. 2.42.

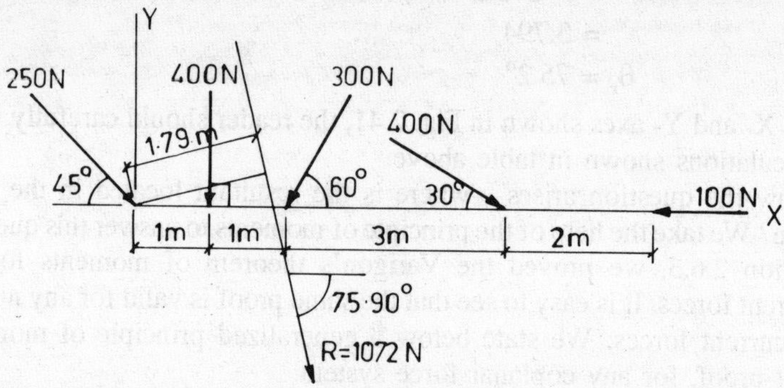

Fig. 2.42

We summarise the procedure outlined above for ready reference as under:

(i) Select suitable co-ordinate axes and determine the *X*- and *Y*- components, carefully assigning proper signs to the components. A component is positive when it is along the positive direction of the axes.

(ii) Determine the magnitude of the resultant by the formula $R = \sqrt{(\Sigma F_x)^2 + (\Sigma F_y)^2}$, where ΣF_x and ΣF_y are the algebraic sum of the components of forces along the X- and Y- axes.

(iii) Determine the direction of the resultant by the formula $\tan \theta_x = \dfrac{\Sigma F_y}{\Sigma F_x}$ or $\tan \theta_y = \dfrac{\Sigma F_x}{\Sigma F_y}$, where θ_x and θ_y are the acute angles the resultant makes with the *X*- and *Y*- axes respectively.

(iv) The sense of the resultant has to be obtained from the signs of the total components ΣF_x and ΣF_y. Fig. 2.43(a) explains the method of determining the sense.

(v) Choose any moment centre and determine the algebraic sum of the moments (ΣM) of the forces about it. This moment divided by the magnitude of the resultant (*R*) gives the perpendicular distance (*d*) of the resultant from the moment centre.

$$d = \frac{\Sigma M}{R}$$

(vi) By knowing the nature of the moment, determine the resultant in the space diagram. Fig. 2.43(b) shows the positioning of the resultant for M clockwise and anti-clockwise, and for different signs of the components F_x and F_y. This is given only to clarify the explanation and is *not* to be committed to memory.

We have outlined the general approach, both graphical and analytical, to the determination of the resultant of a non-concurrent force system. We discuss below particular cases; first the simpler parallel force systems, and then the non-parallel systems.

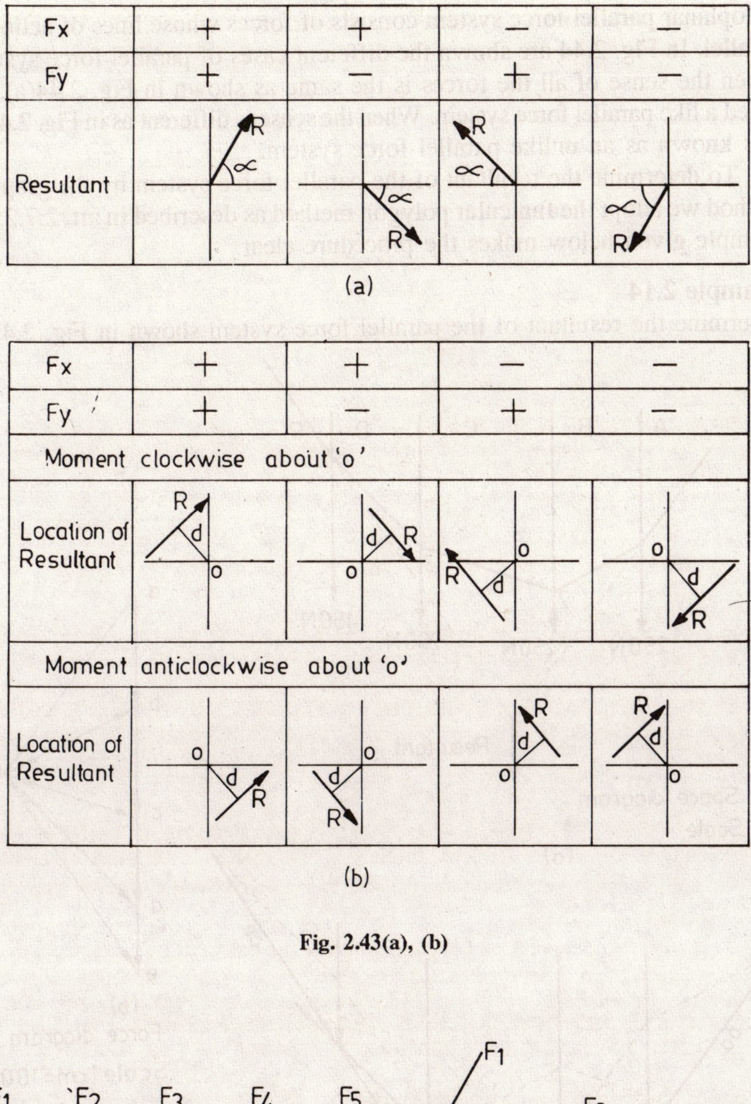

Fig. 2.43(a), (b)

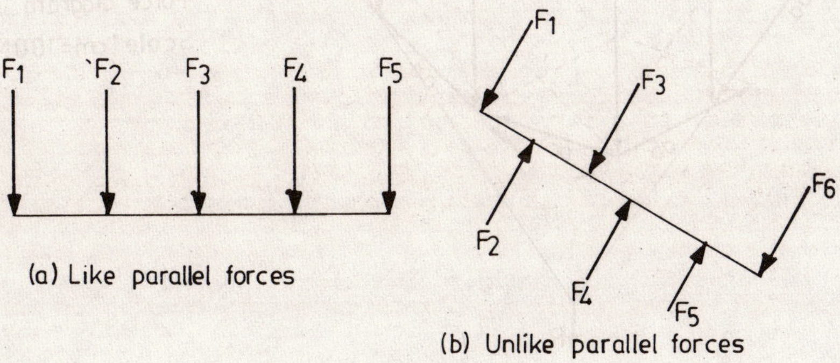

Fig. 2.44

2.7.4 *Resultant of Parallel Force System by Graphical Method*

A coplanar parallel force system consists of forces whose lines of action are parallel. In Fig. 2.44 are shown the different cases of parallel force systems. When the sense of all the forces is the same as shown in Fig. 2.44(a), it is called a like parallel force system. When the sense is different as in Fig. 2.44(b), it is known as an unlike parallel force system.

To determine the resultant of the parallel force system by the graphical method we adopt the funicular polygon method as described in art. 2.7.2. The example given below makes the procedure clear.

Example 2.14

Determine the resultant of the parallel force system shown in Fig. 2.45(a).

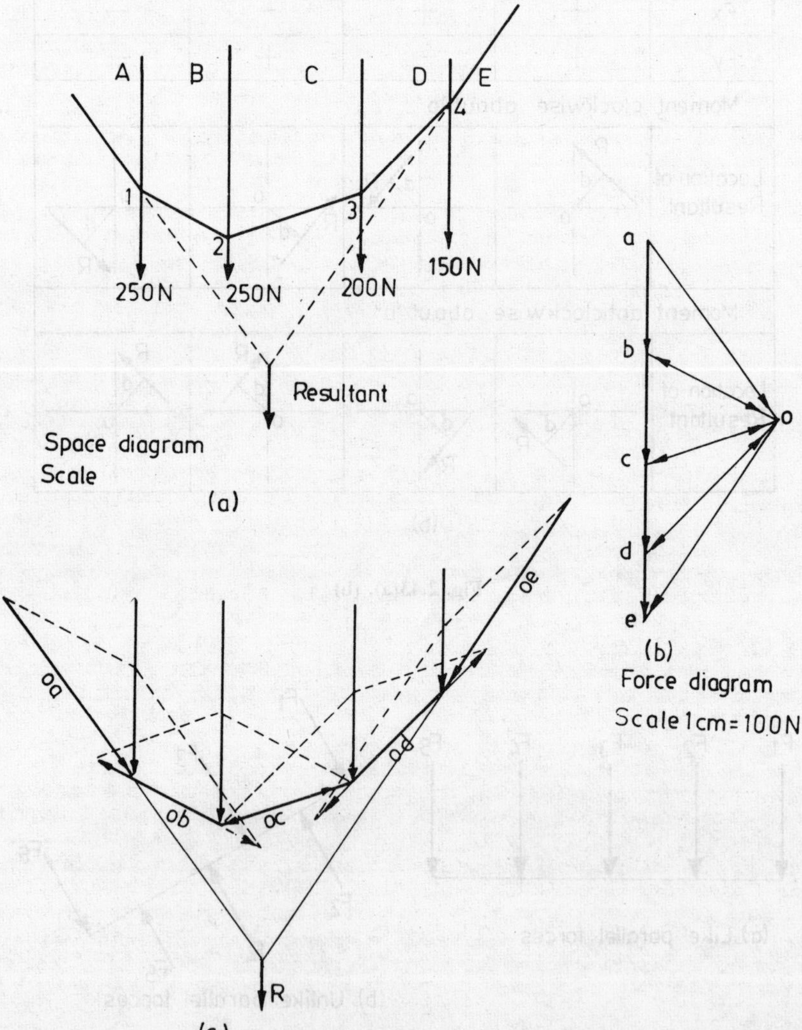

Fig. 2.45

Solution

We start by drawing the space diagram shown in Fig. 2.45(a). The force polygon in this case becomes a straight line, since the lines of action of all the forces are parallel. The force diagram is shown in Fig. 2.45(b). We select a pole and construct the funicular polygon and locate the resultant in space diagram.

The resultant is thus completely determined. From Fig. 2.45, by measurement,

Length of *ae* = 8.5 cm
Magnitude of resultant = 850 N
Direction of resultant = parallel to the given force system
Sense of resultant = from *a* to *e*.

The resultant is drawn in the space diagram in Fig. 2.45(a).

Figure 2.45(c) shows the resolution of forces done by the Funicular polygon showing the unbalanced components *oa* and *oc*.

2.7.5 *Resultant of Parallel Force System by Analytical Method*

To find the resultant of a parallel force system analytically, we apply the general approach described in section 2.5.7 with slight modifications. Since the lines of action of all the forces are parallel, we select the X-axis (or Y-axis) parallel to the lines of action of the forces. The Y-components (or X-components) are then all zero. The resultant is then simply the algebraic sum of all the forces of the system. The principle of moments is applied to locate the resultant in the space diagram. The following example will make the procedure clear.

Example 2.12

Determine analytically the resultant of the coplanar parallel force system shown in Fig. 2.46.

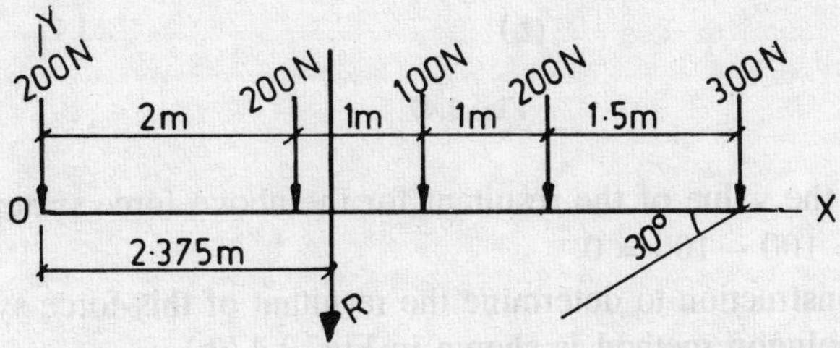

Fig. 2.46

Solution

We select the X- and Y- axes as shown in Fig. 2.46. Since the X-components are all zero, the resultant is the algebraic sum of the forces.

$$\text{Resultant magnitude} = -200 - 200 + 100 + 200 - 300$$
$$= \mathbf{-400\ N.}$$

The resultant magnitude 400 N, has its line of action parallel to that of the given forces and acts in a sense opposite to the positive direction of the Y-axis.

The algebraic sum of the moments of the given forces about '*o*' is calculated as, (clockwise moments +).

$$M = 200 \times 2 - 100 \times 3 - 200 \times 4 + 300 \times 5.5$$
$$= 950 \text{ Nm.}$$

If 'd' is the perpendicular distance of the resultant from point '*o*', then by the principle of moments,

$$R \times d = \Sigma M$$
$$400 \times d = 950$$
$$d = \frac{950}{400}$$
$$= \mathbf{2.375\ m.}$$

2.7.6 *Special Case of Two Unlike, Equal Parallel Forces*

In Fig. 2.47 is shown two equal, unlike parallel forces, acting at a distance '*d*' apart.

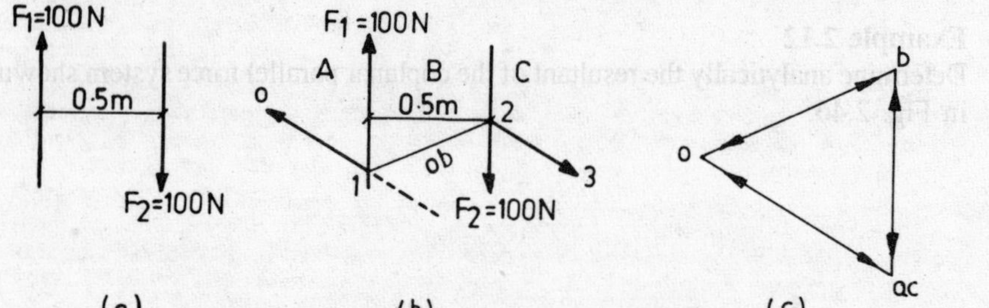

Fig. 2.47

We determine the value of the resultant for the above force system as,

$$R = 100 - 100 = 0$$

A graphical construction to determine the resultant of this force system by the funicular polygon method is shown in Fig. 2.47(b).

We draw the force polygon *abc*, which is a straight line. The points *a* and *c* coincide indicating that the resultant is zero. But the funicular polygon, when drawn shows that the first and last lines are parallel and do not intersect.

What conclusions can we draw from the above results? By the funicular polygon construction, the lines 1-0 and 2-3 are two forces with directions as shown. This again gives a situation of two unlike parallel forces as the given force system.

Such a force system is known as a couple. *A couple consists of two equal, parallel, opposite and non-collinear forces whose resultant force is zero, but has a moment.*

The force system is again shown in Fig. 2.48. Let us determine the moment of this force system about points P, Q, and R shown.

$$\Sigma M \text{ about } P = 100 \times 1.5 - 100 \times 1$$

$$= \mathbf{50\ Nm.}\ \text{Clockwise}$$

$$\Sigma M \text{ about } Q = 100 \times 2.5 - 100 \times 2 = 50 \text{ Nm clockwise}$$

$$\Sigma M \text{ about } R = 100 \times (x + 10.5) - 100 \times x = 50 \text{ Nm clockwise}$$

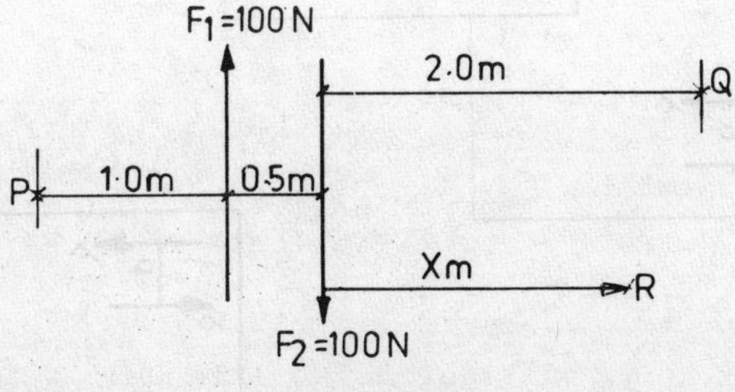

Fig. 2.48

It is evident that the moment about all the points are the same. From the moment about point R, which is at an arbitrary distance 'x', it is clear that the moment about any point will be equal to 50 Nm since the moment about R is independent of 'x'.

Another important conclusion can be drawn from the funicular polygon constructed in Fig. 2.47. We have seen that the funicular polygon gives us two forces represented by 1 - 0 (= *ao* in the force polygon) and 2 - 3 (= *oa* in the force polygon) which are again equal, opposite, non-collinear and parallel. This again is a couple. If we find the moment of this couple, by taking moments about a point on the line of action of 1 - 0, we get

$$\text{Moment} = ao \times d$$

$$= (2.2 \times 50) \times \left(\frac{1.8}{4}\right)$$

$$= 49.5 \text{ Nm} \approx 50 \text{ Nm.}$$

From the above discussion, the following ***characteristics of a couple*** can be arrived at.

The external effect of a couple is one of pure rotation (turning effect). The effect of a couple is dependent upon:

(i) the magnitude of the moment of a couple
(ii) sense of the couple (clockwise or anti-clockwise moment), and
(iii) the plane of the couple.

The above three attributes are known as the characteristics of a couple.

It should be noted that the location of a couple in its plane is not important since the couple has the same moment about every point in its plane. The following modifications, illustrated in Fig. 2.49, can be performed without altering the effect of a couple.

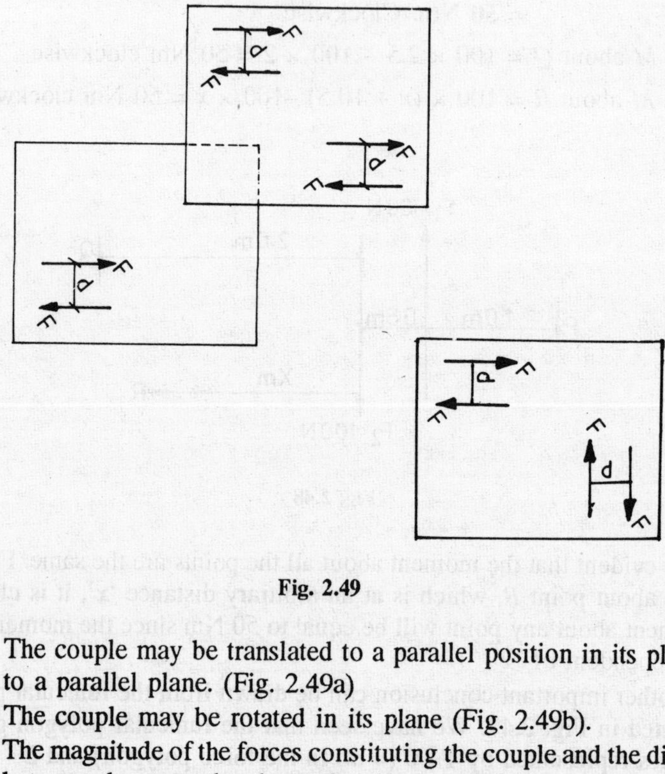

Fig. 2.49

(i) The couple may be translated to a parallel position in its plane or to a parallel plane. (Fig. 2.49a)
(ii) The couple may be rotated in its plane (Fig. 2.49b).
(iii) The magnitude of the forces constituting the couple and the distance between them may be changed so long as the moment of the couple remains the same in magnitude and sense.

2.7.7 *Resolution of a Force into a Force and a Couple*

In many physical situations, it is very convenient to represent a force acting along a particular line of action into a force acting along a different, but parallel line of action. This is done using the concept of a couple and the two fundamental axioms stated below:

Axiom (i) Two forces acting on a body are in equilibrium when they are collinear, equal and opposite.

Axiom (ii) The effect of a force system 'F' on a body remains unchanged if a force system in equilibrium is added to or subtracted from 'F'.

Using the above, we can translate a force parallel to its line of action as shown in Fig. 2.50.

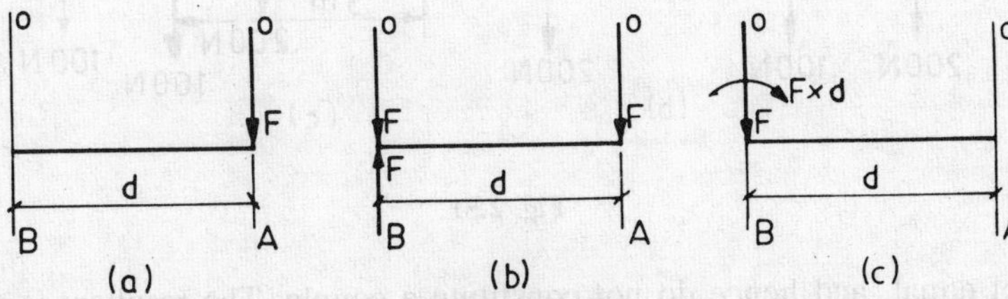

Fig. 2.50

Fig. 2.50(a) shows a body acted upon by a force F acting along the line OA. We want to shift the force F to a parallel line of action OB.

To achieve this, we apply two equal and opposite forces 'F' along the line OB as shown in Fig. 2.50(b). According to axiom (i) the two forces acting along OB are in equilibrium. According to axiom (ii) the addition of these two forces does not alter the effect of force F on the body.

Consider the force F acting along OA and the force equal and opposite to it acting along OB. Obviously, this constitutes a couple, and the force acting downwards along OB is the same as if the force acting along OA is translated to OB. We represent this situation in Fig. 2.50(c), where the curved arrow indicates a couple.

We may, therefore, conclude that whenever we have to translate a force parallel to its line of action, this can be achieved as follows:

(i) *We translate the force parallel to its line of action to the new position,* and

(ii) *Introduce a couple equal in moment to the moment of the force about any point on its new line of action.*

It is to be noted that this is different from the principle of transmissibility studied in chapter I. There the force was shifted along its line of action and this can be done by moving the force along its line of action at will, when the body is rigid as is assumed in Statics. We give below a number of examples to illustrate the principles enunciated so far.

Example 2.13

Determine the resultant of the parallel force system shown in Fig. 2.51(a) and (b).

Solution

We note in Fig. 2.51(a) that these are two unlike parallel forces but they

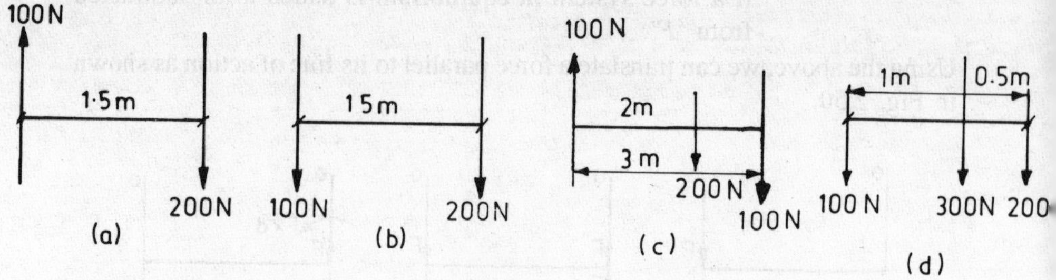

Fig. 2.51

are not equal, and hence do not constitute a couple. The resultant is to be determined considering it as a parallel force system.

$$\text{Resultant} = 200 - 100$$
$$= \mathbf{100\ N} \text{ (acting downwards)}$$

The resultant is located by applying the principle of moments. Taking moments about point '*A*' on the line of action of the 100 N force,

$$100 \times d = 200 \times 1.5$$
$$d = \frac{300}{100} = \mathbf{3\ m.}$$

The resultant, therefore, acts at a distance of 3 m from *A* and is acting downwards as shown in Fig. 2.51(c).

In Fig. 2.51(b) we have two like parallel forces.

Resultant is, therefore, given by

$$\text{Resultant} = 100\text{ N} + 200\text{ N}$$
$$= 300\text{ N}.$$

Applying the principle of moments and taking moments about point *B*, we have

$$300 \times d = 200 \times 1.5$$

or $\quad d = 1$ m

The resultant, therefore, acts as shown in Fig. 2.51(d).

Example 2.14

Shift the force *F* acting at point *A* to a parallel line of action at point *B* shown in Fig. 2.52(a).

Solution

Since the distance between *A* and *B* is $\sqrt{2^2 + 1^2} = \sqrt{5}$, we can shift the force to a parallel line of action at '*B*' as shown in Fig. 2.52(b). In addition a couple is introduced at *B*. The moment of the couple is equal to:

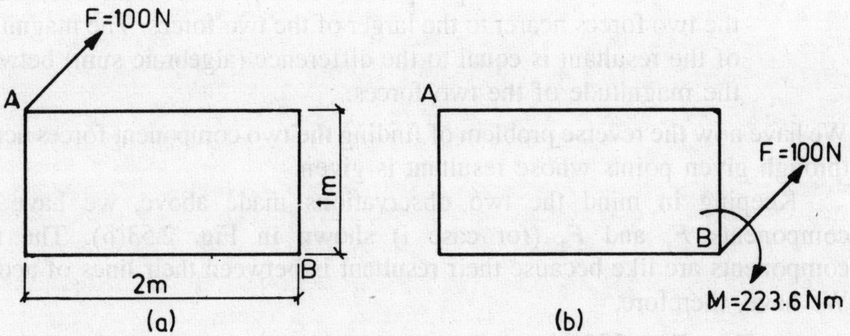

Fig. 2.52

$$\text{Moment of couple} = 100 \times \sqrt{5}$$
$$= \mathbf{223.6\ Nm.}$$

Example 2.15

Resolve the force of 300 N shown in Fig. 2.53(a) into two parallel components (i) at *B* and *C* and (ii) at *C* and *D*.

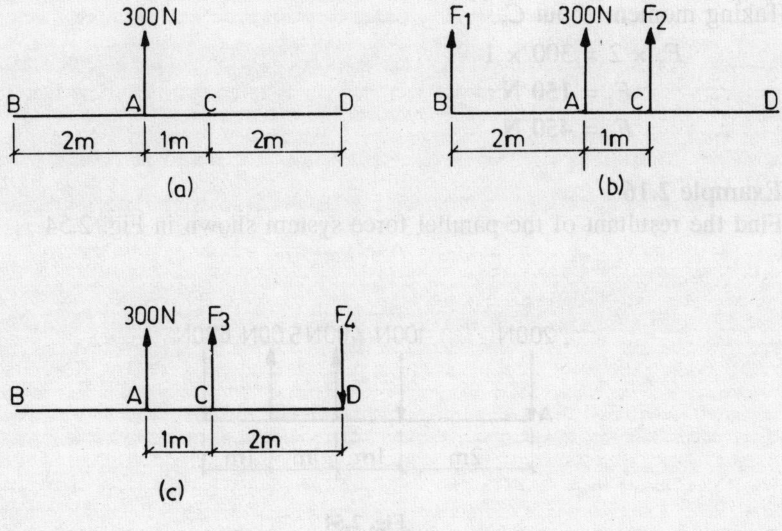

Fig. 2.53

Solution

From example 2.13, we arrive at the following conclusions:

(i) that for two like parallel forces, the resultant lies between the twc forces. The magnitude of the resultant is the sum of the 2 forces.

(ii) that for two unlike unequal parallel forces, the resultant lies outside the two forces nearer to the larger of the two forces. The magnitude of the resultant is equal to the difference (algebraic sum) between the magnitude of the two forces.

We have now the reverse problem of finding the two component forces acting through given points whose resultant is given.

Keeping in mind the two observations made above, we have the components F_1 and F_2 (for case i) shown in Fig. 2.53(b). The two components are like because their resultant is between their lines of action. We have, therefore,

$$F_1 + F_2 = 300 \text{ N}$$

By the principle of moments, taking moment about a point on the line of action of F_1.

$$300 \times 2 = F_2 \times 3$$

$$\therefore \quad F_2 = \mathbf{200\ N}$$

$$\text{and} \quad F_1 = 300 - 200$$

$$= \mathbf{100\ N}$$

For case (ii), the situation is shown in Fig. 2.53(c). Note carefully that components are unlike since their resultant does not lie in between their lines of action. Further, we have marked the force F_3 acting upwards, at C.

Taking moment about C,

$$F_4 \times 2 = 300 \times 1$$

$$\therefore \quad F_4 = \mathbf{150\ N}$$

$$\therefore \quad F_3 = \mathbf{450\ N}$$

Example 2.16

Find the resultant of the parallel force system shown in Fig. 2.54.

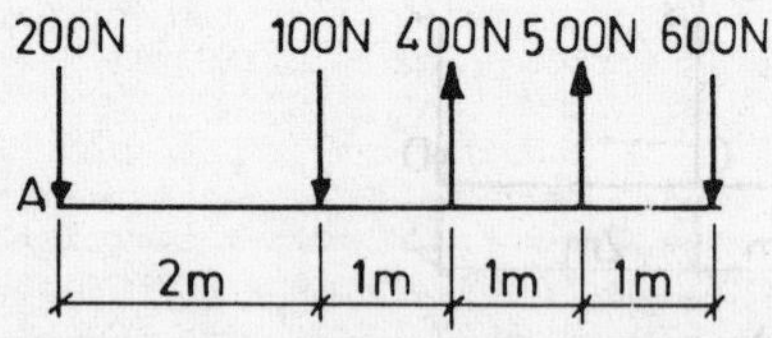

Fig. 2.54

Solution

(a) Analytical solution: The magnitude of the resultant is the algebraic sum of the forces. Taking downward forces as positive,

$$\therefore \quad R = +200 + 100 - 400 - 500 + 600$$

$$R = 0$$

There is no resultant force. The possibility of the system forming a couple should be checked in the case of an unlike parallel force system. This is so because, since R = 0, the sum of the upward forces is equal to the sum of the downward forces. If the resultants of all the upward forces and that of the downward forces are not collinear, they will form a couple when R = 0. If the force system forms a couple, it can be checked by taking moments about any point, since the couple has the same moment about all points in its plane. Taking moments about A, considering clockwise moments +,

$$M \text{ about } A = +\,100 \times 2 - 400 \times 3 - 500 \times 4 + 600 \times 5$$
$$= 0$$

Therefore, the given force system has no resultant force, and does not form a couple.

(b) Graphical solution: The graphical solution is shown in Fig. 2.55. The force polygon *abc def* is drawn to scale. The points *a* and *f* coincide indicating zero resultant force. To check whether the force system forms a couple, we draw the funicular polygon 0123456. The lines 0-1 and 5-6 are parallel and on producing them, they are also collinear. Hence these two unlike forces do not form a couple, confirming the result obtained in the analytic solution.

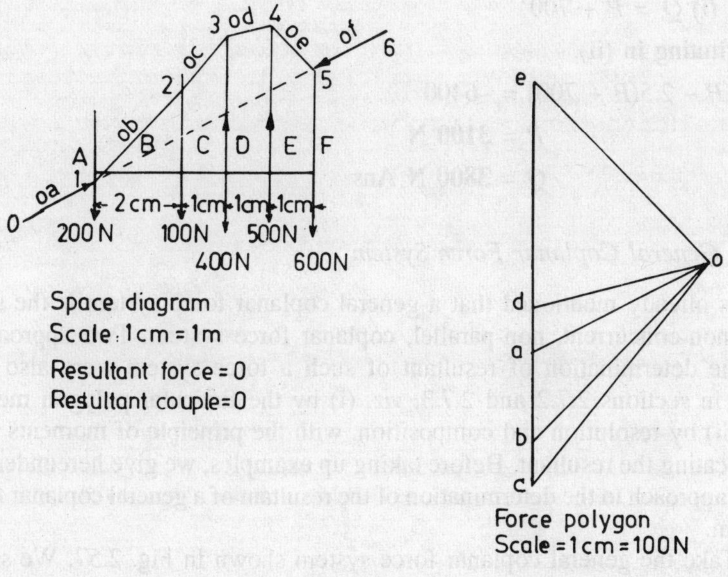

Fig. 2.55

Example 2.17

The four forces shown in Fig. 2.56 have an upward resultant of magnitude 300 N which acts through a point 2 m to the right of the 100 N force. Determine the value of the forces P and Q.

Solution

Since the resultant is a 300 N upward force, we have

$$2000 - Q + P - 1000 = 300$$

$$P - Q = -700 \quad \text{... (i)}$$

Applying the principle of moments about a moment centre on the line of action of the 1000 N force, taking anti-clockwise moments as positive,

$$P \times 1 - Q \times 2.5 + 2000 \times 3.5 = 300 \times 2$$

$$P - 2.5\,Q = -6400 \quad \text{... (ii)}$$

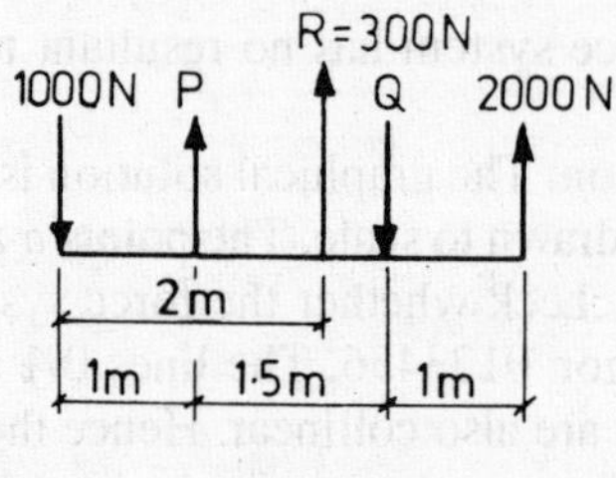

Fig. 2.56

From (i) $Q = P + 700$

Substituting in (ii),

$$P - 2.5(P + 700) = -6400$$

$$P = \mathbf{3100\ N}$$

$$Q = \mathbf{3800\ N} \text{ Ans.}$$

2.7.8 *General Coplanar Force System*

It was already mentioned that a general coplanar force system is the same as a non-concurrent, non-parallel, coplanar force system. Two approaches for the determination of resultant of such a force system were also outlined in sections 2.7.2 and 2.7.3, *viz*, (i) by the funicular polygon method and (ii) by resolution and composition, with the principle of moments used for locating the resultant. Before taking up examples, we give hereunder one more approach to the determination of the resultant of a general coplanar force system.

Take the general coplanar force system shown in Fig. 2.57. We select X- and Y- axes conveniently as shown.

From the discussion given in section 2.5.7 on translating a force parallel to its line of action, we can translate each of the forces shown in Fig. 2.57 to some point, say the origin. This has been done in Fig. 2.57(b). We have, therefore, an equivalent force system consisting of a concurrent force system and a couple, which is the algebraic sum of the couples shown against each

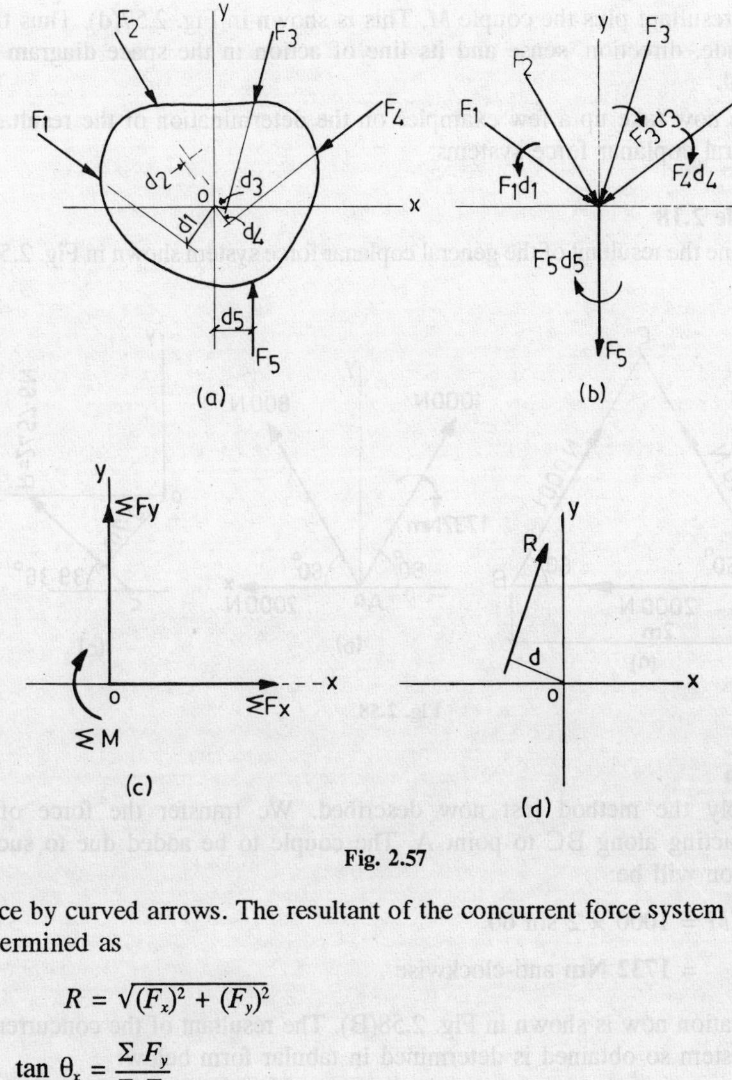

Fig. 2.57

force by curved arrows. The resultant of the concurrent force system can be determined as

$$R = \sqrt{(F_x)^2 + (F_y)^2}$$

$$\tan \theta_x = \frac{\Sigma F_y}{\Sigma F_x}$$

where,

R is the resultant

ΣF_x is the algebraic sum of the X-components

ΣF_y is the algebraic sum of the Y-components

and θ_x is the acute angle made by the resultant with the X-axis.

The system shown in Fig. 2.57(c) is equal to that shown in Fig. 2.57(b). We may now translate the resultant by a distance 'd' given by

$$d = \frac{\Sigma M}{R}$$

We translate the resultant parallel to its line of action by the distance '*d*' from the origin such that when it is translated back to the origin, we get the resultant plus the couple *M*. This is shown in Fig. 2.57(d). Thus the magnitude, direction, sense and its line of action in the space diagram is obtained.

We now take up a few examples on the determination of the resultant of general coplanar force systems.

Example 2.18

Determine the resultant of the general coplanar force system shown in Fig. 2.58.

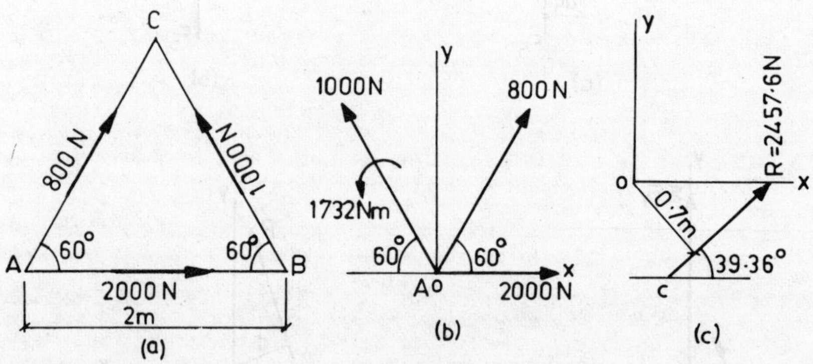

Fig. 2.58

Solution

We apply the method just now described. We transfer the force of 1000N acting along BC to point A. The couple to be added due to such translation will be:

$$M = 1000 \times 2 \sin 60$$

$$= \mathbf{1732\ Nm} \text{ anti-clockwise}$$

The situation now is shown in Fig. 2.58(B). The resultant of the concurrent force system so obtained is determined in tabular form below:

Force	θ_x	*X-component*	*Y-component*
800 N	60°	+ 400	+692.8
1000 N	60°	− 500	+866.0
2000 N	0	2000	0
Total:		+1900	+1558.8

The resultant, *R*, is given by

$$R = \sqrt{(1900)^2 + (1558.8)^2}$$

$$= \mathbf{2457.6\ N}$$

The direction of the resultant is given by

$$\theta_x = \tan^{-1}\left(\frac{1558.8}{1900}\right)$$

$$= \mathbf{39.36°}$$

We now shift the resultant by a distance

$$d = \frac{1732}{2457.6}$$

$$= \mathbf{0.70\ m.}$$

The final location of the resultant is shown in Fig. 2.58(c).

Example 2.19

The rectangular framework is acted upon by a general coplanar force system shown in Fig. 2.59(a). Determine the position and magnitude of the resultant force acting on the framework. What is the moment of the force system about point 'O'?

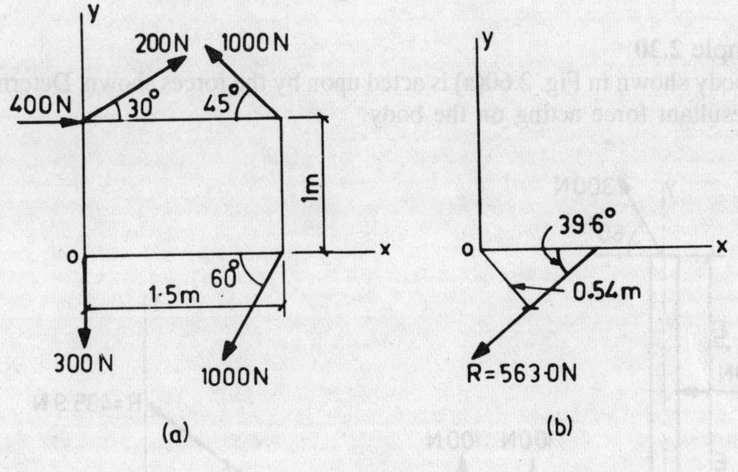

Fig. 2.59

Solution

Select X- and Y-axes as shown. The magnitude of the resultant is determined in tabular form below.

Notice that in the table, we have included a moment column so that we may apply the principle of moment to locate the resultant. Clockwise moments are taken as positive.

Force (N)	*Angle with X-axis*	*X-comp.*	*Y-comp.*	*Moment about 'O' (Nm)*
200	30°	+ 173.2	+ 100	173.2 Nm
1000	45°	– 707.0	+ 707	(– 707 × 1 – 707 × 1.5) = – 1767.5
300	90°	–	– 300	–
1000	60°	– 500.0	– 866	+ 1299.0
600	0	+ 600	–	+ 600.0
Total:		– 433.8	– 359.0	+ 304.7

$$\text{Resultant } R = \sqrt{(-433.8)^2 + (-359.0)^2}$$

$$= \mathbf{563.0\ N}$$

$$\theta_x = \tan^{-1}\left(\frac{-359}{-433.8}\right)$$

$$= \mathbf{39.61°}$$

$$\text{Distance 'd' from 'O'} = \frac{304.7}{563.0}$$

$$= \mathbf{0.54\ m}$$

Position of the resultant is shown in Fig. 2.59(b).

Example 2.20

The body shown in Fig. 2.60(a) is acted upon by the forces shown. Determine the resultant force acting on the body.

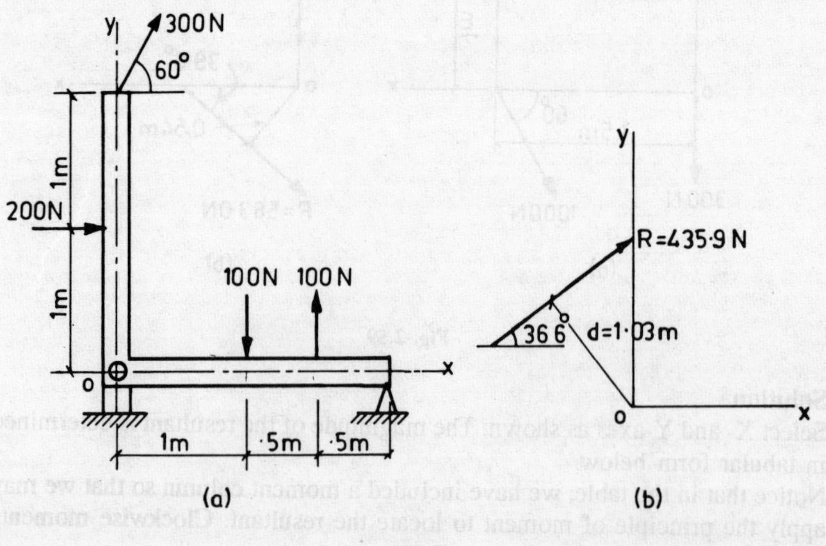

Fig. 2.60

Solution

Analytical: The body is acted upon by a general coplanar force system. We proceed as in Ex. 2.19. The computations are shown in the table below. The X- and Y- axes are selected as shown. Clockwise moments are considered positive.

Force (N)	*Angle with X-axis*	*X-comp.*	*Y-comp.*	*Moment about 'o' Nm ( )*
300	60°	+ 150	+ 259.8	+ 300
200	0	+ 200	–	+ 200
100	90°	–	– 100	+ 100
100	90°	–	+ 100	– 150
Total		+ 350	+ 259.8	+ 450

$$\text{Resultant } R = \sqrt{(350)^2 + (259.8)^2}$$
$$= \mathbf{435.9\ N}$$
$$\theta_x = \tan^{-1}\left(\frac{-259.8}{350}\right)$$
$$= \mathbf{36.58°}$$

$$\text{Perpendicular distance 'd' of Resultant from 'o'} = \frac{450}{435.9}$$
$$= \mathbf{1.03\ M.}$$

The resultant is shown in Fig. 2.60(b).

Graphical: The graphical solution is shown in Fig. 2.61. We lay out the space diagram to scale. The scale chosen is 1 cm = 1 m. The forces are designated by letters ABCDE as shown. The force diagram is drawn to a scale of 1 cm = 100 N. We select a pole '*O*' and draw the funicular polygon 012345. Note carefully the different lines in the funicular polygon. Each line is to be drawn in the space allotted to it. The lines 0-1 and 4-5 intersect at point '*p*'. We draw a line parallel to the resultant *ae* through point '*p*'.

$$\text{Length } ae = 8.5 \text{ cm}$$
$$\text{Resultant} = 8.5 \times 50 = 425 \text{ N}$$

$$\text{Perpendicular distance of resultant from 'O'} = 1.05 \text{ cm}$$
$$= 1.05 \times 1.0 \text{ m}$$
$$= 1.05 \text{ m}$$

You will notice that the results closely agree with the analytical solution given above.

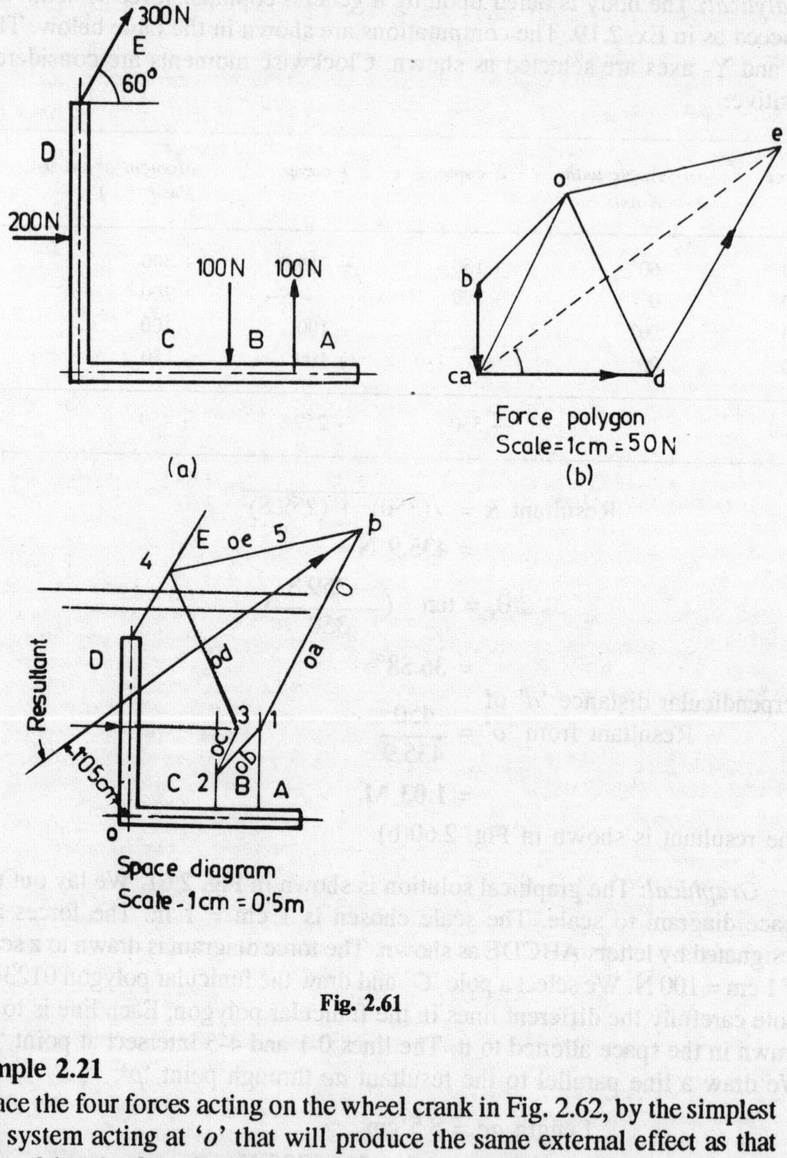

Fig. 2.61

Example 2.21

Replace the four forces acting on the wheel crank in Fig. 2.62, by the simplest force system acting at '*o*' that will produce the same external effect as that produced by the four forces.

Solution

We select the X- and Y- axes as shown. We transfer all the forces to point '*o*' parallel to their line of action. The situation then is as shown in Fig. 2.62(b). We now have a concurrent force system at '*o*' with a couple of 250 Nm. The concurrent force system can be reduced to a single resultant force acting at '*o*' as shown in table on page 80.

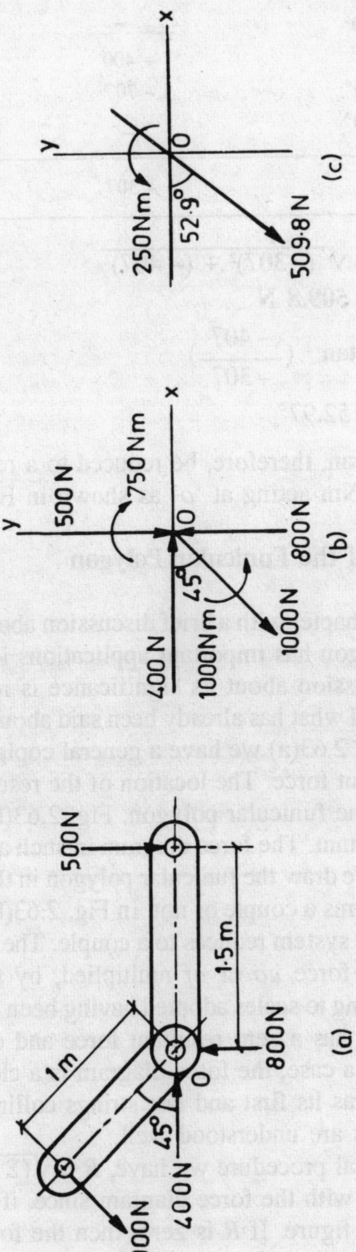
1000N
400N
45°
1m
O
800N
1·5 m
500N
(a)
1000N
1000Nm
400N
45°
500N
750Nm
y
x
O
800N
(b)
509·8 N
52·9°
250Nm
y
x
O
(c)

Fig. 2.62

Force	Angle with X-axis	X- Component	Y- Component
500	90°	–	– 500
400	0	+ 400	–
1000	45°	– 707	– 707
800	90°	–	+ 800
Total		– 307	– 407

$$\text{Resultant } R = \sqrt{(-307)^2 + (-407)^2}$$
$$= \mathbf{509.8\ N}$$
$$\theta_x = \tan^{-1}\left(\frac{-407}{-307}\right)$$
$$= 52.97°$$

The force system can, therefore, be reduced to a resultant force of 509.8 N and a couple 250 Nm acting at '*o*' as shown in Fig. 2.62(c).

2.8 Significance of the Funicular Polygon

We will close this chapter with a brief discussion about the funicular polygon. The funicular polygon has important applications in structural analysis and hence a brief discussion about its significance is relevant at this stage.

To briefly recall what has already been said about funicular polygon, refer to Fig. 2.63. In Fig. 2.63(a) we have a general coplanar force system, which reduces to a resultant force. The location of the resultant is obtained in such cases by drawing the funicular polygon. Fig. 2.63(b) shows a force system that has a zero resultant. The force diagram in such a case is a polygon which is a closed figure. We draw the funicular polygon in this case to check whether the force system forms a couple or not. In Fig. 2.63(b), the funicular polygon shows that the force system reduces to a couple. The magnitude of the couple is obtained by the force *ao* or *of* multiplied, by the distance '*d*', proper conversions according to scales adopted having been done. Fig. 2.63(c) shows a force system that has a zero resultant force and does not form a couple. Notice that in such a case, the force diagram is a closed figure and also the funicular polygon has its first and last strings collinear. It is important that these characteristics are understood well.

By the analytical procedure we have, $R = \sqrt{(\Sigma F_x)^2 + (\Sigma F_y)^2}$. This has a close relationship with the force diagram since, if R is not zero, the force diagram is an open figure. If R is zero, then the force diagram is a closed figure and indicates that $\Sigma F_x = 0$ and $\Sigma F_y = 0$. In the analytical procedure, we locate the resultant by the principle of moments. The corresponding graphical procedure is by drawing the funicular polygon.

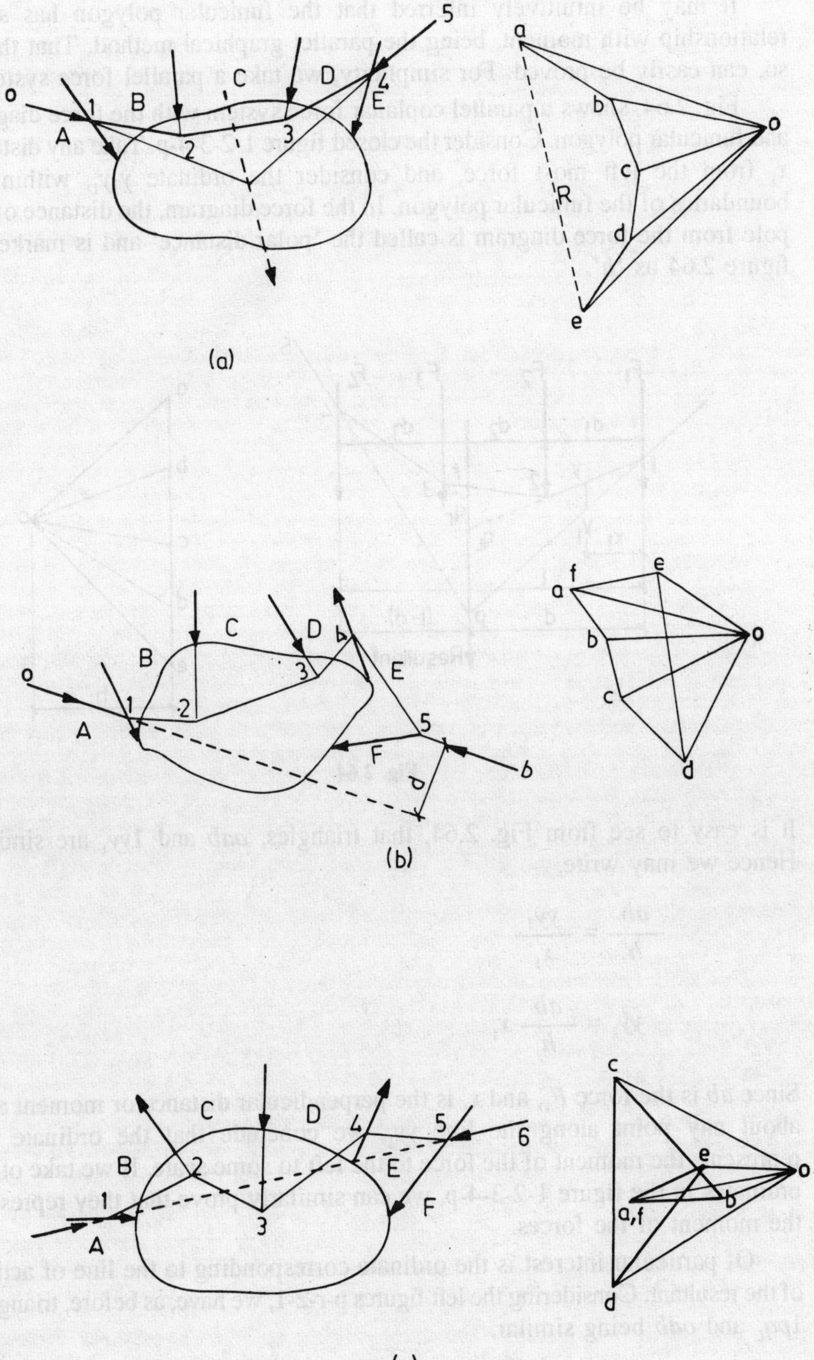

Fig. 2.63

It may be intuitively inferred that the funicular polygon has some relationship with moment, being the parallel graphical method. That this is so, can easily be proved. For simplicity, we take a parallel force system.

Fig. 2.64, shows a parallel coplanar force system with the force diagram and funicular polygon. Consider the closed figure 1-2-3-4-p. Take any distance x_1 from the left most force, and consider the ordinate y-y_1, within the boundaries of the funicular polygon. In the force diagram, the distance of the pole from the force diagram is called the 'polar distance' and is marked in figure 2.64 as '*h*'.

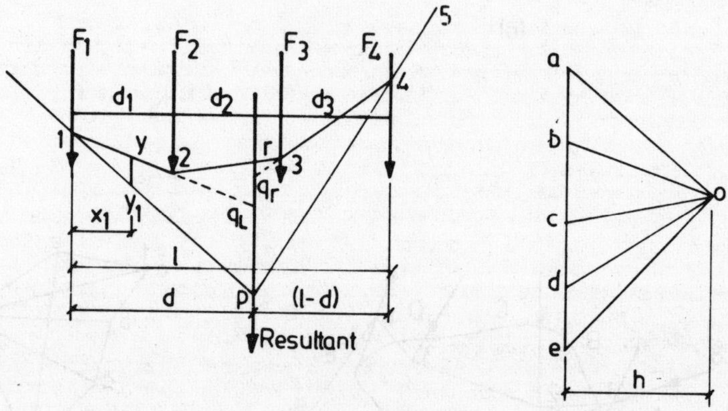

Fig. 2.64

It is easy to see from Fig. 2.64, that triangles, *oab* and $1yy_1$ are similar. Hence we may write,

$$\frac{ab}{h} = \frac{yy_1}{x_1}$$

$$yy_1 = \frac{ab}{h} x_1$$

Since *ab* is the force F_1, and x_1 is the perpendicular distance or moment arm about any point along the line yy_1, we conclude that the ordinate yy_1 represents the moment of the force to the left to some scale. If we take other ordinates in the figure 1-2-3-4-p, we can similarly prove that they represent the moment of the forces.

Of particular interest is the ordinate corresponding to the line of action of the resultant. Considering the left figures p-r-2-1, we have, as before, triangles $1pq_L$ and *oab* being similar,

$$\frac{ab}{h} = \frac{pq_L}{d}, \; pq_L = \frac{ab \times d}{h}$$

Triangles *obc* and $2q_L R$ are also similar.

$$\text{Hence,}\quad \frac{bc}{h} = \frac{q_L\ R}{(d-d_1)},\ Rq_L = \frac{bc \times (d-d_1)}{h}$$

$$\therefore\quad pr = \frac{1}{h}\ (ab \times d + bc\ (d-d_1))$$

= Moment of the forces F_1 and F_2 about a point on the line of action of the resultant to the same scale.

Considering the right figure *p-r·3-4,* we have similarly,

$$\frac{de}{h} = \frac{pq_R}{l-d},\ fq_R = \frac{de\ (l-d)}{h}$$

$$\text{and}\quad \frac{cd}{h} = \frac{rq_R}{(l-d-d_3)},\ rq_R = \frac{cd\ (l-d-d_3)}{h}$$

$$\text{and}\quad \frac{cd}{h} = \frac{rq_R}{(l-d-d_3)},\ rq_R = \frac{cd\ (l-d-d_3)}{h}$$

$$\therefore\ pr = \frac{1}{h}\left[de\ (l-d) + cd\ (l-d-d_3)\right]$$

= Moment of the forces F_3 and F_4 about a point on the line of action of the resultant.

Since the two moments calculated are equal and opposite (one clockwise and the other anti-clockwise), the net moment about a point on the line of action of the resultant of the four forces equal to zero. This is the statement of the principle of moments *when the moment centre is on the line of action of the resultant.*

We can, therefore, conclude the equivalence of analytical and graphical methods as given below:

Analytical	*Graphical*
(i) $R = \sqrt{(F_x)^2 + (F_y)^2}$	The force diagram. When the force diagram closes, $R = 0$ or $\Sigma\ F_x = 0$ and $\Sigma\ F_y = 0$.
(ii) Principle of Moments $M = R \times d$.	The funicular polygon. When $R = 0$ and $\Sigma\ M \neq 0$, we have a couple obtained from funicular polygon. When $R = 0$ and $\Sigma\ M = 0$, we have the first and last lines of the funicular polygon collinear.

We will again discuss these concepts in the next chapter. These concepts are also important in problems relating to structural analysis.

Summary

Force is a vector quantity and requires both magnitude and direction to specify it. Force is measured in unit of Newton as per SI system. Force can be graphically represented by a line with an arrow, the length of the line representing the magnitude, arrow showing the sense and direction of the line showing the direction of the force.

Force systems can be classified as co-planar and non-coplanar (space). Co-planar force systems can further be classified as concurrent, when the lines of action of all the forces meet at a point, and non-concurrent when the lines of action of all forces do not meet at a point. Non-concurrent force systems can further be classified into parallel and non-parallel force systems.

Resultant of a force system is the single force which has the same effect on the body as the given force system. When two concurrent forces are acting, their resultant can be determined by the parallelogram law or the triangle law of forces. When more than two forces act at a point, their resultant is obtained by the closing line of the force polygon whose sides representt the forces in magnitude and direction. The closing line from the starting point to the end point gives the sense of the resultant.

A force can be resolved into two or more components along given directions. When the directions of resolution are at right angles, the components are called rectangular components. To find the resultant of a concurrent force system, the forces are resolved into rectangular components in any two mutually perpendicular directions and then combined to give the resultant.

Moment of a force is the force multiplied by the distance to an axis such that the axis is perpendicular to the plan containing the force. Varignon's Theorem states that the Moment of the resultant of two concurrent forces about any point in their plane is equal to the Algebric sum of the moment of these forces about the same point. This principle applies to non-concurrent forces systems as well.

To find the resultant of non-concurrent force systems, the force polygon is drawn to find the magnitude direction and sense of the resultant. To find the location of the resultant in space, the Funicular Polygon is drawn. The first and last lines of the Funicular Polygon meet at a point through which the resultant passes.

To determine the resultant analytically, resolution into Rectangular components and composition of these components is done. To obtain the location of the resultant in space, the principle of moments is used i.e.

$R = \sqrt{(\Sigma X)^2 + (\Sigma Y)^2}$ and $Rxd = \Sigma F_i \cdot d_i$ about any moment centre in the plane of the forces.

When two equal and opposite parallel forces act, the resultant system is known as a couple. A couple has the same moment about every point in its plane.

The graphical method of force Polygon is equivalent to resolution and composition of the given forces while Funicular Polygon is equivalent to the principle of moments.

Review Questions

1 Define the term "Force".
2 Name the effects of forces on rigid bodies.
3 Define "the line of action of a force".
4 Name the characteristics of a force.
5 Classify the following quantities as scalar and vector : temperature, force, speed, mass, velocity, acceleration, length, momentum.
6 Define "Newton".
7 What does "GN" represent as a force unit multiple? How many Newtons are there in one "GN"?
8 Represent graphically a force of 100 N acting at 45° to the horizontal acting up and towards the right.
9 Sketch diagramatically a Non-concurrent, Non-parallel coplanar force system.
10 Correct the following statement:
The resultant force of a given force system acting on a rigid body is a single force that produces the same internal effect as the given force system.
11 When the magnitude, direction and point of application of two forces are the same, they produce the same effect on a rigid body. Is this true? Explain your answer.
12 Define the term "resultant."
13 A force is to be resolved into two components. How many solutions are possible? Explain.
14 Draw a triangle *abc*. If this is a force triangle, indicate by arrows the force and its components (consider each side as the force alternately).
15 Is it true that the lines of action of any two components of a force intersect on the line of action of the force?
16 Two forces act at a point. State the range of values of their resultant. Explain with a sketch.
17 Can a force have a component in a direction normal to its line of action?
18 Is it possible for a force to have a component larger in magnitude than the force itself? Explain your answer.
19 For a system of concurrent coplanar forces the algebraic sum of their components along X-axis is found to be zero. What can you say about the resultant of these forces?
20 Is the following statement correct? "Rectangular components of a force F are forces acting at right angles to the force F".
21 State the conditions under which a force *does not* have moment about an axis.
22 Give three examples from your everyday experience where the turning effect of a force is involved.
23 Is moment a vector quantity? Explain.
24 State Varignon's theorem.
25 If the algebraic sum of the moments of two concurrent, non-collinear forces is zero about a moment centre in their plane, what conclusions can you draw about their resultant.
26 Illustrate with an example the two statements given below:
(i) A force has no moment about an axis parallel to its line of action.
(ii) A force has no moment about an axis that intersects its line of action.
27 Is it possible to determine the resultant of a general coplanar force system using only the law of parallelogram of forces?
28 What are disadvantages of determining the resultant of a general coplanar force system using only the parallelogram law?
29 What is a funicular polygon? Do the strings of the funicular polygon represent the forces or their components to scale?
30 Explain the construction of the funicular polygon and prove that the intersection of the last two strings is a point in the line of action of the resultant.
31 Why is it not necessary to draw the funicular polygon to locate the resultant in the case of a concurrent, coplanar force system?
32 Can we find the resultant location of a concurrent, coplanar force system by drawing the

funicular polygon? Taking any concurrent coplanar force system, draw the force polygon and funicular polygon and state the conclusions you draw from such a diagram.

33 Is it true that the line of action of the resultant of two parallel forces always passes between the lines of actions of these forces?

34 Does the resultant of unlike parallel forces pass within the lines of action of the two forces?

35 What is a couple? State the characteristics of a couple.

36 When can two equal and opposite forces have zero moment?

37 Is the following statement true: for two equal, opposite and non-collinear parallel forces, the moment of the forces about any point in their plane is the same.

38 The force polygon of a parallel, coplanar force system closes. Which of the following conclusions can be drawn about the force system:

(i) There is no resultant force.
(ii) There is a resultant couple.
(iii) Either of (i) or (ii).

39 Can a couple be balanced by a single force applied to the body?

40 A body is acted upon by a system of two couples. Is the resultant action on the body also a couple? If the system has no resultant, what can you say about the two couples?

41 A system of general coplanar forces consists of five forces. If the resultant of four of the forces is a couple, what conclusions can be drawn regarding the resultant of the five forces?

42 The resultant of a general coplanar force system is a force. The algebraic sum of the moments of all the forces about moment centres A and B are equal but opposite. What conclusion can you then draw regarding the resultant?

43 For a general coplanar force system, it is found that the algebraic sum of the moments of all the forces about a point is zero. What can you say about the resultant of the force system?

44 In determining the resultant of a general coplanar force system what information about the resultant is obtained by applying the principle of moments (a) when the resultant is a force (b) when the resultant is a couple.

45 If the resultant of a general coplanar force system is a couple, in a graphical procedure using the funicular polygon, what would be the effect if the pole is changed?

46 Taking any arbitrary force system, draw the shape of the funicular polygon assuming that the resultant is (i) a force (ii) a couple and (iii) neither a force nor a couple. What can you say about the force polygon in each of the above cases?

Problems

2.1 A force of 320 N acts in a southward direction and another force of 550 N acts in the eastward direction at the same point. Determine the magnitude of the resultant of the two forces graphically.
(Ans. 636 N, 30.2°)

2.2 At a point forces of 60 N and 90 N are acting and the angle between them is 60°. Determine the magnitude and direction of their resultant.
(Ans. 130.8 N, 36.6°)

2.3 A force of 250 N acts at an angle of 50° to a force of 500 N. Determine their resultant.
(Ans. 687.9 N, 41°)

2.4 The resultant of two equal forces P acting at a point is also equal to P. Determine the angle between the two forces.
(Ans. 120°)

2.5 Find the resultant of 2 concurrent forces of 100 N each if the angle between them is (i) 0° (ii) 60° (iii) 90° (iv) 120° and (v) 180°.
(Ans. (i) 200 N (ii) 173.2 N, 30° (iii) 141.4 N, (iv) 100 N, 45° (v) zero)

2.6 A boat is hauled along a canal with parallel banks by horizontal ropes which move along the banks. One rope makes an angle of 20° with the bank and applies a force of 200 N on the boat. The second rope makes an angle of 30° with the bank and applies a force of 160 N. Find the resultant force on the boat and its direction with the bank.
(Ans. 326.7 N, 2°)

2.7 The resultant of two forces P and Q passes through point 'O'. Determine the magnitude of P and the resultant of P and Q if Q = 100 N.
(Ans. 70.7 N, 158 N)

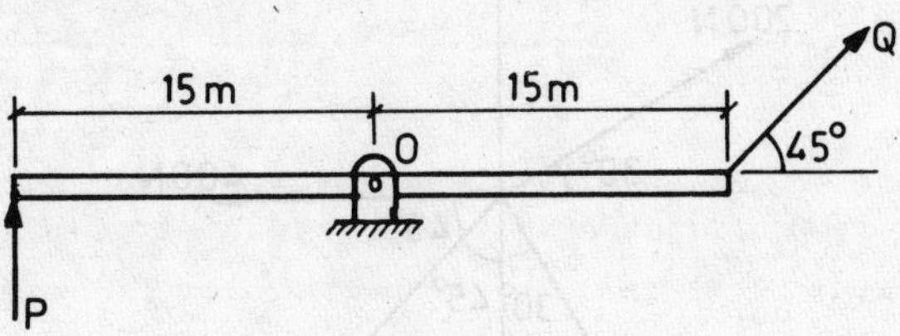

Fig. Prob. 2.7

2.8 A force of 300 N is resolved into two components. If one of the components is equal to 180 N and makes an angle of 30° with the 300 N force, find the other component and the angle between the components.
(Ans. 465N, 18.8°)

2.9 The resultant of the two forces is 200 N force as shown. Find the magnitudes of P and Q.
(Ans. P = 163.3 N, Q = 223 N)

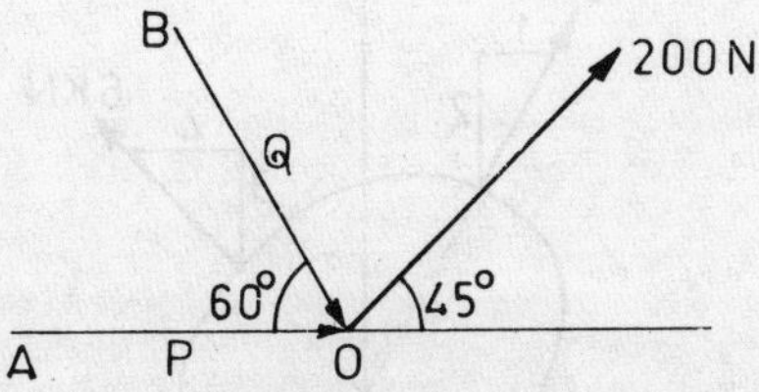

Fig. Prob. 2.9

2.10 P and Q are rectangular components of force F if P = 200 N and makes an angle of 30° with F, find F and Q.
(Ans. F = 231 N, Q = 115.5 N)

2.11 A force F has two rectangular components, F_x and F_y. Find the magnitudes of F and F_y, if F_x = 200 N and makes an angle of 60° with F.
(Ans. 400 N, 346.4 N)

2.12 Determine the resultant of the three forces shown
(Ans. 4.86 N, 109.4°)

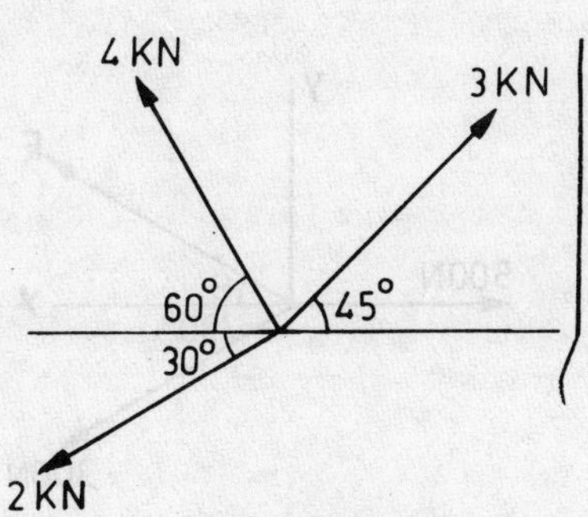

Fig. Prob. 2.12

2.13 Determine graphically the resultant of the concurrent force system shown.
(Ans. 471 N, 127.8°)

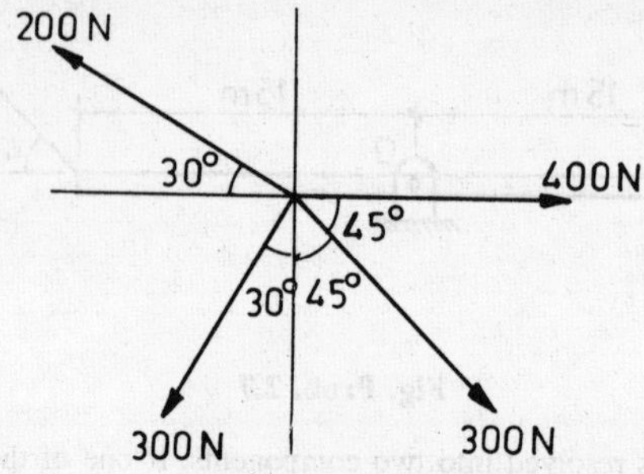

Fig. Prob. 2.13

2.14 Determine the direction and magnitude of the resultant force acting on the body.
(Ans. 12.24 N, 88°)

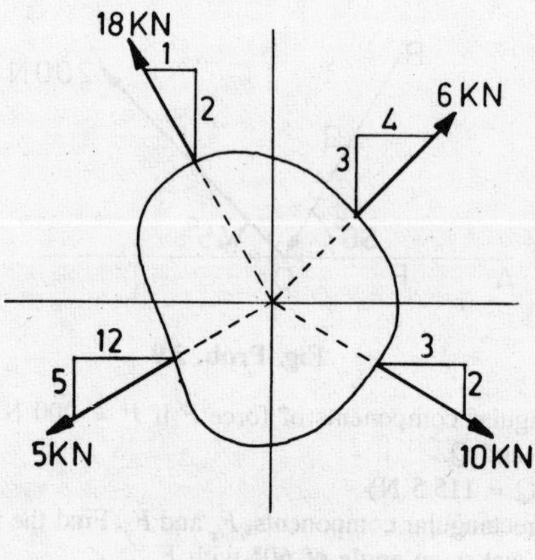

Fig. Prob. 2.14

2.15 The resultant of the concurrent force system shown is a force of 300 N acting up along the Y-axis. Compute the values of *F* and angle θ.
(Ans. 510 N, 62°)

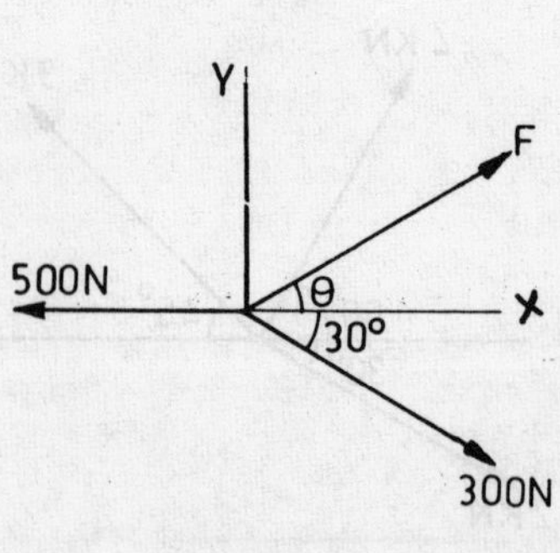

Fig. Prob. 2.15

2.16 Determine graphically the resultant force acting on the hook.
(Ans. 8.13 N, 94°)

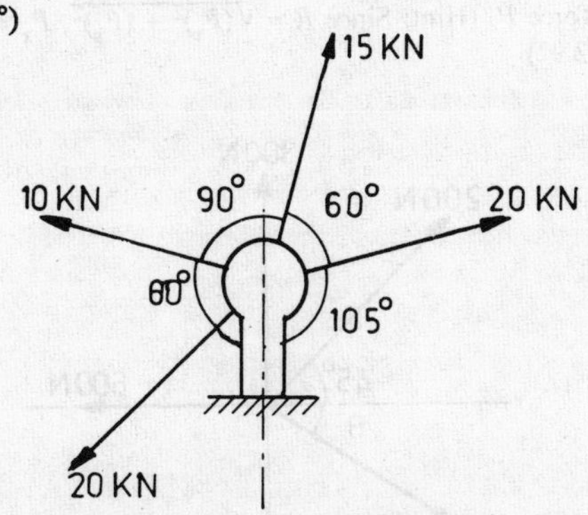

Fig. Prob. 2.16

2.17 The resultant of the five forces shown is a force equal in magnitude to Q and is collinear with and opposite to the force P. Find the magnitudes of the forces P and Q.
(Ans. Q. 38 N, P = 7.29 N)

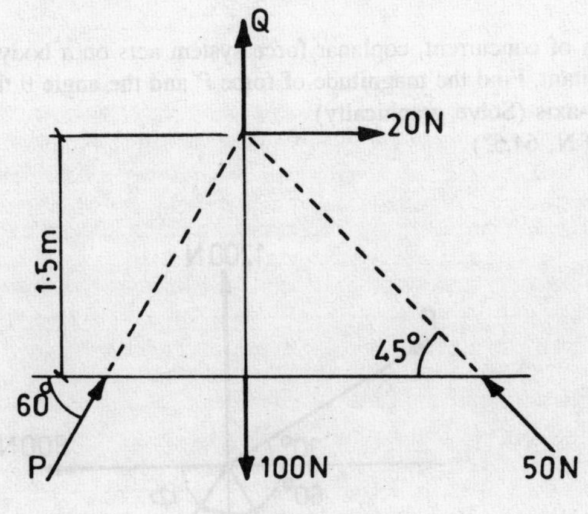

Fig. Prob. 2.17

2.18 The concurrent, coplanar force system shown has a zero resultant. Calculate the magnitude of the forces P and Q.
(Ans. P = 1412.6 N, Q = 437.8 N)

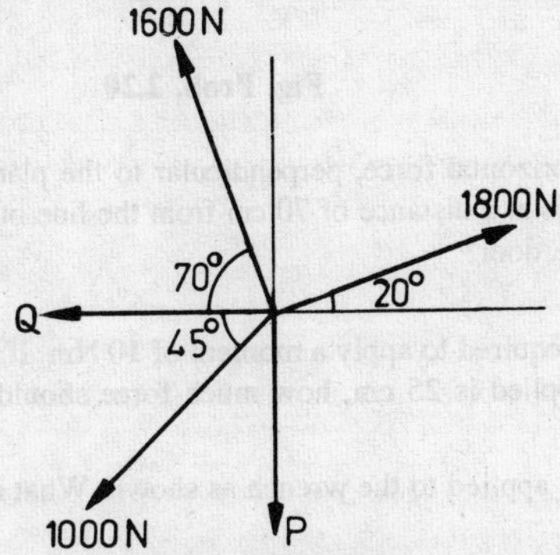

Fig. Prob. 2.18

2.19 The system of concurrent, coplanar forces shown has a zero resultant. Find the magnitude and direction of force P. (Hint: Since $R = \sqrt{(F_x)^2 + (F_y)^2}$, $F_x = 0$ and $F_y = 0$.
(Ans. 636.5 N, 43.9°)

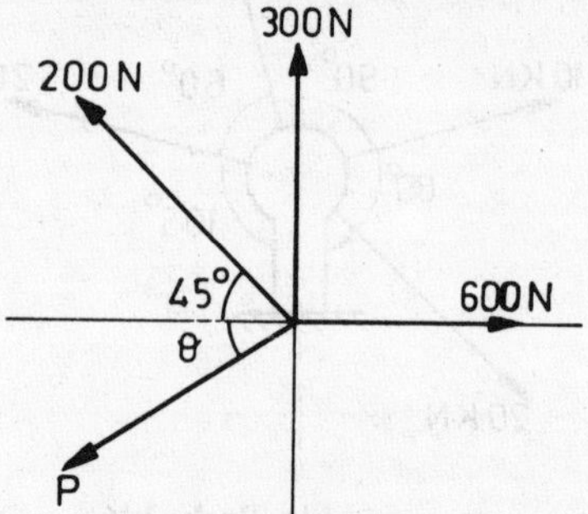

Fig. Prob. 2.19

2.20 The system of concurrent, coplanar force system acts on a body and is known to have a zero resultant. Find the magnitude of force P and the angle θ the 2000 N force makes with the X-axis (Solve graphically)
(Ans. 1560 N, 64.5°)

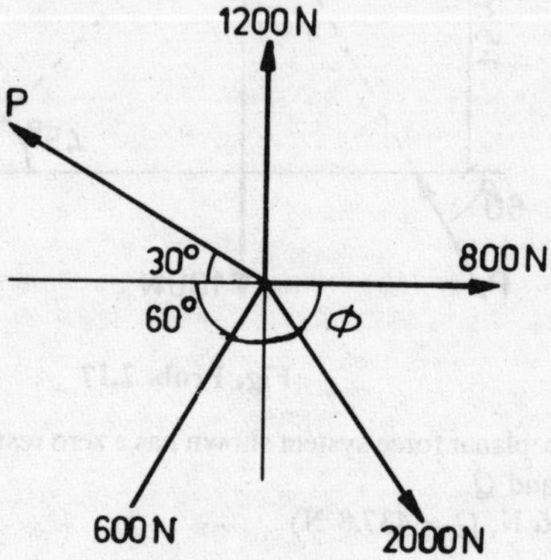

Fig. Prob. 2.20

2.21 A man applies a horizontal force, perpendicular to the plane of the door, of 30 N on a door handle which is at a distance of 70 cm from the line of hinges. What is the moment tending to open the door?
(Ans. 21 Nm)

2.22 To turn a nut, it is required to apply a moment of 10 Nm. If the effective length at which the force can be applied is 25 cm, how much force should be applied to turn the nut?
(Ans. 40 N)

2.23 A force of 160 N is applied to the wrench as shown. What is the turning effect applied?
(Ans. 34.64 Nm)

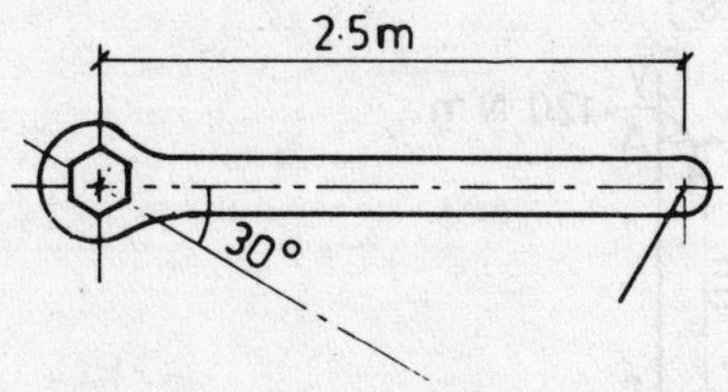

Fig. Prob. 2.23

2.24 Find the total moment of the forces F_1 and F_2 about point '*O*'.
(Ans. 831.66 Nm)

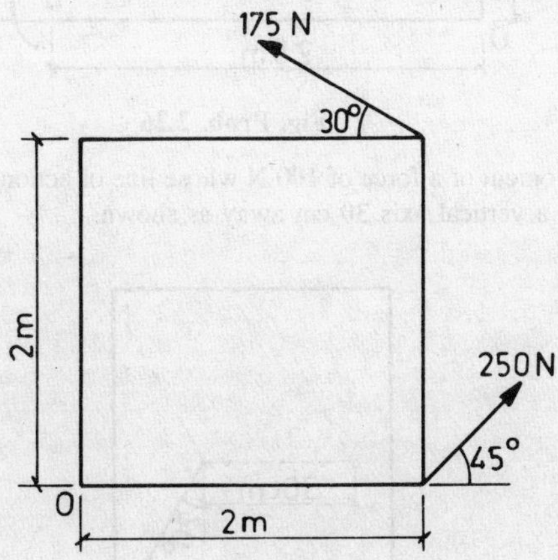

Fig. Prob. 2.24

2.25 Find the total moment of the forces *P* and *Q* about point '*A*' $Q = 100\ N$ & $P = 60\ N$
(Ans. 172.17 Nm)

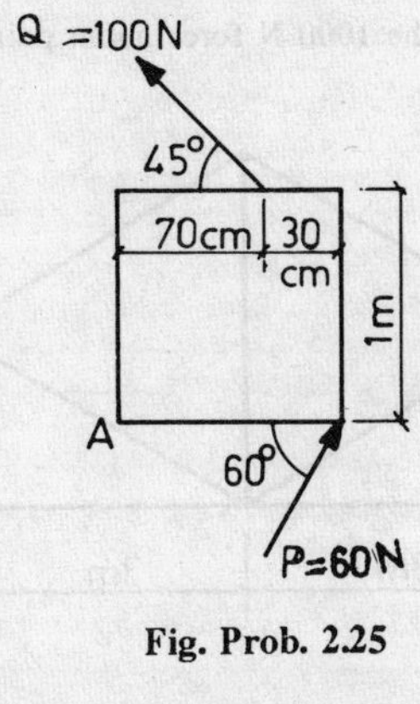

Fig. Prob. 2.25

2.26 The line of action of a force F passes through point 'C'. If the force has a moment of 120 Nm about A and 70 Nm about B both anti-clockwise, calculate the magnitude and direction of the force F.
(Ans. 156.2 N, 39.8°)

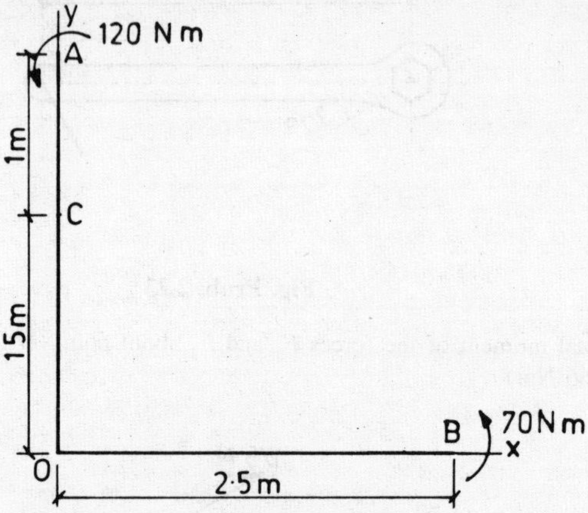

Fig. Prob. 2.26

2.27 Determine the moment of a force of 100 N whose line of action is inclined at 20° to the horizontal about a vertical axis 30 cm away as shown
(Ans. 28.2 Nm)

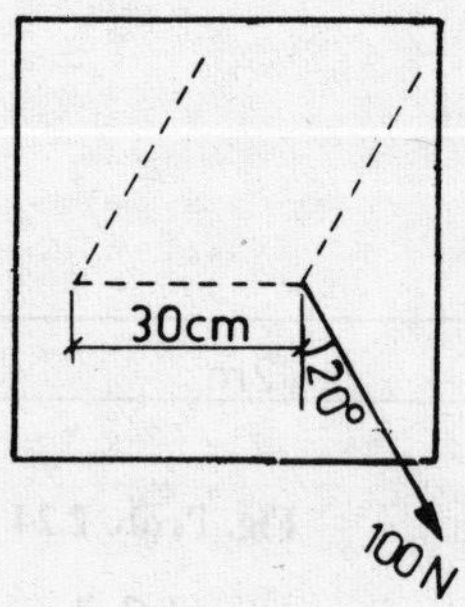

Fig. Prob. 2.27

2.28 Determine the moment of the 1000 N force about point A.
(Ans. 3000 Nm)

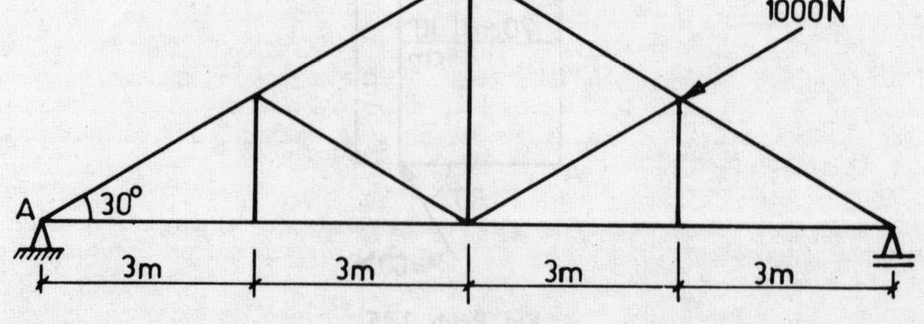

Fig. Prob. 2.28

2.29 Determine the moment of the 100 N force, lying in a plane parallel to the Z-Z plane about (i) Z-axis (ii) about X-axis. (Hint: Resolve the force into components along X- and Z-axis and note that a force has no moment about an axis parallel to its line of action)
(Ans. M_z = 43.3 Nm, M_x = 25 Nm)

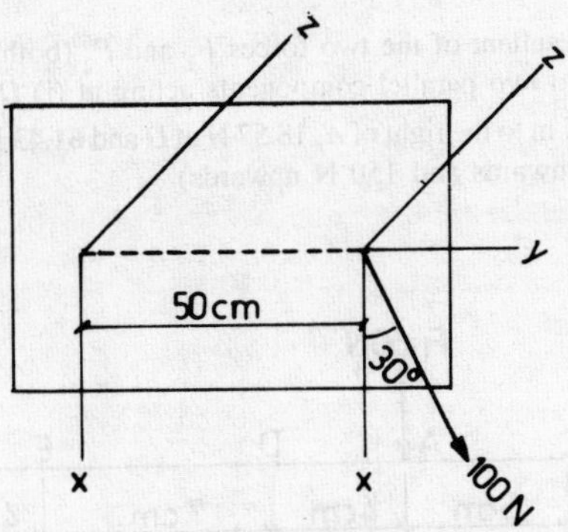

Fig. Prob. 2.29

2.30 Translate the force of 100 N, parallel to its line of action, to point *A*.
(Ans. 100 N, 100 Nm and 100 N, 86.6 Nm)

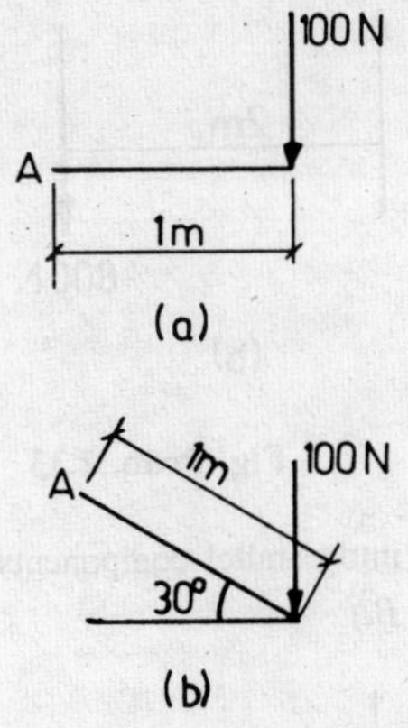

Fig. Prob. 2.30

2.31 Resolve the force of 200 N into (i) parallel components at *O* and *A* (ii) into a force and a couple at '*O*'.
(Ans. 333.3 N at *A*, 133.3 N at *O*)

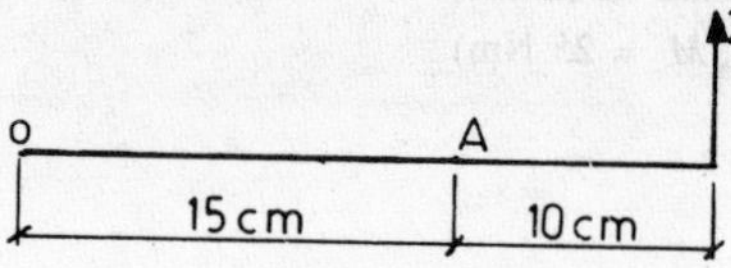

Fig. Prob. 2.31

2.32 Determine the resultant of the two forces F_1 and F_2 (both acting downwards). Resolve this resultant into two parallel components acting at (i) *D* and *E* (ii) *A* and *C*.
(Ans. 80 N, 5.375 m to the right of *A*, 18.57 N at *D* and 61.43 N at *E* both acting downwards. 230 N at *A* downwards and 150 N upwards)

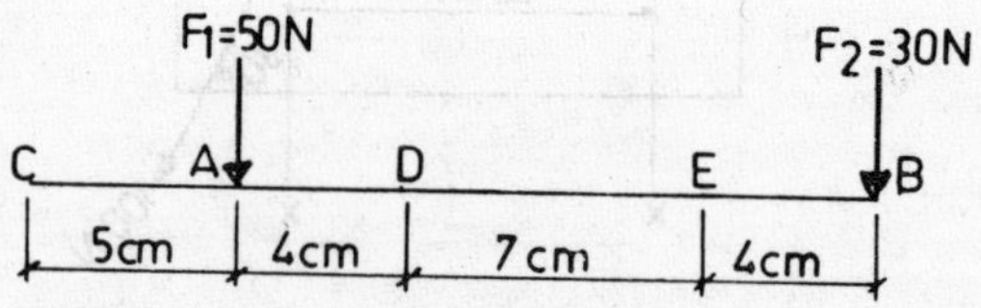

Fig. Prob. 2.32

2.33 Determine the resultant of the parallel forces shown in (a), (b) and (c).
(Ans. (a) 1000 N, 1.2 m (b) 300 N, 3.33 m from 300 N force outside, (c) Couple 400 Nm)

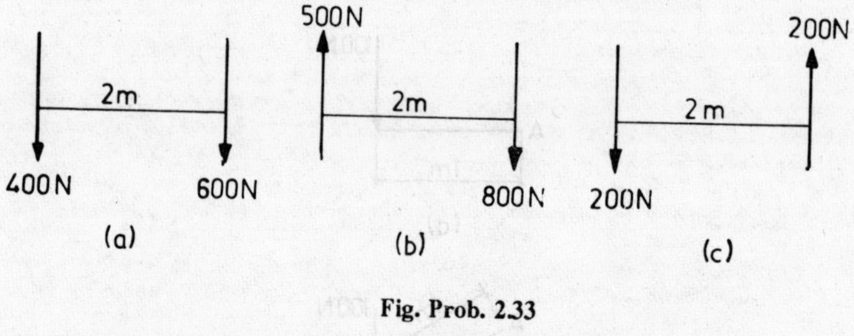

Fig. Prob. 2.33

2.34 Resolve the force of 200 N into parallel components at *A* and *B*.
(Ans. 320 N at *A*, 120 N at *B*.)

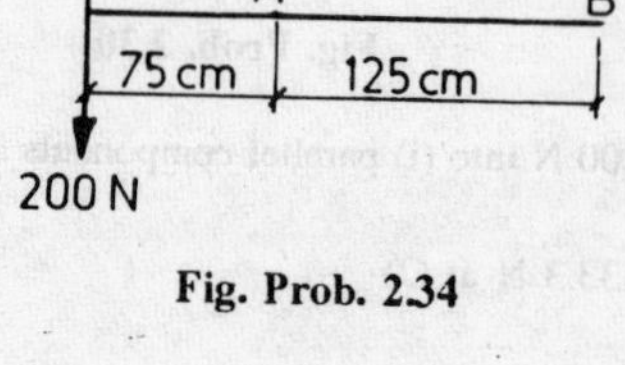

Fig. Prob. 2.34

2.35 Resolve the force of 450 N into parallel components at *D* and *E*.
(Ans. 18 √N at *E*, 270 N at *D*)

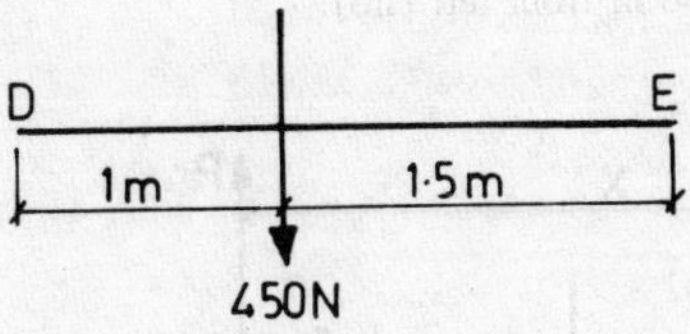

Fig. Prob. 2.35

2.36 The forces *P-Q* and the forces F_1, F_2 produce the same resultant *R* passing through '*O*'
If P = 200 N, find the magnitudes of *Q*, F_1, and F_2.
(Ans. *Q* = 300 N, F_1 = 66.67N F_2 = 133.33 N)

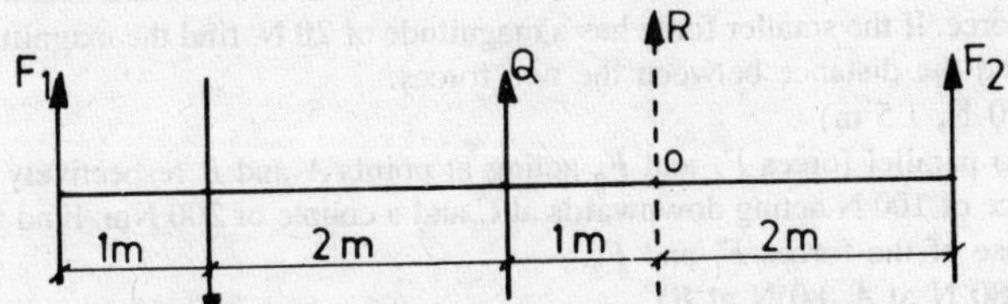

Fig. Prob. 2.36

2.37 Determine graphically the resultant of the I.R.C. class *A* train of vehicles shown.
(Ans. 9.459 m from the left end)

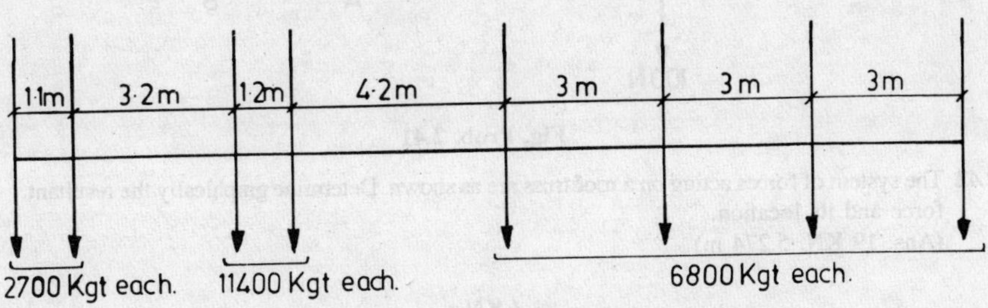

Fig. Prob. 2.37

2.38 Determine graphically the resultant of the parallel force system shown
(Ans. 200 N, 2.75 M from left end)

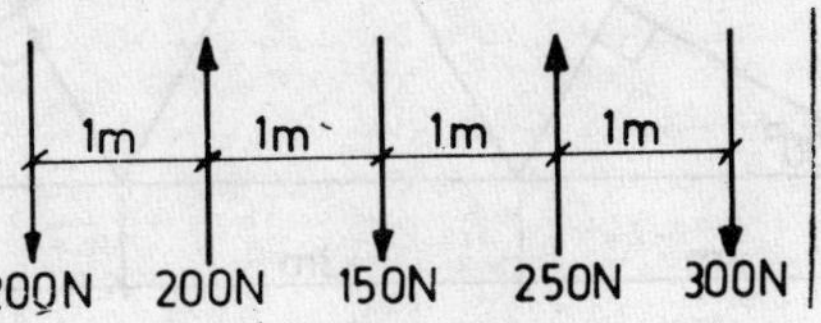

Fig. Prob. 2.38

2.39 For the parallel force system shown, the resultant is a couple of moment 180 Nm clockwise. Determine the magnitude and location of force *P*.
(Ans. 1100 N, 5.38 m from left end)

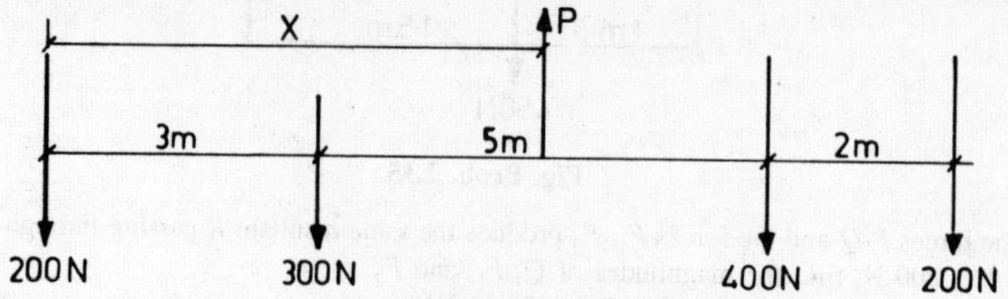

Fig. Prob. 2.39

2.40 The resultant of two unlike parallel forces is a force of 30 N and acts at 1 m from the larger force. If the smaller force has a magnitude of 20 N, find the magnitude of the other force and the distance between the two forces.
(Ans. 50 N, 1.5 m)

2.41 The two parallel forces F_1 and F_2 acting at points *A* and *B* respectively are equivalent to a force of 100 N acting downwards at *C* and a couple of 200 Nm. Find the magnitudes and sense of the forces F_1 and F_2.
(Ans. 180 N at *A*, 80 N at *B*)

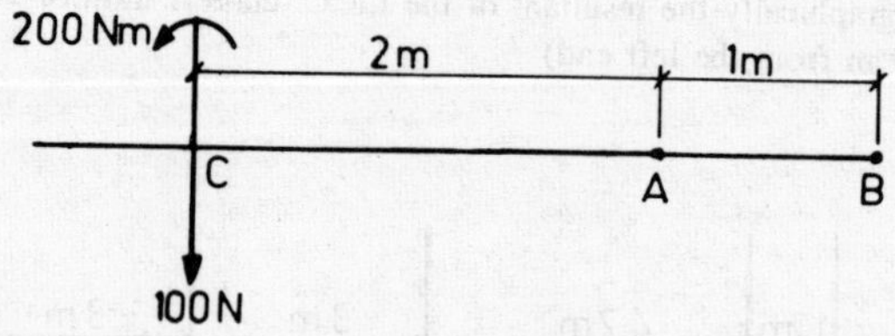

Fig. Prob. 2.41

2.42 The system of forces acting on a roof truss are as shown. Determine graphically the resultant force and its location.
(Ans. 19 KN, 5.274 m)

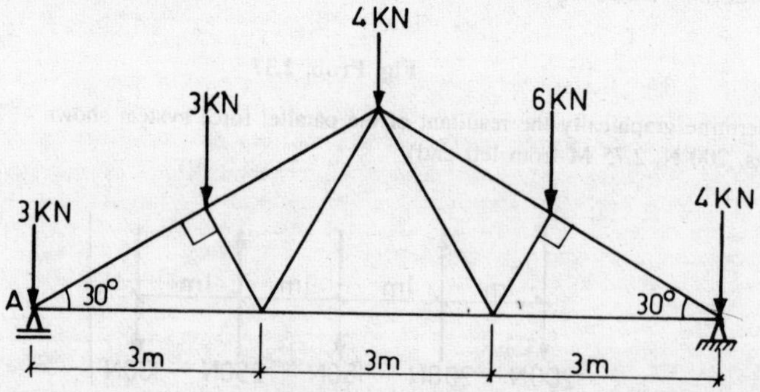

Fig. Prob. 2.42

2.43 The four parallel forces shown have a resultant which passes through a point 2.8 m from point A. Determine the magnitudes of forces P and Q, if the resultant is a force of 200 N acting upwards.
(Ans. P = 180 N, Q = – 380 N)

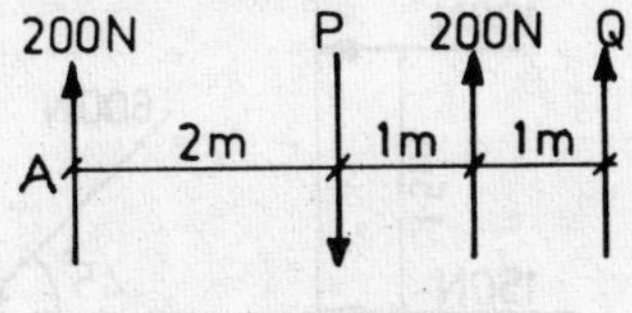

Fig. Prob. 2.43

2.44 If for the parallel force system shown, it is known that there is no resultant force and no resultant couple, find the magnitude and location of force P.
(Ans. P = 800 N, 7.5 m from left)

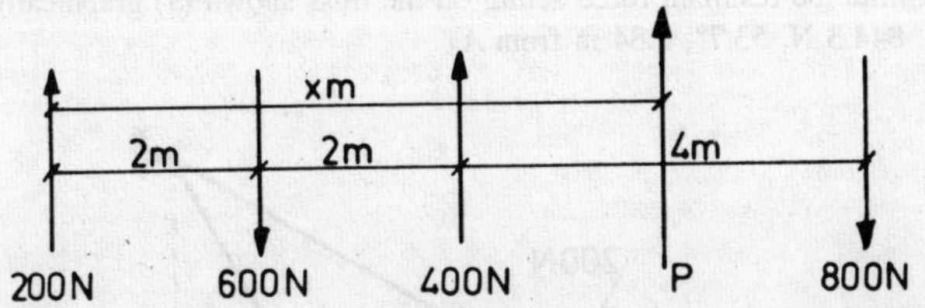

Fig. Prob. 2.44

2.45 If for the five forces shown, the line of action of the resultant (200 N) coincides with that of 100 N force, find the magnitudes of forces P and Q.
(Ans. P = 1200 N, Q = 1100 N)

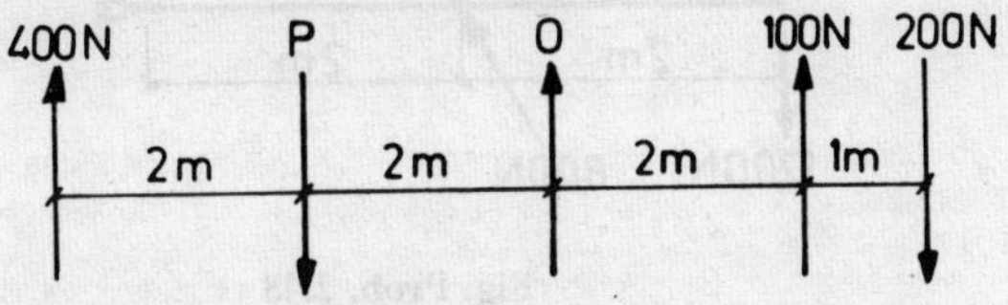

Fig. Prob. 2.45

2.46 Find graphically the resultant of the parallel forces system shown.
(Ans. 900 Nm)

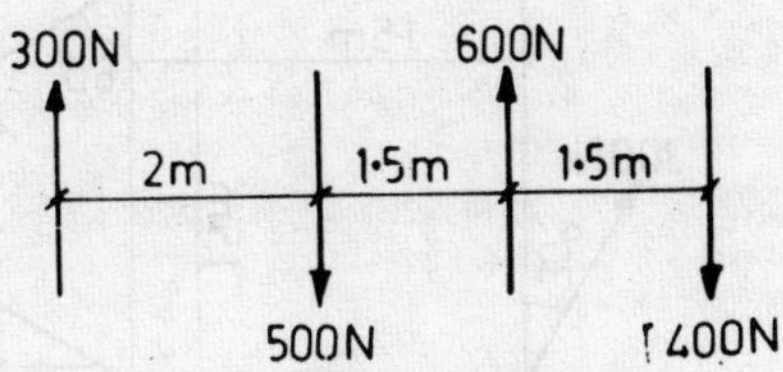

Fig. Prob. 2.46

2.47 Determine the resultant of the general coplanar force system shown.
(Ans. 919 N, 42.8° at 1.086 m from origin)

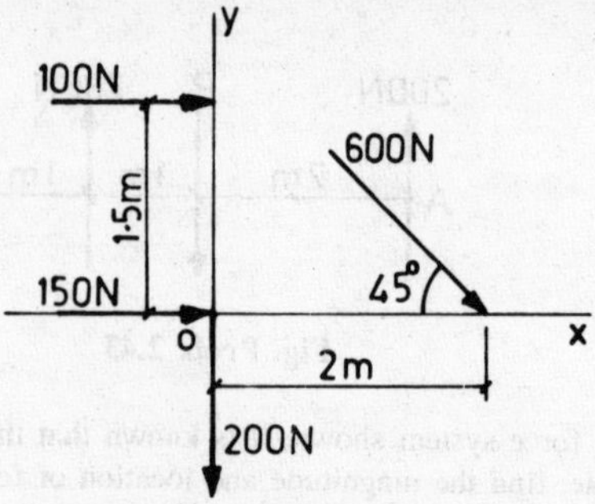

Fig. Prob. 2.47

2.48 Determine the resultant force acting on the truss shown (a) graphically (b) analytically
(Ans. 844.3 N, 53.7°, 2.84 m from A)

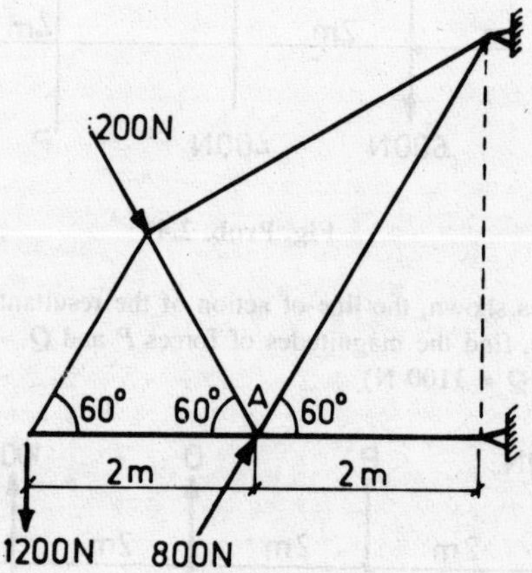

Fig. Prob. 2.48

2.49 Determine completely the resultant of the four forces shown about 'O'
(Ans. 534.6 N, 33.4°, 0.67 m from O)

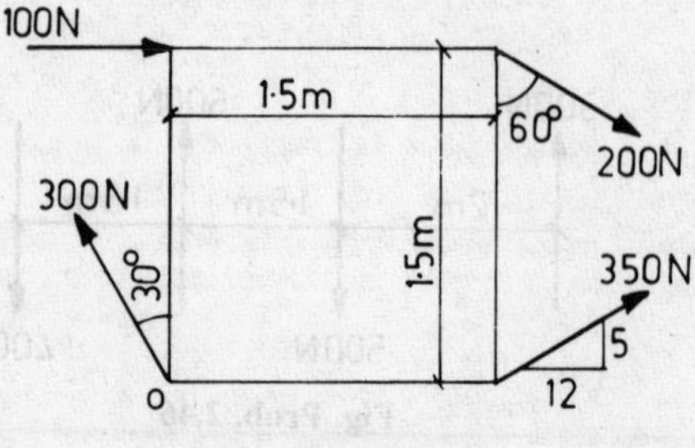

Fig. Prob. 2.49

2.50 The body shown is acted upon by three forces and a couple as shown. Determine the resultant of the force system acting on the body (Hint: A couple does not affect the computation in a force equation, but in a moment equation about any point, it has the same moment). (Ans. 400 N, 26.5°, 0.56 m from *A*)

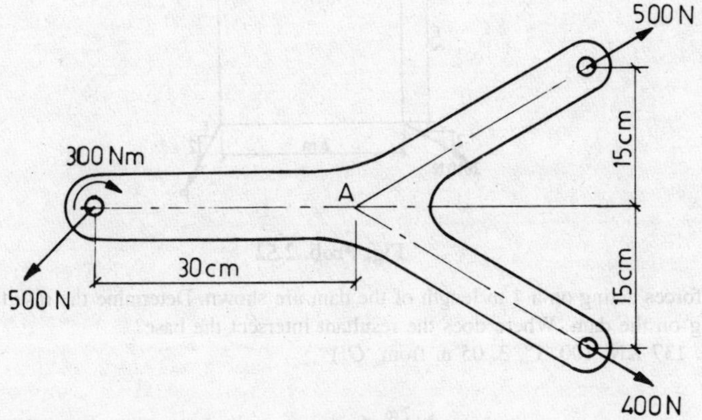

Fig. Prob. 2.50

2.51 The resultant of four forces, three of which are shown, is a couple 4800 Nm anti-clockwise. Determine the fourth force in magnitude, direction and sense.
(Ans. *P* = 2000 N horizontal to the right, 2.06 m below X-axis)

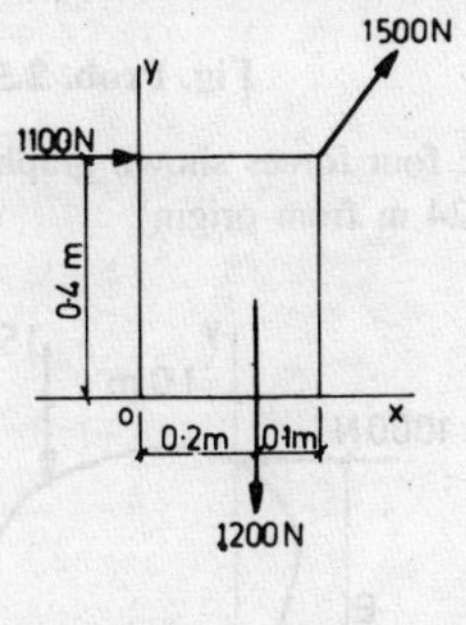

Fig. Prob. 2.51

2.52 The three forces shown have a vertical resultant passing through point *A*. Compute the value of the forces *P* and *Q*.
(Ans. *P* = 2239 N, *Q* = 6332 N)

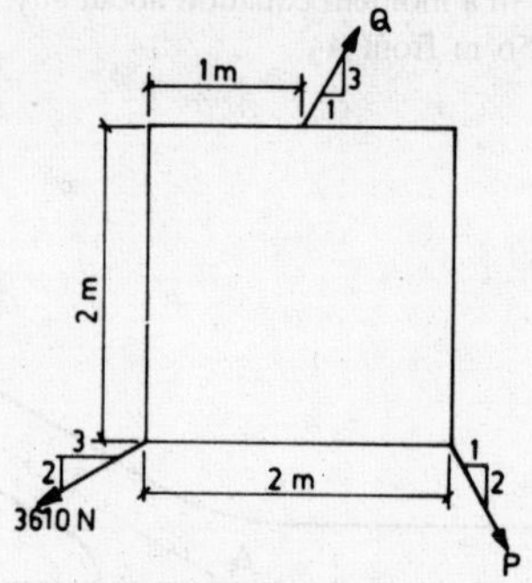

Fig. Prob. 2.52

2.53 The forces acting on a 1 m length of the dam are shown. Determine the resultant force acting on the dam. Where does the resultant intersect the base?
(Ans. 137 KN, 100. 1°, 3. 05 m from '*O*')

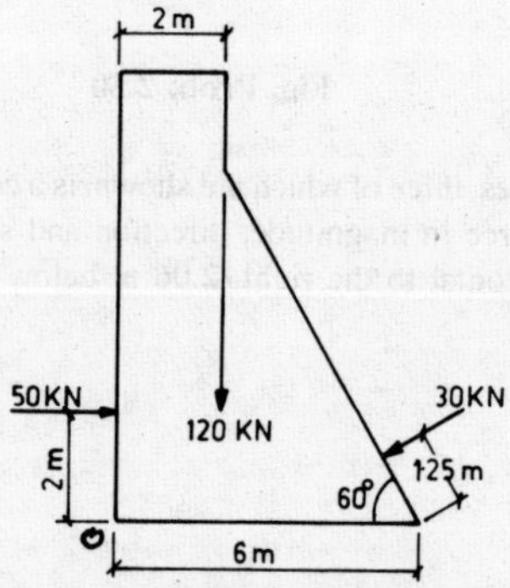

Fig. Prob. 2.53

2.54 Find the resultant of the four forces shown graphically.
(Ans. 2251 N, 12.7°, 1.24 m from origin).

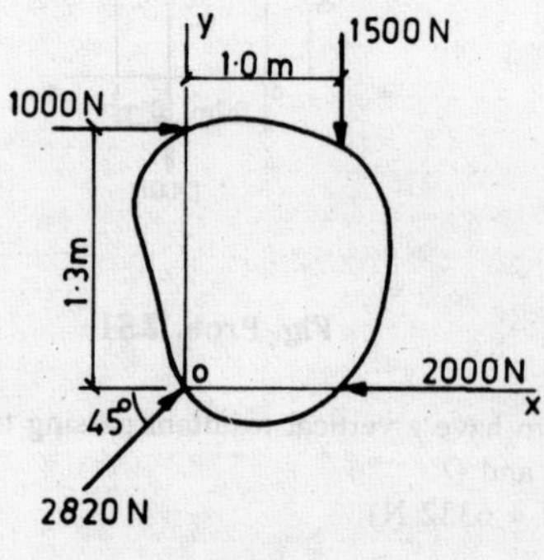

Fig. Prob. 2.54

2.55 Determine the resultant of the four forces acting on the plate graphically.
(Ans. 755.5 N, 74.6°, 3.7 cm from *C*)

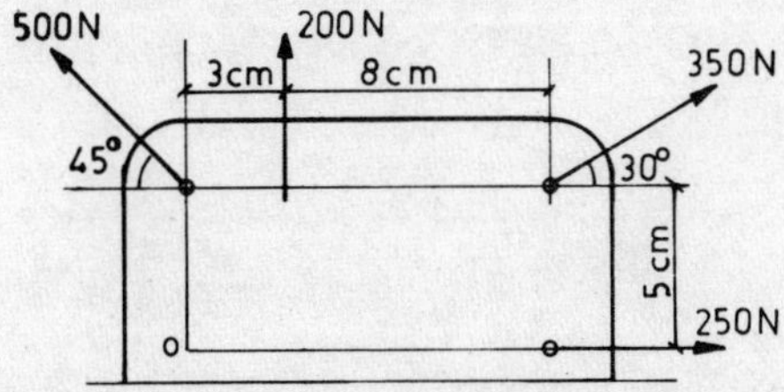

Fig. Prob. 2.55

2.56 Determine the resultant of the three forces acting on the truss graphically.
(Ans. 1000 Nm)

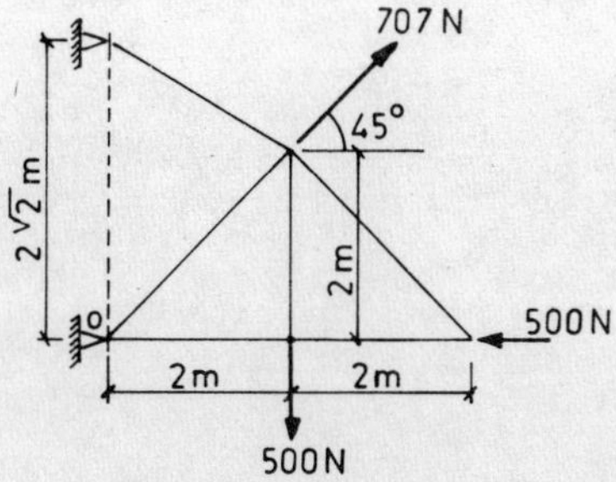

Fig. Prob. 2.56

2.57 The force system shown has neither a resultant force nor a resultant couple. Find the magnitude of the forces *P*, *Q* and *R*.
(Ans. $P = -96$ N, $Q = 559$ N, $R = 1735$ N)

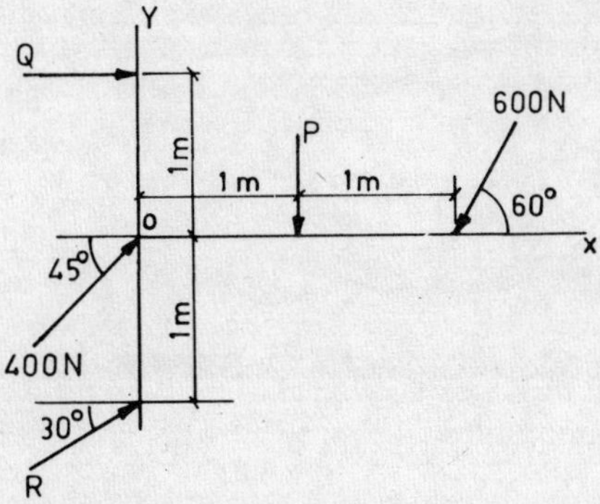

Fig. Prob. 2.57

3. EQUILIBRIUM

OBJECTIVES

After having mastered the content of this chapter, you will be able to:

- Define the term 'Equilibrium'
- Explain the possible displacements of a body subjected to any coplanar force system.
- State the analytical and graphical conditions of equilibrium for a body subjected to any coplanar force system.
- Explain the term body constraint.
- Explain the characteristics of various body constraints used for a planar structure.
- Recognize the body constraints in a given physical situation.
- Explain the concept of free body diagrams.
- Draw free body diagrams for given physical situation involving coplanar forces.
- Select and apply suitable conditions of equilibrium.
- Analyse a physical situation using free body diagrams and conditions of equilibrium to determine unknown reactive forces.

CHAPTER 3

Equilibrium

3.1 Introduction

In the last chapter, we dealt with forces, moments and determination of resultant of coplanar force systems. In calculating the resultant of coplanar force systems, both analytical and graphical methods were dealt with in detail. In this chapter, the problem of equilibrium of bodies acted upon by a system of coplanar forces will be discussed. The principles discussed in the last chapter will be used to develop some important concepts, which help to solve problems of equilibrium.

In the first chapter while discussing the branches of mechanics, it was mentioned that Statics is the science which deals with forces acting on bodies which are at rest. In Statics, the discussion is limited to rigid bodies. The principles developed in this chapter are of the utmost importance in all further studies for the analysis and design of structures, machines etc.

The term equilibrium was also mentioned in the first chapter. This concept will first be examined in detail.

3.2 Concept of Equilibrium

The term 'Equilibrium', in the context of this chapter, essentially means the state of rest of a body. State of rest will mean that the body has a fixed position in space, and there is no motion of the body. As will be explained in a later chapter, a body in motion but without acceleration, is also governed by the same laws as for the state of rest. Therefore, Equilibrium can also, in a general sense, mean a state of uniform motion of the body. However, for the purpose of developing the concepts in this chapter, equilibrium will be taken only as the state of rest of a body.

Figure 3.1 shows a body under the action of a system of coplanar forces. The body itself is three dimensional, but the force system, including the weight. of the body, is coplanar. The plane of the paper is the plane in which all the forces are acting. A system of reference coordinate axes is selected, and any point on the body can be fixed with reference to this system.

A point (1) of the body has the coordinates (x, y) as shown; at a given instant of time the forces acting on the body form a system of non-concurrent, non-parallel, coplanar forces. Equilibrium (State of rest) means that the position of the body or the x-, y- coordinates of the particles of the body remain the same. When the body is not in equilibrium the position of the body continuously changes with reference to X-, Y- axis with time. Therefore, we

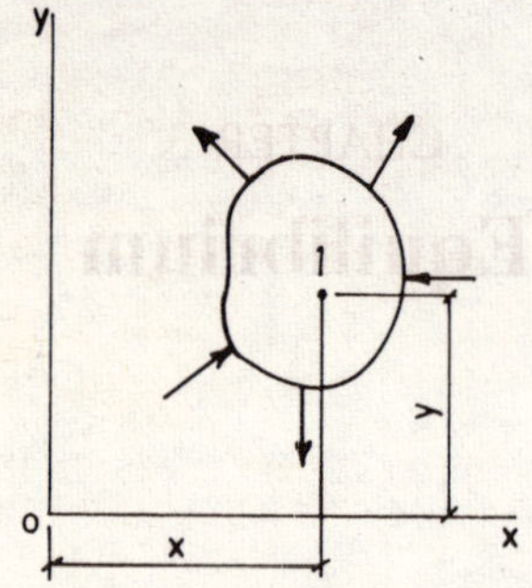

Fig. 3.1

may say that a body in equilibrium has a fixed position with regard to a reference system.

3.3 Possible Displacements of a Body

Before studying the conditions of equilibrium of a body acted upon by a system of coplanar forces, it is advantageous to know the possible displacements of a body. In Fig. 3.2(a) is shown a body acted upon by a system of coplanar forces. Let us assume that this force system can be reduced to a resultant force *R* and a resultant couple *M*. *R* and *M* can be calculated by the methods discussed in Chapter 2.

It has already been mentioned in Chapter 2, that a force has a pushing or pulling effect on the body, and the couple *M* has a rotating effect. Therefore, the resultant force *R* will push the body along its line of action in the sense of the force *R*, and the resultant couple will tend to rotate the body. These conditions are shown hypothetically in Fig. 3.2(b) and 3.2(c).

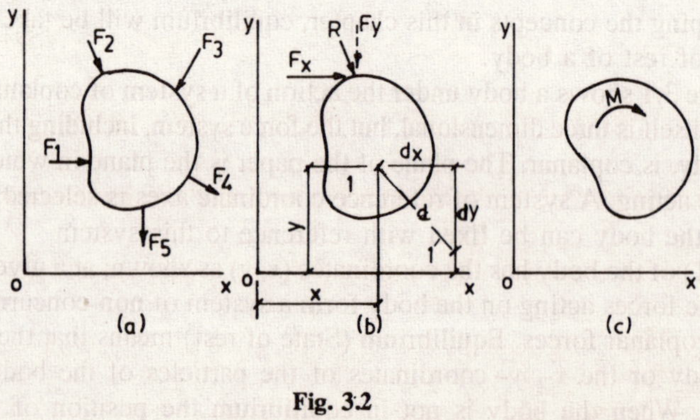

Fig. 3.2

In Fig. 3.2(b) is shown the effect of the resultant force *R*. A particle (1) on the body, let us say, is pushed to position 1' due to the movement of the whole body. This is due to the pushing effect of the force *R*. The distance 1-1', or the displacement, is '*d*'. As shown in Fig. 3.2(b), this displacement *d* may be considered to be the resultant of two displacements d_x and d_y. It is easy to see that d_x is the displacement due to the component F_x parallel the X-axis of *R*, and d_y is due to the component F_y of *R* parallel to the Y-axis.

In general, therefore, we can say that if there is a resultant force acting on the body, the body will have displacements in two perpendicular directions in the same plane as the given force system.

In Fig. 3.2(c), is shown the resultant couple acting on the body. Due to such a couple acting on the body, the body rotates. If the resultant couple *M* is clockwise as shown, the body will rotate in a clockwise direction.

We can now summarise the conclusions from the above discussion as follows: If there is a coplanar force system acting on a body having a resultant force *R* and a resultant couple *M*, then, there are three possible displacements for the body, namely

(i) A linear displacement parallel to the X-axis due to component F_x of *R*.

(ii) A linear displacement parallel to the Y-axis due to component F_y or *R*, and

(iii) A rotation due to resultant couple *M*.

3.4 Conditions of Equilibrium for Coplanar Force Systems

With the concept of displacements given in article 3.3, we can easily set up the conditions under which coplanar force systems are in equilibrium. We will start with the simplest force system consisting of two forces and develop the conditions for the equilibrium of coplanar force systems in the following articles.

3.4.1 *Two Force System*

Axiom : When a body under the action of only two forces is in equilibrium, the two forces must be collinear, equal and opposite.

The above statement is an axiom in mechanics, that is, it is a statement without proof. Fig. 3.3 shows a body under the action of only two forces, and in equilibrium. Force *P* must be equal to *Q* in magnitude, should have

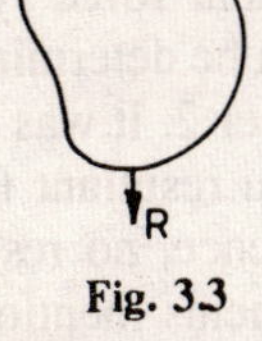

Fig. 3.3

the same line of action and must be opposite in sense to Q. Though there is no analytical proof for this, this fact can be easily verified by experiments.

3.4.2 *Three Force System*

Fig. 3.4 shows a body under the action of three non-parallel forces, and the body is in equilibrium. The condition of equilibrium is that the forces should be concurrent and coplanar.

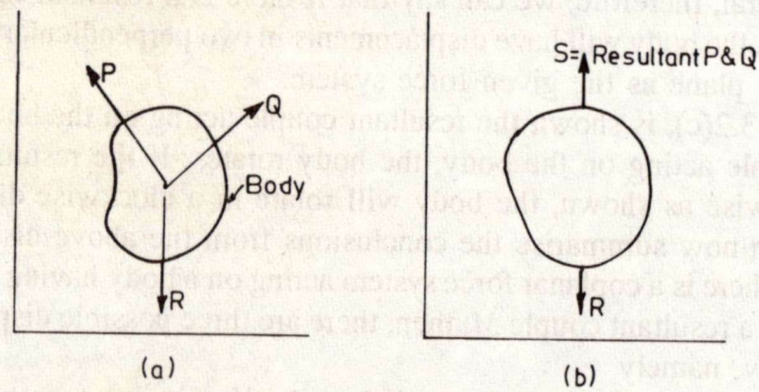

Fig. 3.4

The above statement can be easily proved from the axiom given in art. 3.4.1. Three forces P, Q and R act on the body. If we take any two forces, say P and Q, we can determine their resultant by the parallelogram law. The two forces P and Q can now be replaced by their resultant, say S, as shown in Fig. 3.4(b). The body is now acted upon by two forces S and R. If the body is to be in equilibrium, these two forces should be collinear, equal and opposite. Since force S, which is the resultant of forces P and Q has to pass through the point of intersection of P and Q, R also should pass through the same point. Therefore, if a body is in equilibrium under the action of only three non-parallel forces, the three forces should be concurrent.

The three forces P, Q and R should be coplanar. The plane that contains P and Q, also contains S, because S is their resultant. Since R and S are collinear for equilibrium, R should lie in the same plane as P and Q.

Therefore, for three non-parallel forces, acting on a body, to be in equilibrium, they should be coplanar and concurrent. The above principle is very important and is an useful concept in the solution of problems in equilibrium.

3.4.3 *Concurrent, Coplanar Force System*

Fig. 3.5 shows a concurrent coplanar force system, acting on a body. The resultant of such a force system can be determined analytically or graphically by the methods described in Chapter 2. It was shown in Chapter 2 that such a force system can be reduced to a resultant force, and since all the forces pass through the point of concurrence, no resultant couple is produced.

From the concept of displacements mentioned in article 3.3, we can say

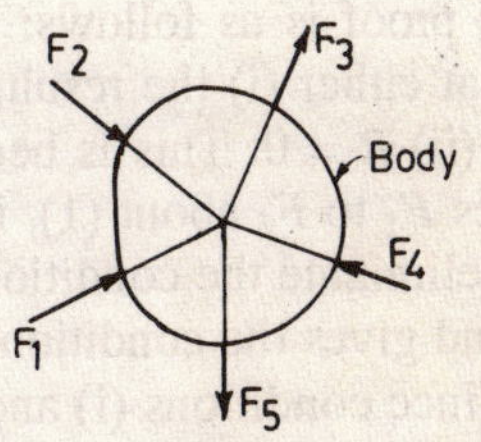

Fig. 3.5

that if the body is to be in equilibrium, i.e. there is no displacement of the body, the resultant force R should be equal to zero, or the total X-components and the total Y-components should each be equal to zero. The conditions of equilibrium are, therefore,

either $R = 0$ 3.1

or $\Sigma F_x = 0$
and $\Sigma F_y = 0$ 3.2

Conditions of Equilibrium in Terms of Moments

Equations 3.1 and 3.2 give the conditions of equilibrium of a concurrent, coplanar force system in terms of the resultant force R, or its components. The conditions can also be written in terms of moments of the forces of the system, as,

$\Sigma M_1 = 0$
and $\Sigma M_2 = 0$ 3.3

where ΣM_1 and ΣM_2 are the algebraic sums of the moments of the forces of the system about moment centres (1) and (2), and the moment centres (1) and (2) are not collinear with the point of concurrency. (Fig. 3.6).

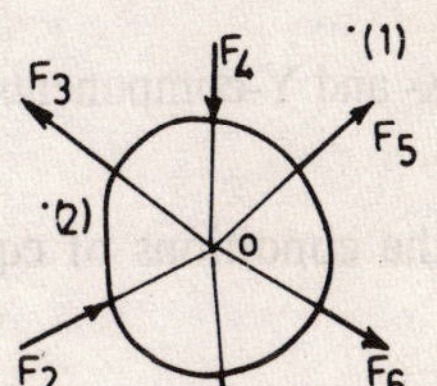

Fig. 3.6

In Fig. 3.6 is shown a system of concurrent, coplanar forces F_1 to F_6, concurrent at point (0); *(1) and (2) are any two moment centres such* that points (0), (1) and (2) are not collinear. Let us assume that ΣM_1 and ΣM_2

are both equal to zero. We can prove that, under these conditions, the body must be in equilibrium. The proof is as follows:
When $\Sigma M_1 = 0$, it means that either (i) the resultant force R of the system passes through point (1), or (ii) $R = 0$. This is because ΣM_1, the algebraic sum of the moments of forces F_1 to F_6 about (1), is equal to the moment of the resultant R about (1). To eliminate the condition (i) above, we take ΣM_2 which is also equal to zero and gives the conditions that either (iii) R passes through point (2) or $R = 0$. Since conditions (i) and (iii) can not be satisfied simultaneously, as points 1 and 2 are not collinear with the point of concurrency, the two conditions $\Sigma M_1 = 0$ and $\Sigma M_2 = 0$, can only mean that $R = 0$, which means the body is in equilibrium.

Graphical Condition of Equilibrium

It was proved in Chapter 2 that the resultant of a concurrent, coplanar force system is given by the closing line of the force polygon drawn. The magnitude of R is given by the length of the closing line and its direction is given by the inclination of the closing line. Now if $R = 0$, the body is in equilibrium. This means that the force polygon should have no closing line, or the polygon should close by itself. Fig. 3.7 shows this condition when the first and last points, that is points (1) and (6), coincide, the resultant is zero and the body is in equilibrium.

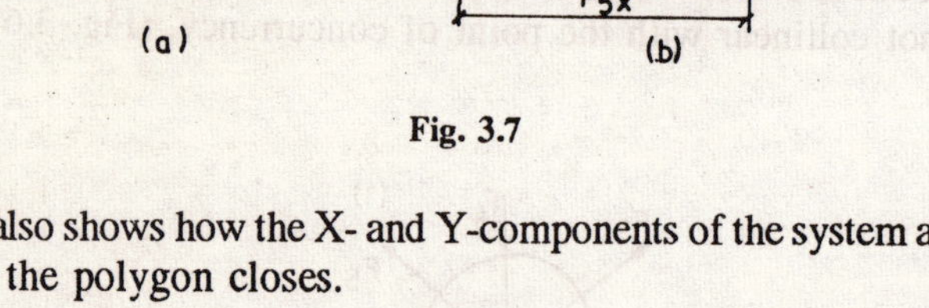

Fig. 3.7

Fig. 3.7(b) also shows how the X- and Y-components of the system also become zero when the polygon closes.

We can now summarise the conditions of equilibrium of a concurrent coplanar force system as,

(i) $R = 0$

or (ii) $\Sigma F_x = 0$

$\Sigma F_y = 0$

or (iii) $\Sigma M_1 = 0$, the moment centres 1 and 2 not collinear with the point of concurrency.

$\Sigma M_2 = 0$

or (iv) The force polygon closes.

Equilibrant

We have seen that a concurrent, coplanar force system can only have a resultant force *R*, and no resultant couple. If we apply a force equal, opposite and collinear with *R* to the body, the body should be in equilibrium. Such a force which is equal, opposite and collinear to the resultant of a concurrent force system is known as the equilibrant of the concurrent force system. Equilibrant, therefore, is the force which, when applied to the body acted upon by the concurrent force system, keeps the body in equilibrium.

Example 3.1
Check whether the concurrent, coplanar force system shown in Fig. 3.8 is in equilibrium. Determine the equilibrant of the given force system, if it is not in equilibrium.

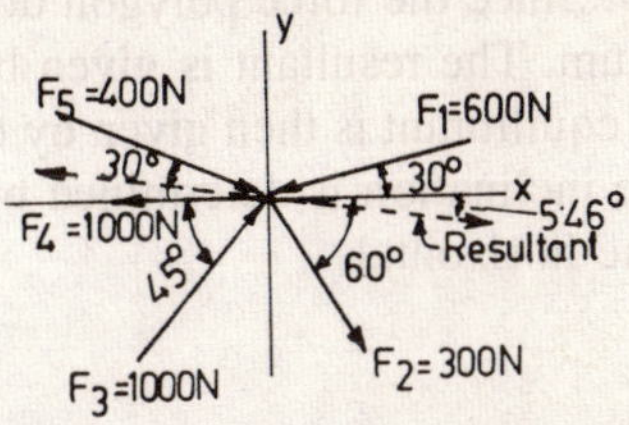

Fig. 3.8

Solution
The problem can be solved both analytically and graphically. Both the solutions are given below:

(a) *Analytical Solution.* Table 3.1 gives the X- and Y-components of all the forces and their sum.

Table 3.1 *Rectangular components of Forces in Fig. 3.8.*

Force	*Magnitude (N)*	*Angle with X-axis*	*X-component (N)*	*Y-component (N)*	*Remarks*
F_1	600	30	– 600 cos 30 = – 519.6	– 600 sin 30 = – 300	
F_2	300	60	+ 300 cos 60 = + 150	– 300 sin 60 = – 259.8	
F_3	1000	45	+ 1000 cos 45 = + 707	+ 1000 sin 45 = + 707	
F_4	100	0	– 100 cos 0 = – 100	– 100 sin 0 = 0	
F_5	400	30	+ 400 cos 30 = 346.4	– 400 sin 30 = – 200	
	Total		+ 583.8	– 52.8	

$$\text{Resultant force } R = \sqrt{(583.8)^2 + (52.8)^2}$$
$$= 586.20 \text{ N}$$
$$\text{Angle with X-axis} = \tan^{-1} \frac{(\Sigma F_y)}{(\Sigma F_x)}$$
$$= \tan^{-1} \left(\frac{52.8}{583.8} \right)$$
$$= 5.46°$$

The force system is not in equilibrium. The equilibrant has a magnitude of 586.2 N inclined at 5.46° to the X-axis, and is opposite to the resultant. Both the equilibrant and the resultant are shown in Fig. 3.8.

(b) *Graphical Solution.* The force polygon 1-2-3-4-5-6 is drawn first to scale as shown in Fig. 3.9. Since the force polygon does not close, the force system is not in equilibrium. The resultant is given by the closing line 1-6 of the force polygon. The equilibrant is then given by the vector 6-1, and can be measured to scale. The inclination is determined by measuring the angle of the equilibrant with the horizontal.

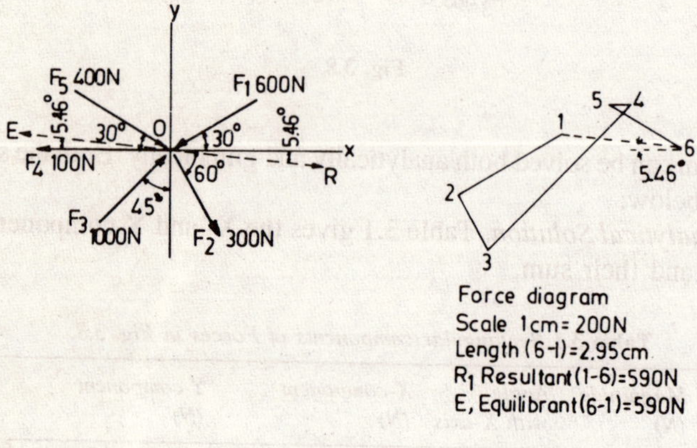

Fig. 3.9

3.4.4 *Coplanar, Parallel Force System*

Fig. 3.10 shows a coplanar, parallel force system. We can determine the

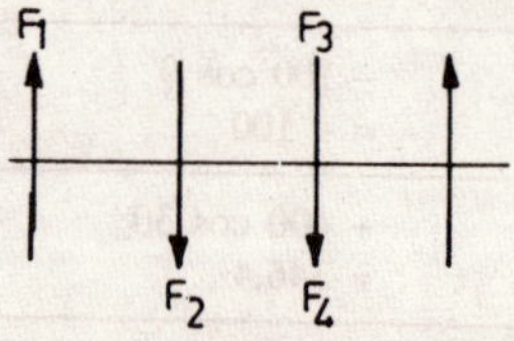

Fig. 3.10

resultant force or couple of such a force system by the methods described in Chapter 2.

The resultant force should be parallel to the lines of action of the forces of the system. It was shown in article 2.7.5, that even if the resultant force of a parallel force system is zero, the system may reduce to a couple.

From the displacement concept given in article 3.3, we can say that the resultant force will push the body in its direction and the resultant couple will rotate the body. If the body is to be in equilibrium, both the linear displacement due to R, and the rotation due to couple M should be zero.

Therefore, the conditions of equilibrium of a coplanar, parallel force system are,

$$\left.\begin{aligned} \Sigma F &= 0 \\ \Sigma M &= 0 \end{aligned}\right\} \qquad \text{... 3.4}$$

For equilibrium, both the resultant forcc R, and the resultant couple M should be zero.

Equilibrium Conditions in terms of Moments

The equilibrium conditions as given by equations 3.4 can also be expressed as

$$\left.\begin{aligned} \Sigma M_1 &= 0 \\ \Sigma M_2 &= 0 \end{aligned}\right\} \qquad \text{... 3.5}$$

where ΣM_1 and ΣM_2 are the algebraic sums of the moments of the forces of the system about moment centres (1) and (2) such that the line joining (1) and (2) is not parallel to the line of action of the forces of the system.

The above statement can be easily proved. If for the force system $\Sigma M_1 = 0$, it means that the system does not reduce to a couple, as a couple has the same moment about any point in the plane. The system may reduce to a resultant force R under these conditions, and ΣM_1 may be zero because R passes through point (1). If ΣM_2 is also equal to zero, it means that $R = 0$, since R cannot pass through moment centres (1) and (2), and we have already concluded that there is no resultant couple. Therefore, for a parallel force system,

$$\Sigma M_1 = 0$$

and $$\Sigma M_2 = 0$$

are the conditions for equilibrium with the condition that moment centres (1) and (2) do not lie in a line parallel to the line of action of the forces of the system.

Graphical Conditions of Equilibrium

In a graphical construction, the force polygon of a parallel force system reduces to a straight line parallel to the lines of action of the forces. The resultant force is given by the line joining the first and last points. If the first and last points of the force polygon (which is a line in this case) coincide, then the resultant force is zero. But we have seen that even if the resultant is zero, there may be a resultant couple, and this has to be zero for equilibrium.

We have also seen in Chapter 2 that for a parallel force system, if the resultant force $R = 0$, then the resultant couple is given by the first or last ray in the force polygon multiplied by the distance between them in the force polygon.

If the resultant couple is to be zero the first and last rays of the funicular polygon should coincide. This condition is shown in Fig. 3.11.

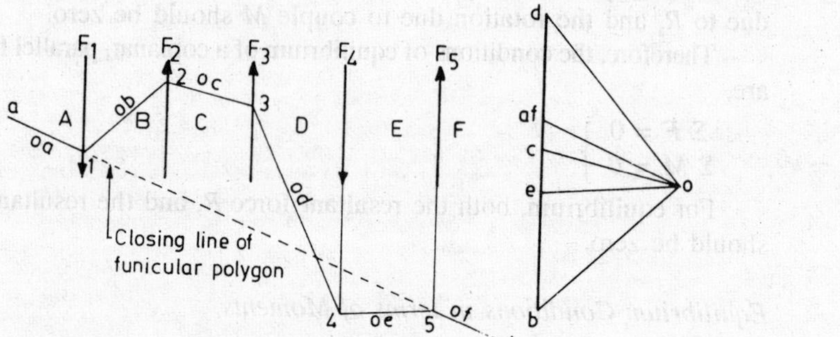

Fig. 3.11

Therefore, the graphical conditions of equilibrium are
(i) the force polygon should close, and
(ii) the funicular polygon should close.

Concluding the above discussion, we can say that a body subjected a system of coplanar parallel forces will be in equilibrium, if

(i) $\Sigma F = 0$
(ii) $\Sigma M = 0$

OR

(i) $\Sigma M_1 = 0$
(ii) $\Sigma M_2 = 0$ } Moment centres (1) and (2) not parallel to the lines of action of the forces.

OR

(i) The force polygon closes
(ii) The funicular polygon closes

Example 3.2

Determine whether the parallel force system given in Fig. 3.12 is in equilibrium. If not, is it possible to keep the system in equilibrium by applying a single force?

Fig. 3.12

Solution

The problem can be solved both analytically and graphically. Both the solutions are presented below:

(a) *Analytical Solution*

The resultant force = + 200 – 800 – 1000 + 1200 + 400 = 0

The resultant force is 0. We have to check whether the system reduces to a couple. For this, we take moment about a point on the line of action of the 200 N force.

$$M = +800 \times 1 + 1000 \times 2 - 1200 \times 2.75 - 400 \times 4.25$$
$$= 800 + 2000 - 3300 - 1700$$
$$= \mathbf{+2200\ Nm}$$

This is a clockwise moment. The given parallel force system, therefore, reduces to a clockwise couple of 2200 Nm. The system is not in equilibrium.

We have already seen that a couple cannot be balanced by a single force. Therefore, we cannot bring the system to equilibrium by applying a single force.

(b) *Graphical Solution*

Fig. 3.13 shows the graphical solution to this problem.

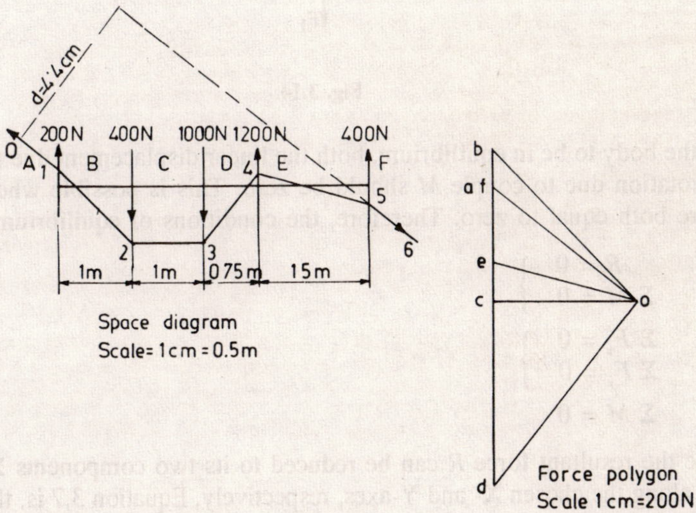

Fig. 3.13

We draw first the space diagram to a linear scale of 1 cm = 0.5 m. The force diagram is then drawn to a scale of 1 cm = 200 N. The force diagram is a straight line *abcdef* in which points *a* and *f* coincide indicating that there is no resultant force. We choose a pole '*O*' and draw the funicular polygon 0-1-2-3-4-5-6. The last lines 5-6 and the first line 0-1 are parallel. Along line 5-6 a force equal to *fo* is acting as shown in the figure and along line 1.0

a force equal to *oa* is acting. These forces constitute a couple. The magnitude of the couple can be determined as

$$M = \text{force } oa \times \text{distance } d$$
$$= (5 \times 200) \times (4.4 \times 0.5)$$
$$= \mathbf{2200\ Nm.}$$

3.4.5 *Non-concurrent, Non-parallel Force System*

Fig. 3.14 shows a general coplanar force system. It was shown in Chapter 2 that such a general coplanar force system may be reduced to a resultant force R and/or a resultant couple M acting in the same plane.

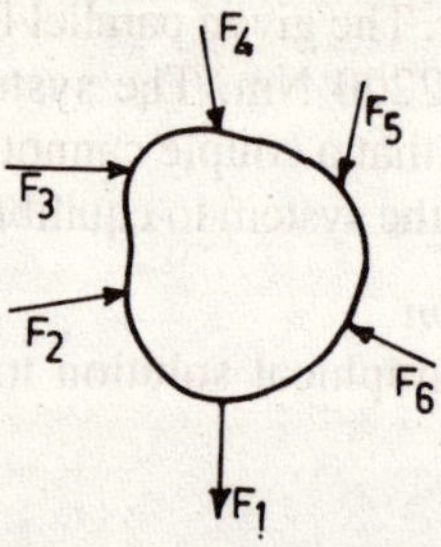

Fig. 3.14

For the body to be in equilibrium, both the linear displacement due to R, and the rotation due to couple M should be zero. This is possible when R and M are both equal to zero. Therefore, the conditions of equilibrium are,

$$\left.\begin{aligned} R &= 0 \\ \text{and} \quad \Sigma M &= 0 \end{aligned}\right\} \quad \text{... 3.6}$$

$$\text{or} \quad \left.\begin{aligned} \Sigma F_x &= 0 \\ \Sigma F_y &= 0 \end{aligned}\right\} \quad \text{... 3.7}$$

$$\text{and} \quad \Sigma M = 0$$

Since the resultant force R can be reduced to its two components ΣF_x and ΣF_y along the chosen X- and Y-axes, respectively. Equation 3.7 is, therefore, the most general condition for the equilibrium of a coplanar force system.

Equilibrium Conditions in terms of Moments

The conditions of equilibrium as expressed by equation 3.7 can be expressed in many ways by moment equations, such as

$$\left.\begin{aligned} \Sigma F_x &= 0 \\ \Sigma M_1 &= 0 \\ \Sigma M_2 &= 0 \end{aligned}\right\} \quad \text{... 3.8}$$

where ΣM_1 and ΣM_2 are the algebraic sums of moments of all the forces about moment centres (1) and (2) which do not lie on a line parallel to Y-axis (Fig. 3.15a).

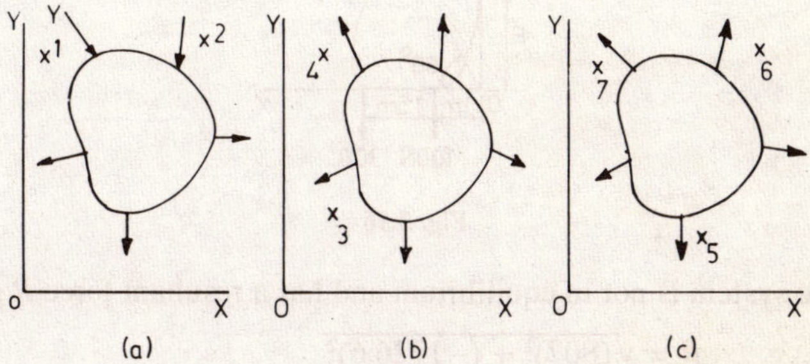

Fig. 3.15

$$\left.\begin{aligned}\Sigma F_y &= 0\\ \Sigma M_y &= 0\\ \Sigma M_4 &= 0\end{aligned}\right\} \quad \text{or} \qquad \ldots\ 3.9$$

where ΣM_3 and ΣM_4 are the algebraic sums of moments of all the forces about moment centres (3) and (4) which do not lie on a line parallel to the X-axis (Fig. 3.15b)

$$\left.\begin{aligned}\Sigma M_5 &= 0\\ \Sigma M_6 &= 0\\ \Sigma M_7 &= 0\end{aligned}\right\} \quad \text{or} \qquad \ldots\ 3.10$$

where ΣM_5, ΣM_6 and ΣM_7 are the algebraic sums of the moments of all the forces about moment centres (5), (6) and (7) such that (5), (6) and (7) do not lie in a straight line (Fig. 3.15c).

The proof for the above statements is on the same lines as was discussed in art. 3.4 and 3.5. It is left as an exercise to the reader to prove the statements in equations 3.8, 3.9 and 3.10.

Example 3.3

Check the equilibrium condition of the force system shown in Fig. 3.16.

Solution

The problem can be solved both analytically and graphically. Both the methods of solution are given below:

(a) *Analytical Method:* The components in the X and Y directions are calculated.

$$\Sigma F_x = 1000 \cos 45 + 400 - 600$$
$$= \mathbf{+\ 807\ N}$$
$$\Sigma F_y = -1000 \sin 45 - 600 \sin 60 - 100 - 400$$
$$= \mathbf{-1726.6\ N.}$$

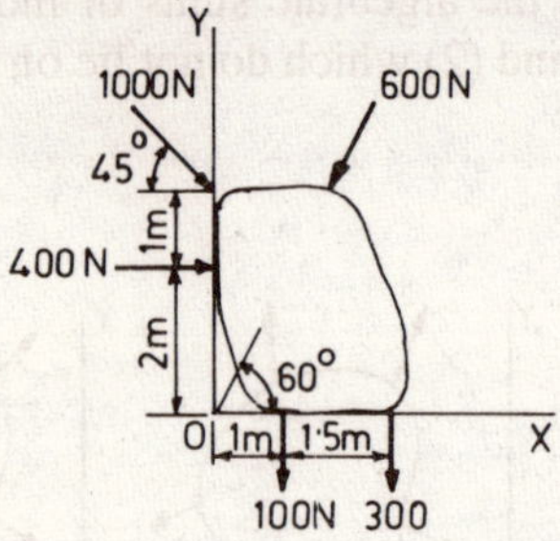

Fig. 3.16

∴ The system is not in equilibrium and has a resultant force R given by

$$R = \sqrt{(807)^2 + (-1726.6)^2}$$

$$= \mathbf{1906\ N\ Ans.}$$

and inclined at $\theta = \tan^{-1} \dfrac{(-1726.6)}{807} = \tan^{-1} (-2.015)$

$$= 65° \text{ with the X-axis.}$$

Taking moments @ '*o*' of all the forces, considering clockwise moments positive,

$$\Sigma M = 1000 \times \cos 45 \times 3 + 400 \times 2 + 100 \times 1 + 400 \times 2.5$$

$$= \mathbf{4021\ Nm.}$$

Distance '*d*' of the resultant from point '*o*' is given by,

$$d = \frac{M}{R} = \frac{4021}{1906} = \mathbf{2.1\ m.}$$

The resultant is shown in Fig. 3.17.

(b) *Graphical Method:* The graphical solution is shown in Fig. 3.17. We lay the space diagram to a scale of 1 cm = 0.5 m. The force diagram is drawn to a scale of 1 cm = 200 N. The resulting diagram, *abcde*, is not closed. The resultant, therefore, is *af* (from *a* to *f*) inclined at 65° to the horizontal. By measurement,

$$af = 9.6 \text{ cm}$$

$$\text{Resultant} = 9.6 \times 200 = \mathbf{1920\ N}$$

To locate the resultant in the space diagram, we select a pole '*o*' and draw the funicular diagram 0-1-2-3-4-5-6. The line 0-1 and 5-6 meet at '*p*' a point on the line of action of the resultant *R*. A line drawn parallel to *a-f* through '*p*' represents the resultant. By measurement, distance of *R* from '*O*'

$$= 4.2 \text{ cm.}$$

$$= 4.2 \times 0.5$$

$$= \mathbf{2.1\ m\ Ans.}$$

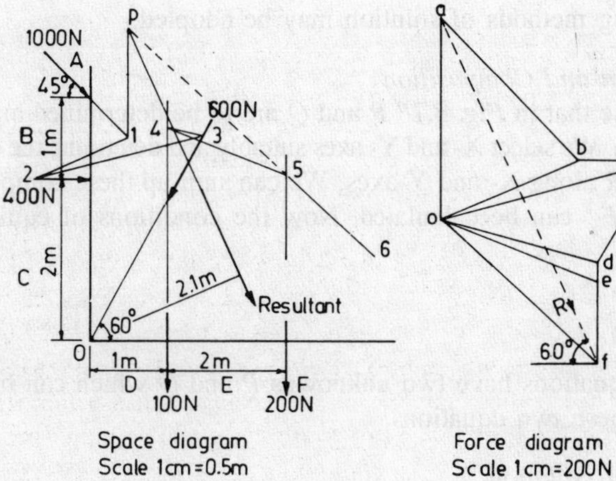

Fig. 3.17

3.4.6 *Problems in Equilibrium Involving Three Concurrent Forces*

It was shown in sec. 3.4.2 that when three non-paralllel forces are in equilibrium, they must be concurrent and coplanar. The above principle is very important and is applied even for a general, coplanar force system by reducing the system to only three forces. Different methods of solution of this problem will be discussed in the ensuing paragraphs.

Fig. 3.18 shows three non-parallel forces keeping the body in equilibrium. It is known that they must be concurrent and coplanar.

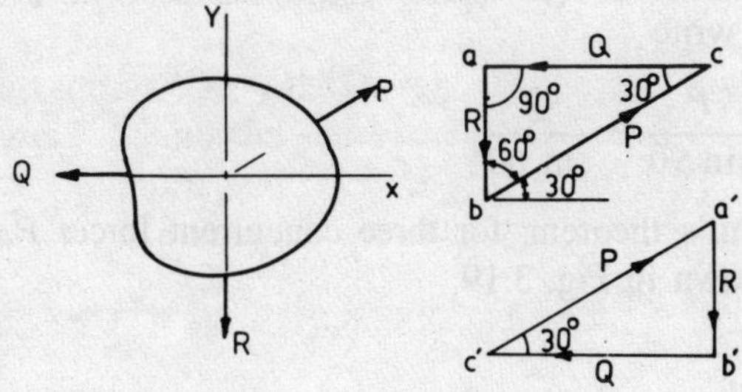

Fig. 3.18

The problem in such cases is to find either (i) One force and its direction or (ii) Magnitude of two unknown forces. It must be remembered that for any concurrent force system of more than two forces, there are two independent conditions of equilibrium, namely $\Sigma F_x = 0$ and $\Sigma F_y = 0$. By applying these conditions, only two unknowns can be determined which are required for analysis and design.

The following methods of solution may be adopted.

(i) *Resolution and Composition*

Let us assume that in *Fig. 3.18* P and Q are to be determined and the value of R is known. We select X- and Y- axes suitably and determine the components of P, Q and R along X- and Y-axes. We can sum up these components and ΣF_x and ΣF_y, can be calculated. Now the conditions of equilibrium are applied as,

$$\Sigma F_x = 0$$

$$\Sigma F_y = 0$$

These two equations have two unknowns P and Q which can be evaluated by solving these two equations.

(ii) *Graphical Solution*

The magnitude of R and the directions of all the forces are known. We can draw a force triangle to a selected scale. *ab* is drawn vertically downward representing the force R. A line is drawn horizontally from *a* and a line is drawn at 30° with the horizontal from *b*. These two lines intersect at C completing the force triangle shown in Fig. 3.18. The triangle can also be drawn as a′ b′ c′ as in Fig. 3.18. The values of P and Q can be measured from the force triangle to scale.

Please note carefully that the force triangle can not be completed if there are more than two unknowns, either as magnitudes or directions.

(iii) *Lami's Theorem*

Lami's Theorem is another analytical method of solution for three concurrent forces in equilibrium.

Refer to the force triangle *abc* in Fig. 3.18. From the property of the triangle, we may write

$$\frac{R}{\sin 30} = \frac{P}{\sin 90} = \frac{Q}{\sin 60}$$

According to Lami's theorem, for three concurrent forces P, Q and R in equilibrium as shown in Fig. 3.19

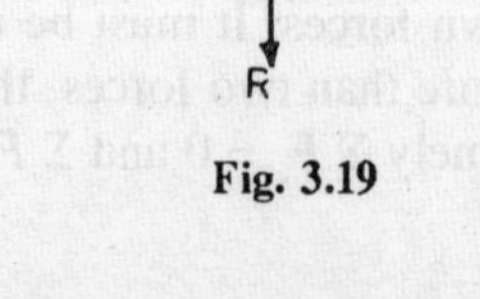

Fig. 3.19

$$\frac{P}{\sin \gamma} = \frac{R}{\sin \beta} = \frac{Q}{\sin \alpha}$$

In the problem shown in Fig. 3.18

$$\alpha = 120°, \beta = 150° \text{ and } \gamma = 90°.$$

$$\therefore \frac{P}{\sin 90} = \frac{R}{\sin 150} = \frac{Q}{\sin 120}$$

Since sin 150 = sin 30 and sin 120 = sin 60.
This can be written as,

$$\frac{P}{\sin 90} = \frac{R}{\sin 30} = \frac{Q}{\sin 60}$$

which is the same as we derived from the force triangle. Lami's theorem is only a method of solution of the force triangle.
Please note carefully that Lami's theorem gives only two equations, namely,

$$\frac{P}{\sin 90} = \frac{R}{\sin 30}$$

$$\frac{P}{\sin 90} = \frac{Q}{\sin 60}$$

from which two unknowns can be calculated. When there are more than two unknowns, though we may write equations as per Lami's theorem, we will not be able to solve the equations for all the unknowns.

The above three methods are frequently employed for solving problems of three concurrent forces.

So far we have derived the conditions of equilibrium for coplanar force systems. These conditions of equilibrium are very useful in analysing bodies in equilibrium and subjected to a set of coplanar forces. The equilibrium conditions, developed so far, are summarised in table 3.1.

In the next sections, we shall see how the conditions of equilibrium are applied to physical situations to solve equilibrium problems in Mechanics.

3.5 Body Constraints and Free Body Diagrams

So far we have applied the conditions of equilibrium together with the concepts developed in Chapter 2 for the solution of problems that did not refer to any physical situations. That is, the problems were presented in such a form that they only needed the application of the formulae developed to give the solution. In fact this is only a part of the problem, that too not the first part, that an analyst has to face. We will now discuss many problems that refer to physical situations from which one has to derive the problem in a proper form for solution.

A physical problem in mechanics is solved essentially by means of mathematical computations and/or graphical methods. For a mathematical solution, any problem has to be represented diagramatically with sketches which help in the visualisation of the physical problem. It is of the utmost fundamental importance to visualise the physical problem, and prepare a mathematical model of the problem for solution. Before taking up the procedure for solution, a detailed discussion on freebody diagrams and body constraints is essential, as these help very much in the solution of problems.

Table 3.1 *Conditions of Equilibrium of Coplanar force system*

Force System	*Conditions of Equilibrium*	
	Analytical	*Graphical*
1. Concurrent force system.	(i) $\Sigma H = 0$; $\Sigma V = 0$ OR $R = 0$ (ii) $\Sigma M_1 = 0$; $\Sigma M_2 = 0$; the moment centres 1 and 2 not collinear with point of concurrency.	Force polygon closes.
2. Three-force system (Non-parallel)	(i) Concurrent & coplanar (ii) Conditions as in 1.	Force polygon (triangle) closes.
3. Coplanar parallel force system.	(i) $\Sigma F = 0$, $\Sigma M = 0$ (ii) $\Sigma M_1 = 0$; $\Sigma M_2 = 0$ line joining moment centre 1 & 2 not parallel to the line of action of forces.	(i) Force polygon (line) closes. (ii) Funicular Polygon closes.
4. General coplanar force system.	(i) $\Sigma F_x = 0$; $\Sigma F_y = 0$; $\Sigma M = 0$ (ii) $\Sigma F_x = 0$; $\Sigma M_1 = 0$, $\Sigma M_2 = 0$ line joining moment centres 1 & 2 not parallel to Y axis. (iii) $\Sigma F_y = 0$; $\Sigma M_3 = 0$, $\Sigma M_4 = 0$. line joining moment centres 3 & 4 not parallel to X axis. (iv) $\Sigma M_5 = 0$; $\Sigma M_6 = 0$, $\Sigma M_7 = 0$ moment centres 5, 6, 7 not collinear.	(i) Force polygon closes. (ii) Funicular polygon closes.

3.5.1 *Body Constraints*

In a physical problem in Statics, the body acted upon by forces can be anything from a photograph hanging on a nail to a multispan bridge carrying a number of vehicles. Take the simple case of a photograph hanging on a nail (Fig. 3.20). Without the nail, the photograph cannot stay on the wall. The nail in this case is known as a body constraint. The nail is a constraint provided to prevent the photograph from falling down. *A body constraint is a contrived support or force provided, such that the body remains in equilibrium. The constraints*

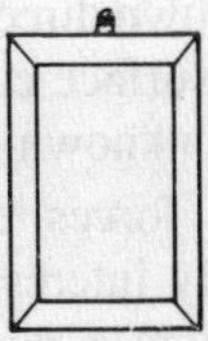

Fig. 3.20

apply either reactive forces and/or couples depending upon the type of constraint provided.

We shall briefly describe a few of the body constraints that are of importance in analysing many structures and machine parts. Only constraints provided for coplanar structures (i.e. subjected to coplanar force systems) are discussed. Let us begin our discussion with a description of surfaces acting as constraints.

(a) Smooth Surface. A smooth surface provides one reactive force which acts normal to the surface. A simple example is a book resting on a table (Fig. 3.21). The table is a constraint for the book because without the table the book will fall to the floor. The table applies one reactive force on the book, and it is normal to the table as well as the book. It may also be noted that of the possible displacements under a coplanar force system, a smooth surface *prevents* one displacement, i.e., a motion normal to the smooth surface. The smooth surface *permits* two displacements, one tangential to it and the other, a rotation.

(b) Rough Surface. A rough surface provides one reactive force in any possible direction. In Fig. 3.22 we have shown a rough surface. Because in a rough surface frictional forces tangential to the contacting surfaces come into play, (friction will be dealt with in detail in Chapter 5) both normal and tangential motions due to applied forces are prevented.

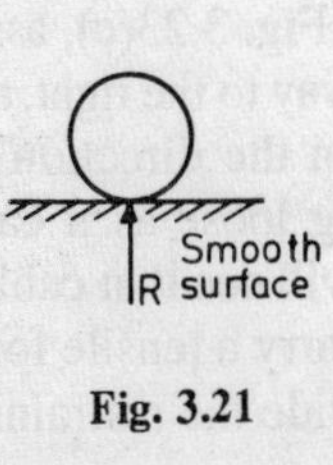

Fig. 3.21

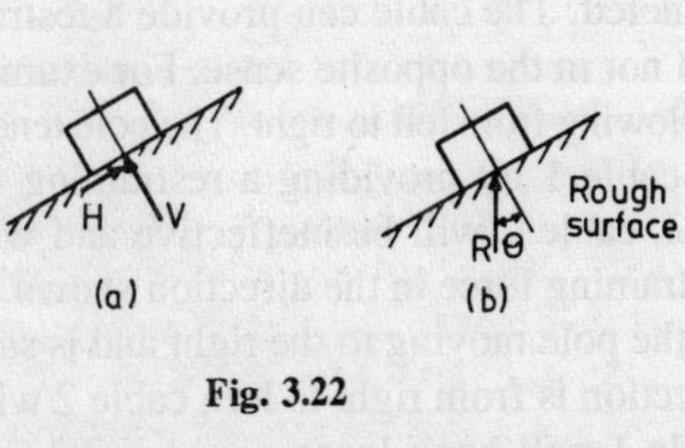

Fig. 3.22

We can, in general, say that a rough surface provides two reactive forces—one normal and one tangential to the surface. Alternatively, the two component reactive forces, normal and tangential can be thought of as one single reactive force with an unknown direction. This is shown in Fig. 3.22(b). The reactive forces at the rough surface can be thought of as (i) V and H or (ii) one reactive force R in the unknown direction θ.

The fact that two reactive forces exist is important, because in a physical situation one is normally interested in finding the reactive forces at the constraints. Therefore, where a rough surface is the constraint, one should allow for two unknowns at that constraint, namely, V and H, or R and θ.

From the point of view of displacements, we may say that a rough surface prevents two possible displacements, namely, translation along two perpendicular directions. This means that there is *no restraint* against rotation at the rough surface.

(c) A String or Cable Tied to the Body. This is a type of constraint that is commonly used in many physical situations. The wire from which a bulb is hung acts as a cable. Similarly, it is a common practice to use cables as stay wires for electric poles to make them stable (Fig. 3.23).

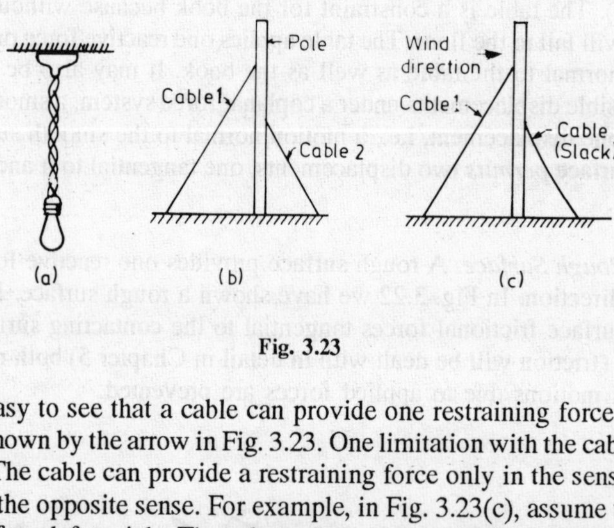

Fig. 3.23

It is easy to see that a cable can provide one restraining force along its direction shown by the arrow in Fig. 3.23. One limitation with the cable should be noted. The cable can provide a restraining force only in the sense shown, and not in the opposite sense. For example, in Fig. 3.23(c), assume that wind is blowing from left to right. The pole tends to sway to the right, and is restrained by cable 1 by providing a restraining force in the direction shown. In this case, cable 2 will be ineffective and will hang loose as it cannot provide a restraining force in the direction shown. You will see that cable 1 is stretched by the pole moving to the right and is said to carry a tensile force. If the wind direction is from right to left, cable 2 will provide the restraining force while cable 1 will hang loose.

We can, therefore, say that the cable support provides a restraining force along the direction of the cable and the cable is effective only when it is stretched. From the point of view of displacements, this is a constraint against translation along the direction of the cable, and no rotational constraint is provided. For example, in Fig. 3.23(a), there is no restraint against the bulb swaying horizontally, the only restraint being against its vertical, downward movement. In Fig. 3.23(b), if the wind is blowing perpendicular to the plane of the paper, you can easily visualize that the two cables provide no restraint against the wind forces.

(d) *Roller Support.* This type of support is shown in Fig. 3.24. The body is attached to the support at the pin which by its very nature does not prevent the rotation of the body about the pin. The rollers can move over the firm support which means that the support cannot provide any reactive force parallel to the plane on which the roller moves. It is assumed that the roller support is as shown in Fig. 3.24(c), which can provide one reactive force in either direction.

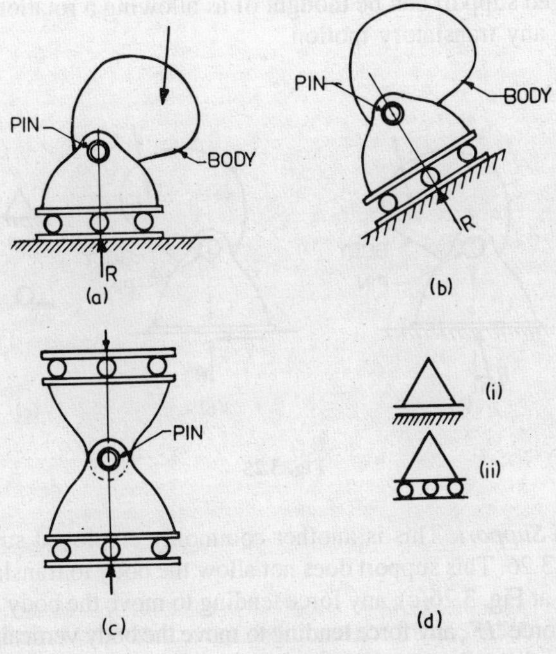

Fig. 3.24

In actual practice, there will be some resistance to the free movement of the rollers, and some tangential force parallel to the support bed is also developed. This is neglected in the analysis.

Such supports are commonly used in bridges. This support is symbolically represented as shown in Fig. 3.24(d).

From the point of view of permitted displacements, this support is similar to a smooth surface. The support *prevents* movement perpendicular to the plane on which the rollers move and *allows* translation (in a parallel direction) and rotation.

(e) *Hinged or Pinned Support.* This type of support is shown in Fig. 3.25. The body is connected by a pin to the support which allows free rotation of the body. The support is firmly attached to the ground, and thus prevents any translatory motion of the body. Thus if forces act on the body tending to move the body in any direction, reactive forces are developed to prevent such translatory motion.

The reactive force provided by the support can be thought of as a force in any direction, or as two components in any two perpendicular directions (Fig. 3.25b). Usually it is convenient to consider two reactive forces *H* and *V* in perpendicular directions.

The symbolic representation for such a support is shown in Fig. 3.25(c).

The hinged support can be thought of as allowing a rotation of the body and resisting any translatory motion.

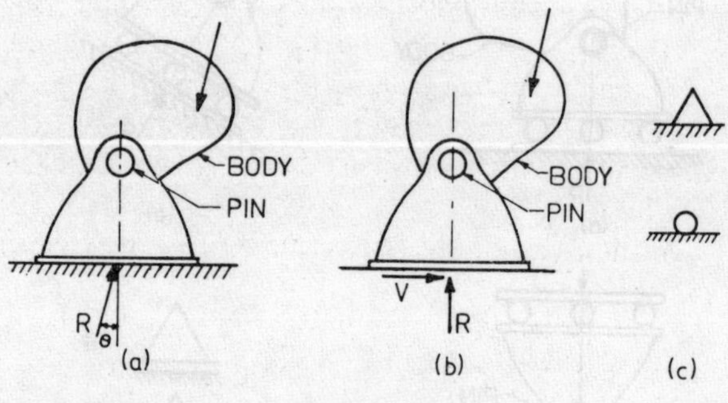

Fig. 3.25

(f) *Fixed Support.* This is another commonly employed support and is shown in Fig. 3.26. This support does not allow the body to translate or rotate. Thus, looking at Fig. 3.26(c), any force tending to move the body horizontally is resisted by force '*H*', any force tending to move the body vertically is resisted by force '*V*', and any force tending to rotate the body is resisted by the couple formed by the forces *r* and *r'*. Thus such a support completely restrains a body acted upon by a coplanar force system.

This support is symbolically represented as shown in Fig. 3.26(d).

The roller, hinge and fixed supports as explained here are only ideal and can be attained in practical construction only to some extent. However, such assumption of ideal supports makes the analysis easier, and is sufficiently accurate for practical purposes.

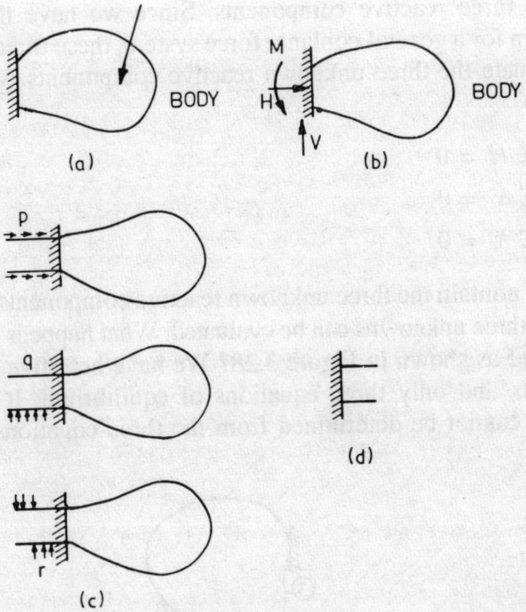

Fig. 3.26

3.5.2 *Body Constraints and Statical Determinancy*

In restraining a body, one has to use a support, or a combination of supports, as required. The fixed support alone can restrain a body subjected to a coplanar force system. A combination has to be used, otherwise, for complete restraint and equilibrium. Thus in Fig. 3.27(a), hinge and roller support is used to restrain the body. In Fig. 3.27(b), a hinge and a cable are used in combination. In Fig. 3.27(c), the fixed support alone can completely restrain the body.

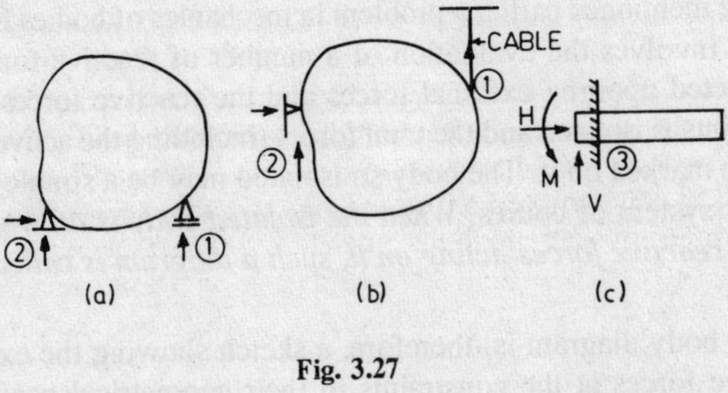

Fig. 3.27

In each of the cases shown in Fig. 3.27, the number of reactive forces are shown against each constraint provided. You will find that each body has a total of three reactive components. Since we have three conditions of equilibrium for a general coplanar force system, these three conditions enable us to evaluate the three unknown reactive components. We can write three equations,

(i) $\Sigma H = 0$

(ii) $\Sigma V = 0$

(iii) $\Sigma M = 0$

which will contain the three unknown reactive components. From these three equations, three unknowns can be evaluated. What happens when the structure is supported as shown in Figure 3.28? We have here five unknown reactive components and only three equations of equilibrium. It is clear that five unknowns cannot be determined from the three equations.

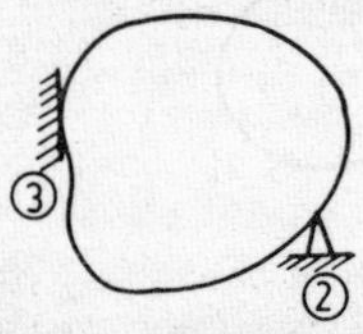

Fig. 3.28

The bodies shown in Fig. 3.27 are known as *statically determinate* because the unknown reactive forces that restrain them can be evaluated from the (statical) conditions of equilibrium. The body shown in Fig. 3.28 is statically *indeterminate* as the five reactive forces *cannot be determined* from the three conditions of equilibrium.

We shall be dealing with only bodies that are statically determinate.

3.5.3 *Free Body Diagrams*

As we have mentioned earlier a problem in mechanics of bodies in equilibrium essentially involves the evaluation of a number of reactive forces. For this, the body acted upon by external forces and the reactive forces provided by the constraints is isolated and the total forces (including the active and reactive forces), are marked on it. The body so isolated may be a simple one or a part of a bigger system of bodies. *When the isolated body is drawn, with all the forces and reactive forces acting on it, such a diagram is called a free body diagram.*

A free body diagram is, therefore, a sketch showing the external forces and reactive forces at the constraints in their geometrical positions. Please note that we remove the constraints and draw the restraining forces provided by it.

Free body diagrams provide an excellent method of book keeping in a mechanics problem. By drawing the free body diagram, we can easily account for all the forces acting on the body and no forces are omitted. It is necessary to carefully identify the active and reactive forces, as a mistake in this, may lead to a mistake in the solution.

Free body diagrams in addition help us to prepare a mathematical model of the physical situation. Once the free body diagram is drawn, and we mark the active and reactive forces on it, we are in a position to apply the conditions of equilibrium to these forces to evaluate the unknowns.

From the above it is clear that free body diagram is the intermediate stage between the physical problem and its mathematical solution. The importance of drawing the free body diagram cannot be over-emphasized. We have already discussed the body constraints and the reactive forces provided by them. A number of examples to illustrate the method of drawing the free body diagrams are given below.

Example 3.4

A book weighing 30 N is resting on a smooth table (Fig. 3.29). Draw the free body diagram of (i) the book and (ii) the table.

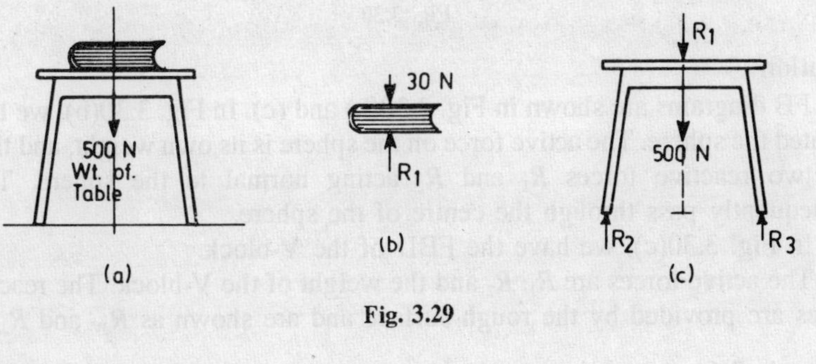

Fig. 3.29

Solution

The free body diagrams are shown in Fig. 3.29(b) and (c). In Fig. 3.29(b), we have isolated the book as a free body.

There are only 2 forces acting on the book:

(i) The weight of the book, and

(ii) The reactive force provided by the table. Thus 30 N is the active force, and R_1 is the reactive force on the book.

In Fig. 3.29(c) we have isolated the table. In this case we use the fundamental principle enunciated in Newton's third law—action and reaction are equal and opposite. Thus, as the table provides a reactive force of R_1 upwards on the book, the book will exert a force of R_1 downwards on the table. The other active forces acting on the table is the weight of the table at the point shown. The reactive forces are provided by all the four legs of the table. We have marked two reactive forces R_2 and R_3, R_2 being the total reactive force supplied by the left legs and R_3 that by the right legs.

Please note that in the free body diagram of the book, we have a two force system, and in the free body diagram of the table, we have a parallel force system. Equilibrium conditions can now be applied to evaluate the constraints.

Example 3.5

A smooth sphere of dia 30 cm and weight 300 N is resting on a V-block, which in turn rests on a rough surface as shown in Fig. 3.30(a). Draw the free body diagrams of the sphere and the V-block.

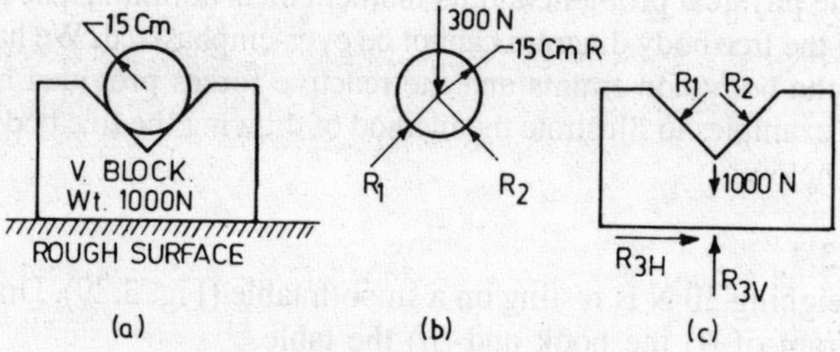

Fig. 3.30

Solution

The FB diagrams are shown in Fig. 3.30(b) and (c). In Fig. 3.30(b), we have isolated the sphere. The active force on the sphere is its own weight, and there are two reactive forces R_1 and R_2 acting normal to the sphere. They consequently pass through the centre of the sphere.

In Fig. 3.30(c), we have the FBD of the V-block.

The active forces are R_1, R_2 and the weight of the V-block. The reactive forces are provided by the rough surface and are shown as R_{3V} and R_{3H}.

Example 3.6

Two pipes of weights 2000 N and 3000 N, and diameters 1 cm and 2 cm respectively, rest in a cutting as shown in Fig. 3.31. The section through the centre at which these weights act is shown in the figure. Draw the free body diagrams of the two pipes.

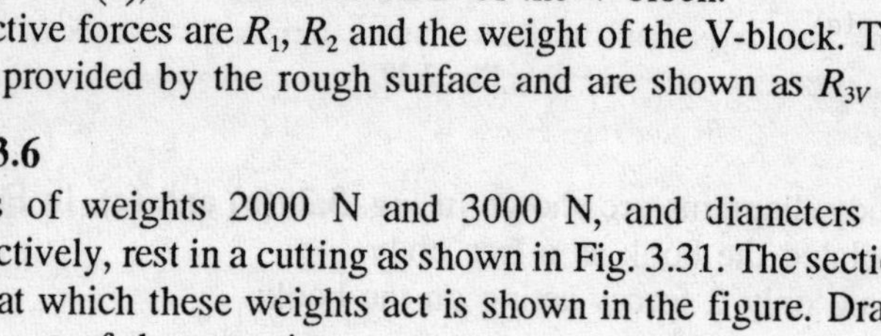

Fig. 3.31

Solution

The free body diagrams are shown in Fig. 3.31(b) and (c). Please note that in drawing these diagrams, the pipes are assumed to be smooth and the surfaces of the cutting as rough. Thus in Fig. 3.31(b), in the free body diagram of the larger cylinder, we have on reactive component R_2 due to the smaller cylinder, and two components R_{1H} and R_{1V} due to the surface of the cutting. In Fig. 3.31(c), we have R_2 marked acting on the smaller cylinder as reaction from the larger cylinder, equal and opposite to R_2 shown in Fig. 3.31(b). The smaller cylinder is supported at two points on the rough surface of the cutting and we have R_{3H} and R_{3V} at one point and R_{4H} and R_{4V} at the other. It can be easily seen that by drawing the F.B. diagrams, we are splitting the problem into smaller parts.

In Fig. 3.32, we have taken two cylinders together and drawn the F.B. diagram. Please note in this case that the reaction R_2 vanishes, being equal and opposite, and reactive forces act only at the points of contact with the cutting.

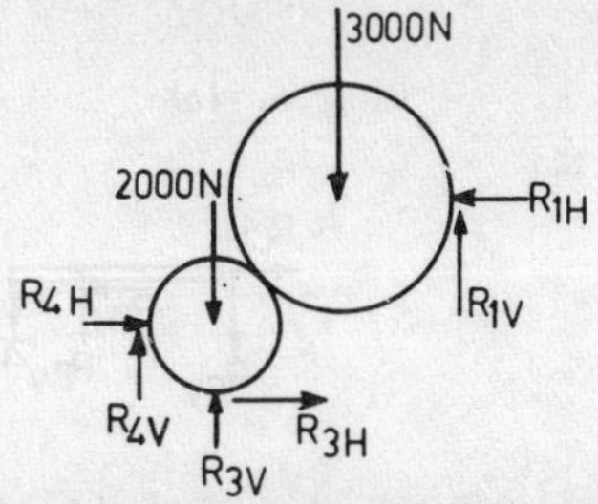

Fig. 3.32

Thus, we have to draw free body diagrams depending upon the need of the problem. If the reaction R_2 is to be calculated, then the free body diagrams shown in Fig. 3.31 have to be drawn. But if we need compute only the reactions at the points of contact with the surface of the cutting, the free body diagram as given in Fig. 3.32 is enough, though it will be necessary to draw the split FBDs to evaluate the unknowns as these are more than 3.

Example 3.7

Draw the free body diagram of the rod AB shown in Fig. 3.33(a).

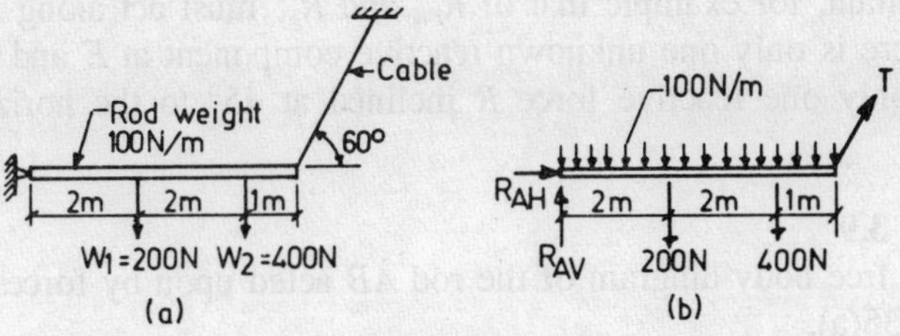

Fig. 3.33

Solution

The free body diagram is shown in Fig. 3.33(b). We first draw the rod and mark the active forces. These are (i) the weight of the rod uniformly distributed over its whole length, (ii) weight W_1 = 200 N, and (iii) weight W_2 = 400 N. Now we mark the reactive forces. The constraint at the left is a hinged support, and at B, it is a cable. Hence, we have the reactive forces R_{AH} and R_{AV} at A and the tension T in the cable at B.

Example 3.8

Draw the free body diagram of each of the rods *AB*, *CD* and *EF* shown in Fig. 3.34(a).

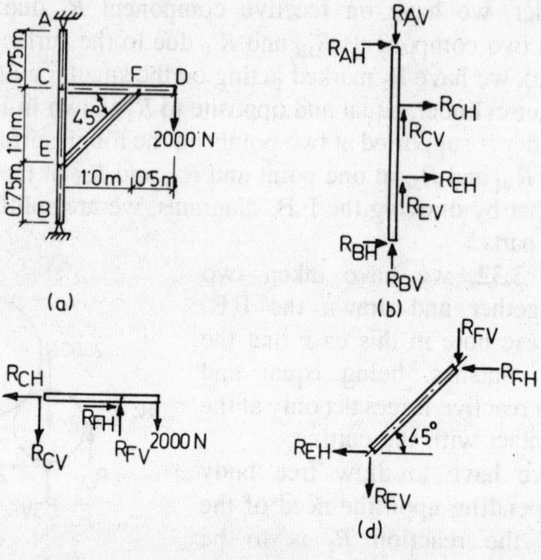

Fig. 3.34

Solution

The free body diagrams shown in Fig. 3.34(b), (c) and (d) can be easily drawn once it is realised that the supports at *A* and *B* are hinge supports, and the constraints at *C*, *E* and *F* are again hinges transferring forces to the other bars. However, one should notice the restriction at the hinges *E* and *F*. Both at *E* and *F*, the horizontal and vertical components are not independent, since their resultant, for example that of R_{FH} and R_{FV} must act along *EF*. In this sense, there is only one unknown reactive component at *E* and *F*. That is, there is only one reactive force *R* inclined at 45° to the horizontal at *E* and *F*.

Example 3.9

Draw the free body diagram of the rod *AB* acted upon by forces as shown in Fig. 3.35(a).

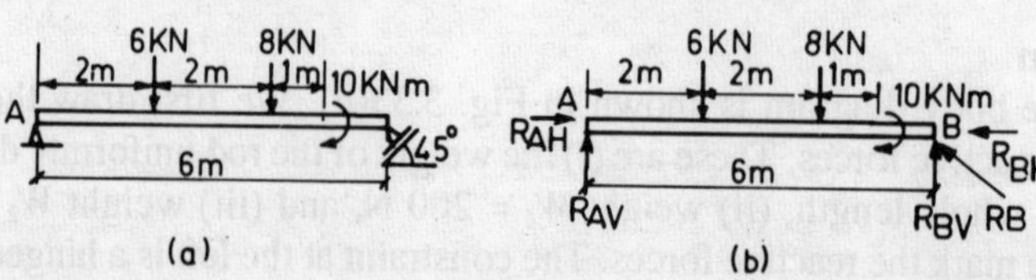

Fig. 3.35

Solution

The free body diagram is shown in Fig. 3.35(b). The support at the right is a hinge and at the left a roller. The importance of the free body diagram can be emphasized from this example. Since only vertical forces are acting, one should not assume that the reactive forces also will be vertical and miss the reactive force R_{AH} at A. In fact, R_{AH} is not equal to zero, and this should be drawn in the free body diagram.

Example 3.10

Draw the free body diagram of the cantilever AB loaded as shown in Fig. 3.36.

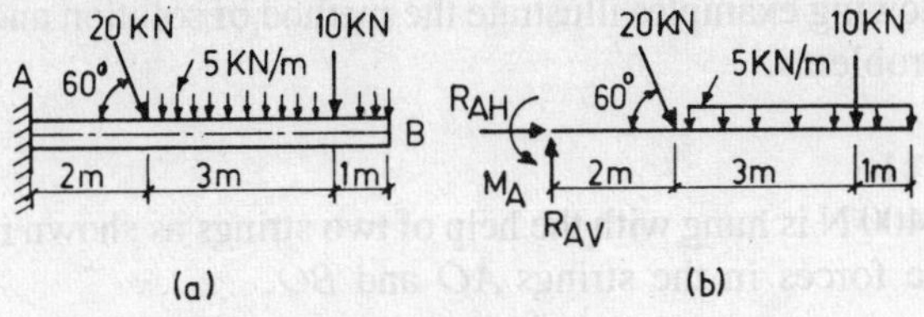

Fig. 3.36

Solution

The free body diagram is shown in Fig. 3.36(b). It is again emphasized here that since the support at A is a fixed support, three reactive components, R_{AH}, R_{AV} and M_A should be shown there.

Once the constraints are identified and free body diagrams drawn as discussed above, the solution of the problem is straight forward with the application of the conditions of equilibrium.

3.5.4 *Procedure for the Solution of Problems*

As already mentioned, a problem in equilibrium essentially involves the evaluation of the forces at the constraints. Conditions of equilibrium, body constraints and free body diagrams are essential components in the solution of all problems. For any particular problem there can be more than one method of solution, and no hard and fast rule can be stated regarding the procedure for a solution. However, the following steps are quite general and have to be gone through the solution of all problems.

(i) Draw a diagram or sketch representing the physical problem. This enables one to visualise and analyse the number and type of constraints involved.

(ii) Analyse the sketch drawn in (i) and from the statement of the problem, identify the known and unknown quantities. A list of the unknowns to be evaluated may be of great help.

(iii) Formulate a strategy for the solution. This will involve the determination of the number of free body diagrams to be drawn, the number of unknowns in each, and the order in which the free body diagrams are to be tackled.

(iv) Draw free body diagram (s) as determined and properly name and identify the unknowns. Also keep in mind that action and reaction are equal and opposite.

(v) Apply the conditions of equilibrium to the free bodies depending upon the force system involved.

(vi) Solve the equations so obtained to evaluate the unknowns.

(vii) If a graphical solution is preferred, the free body diagrams should be drawn to scale. A force diagram and/or funicular polygon as may be required has to be then drawn to complete the solution. The values of the unknowns can be scaled off from the *diagrams.*

The following examples illustrate the method of solution and cover a wide variety of problems.

Example 3.11

A weight of 400 N is hung with the help of two strings as shown in Fig. 3.37(a). Compute the forces in the strings *AO* and *BO*.

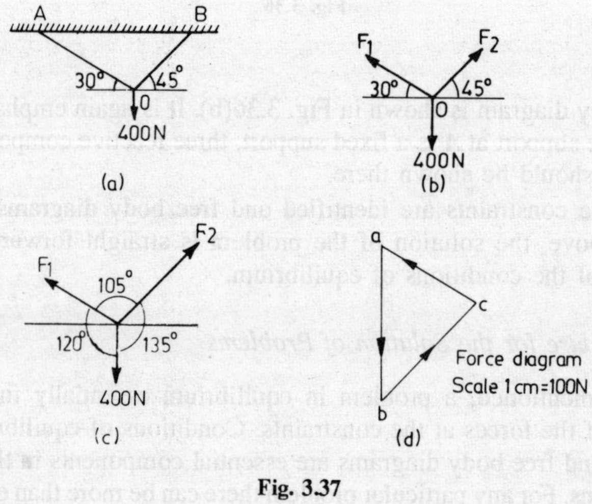

Fig. 3.37

Solution

We draw a free body diagram of the point '*O*' where the two strings and the line of action of the weight meet. The free body diagram of point '*O*' is shown in Fig. 3.37(b). The direction of the two forces in the strings will be along the strings themselves and their sense is marked by intuition only. Since the 400 N weight acts down and stretches the strings, the forces in the strings will be opposed to the direction of stretch. They can as well be marked in opposite direction. If the assumed direction is wrong, the result will come out to be negative, which means that the force actually acts in the opposite direction.

In the free body diagram, we have a concurrent force system of three forces in which two of the forces are known in direction, but have unknown magnitudes. There are many methods in which this can be solved, but we will indicate only two methods of solution.

(i) *By Lami's Theorem.* In Fig. 3.37(c), the three forces are shown with the angle between the forces calculated from the given directions. Hence,

$$\frac{F_1}{\sin 135} = \frac{F_2}{\sin 120} = \frac{400}{\sin 105}$$

There are two equations here and two unknowns, and hence the unknowns can be evaluated.

$$F_1 = \frac{400}{\sin 105} \times \sin 135 = \mathbf{292.8\ N} \text{ (tension in the string } OA)$$

and $$F_2 = \frac{400}{\sin 105} \times \sin 120 = \mathbf{358.6\ N} \text{ (tension in the string } OB)$$

Since both the results have positive signs, the assumed directions are correct.

(ii) *Graphical Solution.* The three concurrent forces in equilibrium should form a closed force triangle. We mark a line *ab* downward to represent the force of 400 N to a scale of 1 cm = 100 N (Fig. 3.37d). From the ends *a* and *b*, we draw two lines inclined at 30° and 45° to the horizontal. They intersect at point *c*. Since *ab* is downward, the sense of the other two forces must be *b* to *c* and *c* to *a*. These can now be measured off to the scale. From Fig. 3.37(d) by measurement,

bc = **360 N** (tension in string *OB*)

ca = **290 N** (tension in string *OA*)

Note. To illustrate what happens if the directions of the unknown forces are assumed wrongly let us say that in Fig. 3.37(b), we have assumed the direction of F_2 as in Fig. 3.38(a). We again draw the diagram as in Fig. 3.38(b) and apply Lami's theorem.

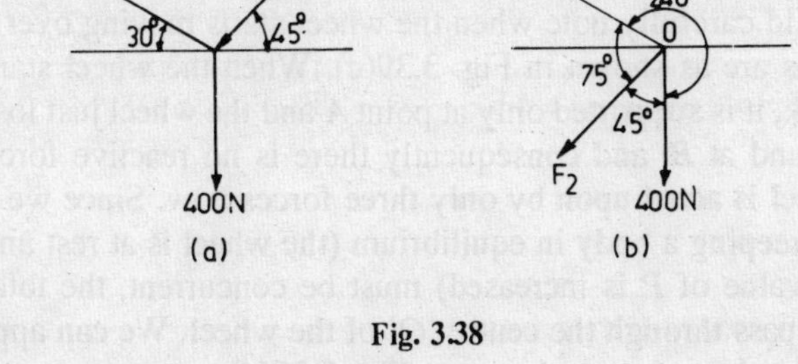

Fig. 3.38

$$\frac{F_1}{\sin 45} = \frac{F_2}{\sin 240} = \frac{400}{\sin 75}$$

$$F_1 = \frac{400}{\sin 75} \times \sin 45 = \mathbf{292.8\ N}$$

and $$F_2 = \frac{400}{\sin 75} \times \sin 240 = \mathbf{-358.6\ N}$$

The negative sign indicates that the direction of F_2 is opposite to that shown in Fig. 3.38(a).

Example 3.12

A wheel 25 cm radius and weighing 5 kN is resting on a rough surface as shown in Fig. 3.39(a). The force *P* is applied to start the wheel over the obstruction 10 cm high.

(a) If *P* is applied horizontally, determine the minimum value of '*P*' needed to start the wheel over the obstruction, and

(b) If *P* is applied inclined at an angle α to the horizontal, find the minimum value of *P*, the angle α and the reaction at the ground.

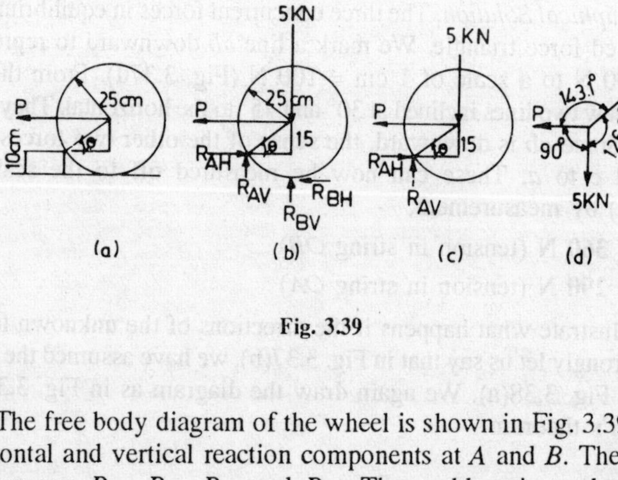

Fig. 3.39

Solution

(a) The free body diagram of the wheel is shown in Fig. 3.39(b). There are horizontal and vertical reaction components at *A* and *B*. There are thus four unknowns, R_{AH}, R_{AV}, R_{BH} and R_{BV}. The problem is to determine the minimum value of *P* at which the wheel starts moving over the obstruction.

We should carefully note when the wheel starts moving over the block, the conditions are as shown in Fig. 3.39(c). When the wheel starts moving over the block, it is supported only at point *A* and the wheel just loses contact with the ground at *B*, and consequently there is no reactive force at *B*.

The wheel is acted upon by only three forces now. Since we know that three forces keeping a body in equilibrium (the wheel is at rest and will just move if the value of *P* is increased) must be concurrent, the total reaction R_A at *A* must pass through the centre '*O*' of the wheel. We can apply Lami's theorem to the three forces shown at Fig. 3.39(d).

$$\text{Angle } \theta = \sin^{-1}\left(\frac{15}{25}\right) = \mathbf{36.9°}$$

The angles between the forces can now be calculated and are shown in Fig. 3.39(d). We have, by Lami's theorem

$$\frac{P_A}{\sin 90^\circ} = \frac{P}{\sin 126.9^\circ} = \frac{W}{\sin 143.1^\circ}$$

$$P = \frac{5 \times \sin 126.9}{\sin 143.1} = \mathbf{6.66\ kN}$$

$$R_A = \frac{5 \times \sin 90}{\sin 143.1} = \mathbf{8.33\ kN}$$

This is the minimum value of P needed to raise the wheel over the 10 cm block.

Alternatively a graphical solution can be used. We draw to scale the space diagram shown in Fig. 3.40(a). We draw the triangle of forces by first drawing ab = 5 kN downwards. We then draw ac horizontal and bc at an angle 36.9° at b. These two lines intersect at c. We measure ac which gives the value of P to scale, and 'bc' which gives R_A to scale. From Fig. 3.40(b) by measurement,

$$ac = 6.7 \text{ cm} = \mathbf{6.7\ kN}$$

$$bc = \mathbf{8.3\ kN}$$

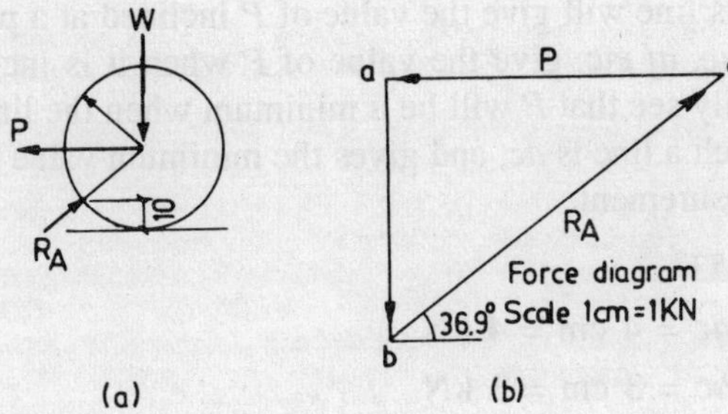

Fig. 3.40

The reader should again carefully study how the arrows in the force diagram are put. Knowing the direction of ab from a to b, we put the arrows in the same sense on the other two sides, from b to c and from c to a.

(b) If P is inclined at an angle α, then the free body diagram when the wheel is about to move will be as shown in Fig. 3.41(a). Again applying Lami's theorem,

$$\frac{R_A}{\sin (90 + \alpha)} = \frac{P}{\sin 126.9^\circ} = \frac{W}{\sin (143.1 - \alpha)}$$

$$P = \frac{5 \times \sin 126.9^\circ}{\sin (143.1 - \alpha)}$$

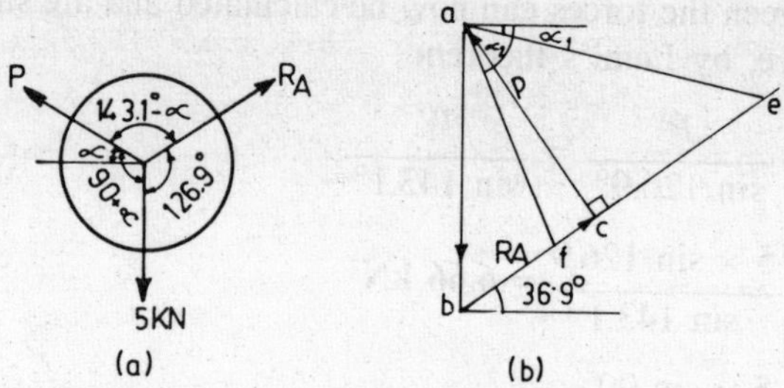

Fig. 3.41

The value of P in this equation will be a minimum when the value of sin (143.1 – α) is maximum. One can eassily find that the value of α must be equal to 53.1°, which makes sin(143.1 – α) = sin 90 = 1.0, which is the maximum value of the sine of any angle.

$$P = 5 \times \sin 126.9 = \mathbf{4\ kN}$$

$$R_A = \mathbf{3\ kN}$$

The problem can also be solved graphically as shown in Fig. 3.41(b). We lay out ab = 5 kN to scale.

We draw a line inclined at 36.9° to ab at b. Any line drawn from 'a' to a point on this line will give the value of P inclined at a particular angle. Thus the lines ae, af etc. give the value of P when it is inclined at α_1, α_2, etc. We can easily see that P will be a minimum when the line ae is at right angles to bc. Such a line is ac, and gives the minimum value of P, from Fig. 3.41(b), by measurement,

$$\alpha = 53°$$

$$P = ac = 4 \text{ cm} = \mathbf{4\ kN}$$

$$R_A = bc = 3 \text{ cm} = \mathbf{3\ kN}$$

Example 3.13

A box weighing 8 kN is held at rest on a smooth plane inclined at 30° to the horizontal by the force P as shown in Fig. 3.42(a). Determine the value of P and the reaction at the plane.

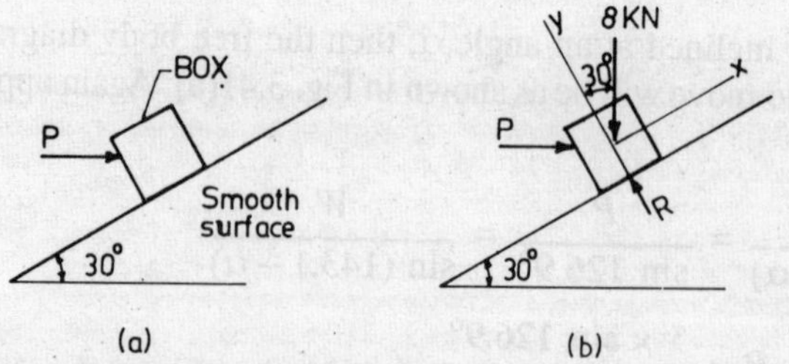

Fig. 3.42

Solution

The free body diagram is shown in Fig. 3.42(b). We use a different approach this time for illustrative purpose. Select X- and Y- axes as shown. We know that for equilibrium, $\Sigma F_x = 0$, and $\Sigma F_y = 0$. We write these two equations as,

$$\Sigma F_x = 0,\ P \sin 60 - 8 \sin 30 = 0$$

$$P = \frac{8 \sin 30}{\sin 60^\circ} = \mathbf{4.62\ kN}$$

$$\Sigma F_y = R - P \cos 60 - 8 \cos 30 = 0.$$

$$R = P \cos 60 + 8 \cos 30$$

$$= 4.62 \times \frac{1}{2} + 8 \times 3/2$$

$$= \mathbf{9.24\ kN}$$

Example 3.14

The cable and the bar shown in Fig. 3.43 support a load of 1000 N. Determine the tension in the cable and the reaction at the hinge *A*.

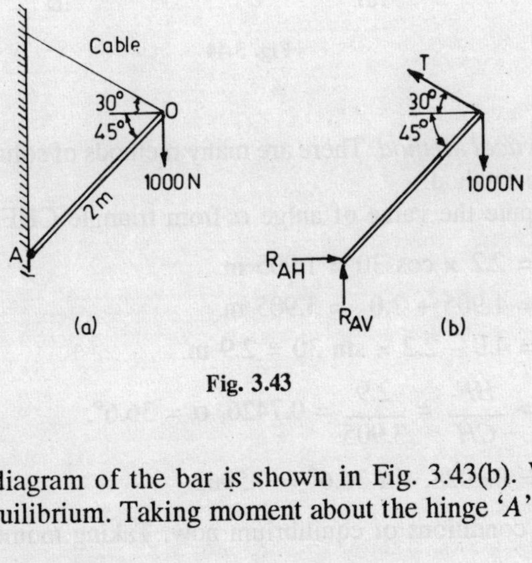

Fig. 3.43

Solution

The free body diagram of the bar is shown in Fig. 3.43(b). We apply the conditions of equilibrium. Taking moment about the hinge 'A' and equating to zero,

$$\Sigma M_A = 0,\ 1000 \times 2 \cos 45 - T \times 2 \sin 75 = 0$$

$$T = \frac{1000 \times 2 \cos 45}{2 \sin 75} = \mathbf{732\ N}$$

$$\Sigma H = 0,\ R_{AH} - T \cos 30 = 0,\ R_{AH} = 732 \cos 30 = \mathbf{634\ N}$$

$$\Sigma V = 0,\ R_{AV} + T \sin 30 - 1000 = 0,\ R_{AV} = 1000 - 732 \sin 30 = \mathbf{634\ N}$$

Reaction at the hinge $R = \sqrt{634^2 + 634^2} = \mathbf{896.6\ N}$ inclined at 45° to the horizontal.

This result could have been anticipated, if we had observed that if we combine R_{AH} and R_{AV} into one force, the beam is acted upon by only three forces, T, 1000 N and R, which should be concurrent.

Example 3.15

A concrete slab weighing 900 kN is being raised into a vertical position by the arrangement shown in Fig. 3.44(a). Calculate the tensions T_1 and T_2 in the cables when angle $\theta = 30°$.

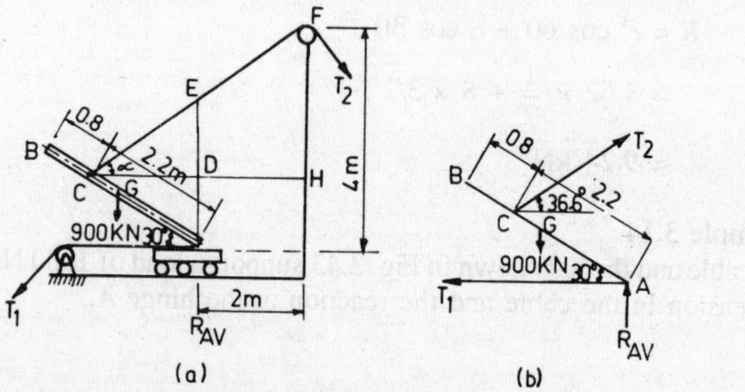

Fig. 3.44

Solution

(a) Analytical Method. There are many methods of solution. We present here one such method.

We first compute the value of anlge α from triangle CHF.

$$CD = 2.2 \times \cos 30 = 1.905 \text{ m.}$$

$$CH = 1.905 + 2.0 = 3.905 \text{ m}$$

$$HF = 4.0 - 2.2 \times \sin 30 = 2.9 \text{ m.}$$

$$\tan \alpha = \frac{HF}{CH} = \frac{2.9}{3.905} = 0.7426, \ \alpha = 36.6°.$$

$$\cos 36.6° = 0.8028, \ \sin 36.6° = 0.5962$$

We apply the conditions of equilibrium now. Taking moments about A, in Fig. 3.44(b),

$$\Sigma M_A = 0, \ T_2 \cos \alpha \times 1.1 + T_1 \sin \alpha \times 1.905 - 900 \times 1.299 = 0$$

$$T_2 = \mathbf{579 \ N}$$

$$\Sigma H = 0, \ T_2 \cos \alpha - T_1 = 0$$

$$T_1 = T_2 \cos 36.6° = 579 \times 0.8028 = \mathbf{465 \ N}$$

The third condition, $\Sigma V = 0$, can be used to calculate the value of R_{AV}.

(b) Graphical Method. The problem can be solved very easily graphically.

We lay out the space diagram to a scale of 1 cm = 0.5 m as in Fig. 3.45(a). If we combine the reaction components T_1 and R_{AV} at A, the slab is acted upon by only three forces T_2, W and R. These three forces should be concurrent. We produce the line of action of W upwards to meet the line of action of T_2 at D and join AD which is the line of action of R. We now draw the force triangle *abc* shown in Fig. 3.45(b) to a scale of 1 cm = 100 N. *ab* represents W, *bc* gives R and *ca* gives T_2. T_2 can be directly scaled off.

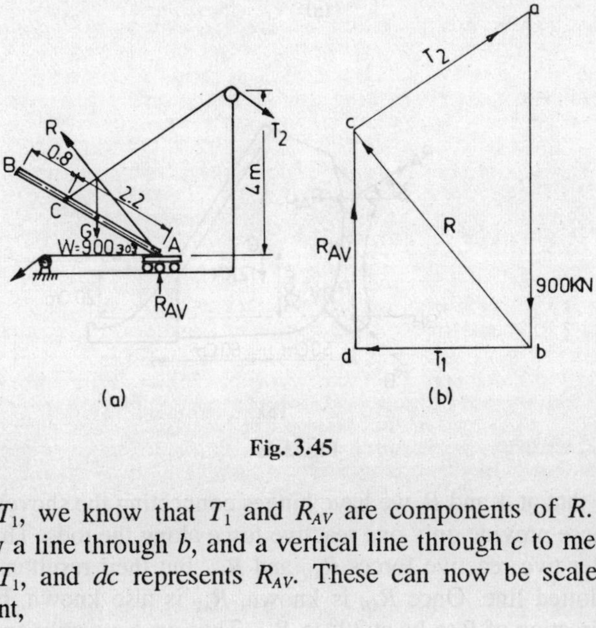

Fig. 3.45

To obtain T_1, we know that T_1 and R_{AV} are components of R. We draw horizontally a line through b, and a vertical line through c to meet at d. *bd* represents T_1, and *dc* represents R_{AV}. These can now be scaled off. By measurement,

$$T_2 = ca.$$
$$= 5.8 \text{ cm}$$
$$= \mathbf{580\ N}$$

$$T_1 = bd$$
$$= 4.7 \text{ cm}$$
$$= \mathbf{470\ N}$$

Example 3.16

Figure 3.46 shows a shovel used to lift earth. In the position shown, the weight of the shovel and the earth is 12 kN acting at G. The magnitude of the force F is 5 kN. Calculate the values of the reactions at A and B when the shovel is in the position shown.

Solution

The free body diagram is shown in Fig. 3.46(b). Here we have a general co-planar force system. We have three unknowns R_A, R_B and the angle θ.

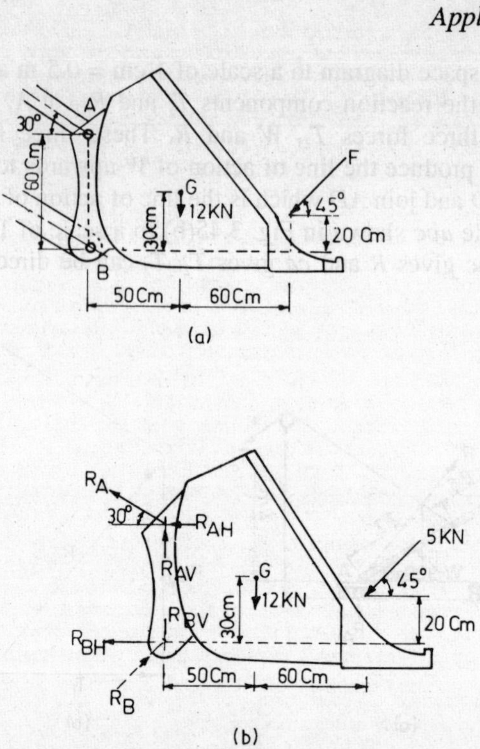

Fig. 3.46

Please note that at A and B, we have hinges connecting the shovel to the two rods; but these provide only one reactive force along the rods. Thus at A, we may consider two reactive forces R_{AH} and R_{AV}, but their resultant should be along the dotted line. Once R_{AH} is known, R_{AV} is also known, because we know the direction of R to be at 30° to R_{AH}. Thus we have only one unknown reactive force at A and two at B. At B, the unknowns can be considered as either R_B and θ, or as R_{BH} and R_{BV}.

The three unknowns can be evaluated by applying the three conditions of equilibrium. Thus, taking moments about B,

$$\Sigma M_B = 0,\ 12 \times 50 - 5 \cos 45 \times 20 + 5 \sin 45 \times 110 - T_A \times \cos 30 \times 60 = 0$$

R_A = **17.66 kN** at 30° to the horizontal

$$\Sigma H = 0,\ -R_{BH} - R_A \cos 30 - 5 \cos 45 = 0$$

R_{BH} = **−18.83 kN** (Sign of R_{BH} is reversed in 3.4.6 (b))

$$\Sigma V = 0,\ R_{BV} + R_A \sin 30 - 12.5 \sin 45 = 0$$

$$R_{BV} = 6.705 \text{ kN}$$

$$R_B = \sqrt{6.705^2 + (-18.83)^2}$$

= **20 kN.**

$$\theta = \tan^{-1}\left(\frac{6.705}{-18.83}\right)$$

= **19.6°**

The directions of the reactions are shown in Fig. 3.46(b).

Graphical Solution. The space diagram is drawn to a scale of 1 cm = 20 cm in Fig. 3.47(a). There are two graphical solutions possible, *viz.* one using the three-force system approach, and the other based upon the funicular polygon. We illustrate the funicular polygon approach here.

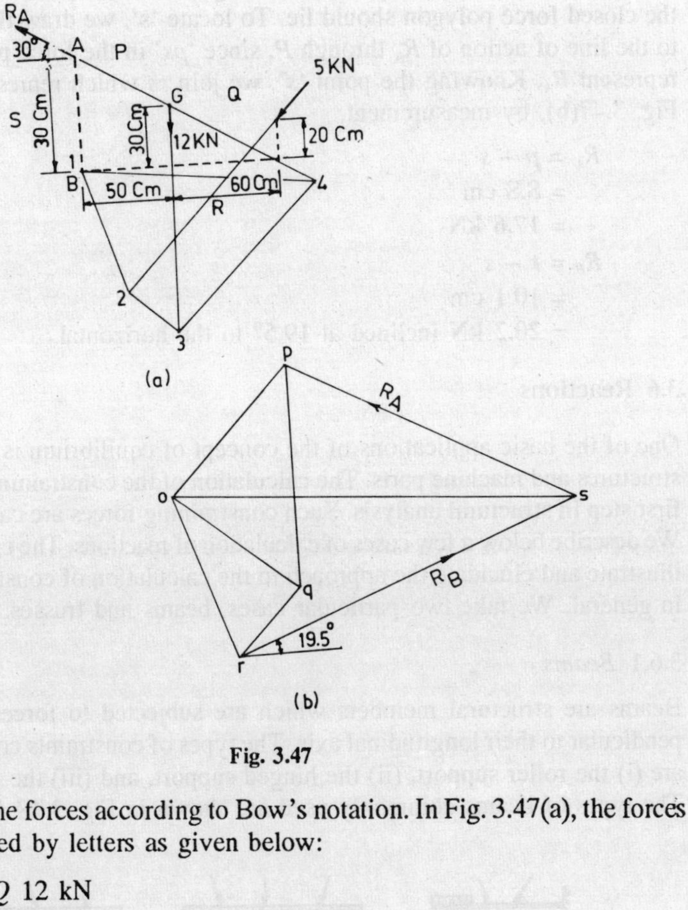

Fig. 3.47

We number the forces according to Bow's notation. In Fig. 3.47(a), the forces are represented by letters as given below:

$P - Q$ 12 kN
$Q - R$ 5 kN
$R - S$ R_B direction and magnitude unknown
$S - P$ R_A direction known and magnitude unknown

We identify three unknowns here, one at *A* and two at *B*.

A force polygon *pqr* is drawn to a scale of 1 cm = 2 kN in Figure 3.47(b). We select a pole '*O*', and start the funicular polygon. Here one has to be careful. We start the funicular polygon at *B*, draw the line parallel to $\overline{or}$ from *B*. This line is 1-2. The line parallel to $\overline{or}$ should lie in space *R*, which is bounded ($\overline{or}$ means the line from o to r, the vector from o to r) by the lines of action of R_B and the 5 kN force. R_B should pass through B, and hence the line 1-2 lies in space R. We continue the funicular polygon by drawing the line 2-3 parallel to $\overline{oq}$, and 3-4 parallel to $\overline{op}$. Since space B is bounded by the lines of action of 12 kN force and R_A, we produce R_A to intersect the line 3-4 at 4. We now close the funicular polygon by joining 1-4.

We draw a line parallel to 1-4 through 's'. On this line the point 's' of the closed force polygon should lie. To locate 's', we draw the line parallel to the line of action of R_A through P, since '*ps*' in the force polygon should represent R_A. Knowing the point '*s*', we join *rs* which represents R_B. From Fig. 3.47(b), by measurement,

$$R_A = p - s$$
$$= 8.8 \text{ cm}$$
$$= \mathbf{17.6 \text{ kN}}$$
$$R_B = r - s$$
$$= 10.1 \text{ cm}$$
$$= \mathbf{20.2 \text{ kN}} \text{ inclined at } 19.5° \text{ to the horizontal.}$$

3.6 Reactions

One of the basic applications of the concept of equilibrium is in analysis of structures and machine parts. The calculation of the constraining forces is the first step in structural analysis. Such constraining forces are called reactions. We describe below a few cases of calculation of reactions. The examples given illustrate and elucidate the approach to the calculation of constraining forces in general. We take two particular cases, beams and trusses.

3.6.1 *Beams*

Beams are structural members which are subjected to forces mainly perpendicular to their longitudinal axis. The types of constraints commonly used are (i) the roller support, (ii) the hinged support, and (iii) the fixed support. The types of beams commonly used are shown in Fig. 3.48. These depend

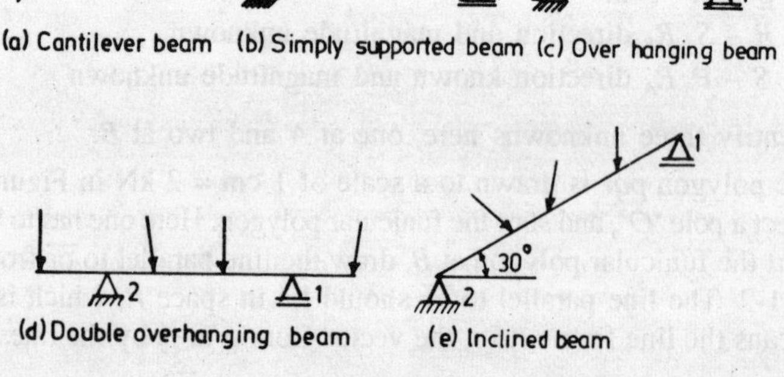

(a) Cantilever beam (b) Simply supported beam (c) Over hanging beam

(d) Double overhanging beam (e) Inclined beam

Fig. 3.48

only upon the number and type of constraints used. You will notice that all the types of beams shown have only three constraining forces. This makes them statically determinate and only such cases can be solved by methods of this chapter. The beams are assumed to be rigid for the purpose of calculating the reactions, although in actual practice they are deformable. The concepts developed are applicable only when the bodies are rigid. The application of these concepts is justified only because the deformations are small and can be neglected.

A general approach to problems concerning beams is discussed in this article and is shown in Fig. 3.49. We describe below three methods of solution, two graphical methods and one analytical method.

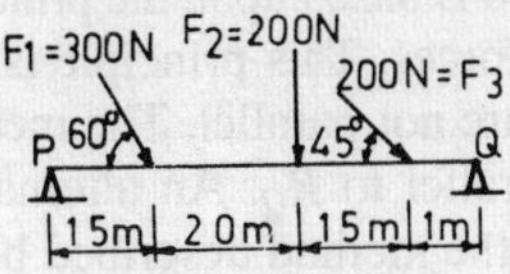

Fig. 3.49

Method I. Considering the forces F_1, F_2 and F_3, we can determine their resultant by the methods developed in Chapter 2. This is done in Fig. 3.50(a) and (b). The force diagram is then drawn to scale and the resultant is obtained by scaling off $\overline{ad}$. The funicular polygon is constructed as usual and the resultant is located in space diagram.

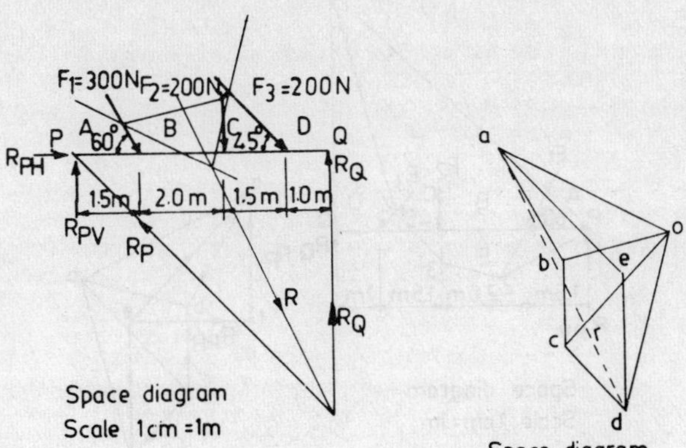

Fig. 3.50

The forces F_1, F_2 and F_3 acting on the beam can be replaced by their resultant R. The beam is now acted upon by only three forces R_P, R and R_Q. Since the beam is in equilibrium, these three forces should be concurrent.

Since R_Q is known to be vertical, we produce R to intersect R_Q at 'S' and join points P and S. $\overline{PS}$ is the direction of the reaction at P. A force diagram can now be constructed for R_P, R and R_Q. All the three directions are known and also the magnitude of R. This may be done separately or on the line ad itself as shown in Fig. 3.50(b). $\overline{de}$ is the reaction at P and $\overline{ea}$ is the reaction at Q. The values can be scaled off from the triangle of forces.

Reaction $R_Q = \overline{ea}$ = 3 cm = **300 N**

$R_P = \overline{de}$ = 4.2 cm = **420 N** (inclined at 46° to horizontal)

If the horizontal and vertical components of the reaction at P are required, these can be obtained by drawing vertical and horizontal lines through e and a respectively.

Method II. The above method is based upon the principle of reducing the forces on the beam to only three forces. This principle is applicable only when the forces acting on the beam are not parallel. The method fails when the forces acting on the beam are parallel to R_Q. An alternative graphical solution is essential in such cases and the method described below is more general, and is applicable in all cases.

In Fig. 3.51 we have drawn the space diagram to scale. Note carefully the way the forces are named. *DE* represents the reaction at *Q* which is vertical and *EA* represents the reaction at *P* whose line of action is not known. We are considering the total reaction R_P at *P* and *not* the components R_{PH} and R_{PV}.

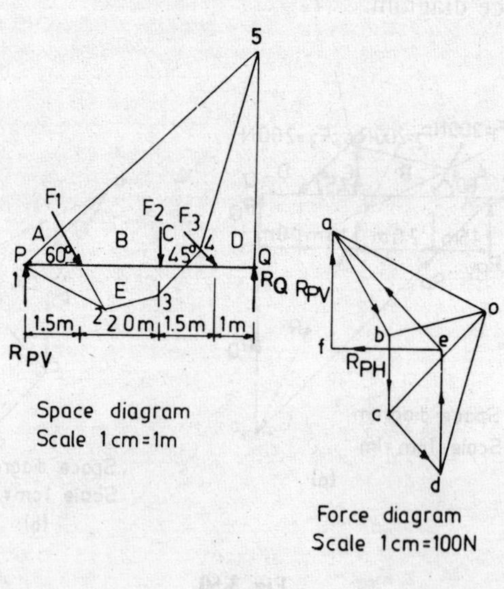

Fig. 3.51

Principle. The five forces F_1, F_2, F_3, R_Q and R_P form a general coplanar force system in equilibrium. The graphical conditions of equilibrium require that the force polygon of the five forces should close and the corresponding funicular polygon should also close.

We draw first the force diagram $\overline{abcd}$ of the known forces F_1, F_2 and F_3. Selecting a pole '*O*' we draw the funicular polygon. An important point to note is that we start the line parallel to $\overline{oa}$ from point '*P*'. The reason is obvious. Any line in the funicular polygon should be between the lines of actions of two forces defining the space between them. The line of action of R_P is not known, but we do know that it should pass through point *P*. Note carefully that we cannot start from line of action of R_Q as we are sure to be held up without knowing the line of action of R_P.

Having drawn the lines $\overline{oa}$, $\overline{ob}$, $\overline{oc}$ and $\overline{od}$ of the funicular polygon, we notice that the line parallel to $\overline{od}$ of the funicular polygon should lie between the lines of action of F_3 and R_Q. We, therefore, produce R_Q to intersect the line parallel to $\overline{od}$ at '5'.

We know that the funicular polygon should close. This means that the line in space (for the reaction *EA*) from point '1' and the line in space *E* (for the reaction *ED*) from point '5' should be collinear. We, therefore, join points 1 and 5. We draw a line parallel to line 1-5 through the pole '*o*'.

The point '*e*', representing the space *E* should lie on this line. This point can be easily located, since we know that the reaction *DE* should be vertical. We draw a vertical line through '*d*' in the force polygon to intersect the line drawn through '*O*' (parallel to 1-5) at *e*. We join *ea*, which gives the magnitude, direction and sense of the reaction R_P. $\overline{de}$ gives the reaction R_Q . Notice that the force polygon closes, a-b-c-d-e-a and the funicular polygon closes, 1-2-3-4-5-1.

From Fig. 3.51(b), by measurement,

de = 3 cm = 3 × 100 N = **300 N** (vertical)

ea = 4.2 cm = 4.2 × 100 = **420 N** (inclined at 46° to horizontal)

Method III. In Fig. 3.52, we have drawn the free body diagram of the beam. There are three unknowns, R_{PH}, R_{PV} and R_Q.

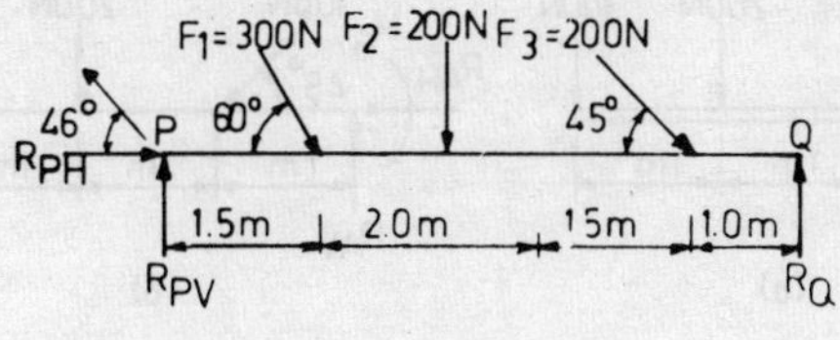

Fig. 3.52

Principle. Since the beam is in equilibrium, the force system acting on the beam (F_1, F_2, F_3, R_{PH}, R_{PV} and R_Q) should satisfy the three conditions of equilibrium

(i) $\Sigma H = 0$

(ii) $\Sigma V = 0$

(iii) $\Sigma M = 0$

These three conditions will yield three equations, from which the three unknowns can be evaluated.

Note that the three conditions need not be applied in the order given. Further, $\Sigma M = 0$ about any point in the plane. Hence, by suitably selecting the moment centre, we can ensure that only one unknown appears in the equation. This facilitates the solution being arrived at quickly.

We apply the conditions of equilibrium to the free body in Fig. 3.52.

Taking moments about point *P*,

$$\Sigma M_p = 0, 300 \sin 60 \times 1.5 + 200 \times 3.5 + 200 \sin 45 \times 5 - R_Q \times 6 = 0.$$

$$\mathbf{R_Q = 300\ N}$$

$$\Sigma_V = 0, + R_{PV} + R_Q - 300 \sin 60 - 200 - 200 \sin 45 = 0$$

$$R_{PV} = 601.2 - 300 = \mathbf{301.2\ N}$$

$$\Sigma_H = 0, R_{PH} + 300. \cos 60 + 200 \cos 45 = 0$$

$$R_{PH} = -150 - 141.4 = \mathbf{-291.4\ N}$$

The sign of R_{PH} indicates that R_{PH} actually acts left wards

$$R_P = \sqrt{(301.2)^2 + (-291.4)^2}$$

$$= \mathbf{419\ N}$$

$$\theta = \tan^{-1}\left(\frac{301.2}{-291.4}\right)$$

$$= \mathbf{46°}$$

The direction and sense of R_P are shown in Fig. 3.52.

The following examples illustrate the procedure.

Example 3.17

Determine the reactions at the support of the cantilever beam carrying the loads shown in Fig. 3.53(a).

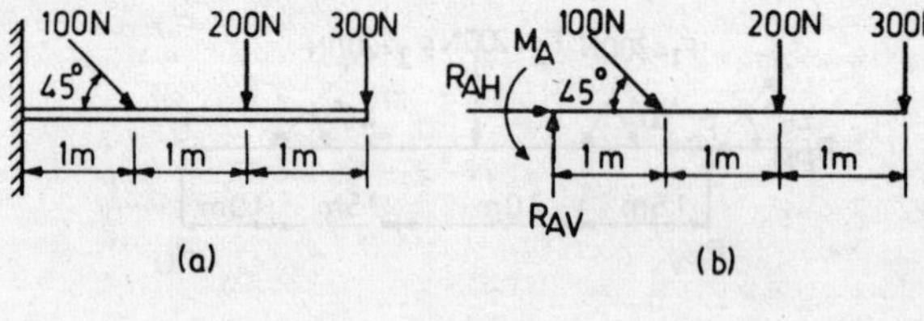

Fig. 3.53

Solution

The free body diagram is shown in Fig. 3.53(b). Please note that there are three reactive components, R_{AH}, R_{AV} and couple M_A at support *A*. The three components can be determined by the three conditions of equilibrium.

Taking moments about A, $\Sigma M = 0$ about A.

$$\Sigma M_A = 0,\ 300 \times 3 + 200 \times 2 + 100 \sin 45 \times 1 - M_A = 0.$$

$$M_A = \mathbf{1370.7\ Nm}$$

(Please note that a couple has the same moment about any points in its plane, and hence will appear in the equation wherever the moment centre may be.)

$$\Sigma H = 0,\ R_{AH} + 100 \cos 45 = 0$$

$$R_{AH} = \mathbf{-\ 70.7\ N}$$

The negative sign shows that R_{AH} actually acts from right to left, opposite to the direction shown in Fig. 3.53(b).

$$\Sigma V = 0,\ R_{AV} - 100 \sin 45 - 200 - 300 = 0$$

$$R_{AV} = \mathbf{570.7\ N}$$

Example 3.18

Determine the reactions at the supports A and B for the beam loaded as shown in Fig. 3.54(a).

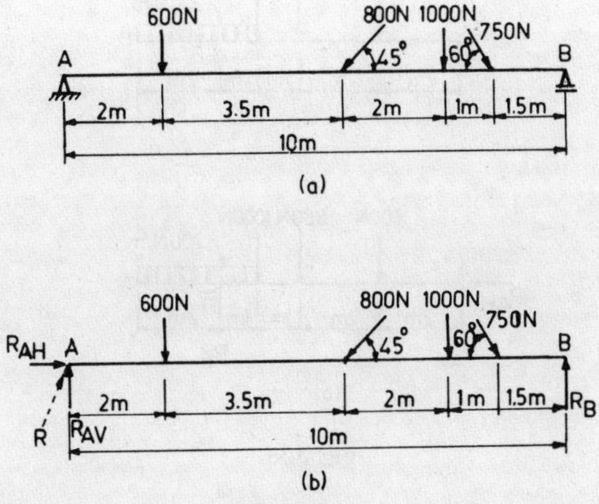

Fig. 3.54

Solution

The free body diagram is shown in Fig. 3.54(b). We can apply the conditions of equilibrium to evaluate R_{AH}, R_{AV} and R_B. Taking moments about A and equating to zero.

$$\Sigma M_A = 0,\ 600 \times 2 + 800 \cos 45 \times 5.5 + 1000 \times 7.5 + 750 \sin 60 \times 8.5 - R_B \times 10 = 0$$

$$R_B = \mathbf{1733.2\ N}$$

$$\Sigma H = 0,\ R_{AH} - 800 \cos 45 + 750 \times \cos 60 = 0$$

$$R_{AH} = \mathbf{190.7\ N}$$

$\Sigma V = 0,\ R_{AV} + R_B - 600 - 800 \sin 45 - 1000 - 750 \sin 60 = 0$

$R_{AV} =$ **1082 N**

Total reaction R at A

$= \sqrt{(190.7)^2 + (1082)^2}$

$=$ **1098.68 N**

Inclination of R to horizontal

$= \tan^{-1}\left(\dfrac{1082}{190.7}\right)$

$\theta =$ **80°**

Example 3.19

Determine the reactions at the supports A and B of the over hanging beam shown in Fig. 3.55.

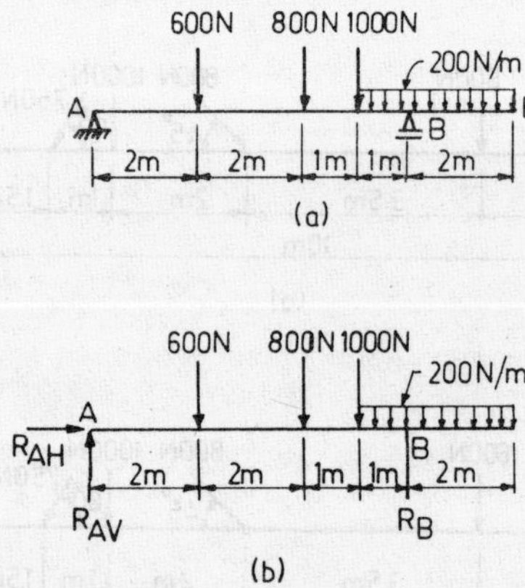

Fig. 3.55

Solution

(a) *Analytical.* The free body diagram is shown in Fig. 3.55(b). Taking moments about A and equating to zero,

$$\Sigma M_A = 0,\ 600 \times 2 + 800 \times 4 + 1000 \times 5 + 200 \times 3 \times 6.5 - R_B \times 6 = 0$$

$$R_B = \frac{13300}{6} = \mathbf{2216.7\ N}$$

Note: In the above equation the uniformly distributed load has a total magnitude of 200 × 3 = 600 N, and acts at the middle of the 3 m length.

$\Sigma H = 0$ for the whole beam

$\therefore R_{AH} = 0.$

Since no other horizontal force, active or reactive, is acting on the beam,

$R_{AH} = 0.$

$\Sigma V = 0, R_{AV} + R_B - 600 - 800 - 1000 - 200 \times 3 = 0$

$R_{AV} =$ **783.3 N**

(b) *Graphical*. We draw the space diagram to scale as shown in Fig. 3.56(a). The force diagram for forces can now be drawn as in Fig. 3.56(b). The only point to be noticed is about the distributed load. For the purpose of finding the reactions, this load can be assumed to act at its centre and has a magnitude of 600 N as shown by the dotted line.

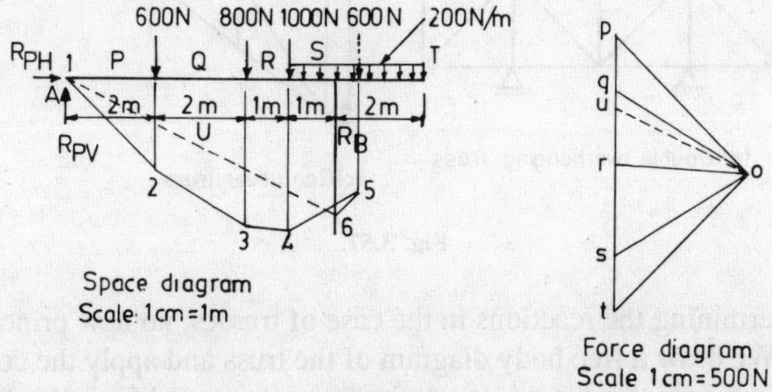

Fig. 3.56

We have applied method II detailed before. The funicular polygon is started from point *A* and is closed. *The only point to be mentioned is the line parallel to ou.*
This line should lie between the 1000 N and 600 N forces. The line parallel to *ot*, the line 5-6, is then drawn between the 600 N force and R_B. The funicular polygon is now closed by joining points 1 and 6. We draw a line through pole '*O*' parallel to line 1-6 and a vertical line from *t*. In this case since the force diagram itself is a line, the point '*u*' is located on this line.

Reaction $R_B = \overline{tu}$

$= 4.45$ cm.

$=$ **2225 N**

Reaction $R_A = \overline{up}$

$= 1.55$ cm

$=$ **7.55 N**

3.6.2 *Trusses*

Figure 3.57(a) shows a truss. Trusses are commonly used in buildings and bridges. The truss consists of a framework of members, which are assumed to be hinged at their ends. Statically determinate trusses normally have two supports, one hinged and the other on rollers. The classification of trusses depending upon the position of supports is shown in Fig. 3.57.

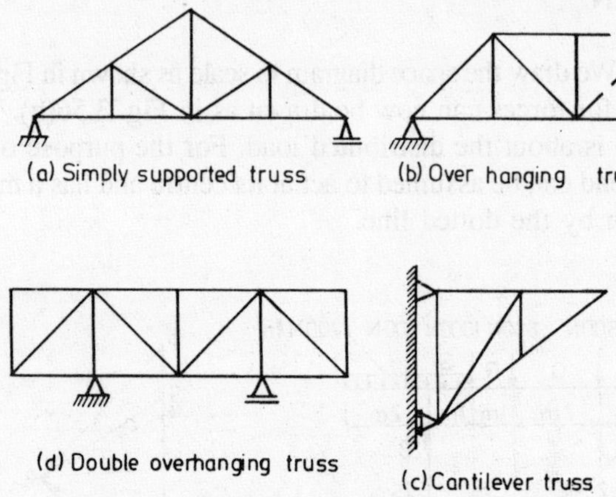

Fig. 3.57

In determining the reactions in the case of trusses, no new principles are involved. We draw a free body diagram of the truss and apply the conditions of equilibrium or seek a graphical solution using force and funicular polygons. The following examples illustrate the method.

Example 3.20
Determine the reactions at the support of the truss carrying loads shown in Fig. 3.58(a).

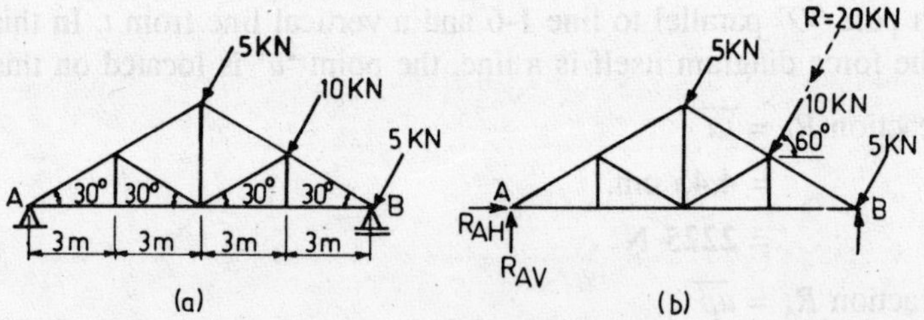

Fig. 3.58

Solution
We have drawn the free body diagram in Fig. 3.58(b). We present below the analytical and graphical solutions.

(a) *Analytical.* We can apply three conditions of equilibrium to the free body diagram. We observe that the three forces acting on the truss have a resultant of 20 kN acting along the line of action of the 10 kN force shown by dotted lines.

$\Sigma M_A = 0$, $20. \sin 60 \times 9 - 20 \cos 60 \times 3 - R_B \times 12 = 0$

$R_B = \mathbf{11.55\ kN}$

$\Sigma V = 0$, $R_{AV} + R_B - 20 \sin 6° = 0$

$R_{AV} = \mathbf{5.77\ kN}$

$\Sigma H = 0$, $R_{AH} - 20 \cos 60 = 0$

$R_{AH} = \mathbf{10\ kN}$

(b) *Graphical.* The graphical solution is shown in Fig. 3.59. We draw the force diagram *p-q-r-s* as shown in Fig. 3.59(b). The funicular polygon in Fig. 3.59(a) should be started from point *A* as explained earlier. The rest of the solution is simple, and Fig. 3.59 is self explanatory. By measurement,

$R_A = \overline{tp}$

$= 2.3$ cm = **11.5 kN**

Inclination of R_A = **30°** to horizontal.

$R_A = \overline{st}$

$= 2.3$ cm = **11.5 kN (vertical)**

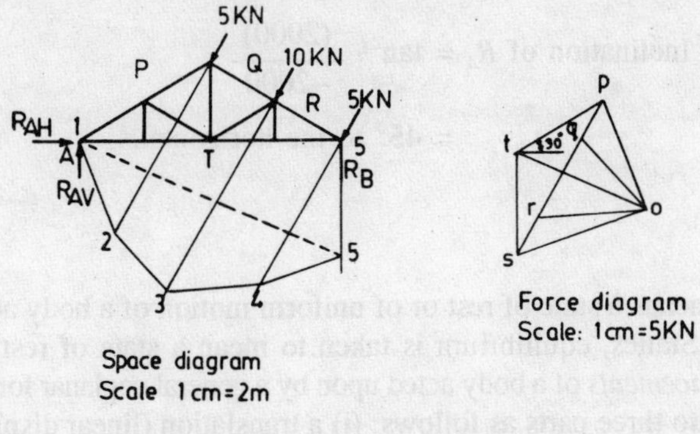

Fig. 3.59

Example 3.21

Compute the reactions in the truss loaded as shown in Fig. 3.60.

Solution

We draw the free body diagram as in Fig. 3.60(b).

$M_A = 0$, $1000 \times 4 + 1000 \times 2 - R_B \times 3 = 0$

$R_B = \mathbf{2000\ N.}$

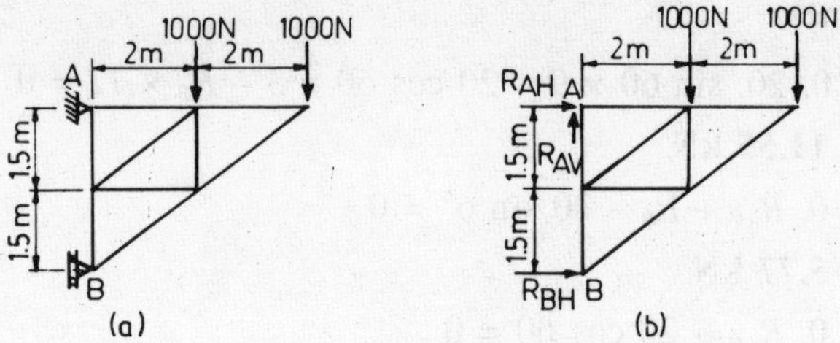

Fig. 3.60

$$H = 0, R_{AH} + R_B = 0$$

$$R_{AH} = \mathbf{-2000\ N}$$

The sign shows that R_{AH} acts in a direction opposite to that shown in Fig. 3.60(b).

$$V = 0, R_{AH} - 1000 - 1000 = 0$$

$$R_{AH} = \mathbf{2000\ N}$$

$$R_A = (2000)^2 + (-2000)^2$$

$$= \mathbf{2000.2\ N}$$

$$= \mathbf{2828\ N}$$

$$\text{Inclination of } R_A = \tan^{-1} \frac{(2000)}{-2000}$$

$$= \mathbf{45°} \text{ to the horizontal.}$$

Summary

Equilibrium means a state of rest or of uniform motion of a body acted upon by forces. In Statics, equilibrium is taken to mean a state of rest.

The *displacements* of a body acted upon by a general coplanar force system can be split into three parts as follows: (i) a translation (linear displacement) along a selected direction (axis), (ii) a translation along another direction perpendicular to that in (i) and (iii) a rotation. Depending upon the type of coplanar force system the above displacements can occur individually or simultaneously.

For a body in equilibrium under a coplanar force system, all the three displacements mentioned above are zero.

For a body acted upon by only a concurrent, coplanar force system, rotation is always zero. In the case of a non-concurrent force systems, rotation is a possible displacement.

The *conditions of equilibrium* of coplanar force systems are as follows:

(i) General coplanar force system	(a) $R = 0$ and (b) $M = 0$ or
(ii) Parallel force system	(a) $F = 0$ and (b) $M = 0$
(iii) Concurrent force system	(a) $R = 0$, or (b) $F_x = 0$ and $F_y = 0$.

The graphical conditions of equilibrium are

(i) General force system	(a) The force polygon should close.
(ii) Parallel force system	(b) the funicular polygon should close.
(iii) Concurrent force system	(c) The force polygon should close.

Where *three non-parallel forces* keep a body in equilibrium, the three forces should be concurrent and coplanar.

Body constraints, or supports are provided to a body which offer reactive forces that keep the body in equilibrium.

Free Body Diagrams, are sketches showing the body or parts of the body acted upon by all the external forces and the reactive forces provided by the constraints. The free body diagram helps in accounting for all the forces acting on the body and also helps in splitting up a problem into small parts.

The evaluation of unknown reactive components is a basic problem in the analysis and design of structures, machine parts, etc. The conditions of equilibrium help us to evaluate these unknown reactive forces.

Review Questions

1 Define the term 'equilibrium'.
2 Distinguish between the terms 'equilibrium' and 'rest.'
3 State the possible displacements of a body acted upon by a coplanar (i) concurrent force system (ii) parallel force system and (iii) general force system.
4 A body is acted upon by only a concurrent coplanar force system. Is it possible for the force system to produce a rotation of the body?
5 State the conditions of equilibrium of a concurrent coplanar force system in terms of moments only.
6 Correct the following statement. If a body is in equilibrium under the action of three forces, the three forces should be concurrent and coplanar.
7 If a general coplanar force system is in equilibrium, what graphical conditions should be satisfied? How do these concur with the analytical conditions?
8 For a general coplanar force system, the conditions of equilibrium are given as (i) $F_x = 0$, (ii) $M_1 = 0$, and (iii) $M_2 = 0$.
   (a) What is the restrictions placed on moment centres 1 and 2?
   (b) If only the first of these conditions is satisfied, what conclusions can be drawn regarding the resultant of the force system?
   (c) If only the condtions (i) and (ii) are satisfied by a force system, what conclusions can we draw about the resultant of the force system?

9 For a general coplanar force system, if the force polygon closes, can you say that the body is in equilibrium? Explain.

10 What are body constraints? Show by means of sketches any five body constraints and the support forces offered by each.

11 Define the term 'free body diagram'. What is the purpose of drawing a free body diagram?

12 State the fundamental principle employed in drawing the free body diagrams for the analysis of bodies in equilibrium.

Problems

3.1 Check whether the force system shown in the figure is in equilibrium. If not, what are the magnitude and direction of the equilibrant force.

(Not in equilibrium; Equilibrant is a force of magnitude 203.37 N at 52.27° with horizontal in the III quadrant).

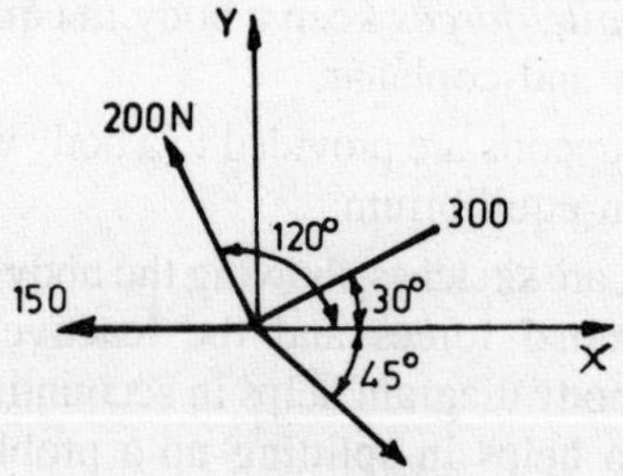

Fig. Prob. 3.1

3.2 Find the magnitudes of the forces F_x & F_y in the figure Prob. 3.2 shown so that the particle *P* is in equilibrium.

(F_x = 3.18 kN towards left, F_y = 4.14 kN upwards)

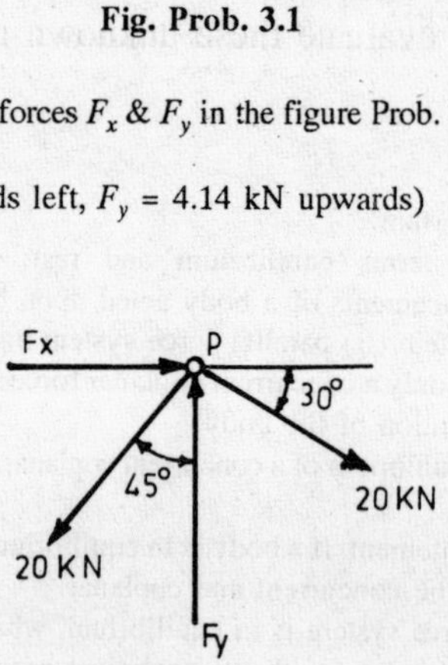

Fig. Prob. 3.2

3.3 For the free body diagram shown in the figure Prob. 3.3, determine the unknown forces for static equilibrium.

(F_{2x} = 3.106 kN; F_{2y} = 1.207 kN; F_1 = 3.586 kN)

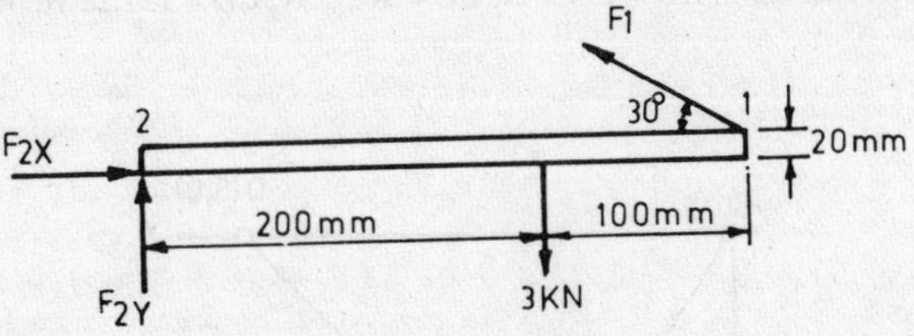

Fig. Prob. 3.3

3.4 For the free body diagram shown in the figure Prob. 3.4 determine the unknown forces for static equilibrium.

(Ans. F_2 = 300 N; F_{1x} = 450 N, F_{1y} = 259.81 N)

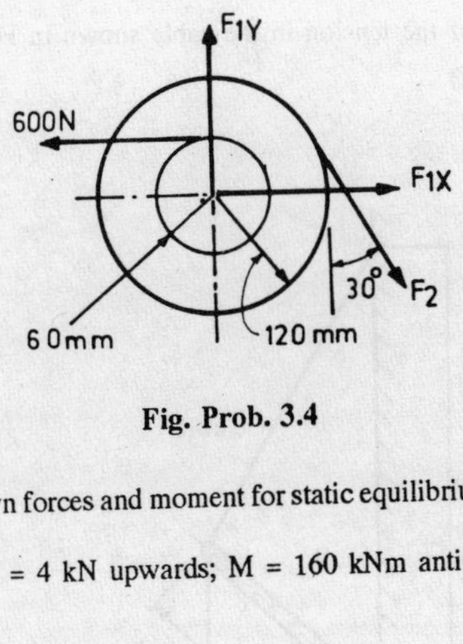

Fig. Prob. 3.4

3.5 Determine the unknown forces and moment for static equilibrium from the given free body diagram.

(Ans. F_x = 0, F_y = 4 kN upwards; M = 160 kNm anti-clockwise).

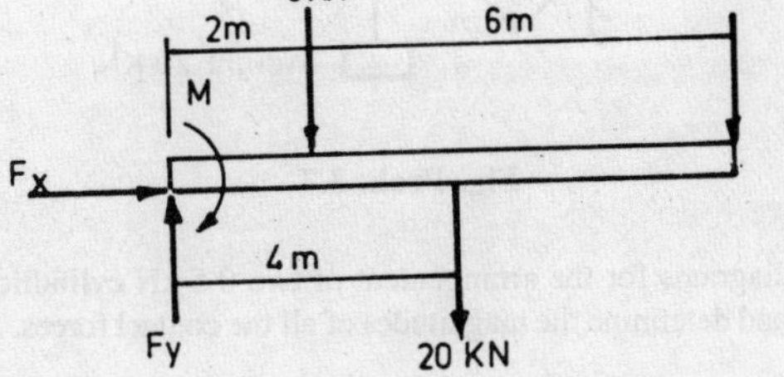

Fig. Prob. 3.5

3.6 A weightless rope is fastened at the points (0, 0) and (9, 0) carries a 100 N weight and an unknown weight *W* as shown. The vertices of the rope polygon lie at the points

(2, 3) and (5, 4). Determine the magnitudes of the forces in the three parts of the rope and the magnitude of the weight *W*.

(Ans. Tensions in *AB* = 154.5 N; *BC* = 90.35 N, *CD* = 121.22 N; *W* = 114.29 N).

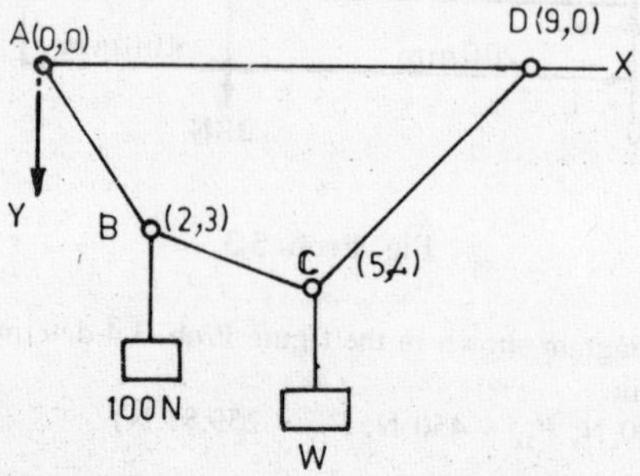

Fig. Prob. 3.6

3.7 Find the magnitude of the tension in the cable shown in Figure Prob. 3.7.

(Ans. 14.736 kN)

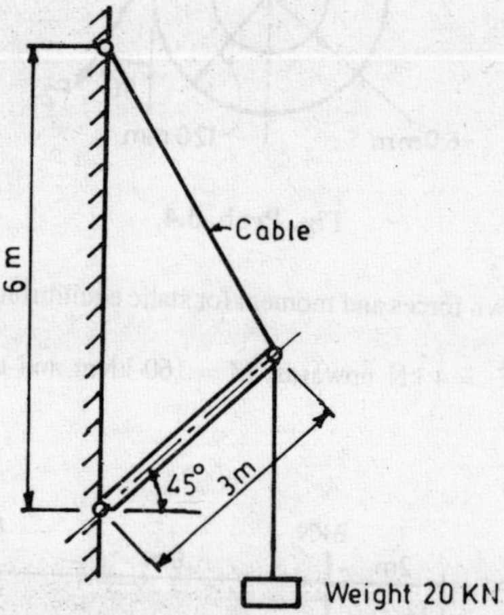

Fig. Prob. 3.7

3.8 Draw freebody diagrams for the arrangement of two 0.5 kN cylindrical rollers shown in Figure Prob. 3.8 and determine the magnitudes of all the contact forces. Assume all surfaces are frictionless.

(Ans. Forces at A = 0.447→
B = 1.0 kN↑
C = 0.447←

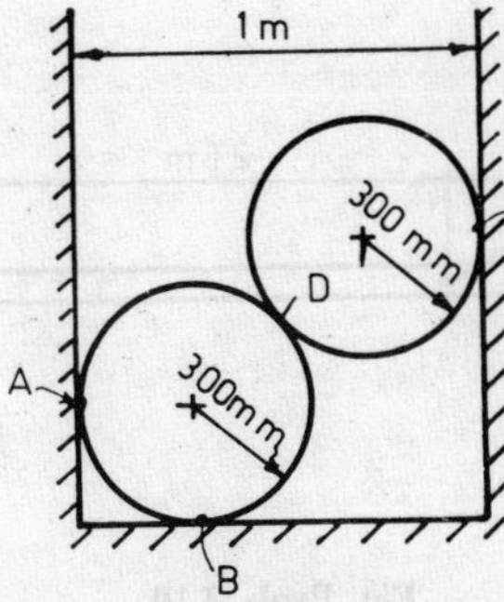

Fig. Prob. 3.8

3.9 The 3 m long truck bed of a dump truck is held in equilbirium in the position shown in Fig. Prob. 3.9. If the truck bed and load weigh 5 kN, draw the free body diagram for the truck bed and determine the magnitude of the reaction at *A* and the force in the hydraulic actuator.

(Ans. F_{AH} = 1.046 kN → F_{AV} = 2.377 kN
Force at point *B* due to actuator = 2.824 kN)

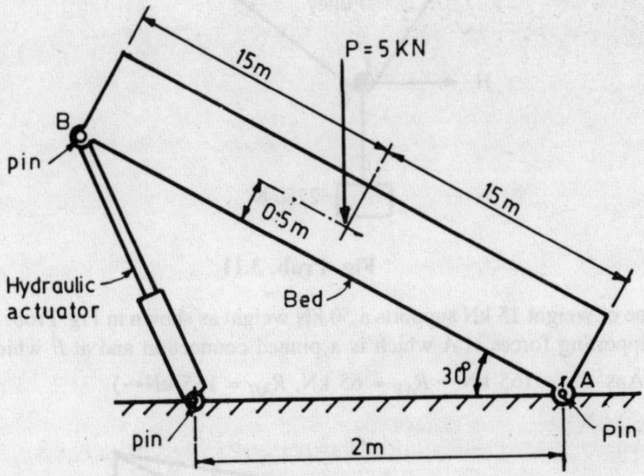

Fig. Prob. 3.9

3.10 A force of 500 N is applied to the bar shown in Figure Prob. 3.10. Draw a free body diagram for the bar and determine the force in the cable if the weight of the bar is negligible.

(Ans. 7000 N)

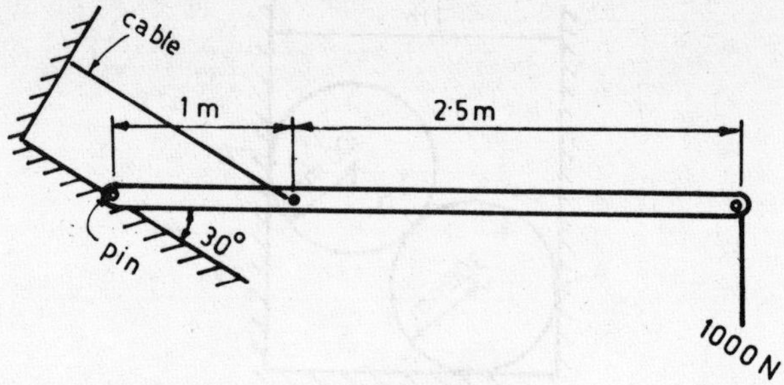

Fig. Prob. 3.10

3.11 What force *H* is needed to keep the mass in the position shown in Fig. Prob. 3.11. Also determine the tension in cables *AC* and *BC*.

(Ans. 321.4 N, 1607 N).

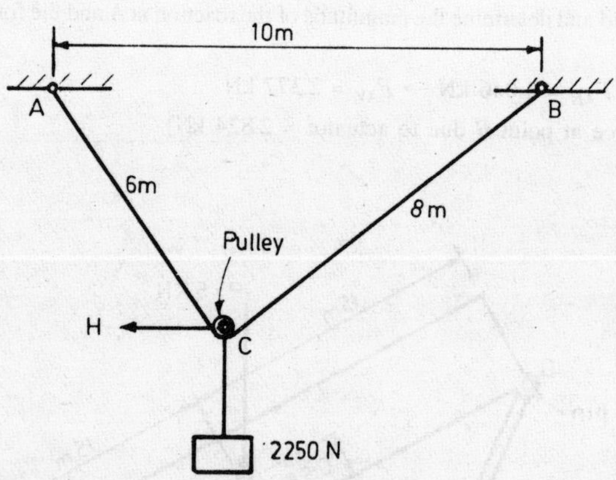

Fig. Prob. 3.11

3.12 A crane of weight 15 kN supports a 50 kN weight as shown in Fig. Prob. 3.12. Determine the supporting forces at *A* which is a pinned connection and at *B* which is a roller.

(Ans. R_B = 165 kN→ R_{AV} = 65 kN, R_{AH} = 165 kN←)

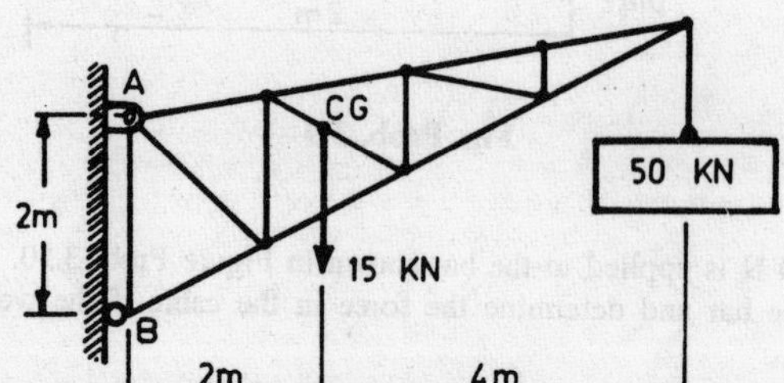

Fig. Prob. 3.12

3.13 Draw free body diagrams for the member AB and for cylinder in Fig. Prob. 3.13. Neglect friction at the contact surfaces of the cylinder. The masses of the cylinder and the member are denoted as W_1 and W_2 respectively.

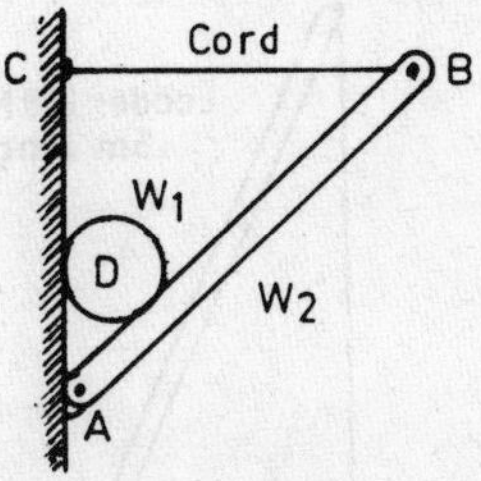

Fig. Prob. 3.13

3.14 Two cantilever beams are pinned together at A in Figure Prob. 3.14. Draw freebody diagrams of each cantilever.

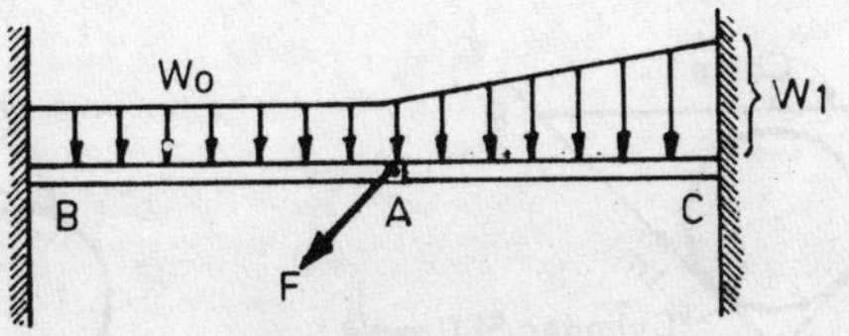

Fig. Prob. 3.14

3.15 Draw freebody diagrams of the parts mentioned in each figure.

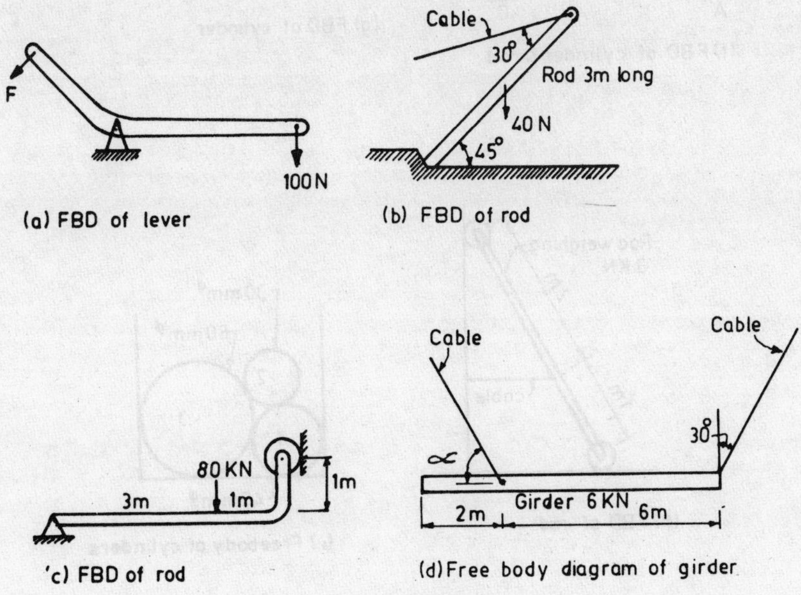

Fig. Prob. 3.15

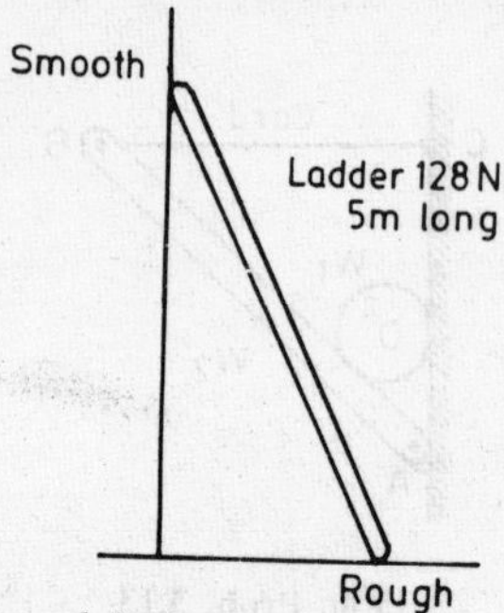

(e) FBD of ladder

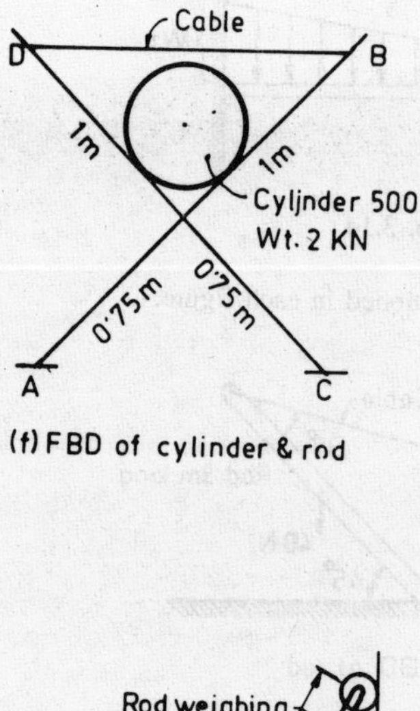

(f) FBD of cylinder & rod

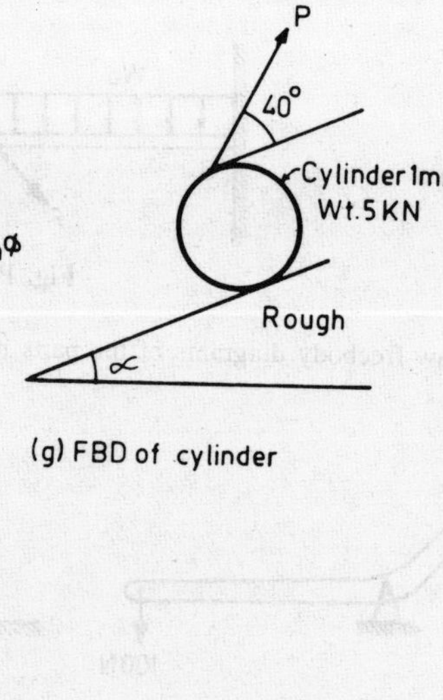

(g) FBD of cylinder

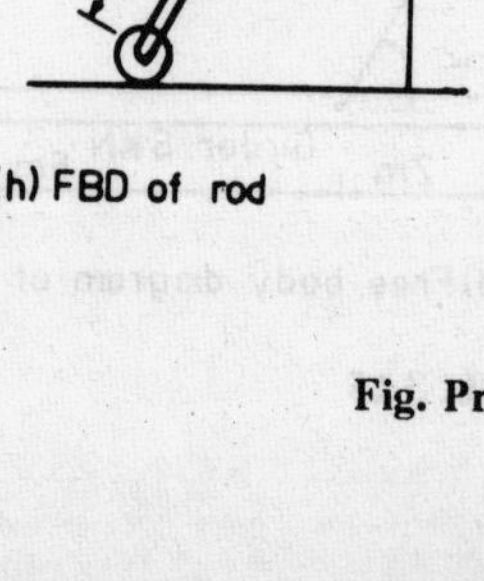

(h) FBD of rod

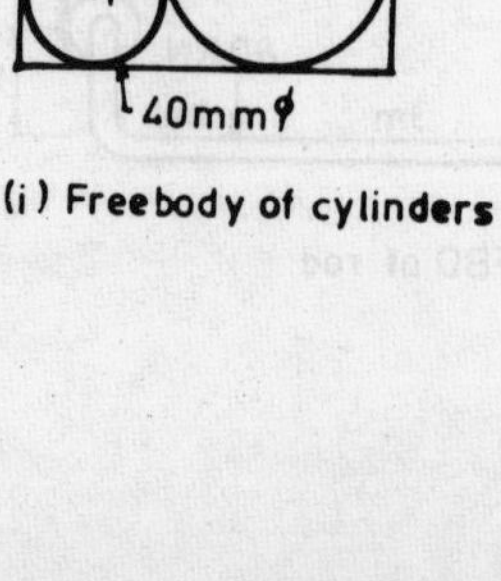

(i) Freebody of cylinders

Fig. Prob. 3.15

3.16 Cylinders *A* and *B* are 300 mm ϕ and have weight of 200 N each in Figure Prob. 3.16 and cylinder *C* is 600 mm ϕ and has a weight of 400 N. Draw free body diagrams for each cylinder and compute all contact forces.

(Ans. Contact force on wall 260 kN; on ground 400 N, between cylinders 328 N).

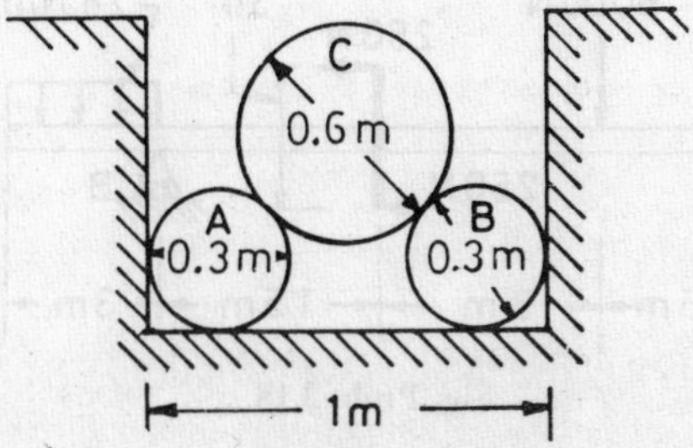

Fig. Prob. 3.16

3.17 Draw the free body diagrams of the bodies shown in Figure Prob; 3.17.

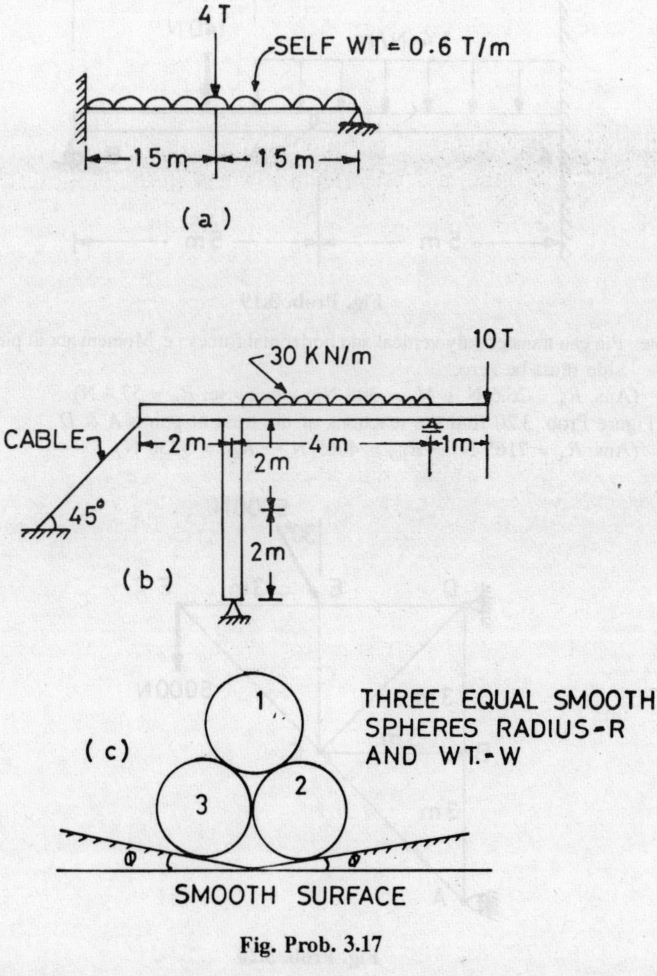

Fig. Prob. 3.17

3.18 In Figure prob. 3.18. find the supporting forces at *A* and *B* on the beam, whose weight is 200 N.

Is the location of the couple significant for this calculation? Why?

(Ans. R_A = 3454 N; R_B = 1630 N)

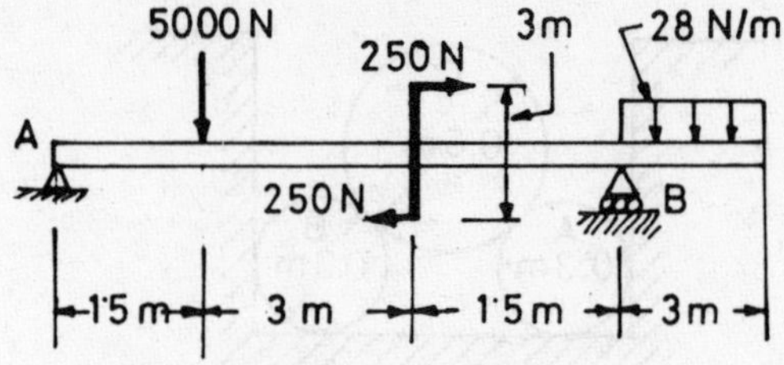

Fig. Prob. 3.18

3.19 Compute the supporting forces for the Cantilever beam and pinned extension shown in Fig. Prob. 3.19.

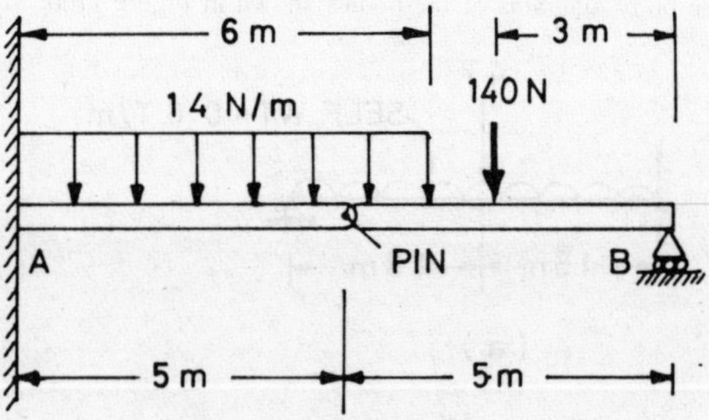

Fig. Prob. 3.19

Note: Pin can transfer only vertical and horizontal forces i.e. Moment about pin from either side must be zero.

(Ans. R_A = 26.6 N ↓ M_A = 308 Nm clockwise; R_B = 57.4 N)

3.20 In Figure Prob. 3.20 find the reactions of the truss at points *A* & *D*.

(Ans. R_A = 7165 N → R_{BH} = 4665 N ← R_{BV} = 9330 N)

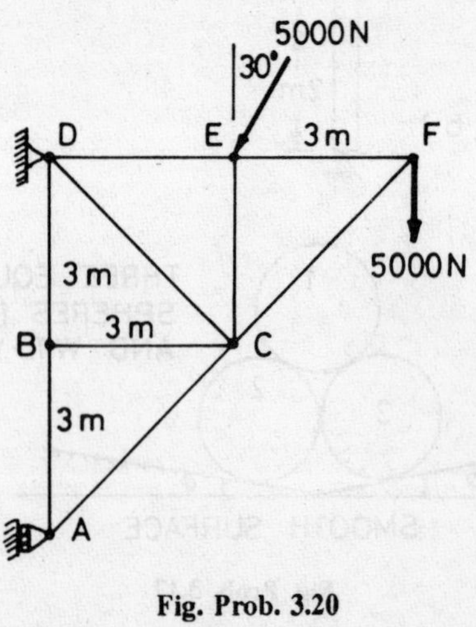

Fig. Prob. 3.20

3.21 In Figure Prob. 3.21, find the reactions of the truss at points *A* & *D*.
(R_{DV} = 15000 N; R_A = 20000 N)

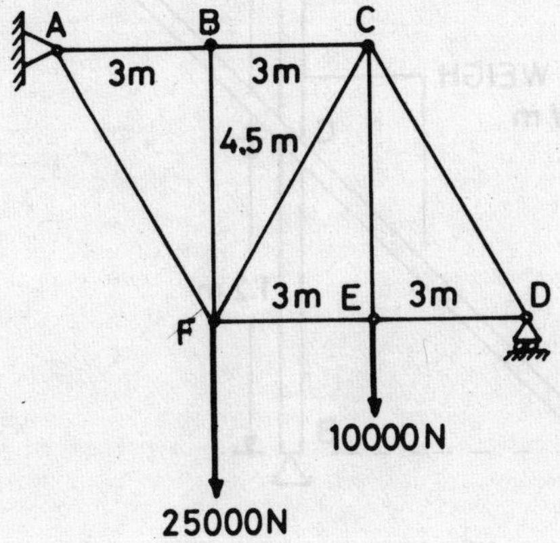

Fig. Prob. 3.21

3.22 In Figure Prob. 3.22, find the reactions of the truss at points *A* & *E*.
(R_A = 16562.5 N; R_{EV} = 18437.5 N R_{EV} = 5000 N←)

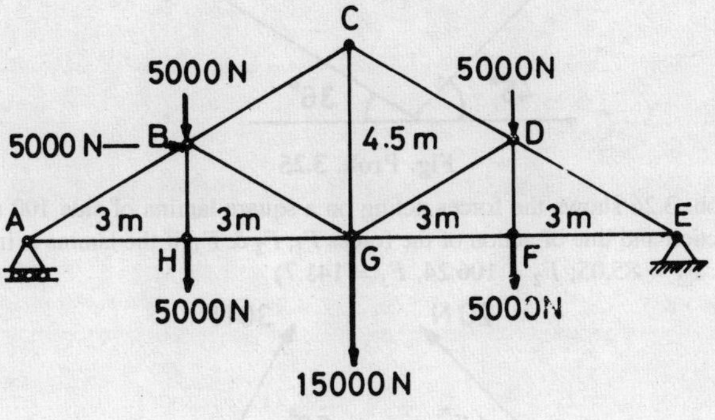

Fig. Prob. 3.22

3.23 Four rods are smoothly joined at their extremities to form a quadrilateral *ABCD* and the opposite corners *A,C* & *B,D*, are joined by tight strings. Prove that if *T*, *T'* denote the tension in *AC* & *BD* then:

$$T\left(\frac{1}{AO}+\frac{1}{OC}\right)=T'\left(\frac{1}{BO}+\frac{1}{OD}\right)$$

3.24 In the set up shown in Figure Prob. 3.24, find the reactions at *A*, *C* and *E*.

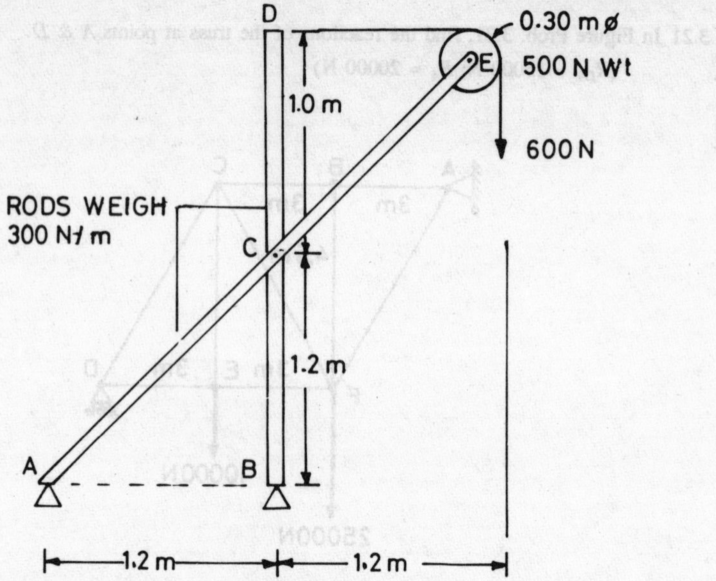

Fig. Prob. 3.24

3.25 In the figure shown, find the distance X so that the 1000 N weight remains in equilibrium. Neglect the wt. of the bar.
(Ans.

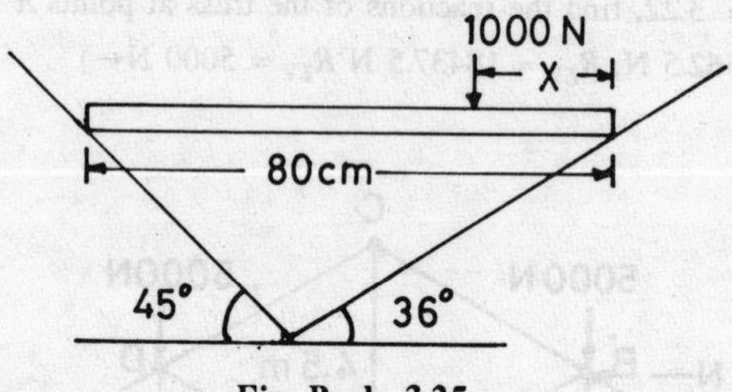

Fig. Prob. 3.25

3.26 Figure Prob. 3.26 shows the forces acting on a square lamina of side 100 mm. Find the magnitude, direction and line of action of the forces F_1, F_2 & F_3, if the lamina is in equilibrium.
(Ans. F_1 = 85.05; F_2 = 106.24, F_3 = 141.7)

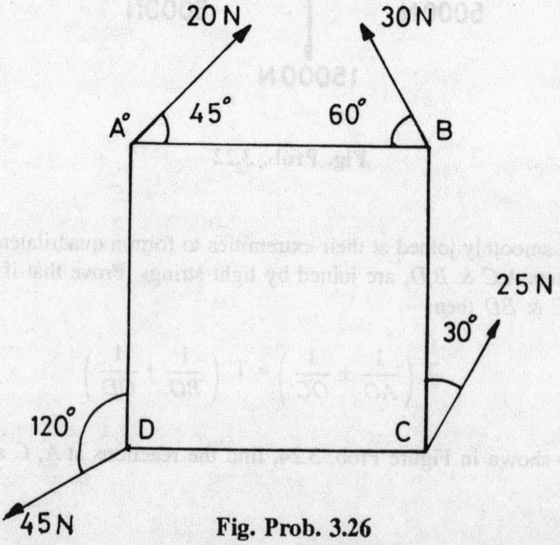

Fig. Prob. 3.26

3.27 Calculate the reactions at the supports in the Figure Prob. 3.27.

(Ans. (a) R_A = 21.5 kN R_B = 21.5 kN (b) R_C = 8.5 kN R_B = 1.5 kN
(c) R_A = 2.7 kN R_B = 6.3 kN (d) R_A = 2650 N R_B = 2800 N
(e) R_A = 10.5 kN R_B = 0 (f) R_A = 1333.33 R_B = 4666.67 N

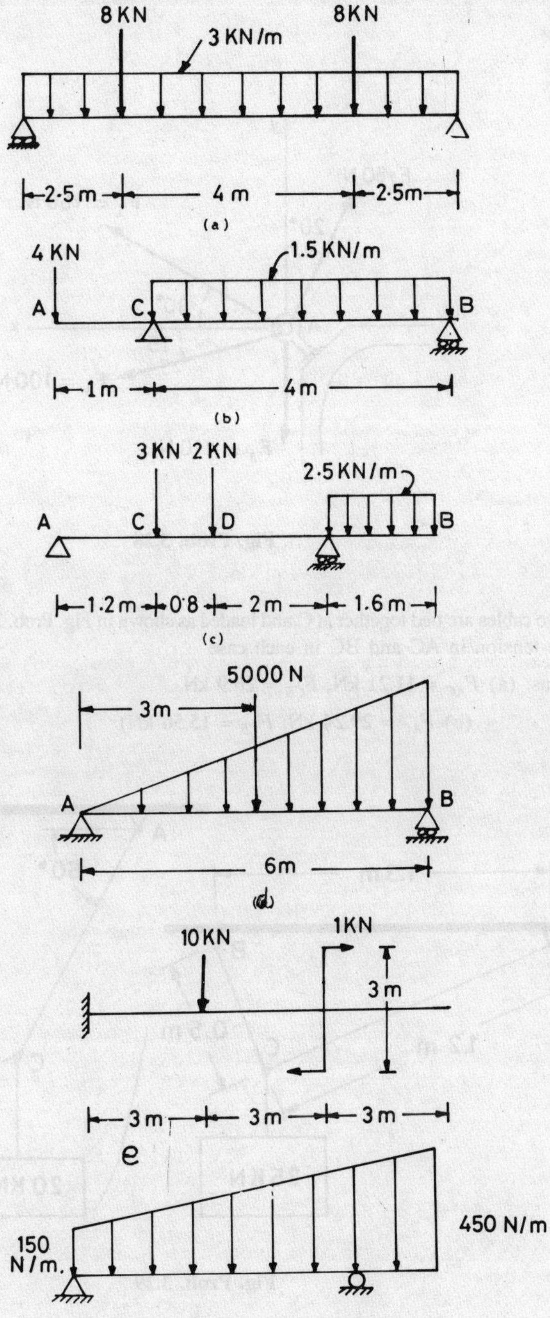

Fig. Prob. 3.27

3.28 Four forces act on an eye bolt *A* as shown in Figure problem 3.28. Determine the forces along X and Y axes which can keep the eye bolt in equilibrium.

(Ans. 199.13 N along X-axis towards left and 14.29 N along Y-axis downwards).

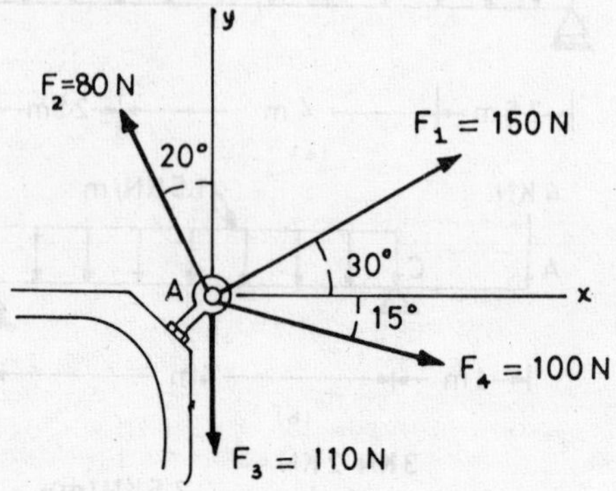

Fig. Prob. 3.28

3.29 Two cables are tied together at C and loaded as shown in Fig. Prob. 3.29 (a) & (b). Determine the tension in AC and BC in each case

(Ans. (a) F_{AC} = 11.21 kN, F_{CB} = 26.9 kN

(b) F_{AC} = 29.24 kN; F_{CB} = 15.56 kN)

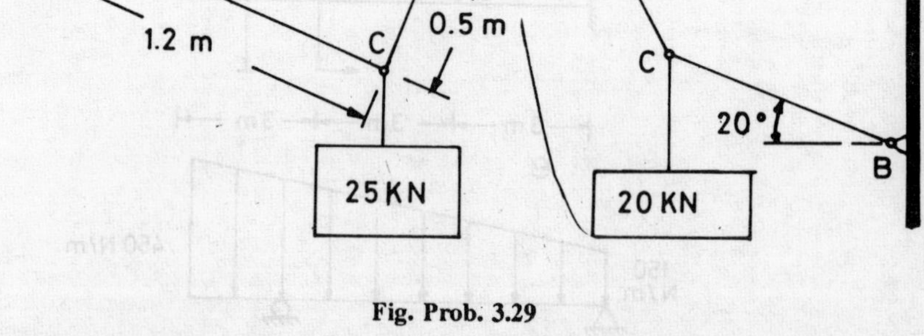

Fig. Prob. 3.29

3.30 A fixed crane has a weight of 10 kN and is used to lift a 24 kN crane. It is held in place by a pin at A and a rocker at B. The centre of gravity of the crane is located at C. Determine the reaction at A and B.

(At B 109.33 kN →; At A 109.33 kN ← and 34 kN↑)

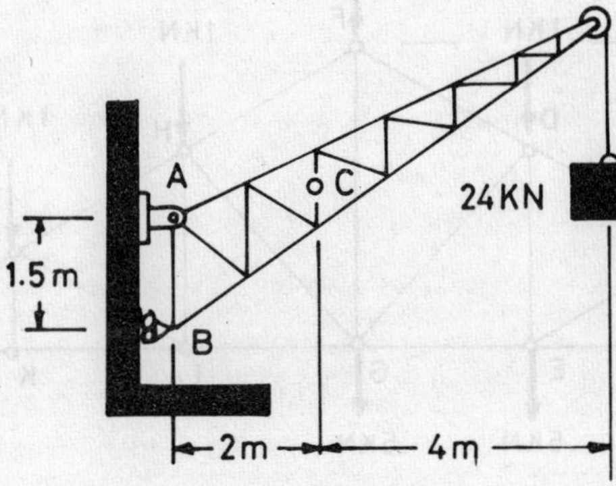

Fig. Prob. 3.30

3.31 Determine the reaction at A & B for the truss and loading shown in Figure Prob. 3.31.

(Ans. R_B = 69.13 kN at 60° with horizontal; At A 5.435 kN ←, 15.13 ↑)

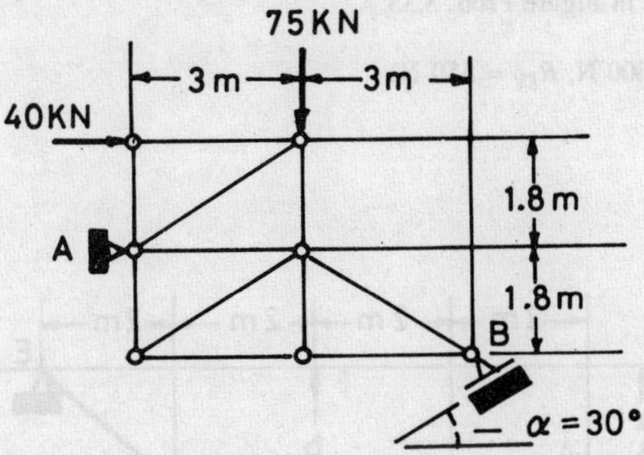

Fig. Prob. 3.31

3.32 Determine the reactions of the roof truss shown in Figure Prob. 3.32

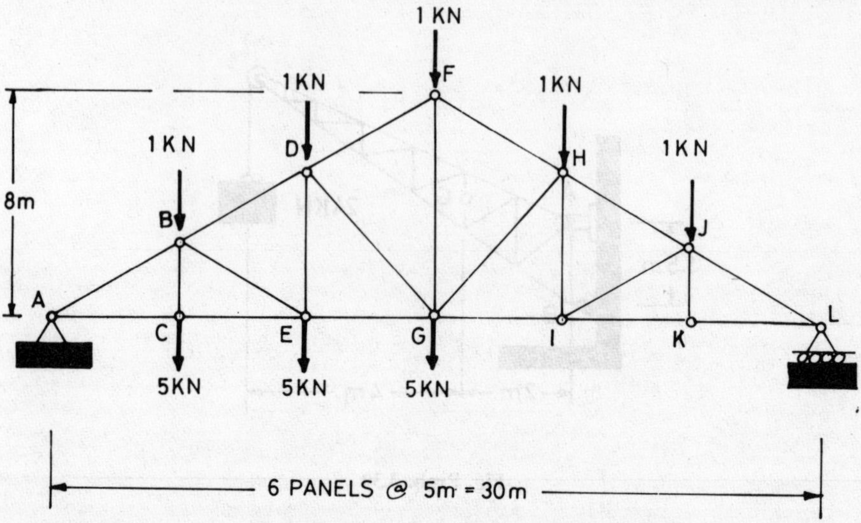

Fig. Prob. 3.32

3.33 If a = 2 m and b = 2.5 m, determine the components of the reaction at *E* for the cable and loading shown in Figure Prob. 3.33.

(Ans. R_{EH} = 600 N, R_{EV} = 550 N).

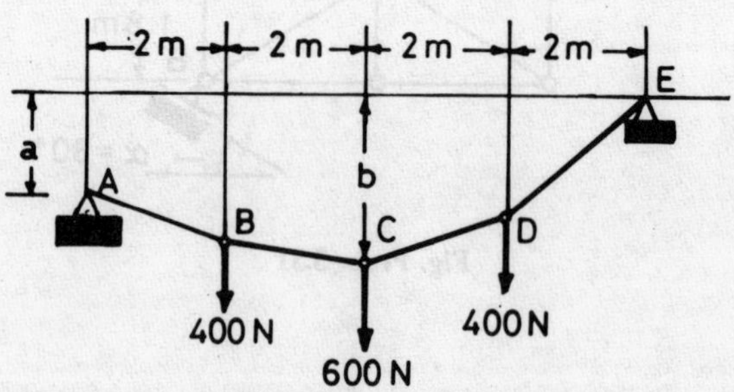

Fig. Prob. 3.33

(Ans. R_B = 20 kN R_E = 60 kN.

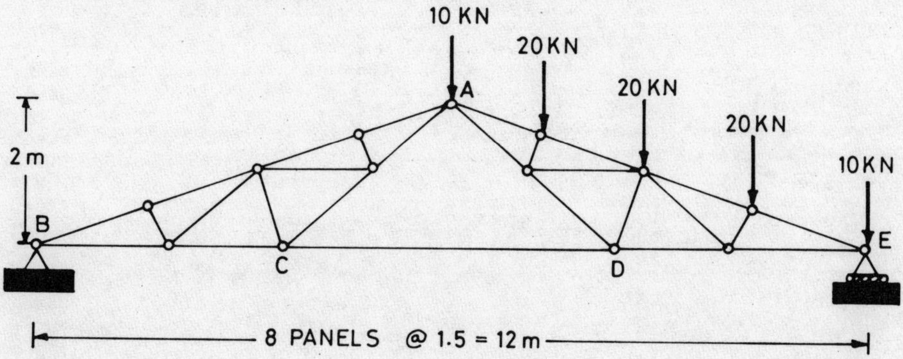

Fig. Prob. 3.34

4. CENTRE OF GRAVITY

After having mastered the content of this chapter, you will be able to:

- Define the term 'centre of gravity'
- Explain the concept of C.G. with illustrative examples.
- State the general principles for the determination of centre of gravity.
- Define the term 'controid'.
- Compare and contrast 'centre of gravity' and 'centroid'.
- State the general principles for locating the centroid of an area.
- Derive the general expression for locating the centroid of an area.
- State the location centroids of a number of regular geometrical shapes.
- Determine, analytically and graphically, the centroids of a given area.
- State the location of C.G. of a given solid composed of a number of regular solids.
- Apply the method of integration to determine the C.G. of a solid with simple curved boundaries.

CHAPTER 4

Centre of Gravity

4.1 Introduction

Let us start our discussion with three simple experiments which the reader can perform very easily.

Experiment No. 1. Take a piece of straight prismatic rod of uniform cross section, say, 20 cm long and of any material. Try to balance it on your finger so that the rod remains horizontal (Fig. 4.1a). It can be seen that the rod does not remain horizontal except when resting with its middle point on your finger. You will also see that in any other position, the rod rotates about your finger downwards to the side where it is longer.

Experiment No. 2. Now try the same experiment by taking a uniformly tapering rod, say, 4 cm square at one end, and 2 cm square at the other and 20 cm long.

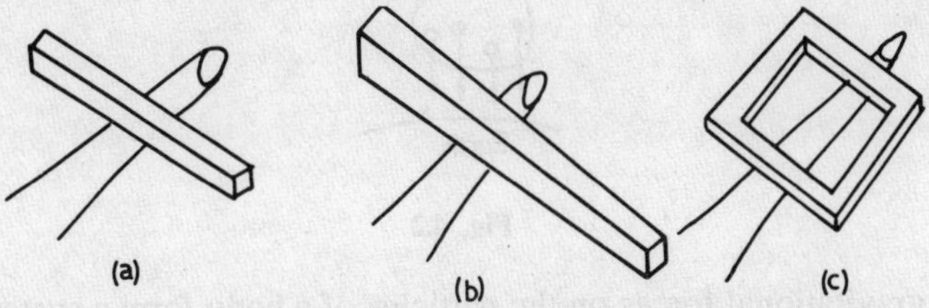

Fig. 4.1(a), (b) and (c)

If you try to balance it on your finger you will see that it does not balance at its mid point and rotates with the 4 cm end going downwards. By repeated trials you can find the point on the rod at which it can be balanced on your finger. This point is nearer to the larger end (Fig. 4.1b).

Experiment No. 3. As a third and more enlightening experiment, cut out a cardboard piece of the shape shown in Fig. 4.1(c). Try to balance this cardboard on your finger such that it remains horizontal. It can be easily seen that it does not balance on any one of its sides. The piece of cardboard can, of course, be balanced and kept horizontal as shown in Fig. 4.1(c).

The reason for the behaviour of the rods and the cardboard lies in what is known as the 'centre of gravity' of the rods and the cardboard. In many engineering situations it is essential to know the position of the 'centre of gravity' of bodies. In this chapter we will discuss the 'centre of gravity', and the methods to locate the same for many bodies.

4.2 Gravity and Gravitational Force

It is well known that all material bodies are attracted by the earth. The attraction of earth on material bodies is called gravity. Due to this attraction, the earth applies a force on all bodies, and this force is called the gravitational force (or the force due to gravity).

This gravitational force is also known as the weight of the body. While studying the type of forces (Chapter 2) we referred to it as the body force. You already know that gravitational force is equal to the product of mass and acceleration due to gravity. The mass of the body is measured in kilogram, and the acceleration due to gravity, which has different values on the surface of the earth depending upon the location, has an average value equal to 9.81 ms^{-2}.

The direction of the gravitational force is towards the mass centre of the earth. Since a body is a collection of small particles, such force of gravity acts on each particle of the body. The gravitational forces on the different mass particles are all directed towards the centre of the earth. Since the bodies, we deal with, are very small compared to the earth, these forces can be assumed to be parallel (Fig. 4.2).

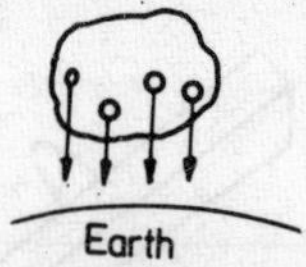

Fig. 4.2

The gravitational forces on the particles of a body form a system of like parallel forces, all acting towards the earth. The resultant of such a parallel force system can be determined by applying the principles developed in Chapter 2. The magnitude of the resultant is equal to the algebraic sum of all the gravitational forces (i.e. $R = \Sigma F$). The point of application of the resultant can be determined from the principle of moments.
The point of application of the resultant of the gravitational forces on a body is known as the 'centre of gravity' (or CG) *of the body.*

4.3 Centre of Gravity

The centre of gravity can, therefore, be defined as follows:
'The Centre of Gravity of a body is the point at which the resultant gravitational force on the body acts, irrespective of the orientation of the body'.

In the above definition, the phrase 'irrespective of the orientation of the body' should be carefully noted. Since the mass particles in a solid body do not change their position relative to each other, whatever be the orientation of the body, the resultant force acts through the same point, the centre of gravity.

4.3.1 *Conclusions from Experiments of Article 4.1*

We can now easily explain the behaviour of the rod in Expt. 1; when the rod is balanced and stays horizontal, only two forces are acting on the rod; the weight of the rod through its centre of gravity and the reaction from the finger. The rod balances at its mid-point obviously because the CG lies there.

At any other position the rod does not balance because the weight of the body and the reactive force, though equal and opposite, are non-collinear producing a couple that rotates the rod. Obviously, this rotation takes the rod downwards in that part of the rod in which CG lies (Fig. 4.3a).

In Expt. 2, the same reasoning applies. The CG lies towards the heavier end of the rod, and balances at that point (Fig. 4.3b).

In Expt. 3, we have an interesting case. The body does not balance at any single point on the cardboard because the CG does not lie there. The CG in fact, lies at the centre of the plan of the figure, at the intersection of the diagonals. The piece of cardboard is balanced on the finger in the way shown because we applied two equal forces whose resultant passed through the CG of the body (Fig. 4.3c)

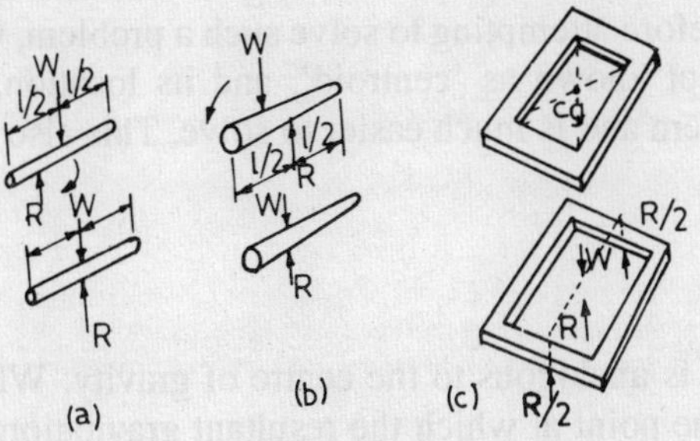

Fig. 4.3

4.3.2 *Location of Centre of Gravity*

In many engineering problems, it is necessary to know the CG of a body, or a group of bodies. We shall now discuss the methods for locating the CG of bodies.

As the CG of a body is the point at which the resultant of the gravity forces act, the problem of locating the CG reduces to one of finding the resultant of a system of parallel forces. In determining the location of the CG, the following principle will simplify the problem in many cases. 'The centre of gravity of a body always lies in a plane of symmetry, if any, of a body'.

A plane of symmetry of a body is a plane through the body such that the mass distribution is symmetrical on either side of the plane. As an example, consider Figure 4.4, which represents a cube. Obviously, the planes *ABCD*, *EFGH* and *IJKL* are planes of symmetry. The CG lies in each of these planes, and can be located as a point common to all the planes (Fig. 4.4).

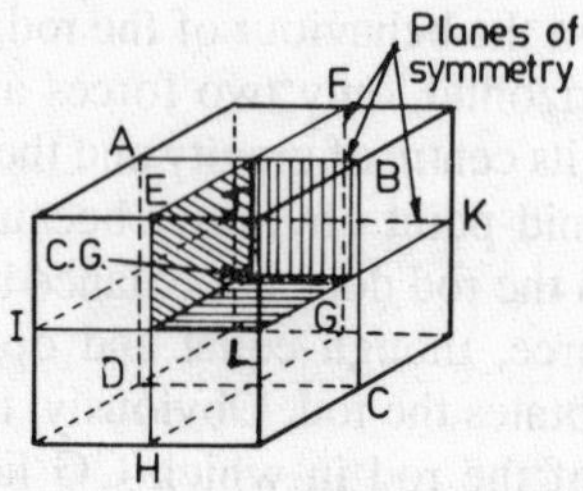

Fig. 4.4

The cube, of course, has three planes of symmetry, and it is easy to locate its CG by inspection. In the case of other bodies, they may or may not have plane(s) of symmetry. In such cases the CG has to be located as the resultant of the parallel forces taking help from the above principle, wherever applicable.

The location of CG of solids is taken up later in art. 4.7. The location of CG is a three dimensional problem, and gives rise to a system of parallel forces in space. Before attempting to solve such a problem, we will first have a look at a concept known as 'centroid', and its location, which is a two dimensional problem and is much easier to solve. This also helps in locating the CG of solids.

4.4 Centroid

The term centroid is analogous to the centre of gravity. While the centre of gravity refers to the point at which the resultant gravitational forces act, the centroid refers to the point in (or outside) an area at which the whole area may be *assumed* to be concentrated, or it is the centre of the area. The concept of centroid is important in many engineering applications, and helps in locating the CG of bodies.

As an example, consider the prismatic rod shown in Fig. 4.5. The rod has a plane of symmetry, at its middle length as shown by the hatched cross section. The CG hence lies in this cross section. Since the X-section of the rod is the same at every point, if we determine the centroid of the X-sectional area, the CG lies on the plane of symmetry at the centroid.

The determination of the centroid of an area is a simpler problem, being two dimensional (an area is a plane figure). We will first study the methods to locate the centroids of plane areas.

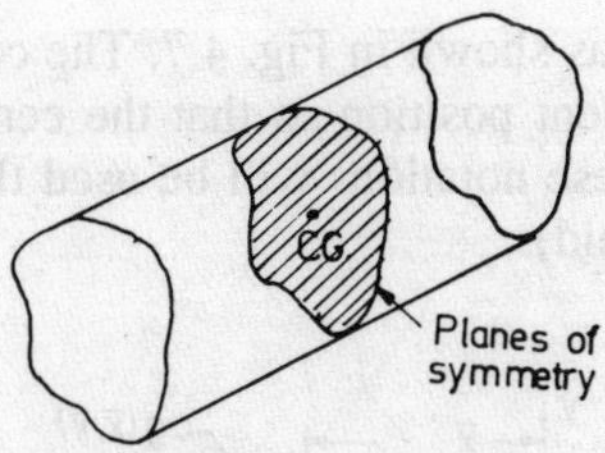

Fig. 4.5

4.4.1 *Location of Centroid*

The centroid of an area is located using the same principles as for the location of CG. Here we assume that the area is composed of a large number of small areas (similar to mass particles), and consider them as a system of parallel forces (areas). We, then, determine the point of application of the resultant force (area).

In locating the centroid, the following principle, similar to the one stated for CG is very helpful.

'The centroid of an area always lies on an axis of symmetry, if any, of an area'.

As an example, take the case of an area in the form of a square as shown in Fig. 4.6. As shown, it is easy to see that the square is symmetrical about axes, *y-y, x-x, p-p* and *q-q*. The intersection of any two of these axes locates the centroid as shown.

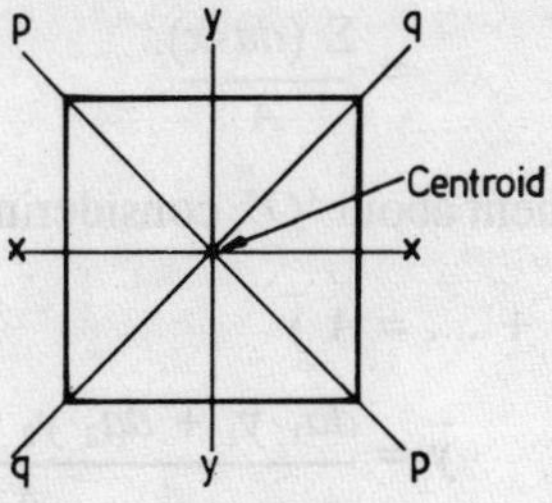

Fig. 4.6

The following three methods are available to locate the centroid of an area

(i) Analytical method

(ii) Graphical method, and

(iii) Experimental method.

Each of the above methods will now be described, and the principle involved will be illustrated with examples later.

4.4.2 *Analytical Method for the Location of the Centroid*

Let us take a plane area as shown in Fig. 4.7. The coordinate axes X and Y are taken in any convenient position so that the centroid is located by two coordinates $\bar{x}$ and $\bar{y}$. (These notations will be used throughout to denote the coordinates of the centroid).

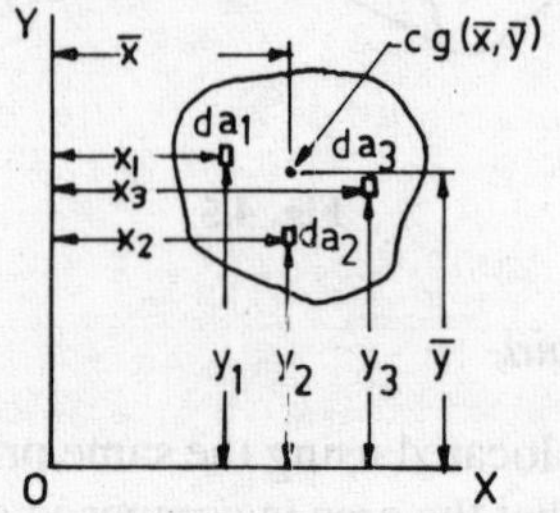

Fig. 4.7

Let da_1, da_2, da_3.... be small elemental areas whose centroids are at distances of x_1, x_2, x_3.... from Y-axis, and y_1, y_2, y_3.... from X-axis. Then, by the principle of moments, the algebraic sum of the moments of these elemental areas about any axis is equal to the moment of the total area '*A*' about the same axis. Thus, taking moment about '*O*', considering forces parallel to *Y*-axis,

$$da_1\, x_1 + da_2\, x_2 + da_3\, x_3 + ... = A\,\bar{x}$$

$$\bar{x} = \frac{da_1\, x_1 + da_2\, x_2 + da_3\, x_3 + ...}{A}$$

$$= \frac{\Sigma\,(da \cdot x)}{A}$$

Similarly, by taking moment about '*O*', considering forces parallel to X-axis,

$$da_1 \cdot y_1 + da_2 \cdot y_2 + da_3 \cdot y_3 + ... = A\,\bar{y}$$

$$\bar{y} = \frac{da_1 \cdot y_1 + da_2 \cdot y_2 + da_3 \cdot y_3 + ...}{A}$$

$$= \frac{\Sigma\,(da.\,y)}{A}$$

where 'Σ' denotes the summation sign. $A = \Sigma\, da.x$ and $\Sigma\, da \cdot y$ can be obtained by integration (see Fig. 4.7).

In the case of regular areas, composed of geometrical shapes which can be split up into a number of simple areas whose centroids are known, this method can be applied by considering $\Sigma\, da \cdot x$ and $\Sigma\, da \cdot y$ as the summation of the moments of simple areas.

4.4.3 *Graphical Method*

The graphical method uses the same principles as the analytic solution. Recalling from Chapter 2 the procedure, using the funicular polygon, for locating the resultant of a system of parallel forces, we arrive at the procedure given below, and illustrated in Fig. 4.8.

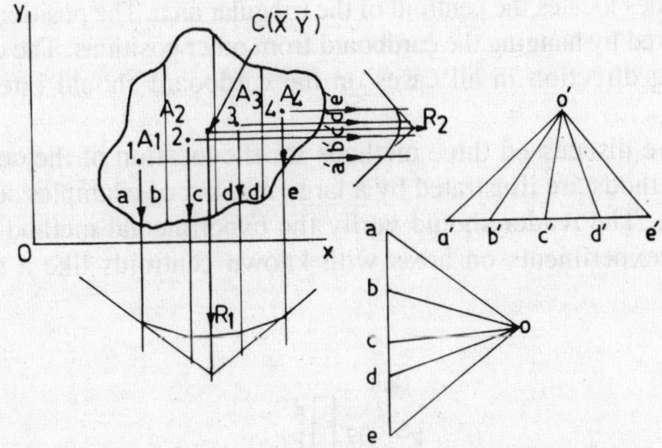

Fig. 4.8

Procedure

(i) Divide the area into a large number of small areas as shown numbered (1), (2), (3) etc.

(ii) Calculate the areas A_1, A_2, A_3 etc. and locate their centroids. (The area is so divided that the centroids of these small areas can be obtained by inspection, or simple formulae).

(iii) Assume that these areas are equivalent to forces whose magnitudes are A_1, A_2, A_3 etc. acting parallel to the Y-axis as shown. They now form a system of parallel forces.

(iv) Determine the line of action of the resultant by drawing the force polygon and funicular polygon. The resultant R_1 is obviously parallel to Y-axis and the distance of the line of action of R_1 from Y-axis gives $\bar{x}$.

(v) Similarly, by assuming these areas to act parallel to the X-axis, and locating the resultant R_2 we obtain the distance $\bar{y}$.

(vi) R_1 and R_2 intersect at point C, which is the centroid with coordinates $(\bar{x}, \bar{y})$.

From the above procedure, it can be easily seen that if the area has one axis of symmetry, only one funicular polygon need be drawn in a direction perpendicular to the line of symmetry. The intersection of this resultant with the line of symmetry locates the centroid.

4.4.4 *Experimental Method*

The method is based upon the experiments illustrated in art. 4.1. When a body is in equilibrium under the action of 2 forces, the two forces must be equal, opposite and collinear. The method is suitable in locating the centroids of irregular areas.

Take for example the irregular area shown in Fig. 4.9. Cut out from a uniformly thick cardboard, a piece exactly the same as that of the irregular area. Then, the cardboard piece can be hung with the help of a string from two points on its periphery. Each time the cardboard piece is hung, we mark on the cardboard the continuation of the line of the string. The intersection of the two lines locates the centroid of the irregular area. The position of centroid can be checked by hanging the cardboard from other positions. The continuation of the string direction in all cases on the cardboard should intersect at the centroid.

We have disccussed three methods for the location of the centroid. The first two methods are illustrated by a large number of examples solved in the next section. The reader should verify the experimental method himself by conducting experiments on areas with known centroids like a semi-circle, triangle etc.

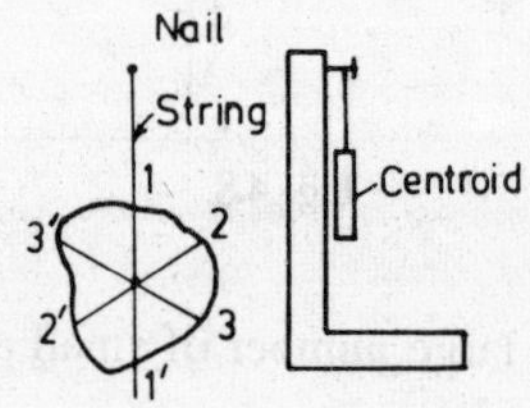

Fig. 4.9

4.4.5 *Centroids of Regular Areas*

We will now take up some simple geometrical areas and locate the centroid mainly by inspection.

We will use the principle stated above, *viz.* the centroid always lies on the axis of symmetry, if any, of an area.

(a) *Rectangle and Square*

Areas in the shape of a rectangle or a square have two axes of symmetry and their intersection locates the centroid. (Fig. 4.10).

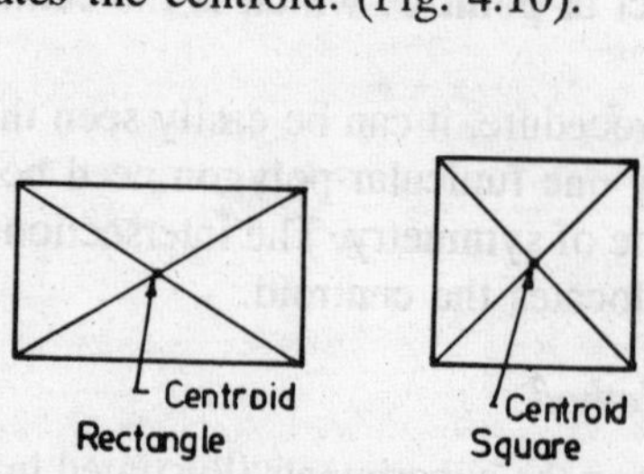

Fig. 4.10

(b) *Parallelogram and Rhombus*

Again there are two axes of symmetry and their intersection locates the centroids (Fig. 4.11).

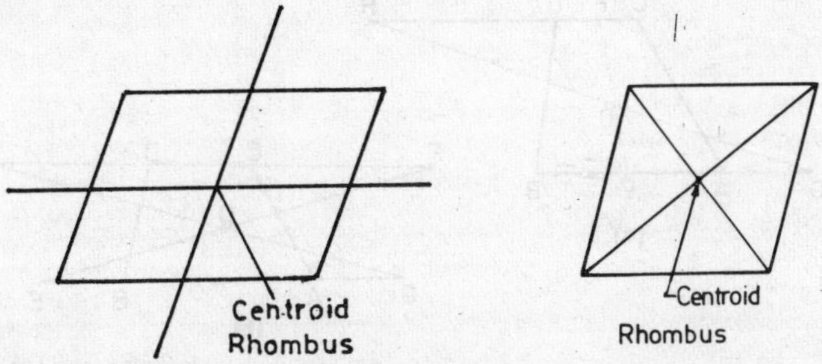

Fig. 4.11

(c) *Triangle*

If we join any vertex of a triangle to the middle point of the opposite side, such a line is known as the median of the triangle (Fig. 4.12). If we divide the triangle into a series of thin strips parallel to this side, it is easily seen that the centroid of each strip lies on the median. Therefore, if we draw two such medians, their intersection locates the centroid of the triangle. The intersection of the medians is at a distance of $\frac{h}{3}$ from the base, where h is the height of the triangle.

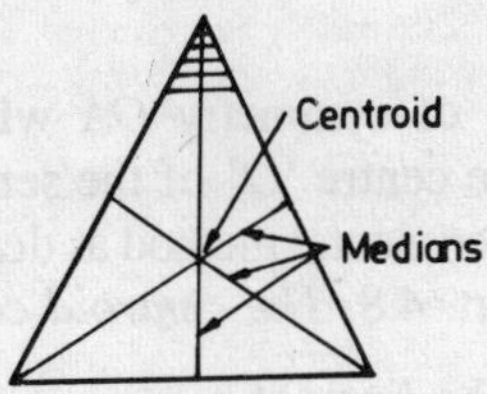

Fig. 4.12

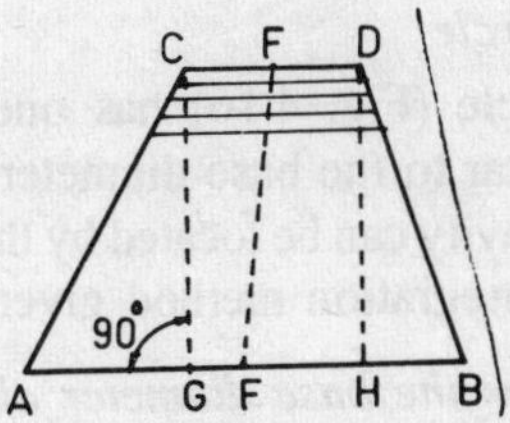

Fig. 4.13

(d) *Trapezium*

The trapezium shown in Fig. 4.13 has two parallel sides *AB* and *CD*. If we join the middle points of parallel sides *AB* and *CD*, the centroid lies on the line *EF* since all elemental strips parallel to *AB* and *CD* have their centroids on this line.

Analytically the centroid can be located by considering the trapezium as composed of 2 triangles *ACG, BDH* and a rectangle *CDHG*. The centroid can now be obtained by taking moments about X-axis and Y-axis and applying the formulae $\bar{y} = \frac{\Sigma\, ay}{A}$ and $\bar{x} = \frac{\Sigma\, ax}{A}$.

The centroid can also be located by the simple geometrical construction shown in Fig. 4.14(a) and (b).

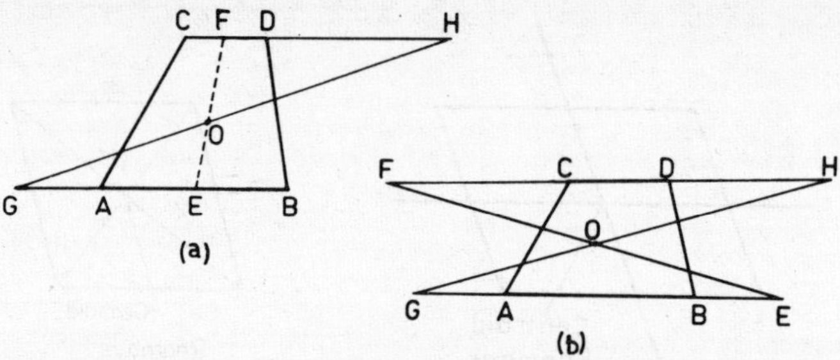

Fig. 4.14

In Fig. 4.14(a) first the centroidal line *EF* is drawn. *BA* is produced to *G* such that *AG* = *CD* and *CD* is produced to *H* such that *DH* = *AB*. *G-H*, is joined, which intersects *EF* at '*O*', the centroid.

In Fig. 4.14(b), the centroidal line is not drawn. Instead *AB* is produced to *E* such that *BE* = *CD* and *DC* is produced to *F* such that *CF* = *AB*. *E-F* is joined. The line *GH* is obtained as in Fig. 4.14(a). *EF* and *GH* intersect at '*O*', which is the centroid.

(e) *Circle*

A circle is symmetrical about all diameters. The intersection of any two diameters, i.e., the centre of the circle is the centroid. (Fig. 4.15).

(f) *Semi-circle*

A semi-circle (Fig. 4.16) has one axis of symmetry *OA* which is a line perpendicular to the base diameter at the centre '*O*' of the semi-circle. The centre of gravity can be located by the experimental method as described earlier or by the integration method given in art. 4.8. *The centroid can be seen to be* $\frac{4r}{3\pi}$ *from the base diameter along the line OA.*

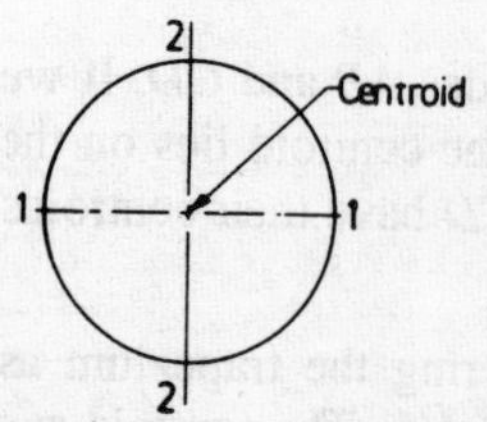

Fig. 4.15

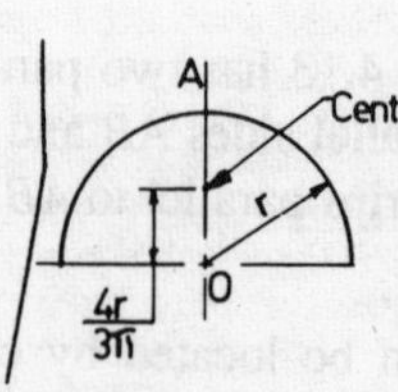

Fig. 4.16

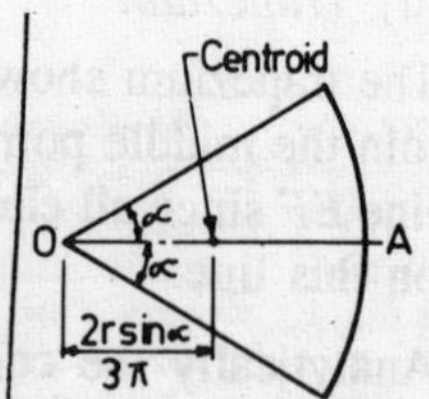

Fig. 4.17

(g) ***Segment of a Circle***

The segment of a circle (Fig. 4.17) has also a line of symmetry *OA*, the line through the centre

and bisecting the angle at the centre. The centroid can be located by the experimental method or by the method of integration. It can be seen to be at a distance of $\frac{2r \sin \alpha}{3\,\pi}$, where r is the radius and 2α is the angle subtended at the centre.

The above results and many additional data regarding centroids of plane geometric shapes are given in Table 4.1. This can be used for ready reference while solving problems.

Table 4.1

S.No.	SHAPE	A	$\bar{x}$	$\bar{y}$	S.No.	SHAPE	A	$\bar{x}$	$\bar{y}$
1.	Square 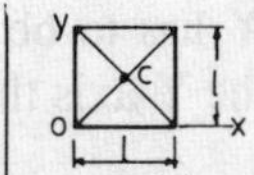	l^2	$\frac{l}{2}$	$\frac{l}{2}$	8	Semicircular area 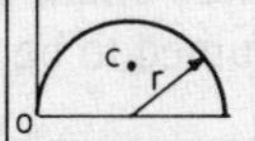	$\frac{\pi r^2}{2}$	r	$\frac{4r}{3\pi}$
2.	Rectangle 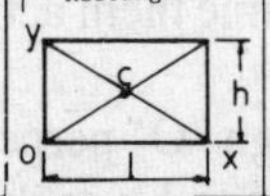	lh	l/2	h/2	9	Arc of a circle 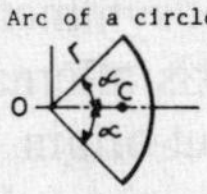	$2\alpha r$ (length)	$\frac{r \sin\alpha}{\alpha}$	0
3.	Triangle 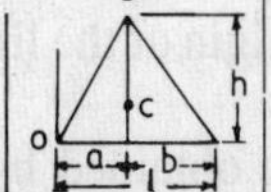	$\frac{lh}{2}$	$\frac{l+a}{3}$	$\frac{h}{3}$	10	Circular sector 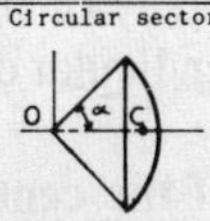	αr^2	$\frac{2r\sin\alpha}{3\alpha}$	0
4.	Trapezium 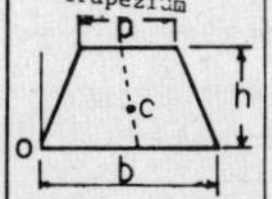	$\frac{h(a+b)}{2}$	$\frac{a+b}{2}$	$\frac{2a+b}{a+b}\,\frac{h}{3}$	11	Ellipse 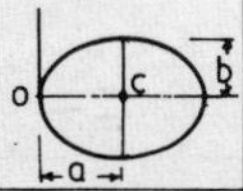	πab	a	0
5.	Parallelogram 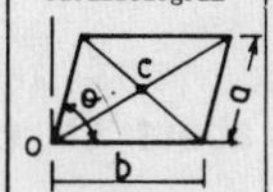	$ab\sin\theta$	$\frac{b+a\cos\theta}{2}$	$\frac{a\sin\theta}{2}$	12	Semi-ellipse 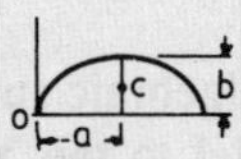	$\frac{\pi ab}{2}$	a	$\frac{4b}{3\pi}$
6.	Circular area 	πr^2	r	r	13	Parabola 	$\frac{4ah}{3}$	a	$\frac{2h}{5}$
7.	Quarter circular area 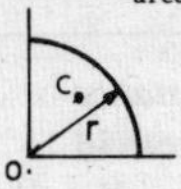	$\frac{\pi r^2}{4}$	$\frac{4r}{3\pi}$	$\frac{4r}{3\pi}$	14	Half Parabola 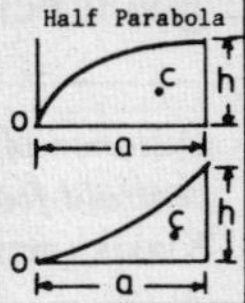	$\frac{2}{3}ah$ $\frac{1}{3}ah$	$\frac{3}{5}a$ $\frac{3}{4}a$	$\frac{3}{8}h$ $\frac{3}{10}h$

c- centroid

4.5 Solution of Problems on Location of Centroid

The following procedure is conveniently adopted in locating centroids:

(i) Determine by inspection whether the area has any axis of symmetry. It can be seen by inspection of the area shown in Fig. 4.18 that a horizontal line at half its height ($\frac{1}{2} \times 300$) is a line of symmetry.

(ii) Divide the area into smaller areas of regular shape whose centroids are known.

The given area can be divided into three rectangles marked (1), (2) and (3) in the Figure. Area (1) is 300 × 20, (2) is 180 × 15 and (3) is 180 × 15, all in sq. mms.

(iii) Select suitable coordinate axes OX and OY (when there is a line of symmetry, one of the axes can be conveniently chosen to coincide with the line of symmetry).

In Fig. 4.10, the X-axis (OX) is selected along the line of symmetry and Y-axis (OY) is selected as the leftmost vertical boundary line of the area. For these axes, $\bar{Y} = 0$ and only $\bar{X}$ has to be calculated.

(iv) Compute areas of each small area into which the figure is divided, locate the centroids of these small areas. Since only $\bar{X}$ has to be calculated, these areas are assumed to be acting parallel to the Y-axis through their centroids.

(If there is no axis of symmetry, we have to assume them acting parallel to X-axis and Y-axis alternately).

(v) Take moments about origin 'O' (an axis through 'O' perpendicular to the plane of the paper) of the areas. This is done by multiplying the areas with the perpendicular distance from the origin of the line of action of these areas.

(vi) The co-ordinates of the centroids $\bar{X}$, $\bar{Y}$, can be obtained by using the formulae,

$$\bar{X} = \frac{\Sigma ax}{A}$$

$$\bar{X} = \frac{\Sigma ay}{A}$$

where Σax is the algebraic sum of the moments about 'O', considering areas acting parallel to the Y-axis,

Σay is the algebraic sum of the moments about 'O', considering areas acting parallel to the X-axis and A is the total area.

In an analytical solution, the computations are best done in tabular form shown in the table given below.

Sr No	*Area*	*Distance of centroid from* X-axis (mm)	*Moment about* *'O' (2 × 3)* (mm^3)	*Distance* Centroid from X-axis	*Moment* about 'O' (2 × 5)
(1)	(2)	(3)	(4)	(5)	(6)
1.	300 × 20 = 6000	10	60000		
2.	180 × 15 = 2700	110	297000		
3.	180 × 15 = 2700	110	297000		
	11400		654000		

$$\bar{X} = \frac{654000}{11400} = 57.4 \text{ mm}$$

(Columns (5) and (6) of the above table are not used in solving the problem shown in Fig. 4.18, because the area has a line of symmetry and $\bar{Y}$ need not be calculated)

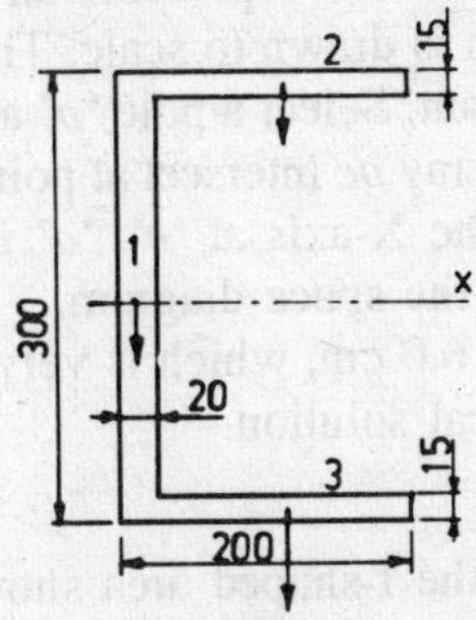

Fig. 4.18

The centroid, therefore, lies on the axis of symmetry (X-axis) at a distance of 57.4 mm from the origin.

Graphical Procedure

If a graphical solution is preferred, the procedure will be as follows: (See Fig. 4.18 for the problem and 4.19 for graphical solution). Steps (i) to (iv) are the same as in the analytical solution.

(v) The plane figure is laid out to a convenient scale. In Fig. 4.19 the space diagram is drawn to a scale of 1 cm = 5 cm.

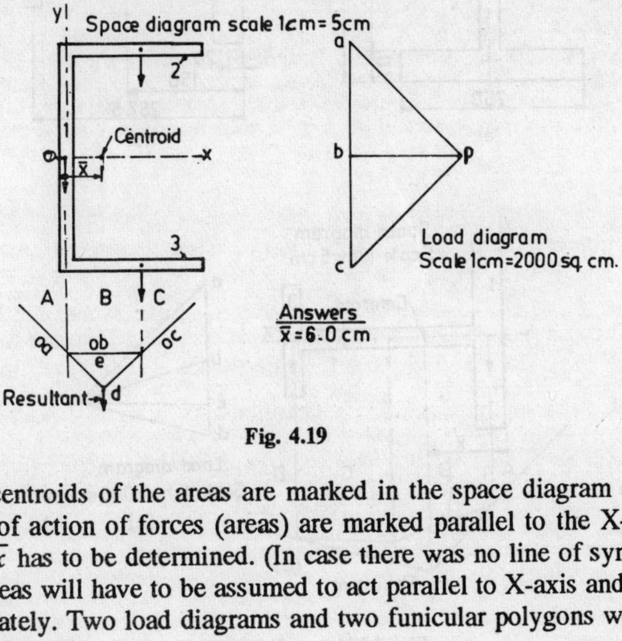

Fig. 4.19

(vi) The centroids of the areas are marked in the space diagram and the lines of action of forces (areas) are marked parallel to the X-axis as only $\bar{x}$ has to be determined. (In case there was no line of symmetry, the areas will have to be assumed to act parallel to X-axis and Y-axis alternately. Two load diagrams and two funicular polygons will have to be drawn).

(vii) The forces are named according to Bow's notation. In this case, the lines of action of areas of parts (2) and (3) have the same line of action, and hence considered as one area of 5400 sq. mm. Thus *AB* represents an area of 6000 sq. mm. and *BC* represents an area of 5400 sq. mm.
(viii) The load diagram is drawn to scale. The scale chosen is 1 cm = 2000 sq. mm. *ac* is the resultant. Select a pole '*p*' and draw the funicular polygon. The first ray *ao* and last ray *oc* intersect at point *d*. A line parallel to *ac* drawn through '*d*' intersects the X-axis at '*e*'. '*o*' is the centroid and distance *oe* gives $\bar{x}$ to the scale of the space diagram.

The value of $\bar{x}$ = 6.0 cm, which is very close to the value of 5.74 cm obtained in the analytical solution.

Example 4.1

Locate the centroid of the I-shaped area shown in Fig. 4.20.

Solution

The X- and Y- axes are selected as shown in Fig. 4.20. For convenience the figure is drawn by turning it through 90°. The X-axis is a line of symmetry. The area is divided into three rectangles as shown in Fig. 4.20(b). The calculations for $\bar{x}$ are shown in the table below.

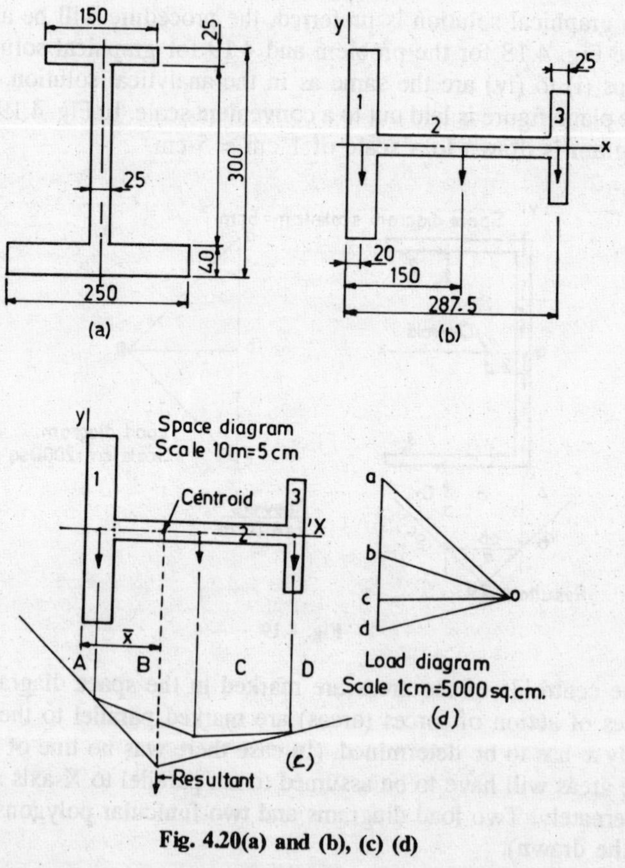

Fig. 4.20(a) and (b), (c) (d)

Sl. No.	*Area*	*Distance of Centroid from Y-axis*	*Moment of area about 'O'*
1.	250 × 40 = 10,000	20	200000.0
2.	235 × 25 = 5875	157.5	925312.5
3.	150 × 25 = 3750	287.5	1078125.0
	Total: 19625	$\bar{x} = \frac{2203437.5}{19625}$ = 112.28	2203437.5

Example 4.2
Locate the centroid of the plane area shown in Fig. 4.21.

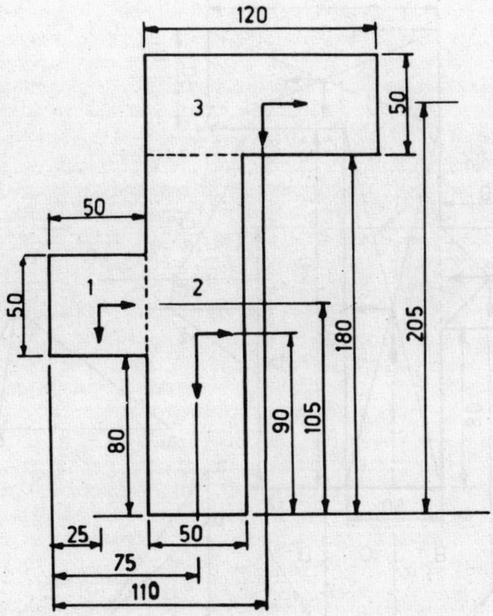

Fig. 4.21

Solution
The X- and Y-axes are selected as shown. There is no line of symmetry in this case. The given area can be divided into three rectangles as shown numbered in the figure. Moments are taken about 'O' of these areas and the centroid located as shown in table below.

Sl. No.	Area	Distance of c from Y-axis.	Moment about 'O'	Distance of c from X-axis.	Moment about 'O'
1.	50 × 50 = 2500	25	62500	105	262500
2.	180 × 50 = 9000	75	675000	90	810000
3.	120 × 50	110	660000	205	1230000
Total:	17500	$\bar{x} = \frac{1397500}{17500}$ = 79.86	139500	$\bar{y} = \frac{2302500}{17500}$ = 131.57	2302500

The same problem has been solved graphically in Fig. 4.22.

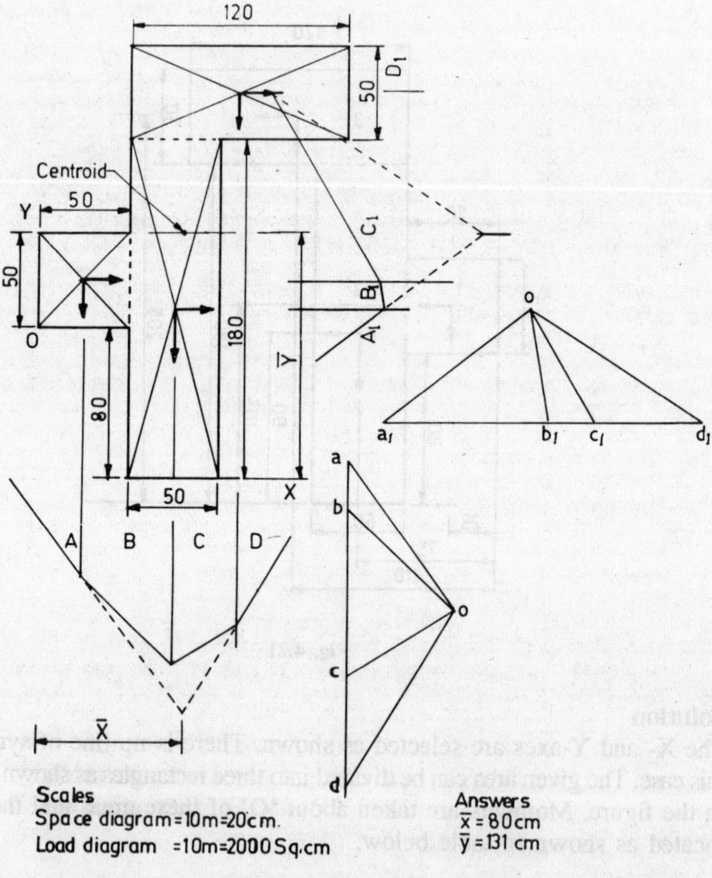

Fig. 4.22

The procedure is exactly same as in Fig. 4.19. In this case, however, there is no line of symmetry. Therefore, two load diagrams, *abcd* parallel to Y-axis, and $a_1b_1c_1d_1$ parallel to X-axis, are drawn, and with two funicular polygons, the lines of action of the resultant in two perpendicular directions are found out. The intersection of these two lines of action locates the point of application of the resultant force which is the centroid.

Example 4.3
Locate the centroid of the plane area shown in Fig. 4.23.

Solution
The X- and Y-axes are selected as shown. There is no line of symmetry

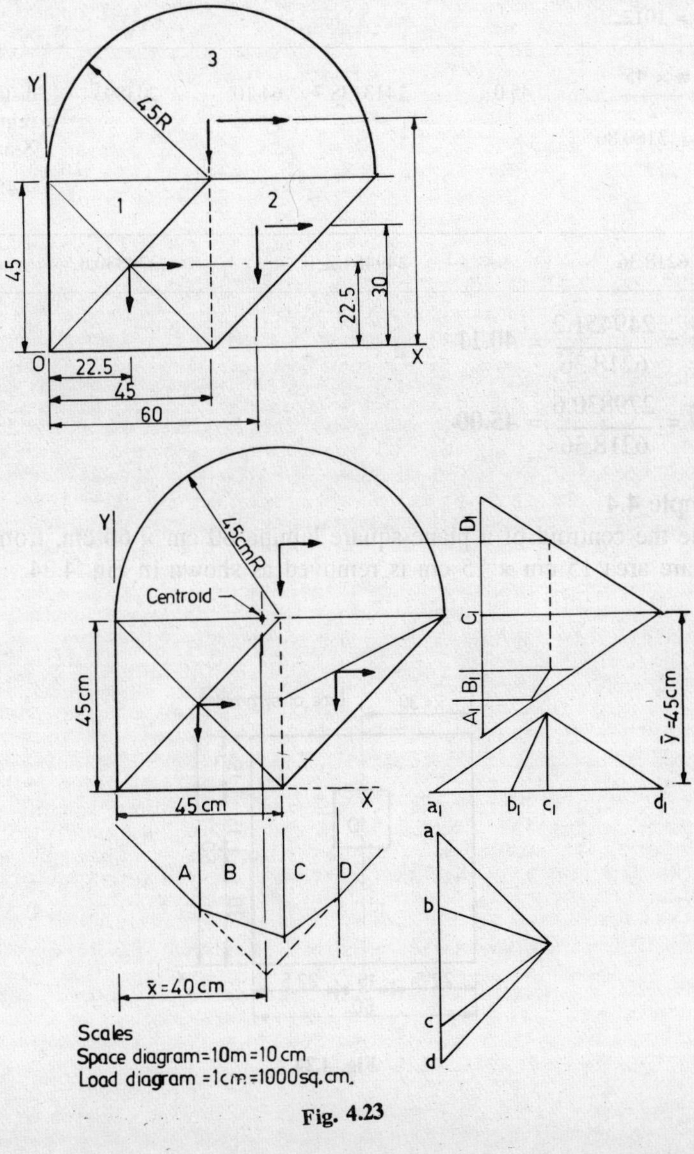

Fig. 4.23

in this case and both $\bar{x}$ and $\bar{y}$ should be computed. The given area is divided into 3 parts, a rectangle, a triangle and a semi-circle, as shown. Moments are taken about the origin and $\bar{x}$ and $\bar{y}$ are calculated as shown in table below.

Sl. No.	*Area*	*Distance* of c from Y-axes.	*Moment* of area about '0'	*Distance* of c from X-axis.	*Moment* of area about '0'	*Remarks*
1.	$45 \times 45 = 2025$	22.5	455625	22.5	455625	
2.	$\frac{45 \times 45}{2} = 1012.5$	60.0	60750	30.00	30375	
3.	$\frac{\pi \times 45^2}{2} = 3180.86$	45.0	1413138.7	64.10	203893	distance of centroid from X-axis $= 45 + \frac{4R}{3\pi}$
Total:	6218.36		249451.2		279830.6	

$$\bar{x} = \frac{249451.2}{6218.36} = \mathbf{40.11}$$

$$\bar{y} = \frac{279830.6}{6218.36} = \mathbf{45.00}$$

Example 4.4

Locate the centroid of a plane square lamina 60 cm × 60 cm, from which a square area 15 cm × 15 cm is removed as shown in Fig. 4.24.

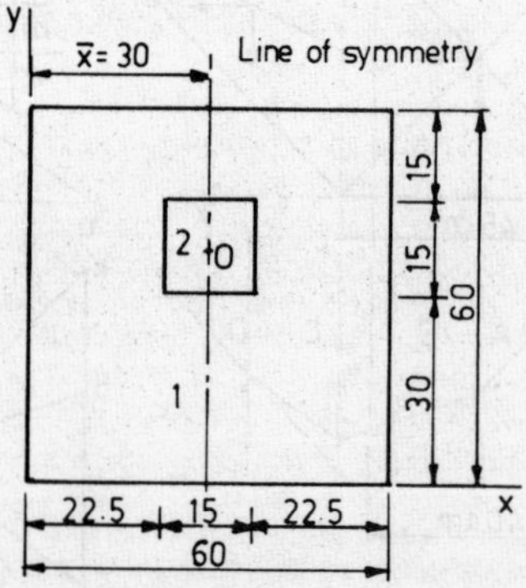

Fig. 4.24

Solution

The X- and Y-axes are selected along the edges of the lamina as shown. The given area is equal to a square area of 60 cm × 60 cm, and a negative area of 15 cm × 15 cm. A line parallel to the Y- axis and 30 cm from it is a line of symmetry, and hence $\bar{x}$ = 30 cm. Only $\bar{y}$ is to be determined which can be obtained from the principle of moments by taking moments about 'O' of areas considered parallel to X-axis.

Areas of lamina 60 cm × 60 cm = 3600 cm^2

Coordinates of its centroid are (30, 30).

Area of lamina 15 × 15 (removed) = –225 cm^2.

Coordinates of centroid are (30, 37.5).

∴ By applying the principle of moments,

$$A\,\bar{y} = A_1 Y_1 + A_2 Y_2$$

$$(3600 - 225)\,\bar{y} = 3600 \times 30 - 225 \times 37.5$$

$$3375\,\bar{y} = 108000 - 8437.5$$

$$\bar{y} = \frac{99562.5}{3375} = \mathbf{29.5\ cm}$$

∴ The coordinates of the centroid of the lamina are (30 cm, 29.5 cm).

Example 4.5

Locate the centroid of the plane lamina shown in Fig. 4.25. A circular hole, 20 cm dia., is cut out of this lamina as shown.

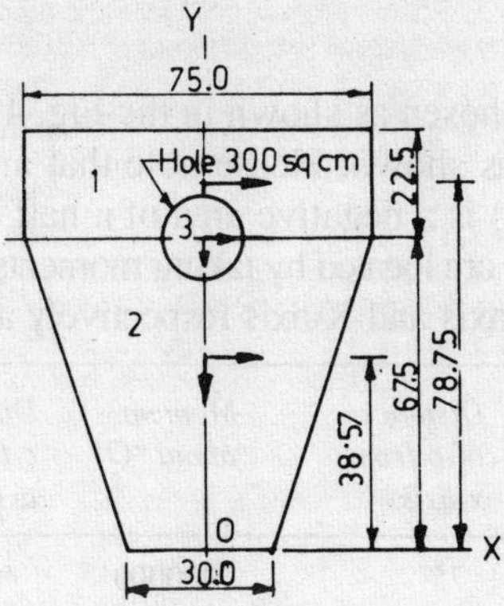

Fig. 4.25

Solution

The X- and Y-axes are selected as shown to make the Y-axis a line of symmetry. The given area is divided into three simple areas, a rectangle, a trapezium and a (negative) circular area. The co-ordinates of the centroids of different areas are:

Area (1) .... (0,78.75)

Area (2) .... (0,38.57) Refer to Table 4.1

$$\bar{y} = \frac{(30 + 2 \times 75)}{(30 + 75)} \times \frac{67.5}{3}$$

$$= 38.57 \text{ cm}$$

Area (3) .... (0,67.5)

The calculations for $\bar{y}$ are shown in the table below.

Sl. No.	*Area (cm²)*	*distance of centroid from X-axis (cm)*	*Moment of area about 'O' (cm³)*
1.	$75 \times 22.5 = 1687.5$	78.75	132890.6
2.	$\frac{(30 + 75)}{2} \times 67.5 = 3543.75$	38.57	136682.4
3.	$\frac{\pi \times 20^2}{4} = -314.16$	67.50	21205.8
Total	4917.09		248367.26

$$\bar{y} = \frac{248367.26}{4917.09} = \mathbf{50.51 \text{ cm}}$$

Example 4.6

Locate the centroid of the plane area shown in Fig. 4.26.

Solution

The X- and Y-axes are chosen as shown in the Fig. 4.26. The area is divided into 4 parts, numbered as shown. Please note that area (1) is a rectangle of 150 × 80 cm and area (4) is a negative area of a half circle of radius 25 mm.

The distances $\bar{x}$ and $\bar{y}$ are located by taking moments about 'O' of these areas considered parallel to Y-axis and X-axis respectively as shown in table below.

Sl. No.	*Area*	*Distance of c from x-axis.*	*Moment about 'O'*	*Distance of c from y-axis.*	*Moment about 'O'*
1.	$150 \times 80 = 12000$	75	900000	40	480000
2.	$90 \times 30 = 2700$	105	283500	95	256500
3.	$\frac{90 \times 60}{2} = 2700$	120	324000	130	351000
4.	$-\frac{\pi}{2} \times 25^2 = -982$	65	−63830	10.62	−10440
Total:	16418	$\bar{y} = \frac{1443670}{16418} = 87.8$	1443670	$\bar{x} = \frac{1087500}{16418} = 65.6$	1077060

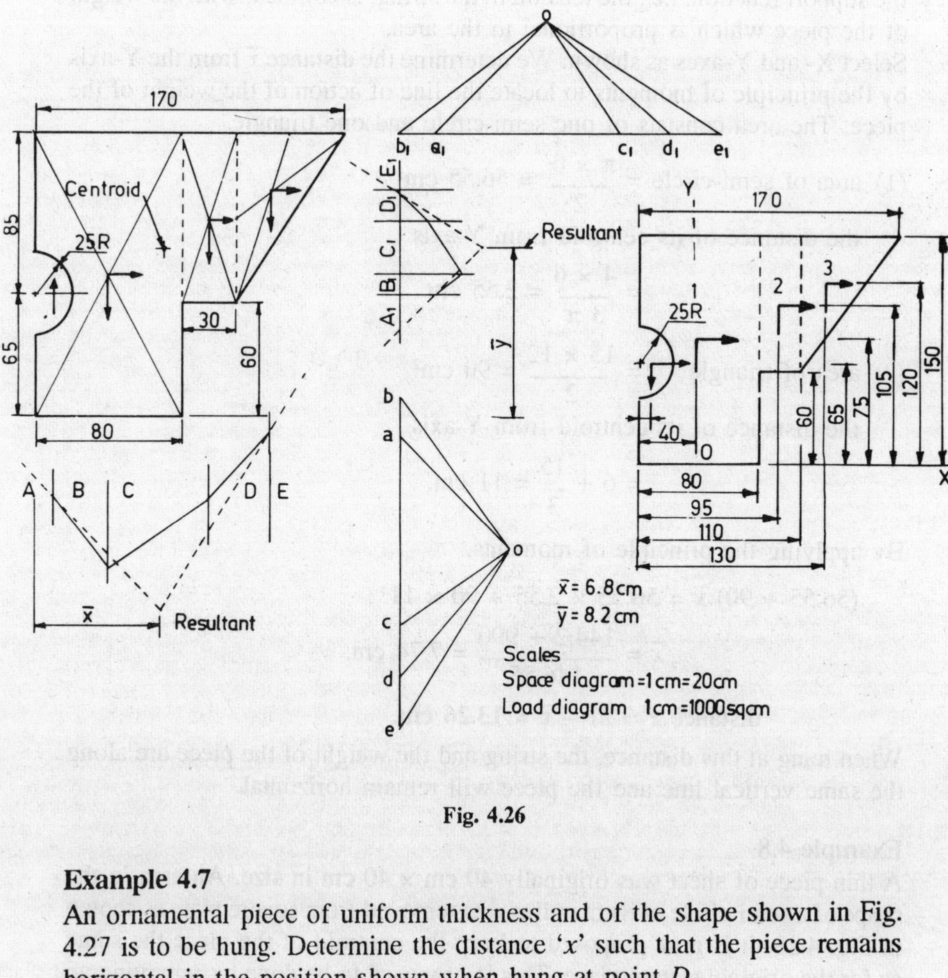

Fig. 4.26

Example 4.7

An ornamental piece of uniform thickness and of the shape shown in Fig. 4.27 is to be hung. Determine the distance 'x' such that the piece remains horizontal in the position shown when hung at point D.

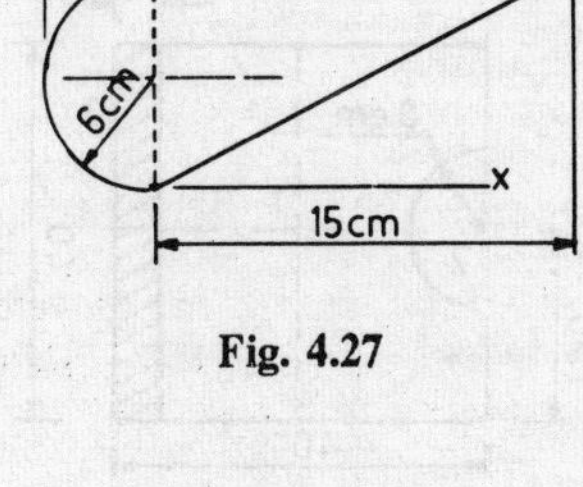

Fig. 4.27

Solution

The piece can be hung in the position shown when the line of action of the support reaction, i.e., the tension in the string, is collinear with the weight of the piece which is proportional to the area.

Select X- and Y-axes as shown. We determine the distance $\bar{x}$ from the Y-axis by the principle of moments to locate the line of action of the weight of the piece. The area consists of one semi-circle and one triangle.

(1) area of semi-circle $= \dfrac{\pi \times 6^2}{2} = 56.55 \text{ cm}^2$

the distance of its centroid from Y-axis

$$= \frac{4 \times 6}{3\pi} = 2.55 \text{ cm.}$$

(2) area of triangle $= \dfrac{15 \times 12}{2} = 90 \text{ cm}^2$

the distance of its centroid from Y-axis

$$= 6 + \frac{15}{3} = 11 \text{ cm.}$$

By applying the principle of moments,

$$(56.55 + 90)\,\bar{x} = 56.55 \times 2.55 + 90 \times 11$$

$$\bar{x} = \frac{144.2 + 990}{146.55} = 7.74 \text{ cm.}$$

$$\text{distance } x = 21 - \bar{x} = \mathbf{13.26 \text{ cm.}}$$

When hung at this distance, the string and the weight of the piece are along the same vertical line and the piece will remain horizontal.

Example 4.8

A thin piece of sheet was originally 40 cm × 40 cm in size. An area in the shape of a semi-circle of 8 cm radius was removed from square plate as shown in Fig. 4.28. It is now proposed to keep the centroid of the piece the same as for the original square piece. This is proposed to be done by removing an area as shown hatched in Fig. 4.28. Calculate the width '*w*' of the cut, for this condition.

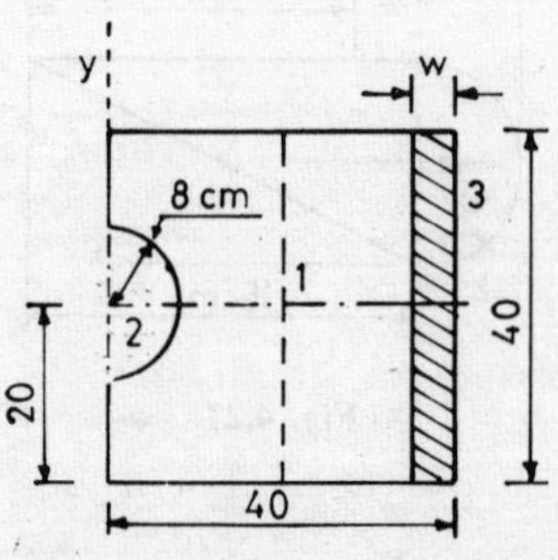

Fig. 4.28

Solution

It is easy to see that $\bar{y}$, the distance of centroid from X-axis shown, does not change by removing the areas. After the hatched area is removed, $\bar{x}$, should be equal to 20 cm. This is a reverse problem, *viz.*, knowing the centroid, we have to determine the area.

The area consists of three parts, numbered as shown in the figure.

(1) A square 40 cm side

$$\text{area} = 40 \times 40 = 1600 \text{ cm}^2$$

centroid is at (20, 20).

(2) Semi-circular area (removed)

$$\text{area} = \frac{-\pi \times 8^2}{2}$$

$$= -100.53 \text{ cm}^2$$

$$\text{Centroid is at } \left(\frac{4 \times 8}{3\,\pi}, 20 \right)$$

(3) Rectangular area (removed)

$$\text{area} = -40 \times w \text{ cm}^2$$

$$\text{centroid is at} \left(40 - \frac{w}{2}, 20 \right)$$

We now apply the principle of moments, for moments about 'O' considering the areas parallel to the Y-axis.

$A\,\bar{x} = A_1\,\bar{x}_1 + A_2\,\bar{x}_2 + A_3\,\bar{x}_3$

$w\bar{x} = dw \cdot x$

(w is proportional to the areas and $\bar{x}$ = 20)

$$(1600 - 100.53 - 40\,w) \times 20 = 1600 \times 20 - 100.53 \times \frac{4 \times 8}{3\,\pi}$$

$$\left[-40\,w \left(40 - \frac{w}{2} \right) \right]$$

$29989.4 - 800\,w = 32000 - 341.3 - 1600\,w + 20\,w^2$

$20\,w^2 - 800\,w + 1669.3 = 0$

$w^2 - 40\,w + 83.465 = 0$

$$w = \frac{40 + 40^2 - 4 \times 83.465}{2}$$

$$= \frac{40 \pm 35.58}{2} = 37.785 \text{ or } 2.21 \text{ cm.}$$

w = 37.785 cm is not possible, therefore,

w = **2.21 cm**

4.6 Centre of Gravity of Regular Solids

In the previous articles, we have discussed about centroids. In the following articles we shall discuss the location of centre of gravity of solids.

Solids are three dimensional bodies. The procedure to be adopted in locating the CG is the same as in the case of centroids, though the addition of the third dimension does involve more calculations. In the case of many regular solids, the CG can be located by inspection. Knowing that the CG lies in the planes of symmetry of the body we shall first take up simple cases and locate the CG.

(a) *Cube and Rectangular Prism*

The cube (Fig. 4.29) is symmetrical about three mutually perpendicular planes. and the CG can be located as the intersection point of these three planes. The CG can be located also by seeing that it lies at the intersection of the two body digonals.

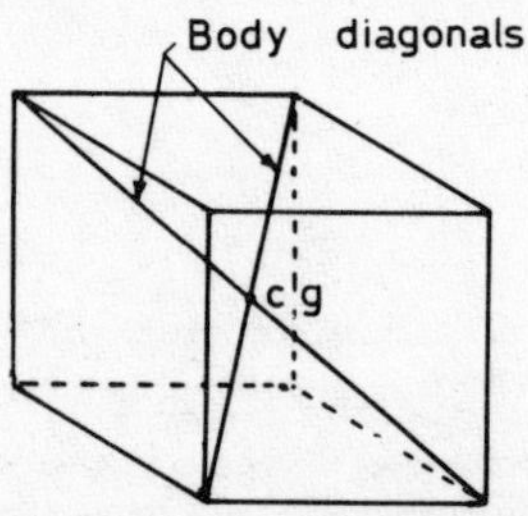

Fig. 4.29

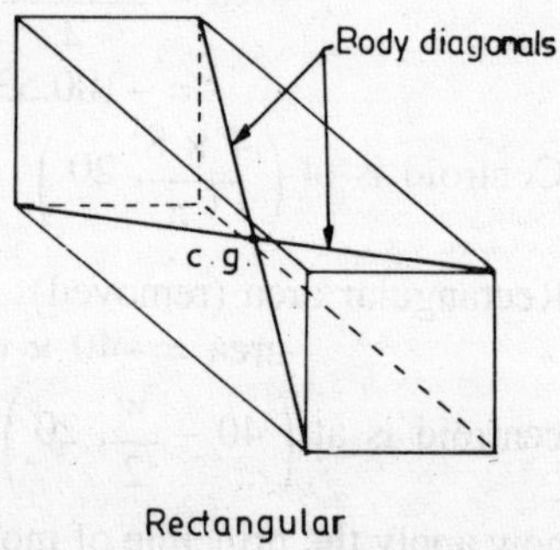

Fig. 4.30

The rectangular prism (Fig. 4.30) is a case similar to the cube. The CG again lies at the intersection of the three mutually perpendicular planes about which it is symmetrical. Alternatively, it can be seen to be at the intersection of the two body diagonals.

(b) *Sphere*

A sphere is symmetrical about all diametrical plane. Hence the CG lies at the centre of the sphere (Fig. 4.31). A hemisphere (Fig. 4.32) is symmetrical about a vertical diameter. The centroid is at $4r/3\pi$ from the base.

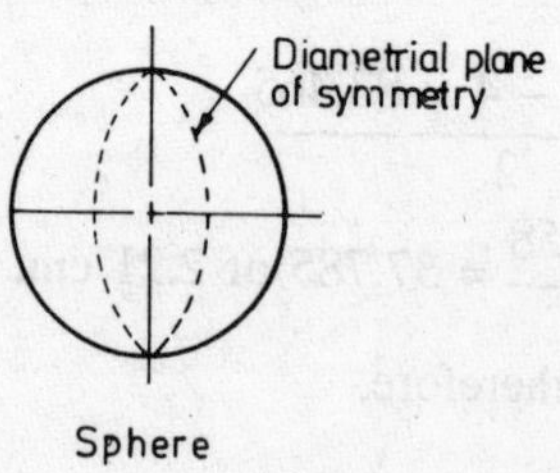

Fig. 4.31

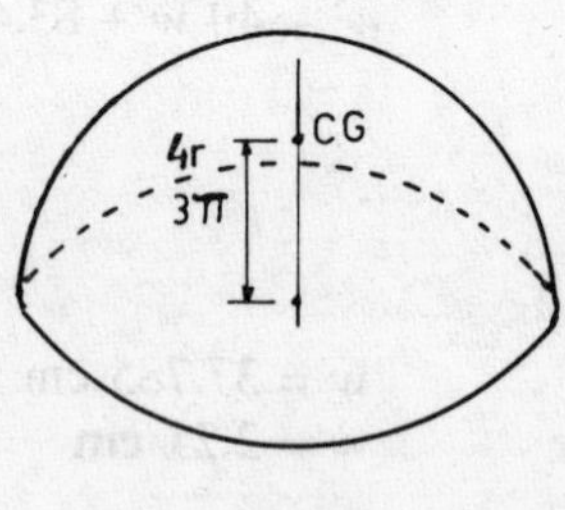

Fig. 4.32

(c) *Cylinder*

A cylinder is symmetrical about its longitudinal axis. It is also symmetrical about a plane perpendicular to the longitudinal axis at its mid point. The intersection of the longitudinal axis and this plane of symmetry locates the CG (Fig. 4.33).

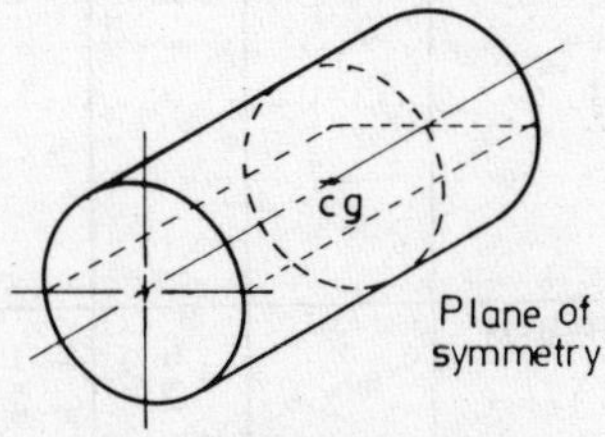

Fig. 4.33

The centre of gravity of many other regular solids are given in Table 4.2. This can be used for ready reference while solving problems.

4.7 Location of CG of Solids

The location of CG of solids is accomplished on the same lines as for the centroid of areas. When a solid can be divided into a number of regular solids, the CG can be located by the procedure outlined below.

(i) Select the coordinate axes such that any plane(s) of symmetry coincides with the coordinate axes.

(ii) Compute the weights and CG of parts of the solids.

(iii) Now the principle of moments can be applied to locate the coordinates of the CG. Let $\bar{x}$, $\bar{y}$ and $\bar{z}$ be the coordinates of the CG. If W is the weight of the body, W_1, W_2, W_3 etc., the weights of individual parts of the body and (x_1, y_1, z_1), (x_2, y_2, z_2), (x_3, y_3, z_3), etc., the coordinates of the CG of the individual parts respectively, then by taking moments about Z, X and Y axes, shown in Fig. 4.35,

$$\bar{x} = \frac{1}{W}(w_1 x_1 + w_2 x_2 + w_3 x_3 + ...)$$

$$\bar{y} = \frac{1}{W}(w_1 y_1 + w_2 y_2 + w_3 y_3 + ...)$$

and $$\bar{z} = \frac{1}{W}(w_1 z_1 + w_2 z_2 + w_3 z_3 + ...)$$

(iv) The above procedure is completely general and can be applied in the case of solids made up of different materials. In cases where the solid is homogeneous, the weights in the above expressions can be replaced by volumes since the weight density is a common factor and cancels out.

Table 4.2

S.No.	SHAPE	Volume	$\bar{x}$	$\bar{y}$	$\bar{z}$
1	CUBE 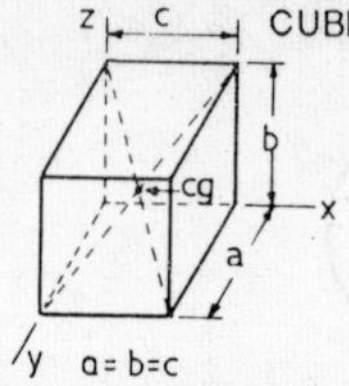	a^3	$\frac{c}{2}$	$\frac{a}{2}$	$\frac{b}{2}$
2	RECTANGULAR PRISM 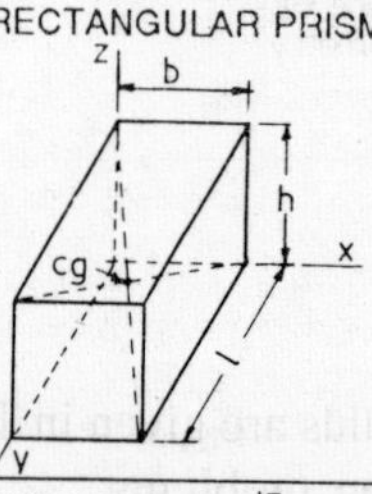	lbh	$\frac{b}{2}$	$\frac{l}{2}$	$\frac{h}{2}$
3	CYLINDER 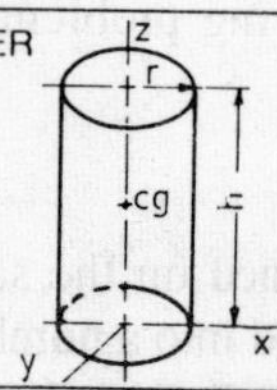	$\pi r^2 h$	0	0	$\frac{h}{2}$
4	SPHERE 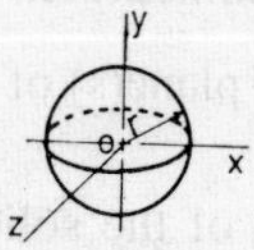	$\frac{4}{3}\pi r^3$	0	0	0
5	HEMI SPHERE 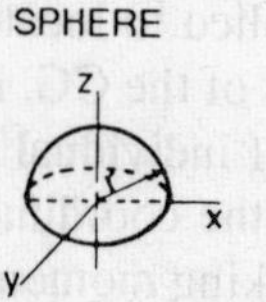	$\frac{2}{3}\pi r^3$	0	0	$\frac{4r}{3\pi}$
6	CIRCULAR PYRAMID (CONE) 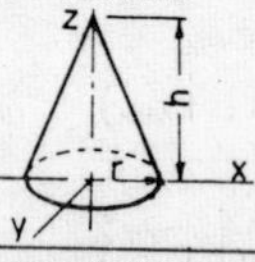	$\frac{1}{3}\pi r^2 h$	0	0	$\frac{h}{3}$
7	RIGHT RECTANGULAR PYRAMID 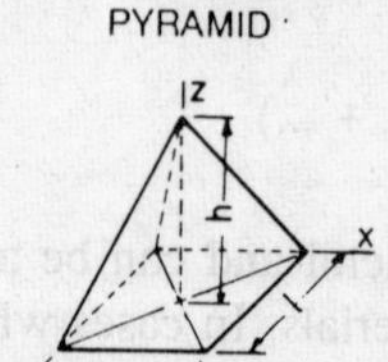	$\frac{1}{3}$ blh	$\frac{b}{2}$	$\frac{l}{2}$	$\frac{h}{3}$

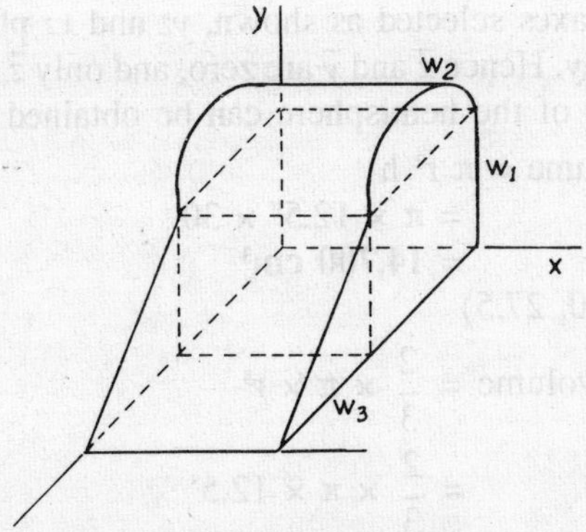

Fig. 4.34

These formulae can also be put in a general form as given below

$$\bar{x} = \frac{\Sigma\, dw \cdot x}{w}$$

$$\bar{y} = \frac{\Sigma\, dw \cdot y}{w}$$

and $$\bar{z} = \frac{\Sigma\, dw \cdot z}{w}$$

suitable for applying the method of integration (see art. 4.8), where dw = elemental weight of an elemental volume having CG at coordinates (x, y, z).

It is desirable to perform the calculations in tabular form as done in the examples given below.

Example 4.9

Locate the CG of the solid, shown in Fig. 4.35. Bottom part is a hemisphere 12.5 cm radius and the top part is a regular cylinder of 25 cm diameter, and 30 cm height.

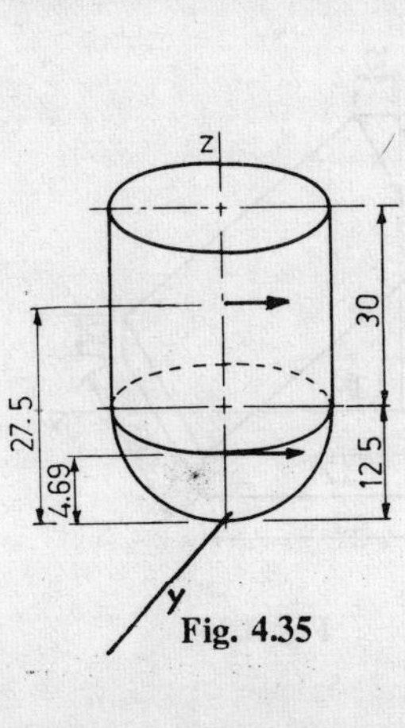

Fig. 4.35

Solution

With the coordinate axes selected as shown, *yz* and *xz* planes are planes of symmetry for the body. Hence $\bar{x}$ and $\bar{y}$ are zero, and only $\bar{z}$ is to be calculated. The volume and CG of the hemisphere can be obtained from Table 4.2.

for the cylinder, volume $= \pi r^2 h$

$$= \pi \times 12.5^2 \times 30$$

$$= \mathbf{14{,}700\ cm^3}$$

and its CG is at (0, 0, 27.5)

for the hemisphere, volume $= \dfrac{2}{3} \times \pi \times r^3$

$$= \frac{2}{3} \times \pi \times 12.5^3$$

$$= \mathbf{4084\ cm^3}.$$

and its CG is at (0, 0, 4.69).

The computations are best done in the tabular form shown below. The tabular form shown is complete even though $\bar{x}$ and $\bar{y}$ need not be calculated.

Sl. No.	*Shape*	*Volume/ Weight cm³*	*Coordinates of CG*			*Moments about axis*		
			X	*Y* *(cm)*	*Z*	*Z* *(w.x)*	*X* *(w.y)*	*Y* *(w.z)*
1.	Cylinder	14,700	0	0	27.5	0	0	404250
2.	Hemisphere	4,084	0	0	4.69	0	0	19154
Total:		18,784 cm³						423404

$$\bar{z} = \frac{423404}{18784} = \mathbf{22.5\ cm.}$$

The co-ordinates of the CG of the solid are (0, 0, 22.5).

Example 4.10

Locate the CG of the block shown in Fig. 4.36.

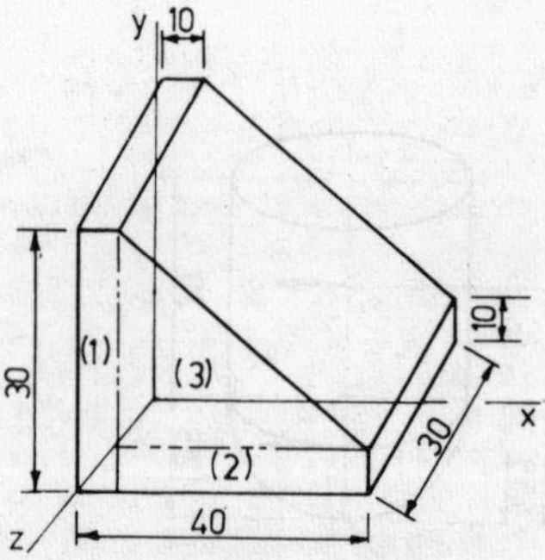

Fig. 4.36

Solution

The solid can be divided into three parts as shown in Fig. 4.36. X-, Y-and Z- axes are selected as shown making X-Y plane a plane of symmetry.

The computations are shown in tabular form below.

Sl. No.	*Volume*	*Coordinates of CG*			*Moment about axes*		
		x	*y*	*z*	*z*	*x*	*y*
1.	$10 \times 30 \times 20$ $= 6000$ cm^3	5	15	0	30,000	90,000	0
2.	$30 \times 10 \times 20$ $= 6000$ cm^3	25	5	0	150,000	30,000	0
3.	$30 \times 20 \times \frac{20}{2}$ $= 6000$ cm^3	20	$\frac{50}{3}$	0	12,000	100,000	0
Total:	18,000				30,000	220,000	0

$$\bar{x} = \frac{300{,}000}{18{,}000}$$

$$= \mathbf{16.66\ cm}$$

$$\bar{y} = \frac{220{,}000}{18{,}000}$$

$$= \mathbf{12.22\ cm} \text{ and } \bar{z} = \mathbf{0}$$

Therefore, the co-ordinates of the centroid are (16.66, 12.22, 0).

Example 4.11

Locate the CG of the assembly shown in Fig. 4.37. The rod weighs 600 N per cm. length and the semi-circular disc and the rectangular plate weigh 1800 N per sq. cm of area.

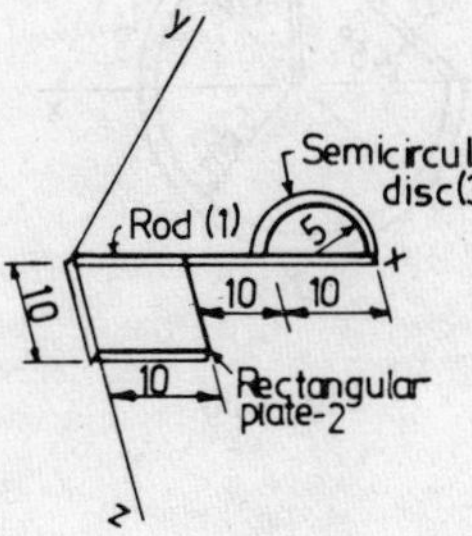

Fig. 4.37

Solution

The assembly consists of three parts, *viz.*, the straight rod (along X-axis) (1), and the rectangular plate in the *xz* plane (2) the semi circular disc in the *y-x* plane (3) Computations are shown in the table below.

Sl. No.	Weight	Coordinates of CG			Moment about axes		
		x	*y*	*z*	*z*	*x*	*y*
1.	30×600 $= 18000$ N	15	0	0	27×10^3	0	0
2.	100×1800 $= 180,000$ N	5	0	5	900×10^3	0	900×10^3
3.	$\pi \times 5^2/2 \times 1800$ $= \dfrac{45000\,\pi \text{ N}}{2}$	25	$\dfrac{20}{3\pi}$	0	1767.15×10^3	150×10^3	0
Total:	268686				2937150	150000	900000

$$\bar{x} = \frac{2937150}{268686} = \mathbf{10.93\ cm}$$

$$\bar{y} = \frac{150000}{268686} = \mathbf{0.50\ cm}$$

$$\bar{z} = \frac{900000}{268686} = \mathbf{3.35\ cm}$$

The CG is, therefore, located at (10.93, 0.50, 3.35)

Example 4.12

Locate the CG of a thin piece of wire bent into the shape shown in Fig. 4.38.

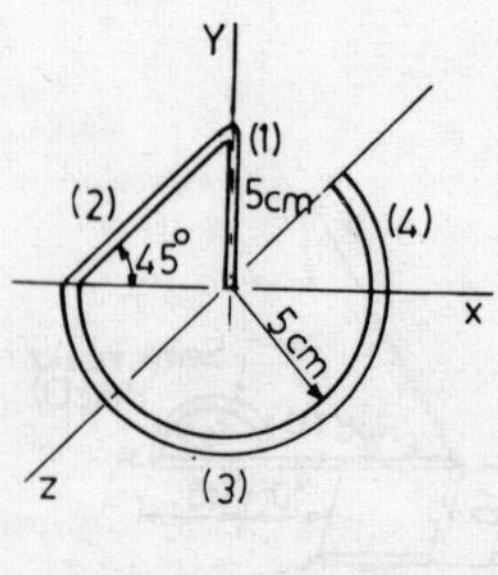

Fig. 4.38

Solution

With the X, Y, Z axes given as shown, the piece of wire can be considered to consist of 4 parts as marked in the figure. We can consider the volume of each part proportional to its length as shown. The circular arc which lies wholly in the X-Z plane is considered in two parts, namely, a semicircular arc (3) of length $\pi \cdot r$, and a quarter circular arc (4) of length $\frac{\pi r}{2}$. The centroids of these axes can be found from Table 4.1. The computations are shown in tabular form below.

Sl. No.	*Length*	*Coordinates of CG*			*Moment about axes*		
		x	*y*	*z*	*z*	*x*	*y*
1.	5	0	2.5	0	0	12.5	0
2.	$\sqrt{5^2+5^2} = 7.07$	–2.5	+2.5	0	–12.5 2	+12.5 2	0
3.	$\pi \times 5 = 15.71$	0	0	$10/\pi$	0	0	50
4.	$\pi \times 2.5 = 7.85$	$10/\pi$	0	$10/\pi$	25	0	25
	Total: 35.63				– 7.32	30.17	+75

$$\bar{x} = \frac{-7.32}{35.63} = \mathbf{-0.2}$$

$$\bar{y} = \frac{30.17}{35.63} = \mathbf{0.85}$$

$$\bar{z} = \frac{75}{35.63} = \mathbf{2.10}$$

The coordinates of the centroid are (– 0.2, 0.85, 2.10).

Example 4.13

Locate the CG of the assembly shown in Fig. 4.39 with reference to the coordinate axes shown.

The cylindrical head weighs 3000 N, the rod BC weighs 2500 N and the rod CD weighs 1000 N.

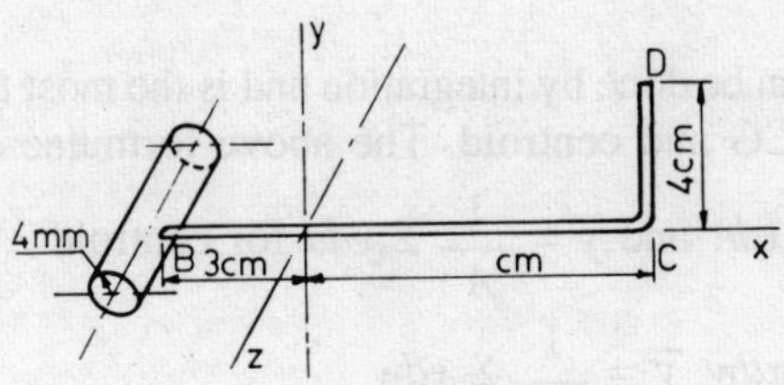

Fig. 4.39

Solution

With reference to the coordinate axes shown, the body is symmetrical with respect to the *xy* plane, and therefore $\bar{z} = 0$. The computations are shown in the tabular form below.

Sl. No.	*Wt. N.*	*Coordinates of CG*			*Moment about axes*		
		x	*y*	*z*	*z*	*x*	*y*
1.	3000	–3.4	0	0	–10200	0	0
2.	2500	2	0	0	5000	0	0
3.	1000	7	2	0	7000	2000	0
Total:	6500				1800	2000	0

$$\therefore \bar{x} = \frac{1800}{6500} = \mathbf{0.28}$$

and $$\bar{y} = \frac{2000}{6500} = \mathbf{0.31}$$

The coordinates of CG, therefore, are (0.28, 0.31, 0.00)

4.8 Method of Integration

It has already been explained that the location of CG or centroid involves a process of summation of the moments of elemental areas, and we have established the formulae

$$\bar{x} = \frac{\Sigma\, x\, dA}{A}$$

and $$\bar{y} = \frac{\Sigma\, y\, dA}{A}$$ for centroids and

$$\bar{x} = \frac{\Sigma\, x\, dW}{W}$$

$$\bar{y} = \frac{\Sigma\, y\, dW}{W}$$

and $$\bar{z} = \frac{\Sigma\, z{\cdot}dW}{W}$$ for centre of gravity.

The summation can be done by integration and is the most fundamental method for locating the CG and centroid. The above formulae can be written as

$$\bar{x} = \frac{1}{A}\,\Sigma_A x dA \text{ and } \bar{y} = \frac{1}{A}\,\Sigma_A y da \text{ for centroid}$$

and $$\bar{x} = \frac{1}{W}\,\Sigma_V x dw,\ \bar{y} = \frac{1}{W}\,\Sigma_V y dw.$$

and $$\bar{z} = \frac{1}{W}\,\Sigma_V z.dw \text{ for CG}$$

Here the signs Σ_A and Σ_V indicate that summation should be done over the whole area for centroids and over the whole volume for centre of gravity for the variables *x*, *y* and *z* appearing in the equations.

The following examples illustrate the use of the integration method.

Example 4.14

Locate the centroid of the area of the triangle shown in Fig. 4.40.

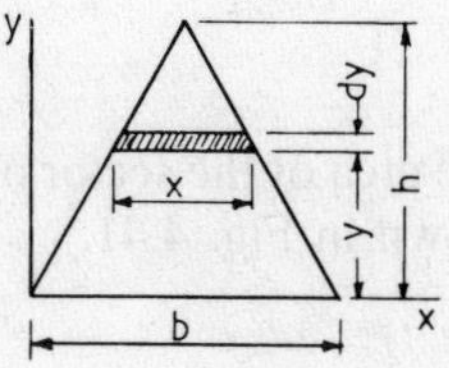

Fig. 4.40

Solution

Select the *x* and *y* axes as shown. We consider an elementary area of width, say '*x*' and height *dy* at a distance of '*y*' from the base. Then area of the elementary strip $= x \cdot dy$

and its moment about *x*-axis $= x \cdot dy \cdot y$

The integration will be done with reference to *y* within limits 0 to *h*, and it is, therefore, necessary to express '*x*' in terms of *y*· This can be done from similar triangles with base '*x*' and '*b*' respectively, and we get

$$x = \frac{b}{h}(h - y)$$

$\therefore$ Moment of shaded strip about Z-axis $= \dfrac{b}{h}(h - y) \cdot y \cdot dy \cdot$

Area of the triangle $= \dfrac{b \cdot h}{2}$

If $\bar{y}$ is the distance of the centroid from the base then,

$$\frac{b \times h}{2}\bar{y} = \int_0^h \frac{b}{h}(h - y)\, y \cdot dy$$

$$\bar{y} = \frac{2}{b \times h} \times \frac{b}{h} \int_0^h (h - y)\, y \cdot dy$$

$$= \frac{2}{h^2}\left| \frac{hy^2}{2} - \frac{y^3}{3} \right|_0^h$$

$$= \frac{2}{h^2}\left(\frac{h^3}{2} - \frac{h^3}{3}\right)$$

$$= \frac{2}{h^2} \times \frac{h^3}{6}$$

$$= \frac{h}{3}$$

The centroid of a triangle is therefore, at a distance of $\frac{h}{3}$ from the base.

Example 4.15

Determine the centroid of the area of the sector of a circle of radius 'R', and subtended angle '2α' as shown in Fig. 4.41.

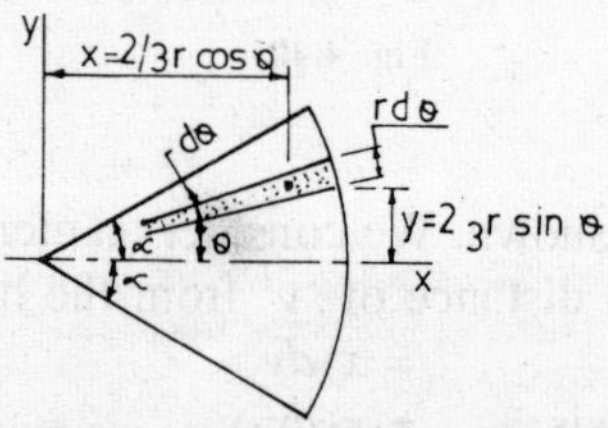

Fig. 4.41

Solution

Select X- and Y-axes as shown. Since X-axis is an axis of symmetry, $\bar{y} = 0$.

The elementary strip selected is shown hatched in the figure. The subtended angle at the centre is $d\theta$ and the length is $r \times d\theta$. As $d\theta$ is very small, the elementary strip can be considered to be a triangle.

The area of the elementary strip $= r \times d\theta \times \frac{r}{2}$

$$= \frac{r^2 \times d\theta}{2}$$

The variable is θ, and varies from $-\alpha$ to $+\alpha$. We can integrate this expression to get the total area.

$$\text{Area of the sector} = \int_{-\alpha}^{+\alpha} \frac{r^2 \cdot d\theta}{2}$$

$$= \left| \frac{r^2}{2}\theta \right|_{-\alpha}^{\alpha}$$

$$= \frac{r^2 \times 2\alpha}{2}$$

$$= r^2 \alpha$$

We are interested in finding the value of $\bar{x}$, the x coordinate of the centroid. We, therefore, take moments about 'O' and apply the principle of moments.

$$\text{Moment of the elementary strip about 'O'} = \frac{r^2}{2} \cdot d\theta \cdot \frac{2}{3} \cdot r \cdot \cos\theta$$

$$= \frac{r^3}{3} \cdot \cos\theta \cdot d\theta$$

Applying the principle of moments, we get

$$\text{Area of the sector} \times \bar{x} = \int_{-\alpha}^{+\alpha} \frac{r^3}{3} \cos\theta \cdot d\theta$$

$$r^2 \cdot \alpha \cdot \bar{x} = \int_{-\alpha}^{+\alpha} \frac{r^3}{3} \cos\theta \, d\theta$$

$$= \frac{r^3}{3} 2 \sin\alpha$$

$$\bar{x} = \frac{r^3}{3} \frac{2\sin\alpha}{r^2 \cdot \alpha}$$

$$= \frac{2}{3} r \frac{\sin\alpha}{\alpha}$$

This is the location of the centroid of any sector. Please note that α must be substituted in radians.

The above result can be used to locate the centroid of a semicircle and quarter of a circle (Fig. 4.42).

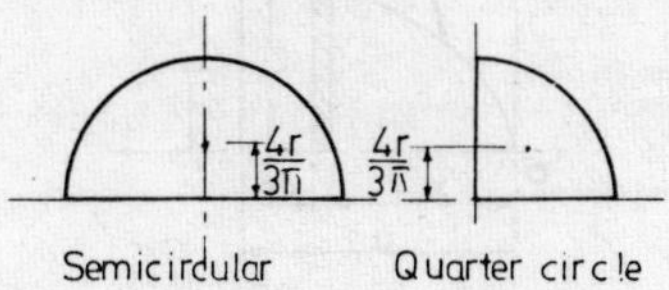

Fig. 4.42

In the case of semicircle, $2\alpha = \frac{2\pi}{2} = 180°$

substituting $\alpha = \frac{2\pi}{2}$ in the above,

$$\text{x-coordinate of the centroid} = \frac{2}{3} \times \frac{r \times \sin \frac{\pi}{2}}{\frac{\pi}{2}}$$

$$= \frac{4r}{3\,\pi}$$

In case of a quarter circle, $2\alpha = \frac{\pi}{2}$ and $\alpha = \frac{\pi}{4}$.

$$\text{x-coordinate of centroid} = \frac{2}{3} \times \frac{r \times \sin \frac{\pi}{4}}{\frac{\pi}{4}}$$

$$= \frac{8r}{3.\,\pi\,\sqrt{2}}$$

The distance from the two bounding radii is obtained as in Fig. 4.42.

$$\bar{x} = \bar{y} = \frac{8\,r}{3\,\pi\,\sqrt{2}} \times \frac{1}{\sqrt{2}}$$

$$= \frac{4\,r}{3\,\pi}$$

Therefore, the distance to the centroid from the base diameter of a semicircle or from the radii of a quarter of a circle are the same.

Example 4.16
Locate the centroid of the area of a parabolic segment as shown in Fig. 4.43.

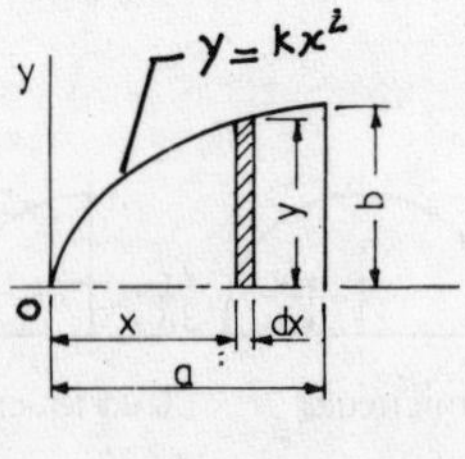

Fig. 4.43

Solution
With reference to the co-ordinate axes shown, we select an elementary strip at a distance x of width dx. The area of the element is $y\,dx$ and the distance of its centroid from Y- and X-axes are x and $\frac{1}{2} \cdot y$.

$$\text{Area of the parabola} = \int_0^a y \cdot dx$$

$$= \int_0^a kx^2 \cdot dx$$

$$= \frac{ka^3}{3}$$

To find $\bar{x}$, we apply the principle of moments about 'O'

$$\frac{ka^3}{3} \cdot \bar{x} = \int_0^a x \cdot ydx$$

$$= \int_0^a k\, x^3\, dx$$

$$= \frac{ka^4}{4}$$

$$\bar{x} = \frac{ka^4}{4} \times \frac{3}{ka^3}$$

$$= \frac{3\, a}{4}$$

To find $\bar{y}$, we take moments about 'O', considering areas parallel to X-axis. Then

$$\frac{ka^3}{3} \cdot \bar{y} = \int_0^a y \cdot dx \cdot \frac{y}{2}$$

$$= \int_0^a \frac{k^2\, x^4\, dx}{2}$$

$$= \frac{k^2 \times a^5}{10}$$

$$\bar{y} = \frac{k^2\, a^5}{10} \times \frac{3}{ka^2}$$

$$= \frac{3\, ka^2}{10}$$

Since $ka^2 = b$,

$$\bar{y} = \frac{3\, b}{10}.$$

Therefore, the centroid of the parabolic segment has coordinates $3a/4$, $\frac{3b}{10}$

Example 4.17

Locate the centre of gravity of a homogeneous hemispherical solid volume as shown in Fig. 4.44.

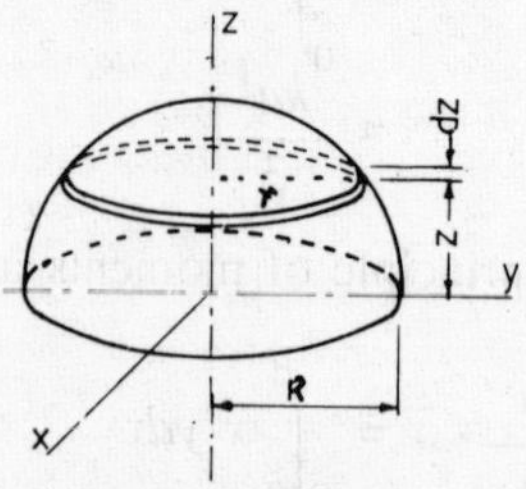

Fig. 4.44

Solution

With the X, Y and Z axes selected as shown in the figure, $\bar{x}$ and $\bar{y}$ are both equal to zero. To locate $\bar{z}$, consider the elementary volume of thickness dz at a distance z, from the xy plane. The elementary volume is in the form of a cylinder of base circle radius

$r^2 = R^2 - z^2$ and height dz.

Elementary volume $= \pi\,(R^2 - z^2)\cdot dz$

The CG of the elementary volume is at a distance z from xy plane.

Moment of the elementary volume about 'O' $= \pi\,(R^2 - z^2)\cdot z dz$.

We need the volume of the hemisphere to calculate $\bar{z}$. We integrate the expression for the elementary volume from limits $z = 0$ to $z = R$ to get the volume of the hemisphere.

$$\text{Volume of hemisphere} = \int_0^R \pi\,(R^2 - z^2)\,dz$$

$$= \pi\left[\left(R^2 z - \frac{z^3}{3}\right)\right]_0^R$$

$$= \frac{2}{3}\,\pi R^3.$$

Then from the principle of moments,

$$\frac{2}{3}\,\pi R^3 \times \bar{z} = \pi \int_0^R (R^2 - z^2)\cdot z\cdot dz$$

$$= \pi\left|\frac{R^2 z^2}{2} - \frac{z^4}{4}\right|_0^R$$

$$= \frac{\pi \times R^4}{4}$$

$$\bar{z} = \frac{\pi R^4}{4} \times \frac{3}{2\pi R^3}$$
$$= (3/8) R$$

Example 4.18
Locate the CG of the right circular cone shown in Fig. 4.45.

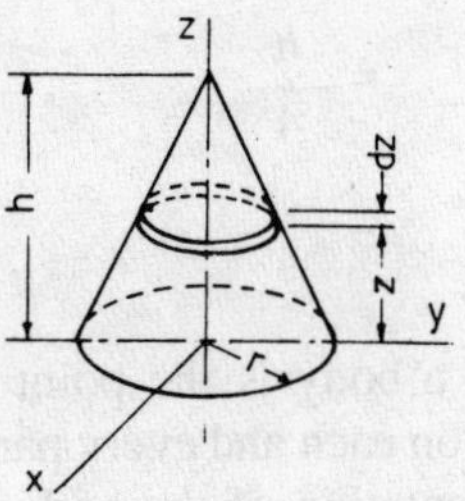

Fig. 4.45

Solution
With the coordinate axes selected as shown, *xz* and *yz* planes are planes of symmetry and, therefore, $\bar{y} = \bar{x} = 0$. Only $\bar{z}$ is to be calculated.
Taking an elementary strip of thickness *dz* at a distance *z* from the x-y plane, radius of strip $= \frac{r}{h}(h - z)$

$$\text{Volume of elementary strip} = \pi \times \frac{r^2}{h^2} (h - z)^2\, dz\cdot$$

$$= \int_0^h \pi \times \frac{r^2}{h^2} (h - z)^2\, dz$$

$$= \frac{\pi r^2}{h^2} \left| \frac{(h - z)^3}{3} \right|_0^h$$

$$= \frac{\pi r^2}{h^2} \times \frac{h^3}{3}$$

$$= \frac{1}{3} \pi r^2 h\cdot$$

To determine $\bar{z}$, we have,

$$\frac{1}{3} r^2\, h\, x\, \bar{z} = \int_0^h x\, \frac{r^2}{h^2} (h - z)^2\, zdz\cdot$$

$$= \frac{x\, r^2}{h^2} \int_0^h (h^2 z + z^3 - 2hz^2)\, dz\cdot$$

$$= \frac{\pi r^2}{h^2} \left| \frac{h^2 z^2}{2} + \frac{z^4}{4} - \frac{2hz^3}{3} \right|_0^h$$

$$= \frac{\pi r^2}{h^2} \cdot \frac{h^4}{12}$$

$$= \frac{\pi r^2 h^2}{12}$$

$$\bar{z} = \frac{\pi r^2 h^2}{12} \cdot \frac{3}{\pi r^2 h}$$

$$= \frac{h}{4}$$

Summary

The Centre of gravity of a body is the point at which the resultant of the gravitational forces acting on each and every particle of a body acts. This point is independent of the orientation of the body.

The Centroid of a plane area or lamina is analogous to the CG of a body. It can be defined as the point at which the whole area may be assumed to be concentrated. The centroid of an area need not lie necessarily within the area.

Location of CG (centroid) of a solid (an area) is done on the same principle as in the determination of the resultant of a system of parallel forces (areas). In locating the centroid the areas can be considered as weights acting along a chosen direction.

The CG of a body always lies in a plane of symmetry, if any, of the body. Similarly the centroid always lies on an axis of symmetry, if any, of an area.

Centroid of composite areas can be located by dividing the area into regular areas whose centroids are known and by determining the point of application of their resultant. This can be done analytically or graphically. Since an area is a plane figure, by determining the line of action of the resultant *areas* in two perpendicular directions, the centroid can be located.

The CG of compound solids can be located by dividing the solid into a number of regular solids whose CGs are known. The resultant of the gravitational forces can then be determined by assuming these forces to act along three mutually perpendicular directions (say, the X, Y and Z axes). The coordinates of CG can thus be located. In the case of homogeneous solids, weights can be replaced by volumes and are proportional to the gravitational forces.

Method of Integration is the most fundamental approach for locating the CG or centroid of many regular areas (volumes). The method is quite general and is the only method available in the case of volumes (areas) with regular but curved boundaries.

Review Questions

1 Give three examples from your every day experience where you have been consciously or unconsciously applying the concept of centre of gravity.

2. Distinguish between centre of gravity and centroid.

3 A plane area has two lines of symmetry. What can you say about its centroid?

4 In determining by the experimental method the centroid of a lamina, what principle of Statics is employed?

5 A body is symmetrical about two mutually perpendicular planes. What can you infer about its CG?

6 Give examples of: (a) an area with one line of symmetry
(b) an area with two lines of symmetry
(c) a body having one plane of symmetry
(d) a body having two planes of symmetry
(e) a body having three planes of symmetry.

In each case state what do you infer about the centroid or CG?

7 State whether the following statements are true or false.
(i) The centroid of a triangle always lies on a median
(ii) If an area has two lines of symmetry, the centroid lies at the intersection of these two lines
(iii) The CG is located by applying the principle of moments to a set of concurrent forces
(iv) The centroid of a plane area always lies on a line of symmetry.
(v) Any line passing through the centroid of an area must be a line of symmetry for that area.

8 Select the right answer from amongst those given in each of the following:
(i) The area shown in figure below is obtained by removing an area equal to the area of the rectangle of sides *a* and *b*. The centroid of this area lies approximately at
(a) point (1)
(b) point (2)
(c) point (3)
(d) point (4)

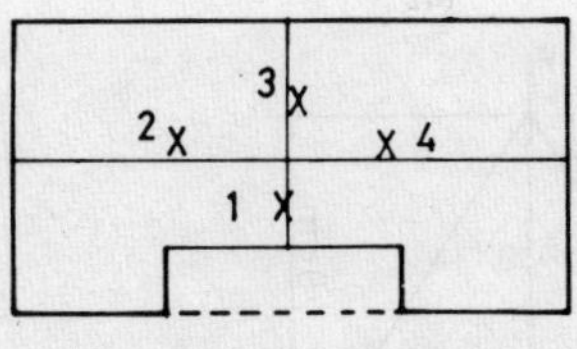

(ii) From a circle an area equal to quarter of a circle is removed as shown in figure below. The centroid of the remaining area is approximately at
(a) point (1)
(b) point (2)
(c) point (3)
(d) point (4).

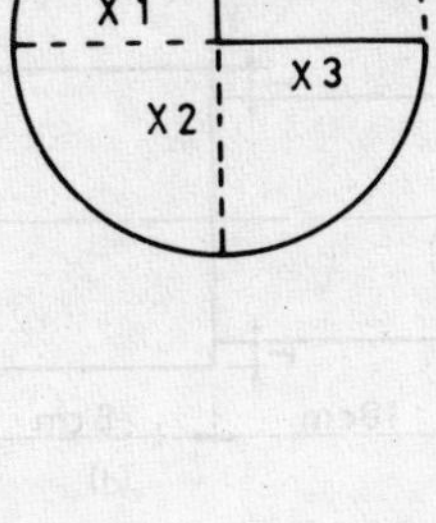

(iii) The CG of a cone is at $\frac{h}{4}$ from the base where h is the height of the cone. If a frustum of a cone is obtained by removing a conical part from the top, the CG of the frustum of the cone will be at

(a) $\frac{h}{4}$ from the base

(b) less than $\frac{h}{4}$ from the base

(c) more than $\frac{h}{4}$ from the base.

Problems

4.1 Locate the centroids of the plane areas shown in the Figure.

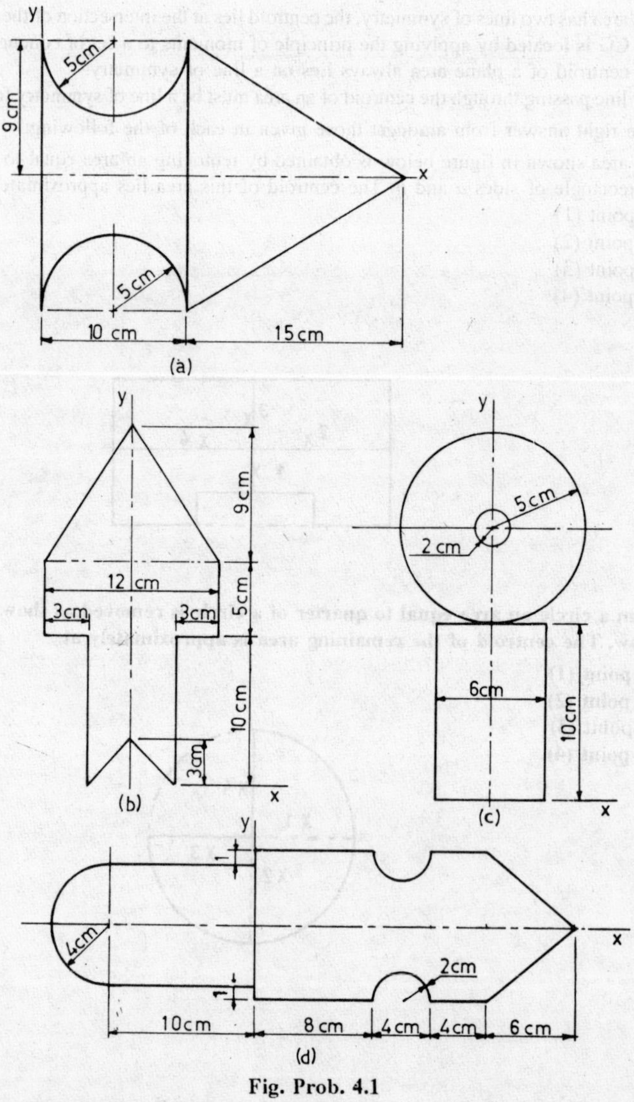

Fig. Prob. 4.1

Ans. (a) ($\bar{x}$ = 10.71 cm, $\bar{y}$ = 0, (b) $\bar{x}$ = 12.2 cm, $\bar{y}$ = 0
(c) $\bar{x}$ = 0, $\bar{y}$ = 10.24 cm (d) $\bar{x}$ = 3.88 cm, $\bar{y}$ = 0.)

4.2 Locate the centroids of the plane areas shown in the figure

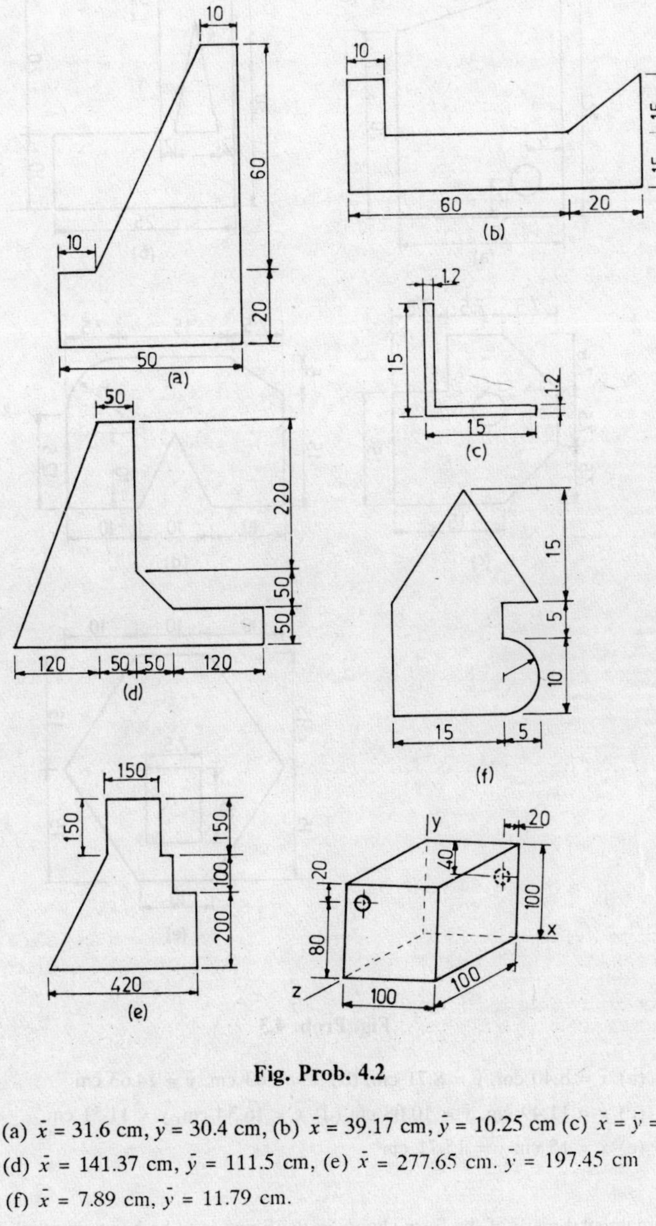

Fig. Prob. 4.2

Ans. (a) $\bar{x}$ = 31.6 cm, $\bar{y}$ = 30.4 cm, (b) $\bar{x}$ = 39.17 cm, $\bar{y}$ = 10.25 cm (c) $\bar{x}$ = $\bar{y}$ = 4.19 cm
(d) $\bar{x}$ = 141.37 cm, $\bar{y}$ = 111.5 cm, (e) $\bar{x}$ = 277.65 cm. $\bar{y}$ = 197.45 cm
(f) $\bar{x}$ = 7.89 cm, $\bar{y}$ = 11.79 cm.

4.3 Locate the centroids of plane areas shown in the figure.

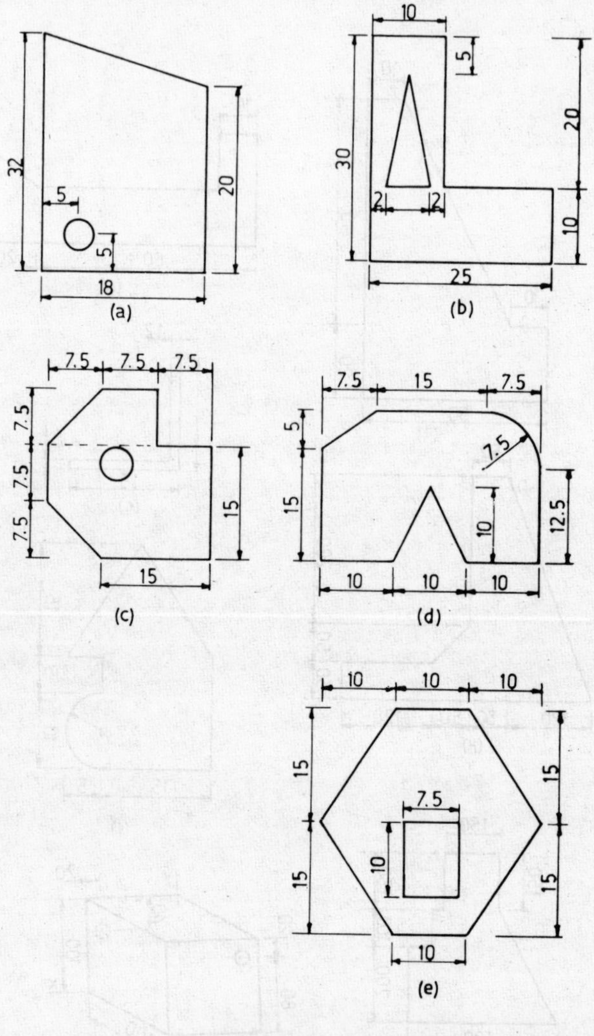

Fig. Prob. 4.3

Ans. (a) $\bar{x} = 8.40$ cm, $\bar{y} = 8.71$ cm, (b) $\bar{x} = 9.63$ cm, $\bar{y} = 14.63$ cm

(c) $\bar{x} = 11.49$ cm, $\bar{y} = 10.08$ cm (d) $\bar{x} = 16.34$ cm, $\bar{y} = 11.51$ cm,

(e) $\bar{x} = 15$ cm $\bar{y} = 15.71$ cm

4.4 An ornamental piece of the form shown in the figure is to be hung on a nail. Locate the point on its top face, where it should be hung if it is to remain horizontal.

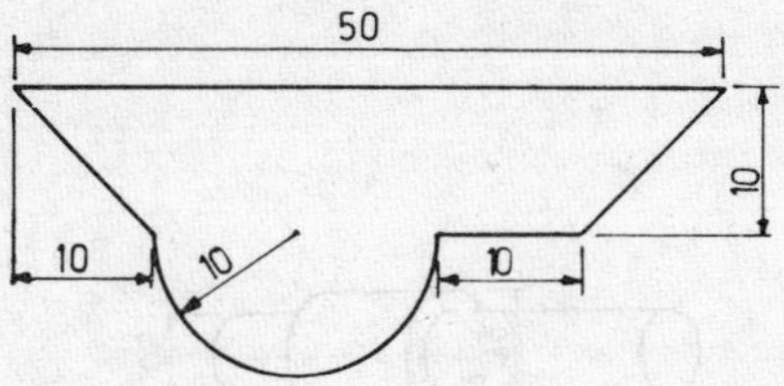

Fig. Prob. 4.4

Ans. 27.46 cm from the left end along the top.

4.5 From a rectangular steel plate, the area of a quarter of a circle has been removed as shown in the figure. Determine the dimensions of piece to be removed (triangle of sides *a* and *b* as shown) such that the centroid of the remainder is at the centre of the original rectangular plate.

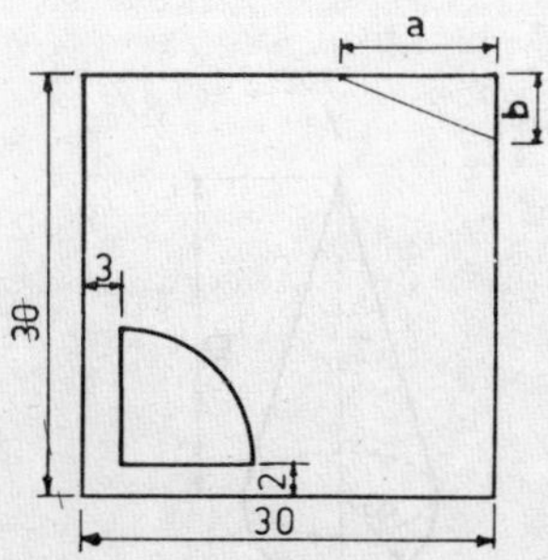

Fig. Prob. 4.5

Ans. $a = b = 2.0125$ cm.

4.6 Locate the CG of a steel rivet having a stem 25 cm in diameter 54 mm long and having a hemispherical head 40 mm in diameter.

Ans. 33.26 mm from the top of head.

4.7 Locate the CG of the body shown in the figure. The L-section is made of steel weighing 78.5 kN/m^3 and the quarter cylinder is made of wood weighing 5.8 kN/m^3.

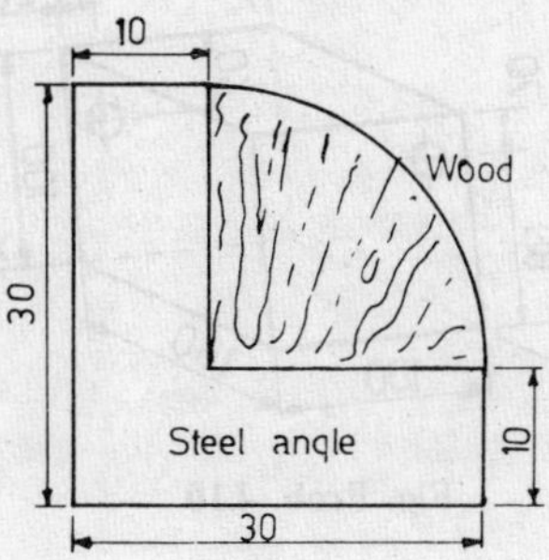

Fig. Prob. 4.7

Ans. $\bar{x} = 11.31$ cm, $\bar{y} = 11.31$ cm.

4.8 On a steel shaft 50 mm dia, a steel hub of 50 mm inner diameter and 80 mm outer diameter is fixed as shown in figure. Calculate the CG of the assembly.

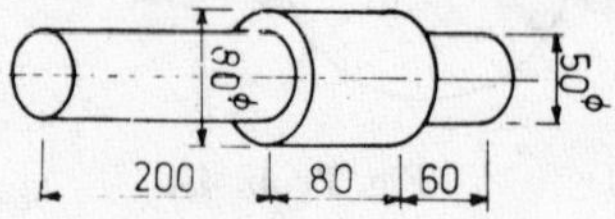

Fig. Prob. 4.8

Ans. $\bar{x}$ = 19.75 cm.

4.9 Locate the centroid of the solid shown in the figure with reference to the coordinate axes shown.

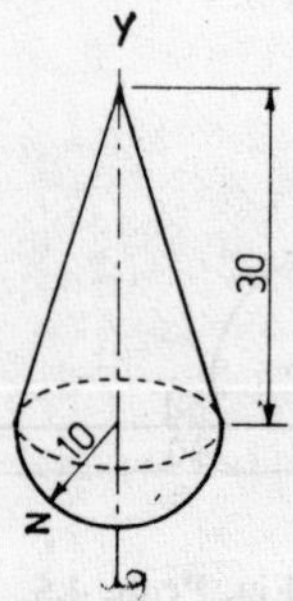

Fig. Prob. 4.9

Ans. $\bar{y}$ = 13 cm.

4.10 Locate the CG of the steel rectangular block shown with respect to the coordinate axes shown.

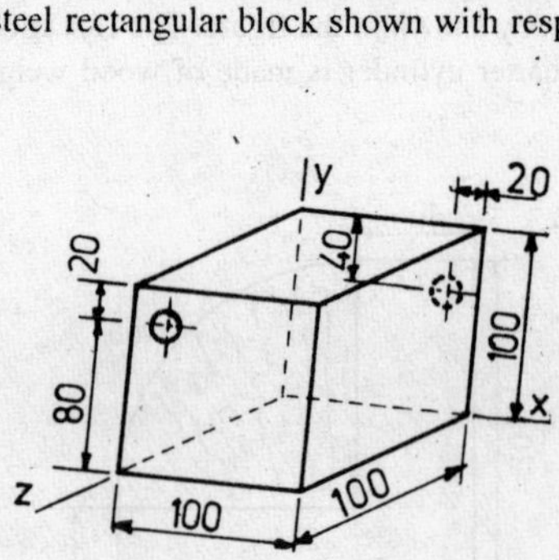

Fig. Prob. 4.10

Ans. $\bar{y}$ = 4 cm, $\bar{z}$ = 3.89 cm.

4.11 A concrete block 3 m long has the cross section shown in figure. It is to be lifted horizontally with a single hook provided in the concrete block. Determine the position of the hook along the X-axis.

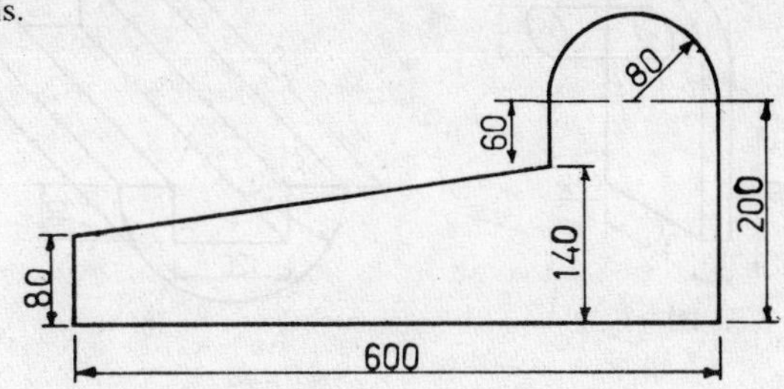

Fig. Prob. 4.11

Ans. $\bar{x}$ = 31.5 cm, $\bar{y}$ = 0, $\bar{z}$ = 9.73 cm.

4.12 Locate the CG of a metal can, 20 cm dia. and 30 cm high, open at the top and of negligible thickness.
Ans. $\bar{y}$ = 12.85 cm.

4.13 Locate the CG of a mallet having a stem 31 cm long and 4 cm diameter, and a head 12 cm dia. and 20 cm long.
Ans. 33.84 cm.

4.14 Locate the CG of the solids shown in figure.

4.15 A cylinder 60 cm dia is mounted on top of a hemispherical base 60 cm dia. Determine the permissible height *h* of the cylinder for stable equilibrium. Use the condition that for stable equilibrium, the CG of the compound body should lie within the hemisphere.
Ans. 23.87 cm.

4.16 Determine the coordinates of the centroid of the area shown in the figure.

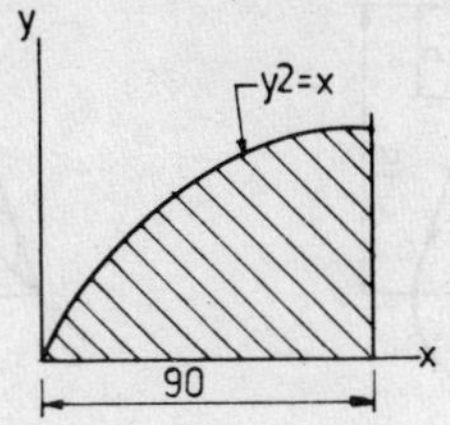

Fig. Prob. 4.16

Ans. (5.4, 1.125)

4.17 Determine the coordinates of the centroid of the area shown shaded in the figure.

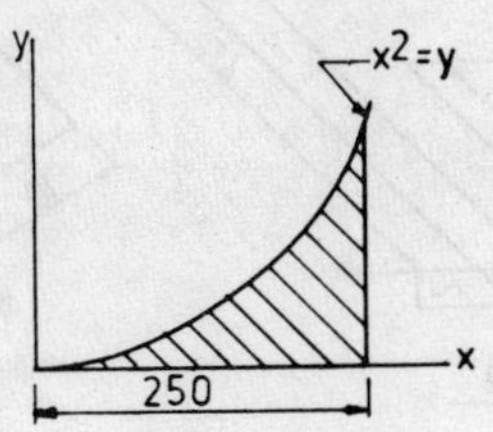

Fig. Prob. 4.17

Ans. (3.75, 7.5)

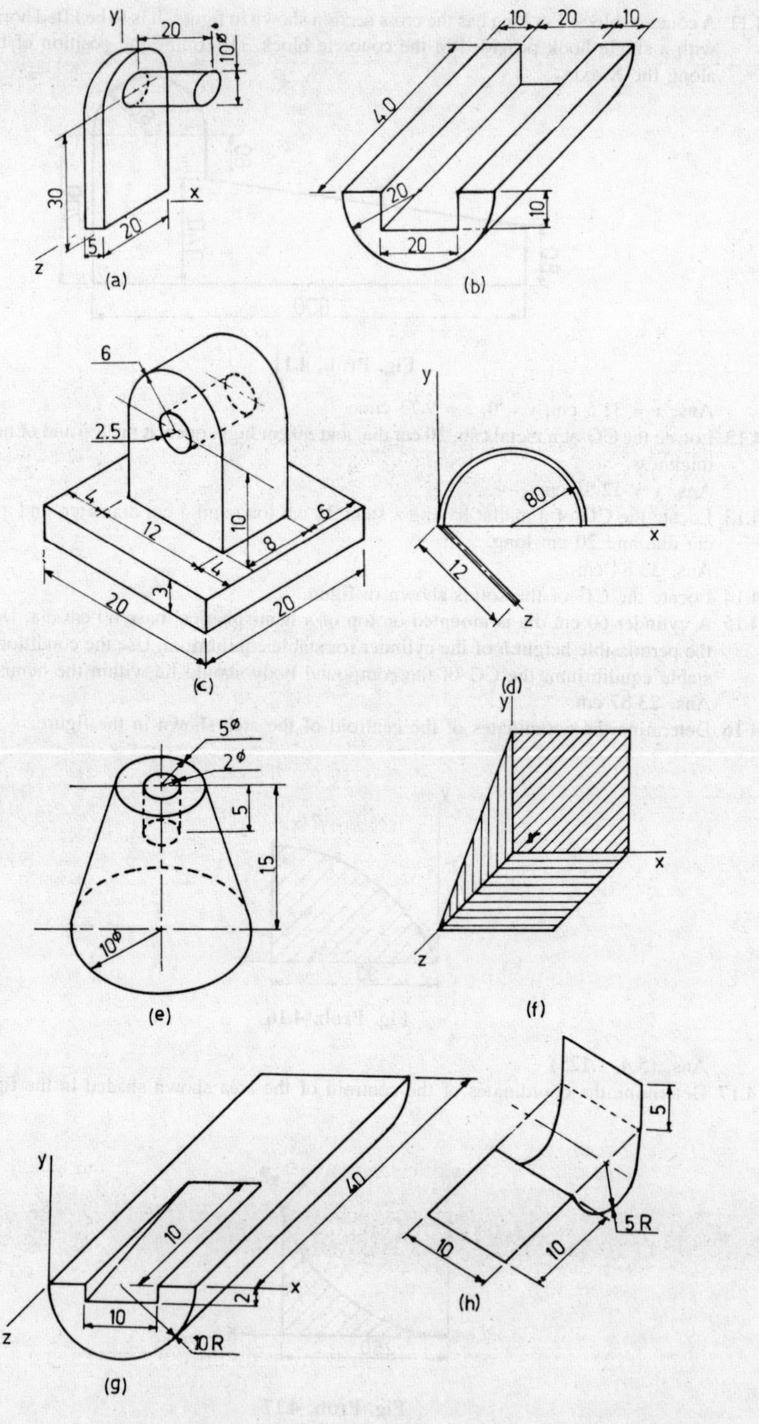

Fig. Prob. 4.14 *(Contd.)*

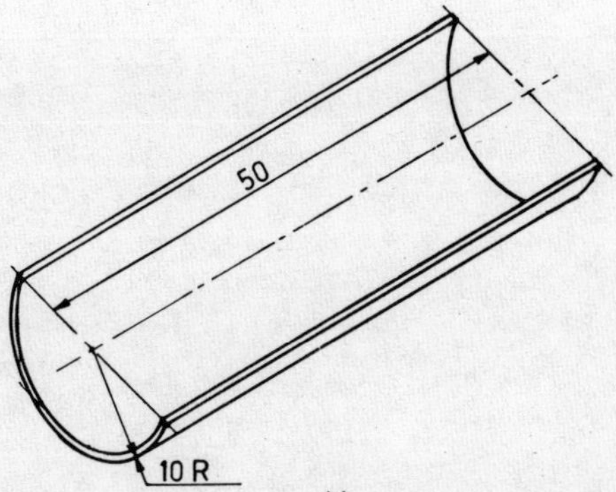

(i)

Fig. Prob. 4.14

Ans. (a) (3.28, 13.84, 0) (b) (0, −10.12, 0) (c) (0, 7.86, 0) (d) (5.41, 3.44, 1.44)
(e) (0, 5.74, 0) (f) (4, 8/3, 8/3) (g) (0, −4.41, 19.33) (h) (0, −1.22, 7.44)
(i) (0, 5.4, 22.73)

4.18 Determine the coordinates of the centroid of the area shown shaded in figure.

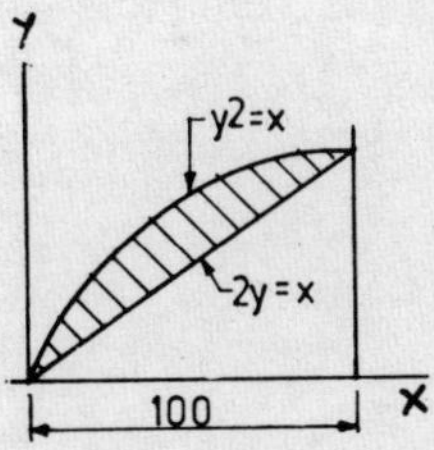

Fig. Prob. 4.18

Ans.

5. FRICTION

After having mastered the contents of this chapter, you will be able to:

- Define the term 'friction'
- Explain, giving illustrative examples, the importance of friction in our daily lives.
- Explain the terms 'fluid friction' and dry friction
- Explain the characteristics of frictional force.
- State and explain the 'laws of friction'
- Differentiate between 'static' and 'Kinetic' friction.
- Explain the concept of 'cone of friction'.
- Explain the term 'coefficient of friction'.
- Apply the concept of friction and solve problems involving frictional forces.

CHAPTER 5

Friction

5.1 Introduction

In the foregoing chapters, while solving many problems, we have assumed the contact surfaces as smooth, and consequently the reactions at points of contact were taken as normal to the surfaces. This assumption can be seen to be quite invalid in actual practice, and there is hardly any surface that can be termed 'smooth'. Let us take a specific example as shown in Fig. 5.1(a). A block of weight '*W*' is resting on a horizontal surface. The free body diagram of this block is shown in Fig. 5.1(b).

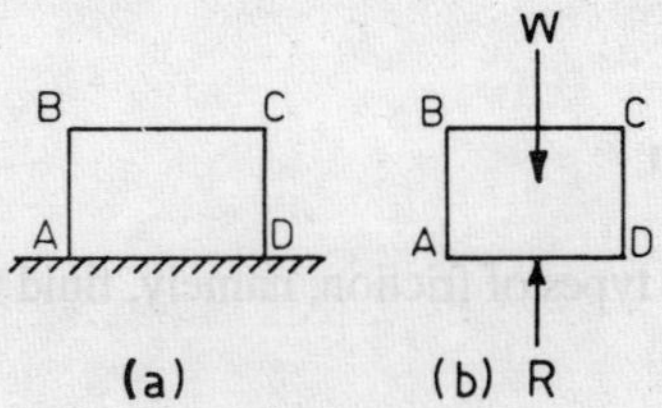

Fig. 5.1 (a) and (b)

No horizontal force is acting on the body as is clear from its free body diagram. In such a case, we may say that any small horizontal force applied to the block should move it (neglecting air resistance, etc.). However, it is everyday experience that a considerable force is needed to move the block. The heavier the block, the larger is the horizontal force needed to move it. The reason for this is that tangential forces are developed at the surface of contact between the block and the horizontal surface. These resist the motion of the block. So long as the body does not move, these forces are equal to the applied horizontal force. At a certain value of the horizontal force, the body moves. This indicates the limitation of the amount of tangential forces that can be developed.

The development of tangential forces between contacting surfaces is due to a phenomenon known as 'friction'. In this Chapter, we will study the phenomenon of friction in detail and solve some interesting problems involving frictional forces.

5.2 Definition of Friction

When two surfaces are in physical contact and there is relative motion between the two, or a tendency to move relatively, it is found that there is a force acting tangential to the contact surfaces. This tangential force always acts in a direction opposed to that of motion. This phenomenon is known as friction, and the force so developed is called the frictional force, or simply friction. We, can, therefore, define friction as follows

Friction is the contact resistance developed between the contacting surfaces when one of the bodies moves, or tends to move, over the other. It always opposes the motion or tendency to move.

5.2.1 *Importance of Friction*

The phenomenon of friction is very important in many engineering applications. Frictional forces exist in nature throughout and, in fact, we would not have been able to even walk without friction.

Friction is rightly called "the necessary evil". In many engineering applications, such as bearings and gears (as in an automobile) and the flow of fluids in pipes, friction is quite undersirable. In many other engineering applications, such as brakes and clutches (as in an automobile) and in belt drives (as in a 'Chakki' for grinding) friction is quite necessary, as their working depends on friction.

5.3 Nature of Friction

There are essentially two types of friction, namely, fluid friction and dry friction.

5.3.1 *Fluid Friction*

When a lubricating fluid is introduced between the contact surfaces of two bodies, fluid friction is developed. When there is motion between the two surfaces, the frictional forces developed depend upon the properties of the fluid. A simple example is the oiling of door hinges to avoid creaking sounds and the oiling of a sewing machine. The fluid is introduced *to reduce* friction and *to aid* motion. In all machine parts, where friction reduces the efficiency or usefulness of a machine, lubrication is employed to decrease friction. An analysis of fluid friction depends upon the fluid properties and is beyond the scope of this book. It is, however, mentioned here due to its importance in many engineering applications.

5.3.2 *Dry Friction*

Dry Friction or Coulomb friction, as it is commonly called, is the phenomenon when the contacting surfaces are dry and there is a tendency for relative motion. The principles of dry friction were first developed by Coulomb in the year 1781, and later by Morin, who published his results in the year 1831.

5.4 Theory of Dry Friction

To understand the theory of dry friction, let us first study a simple experiment which the reader can himself perform with very simple apparatus.

The experimental set up is shown in Fig. 5.2(a). A block of weight '*W*' is resting on a horizontal surface. A horizontal force 'P' is applied to this block and the behaviour of the block is observed as the force 'P' is continuously increased.

The freebody diagram of the block is shown in Fig. 5.2(b). As can be easily seen, so long as the body is in equilibrium, the conditions of static equilibrium require that the force '*P*' and the frictional force between the contacting surfaces should be equal. As the value of '*P*' is continuously increased it will be found that the body will start moving for a particular value of *P*. This value of '*P*', at which the body starts moving, is the minimum horizontal force 'P_s' needed to *just move* the block. We can prove with the same experimental set up that once the body starts moving, a force P_k, smaller than P_s, is sufficient to keep the body moving uniformly.

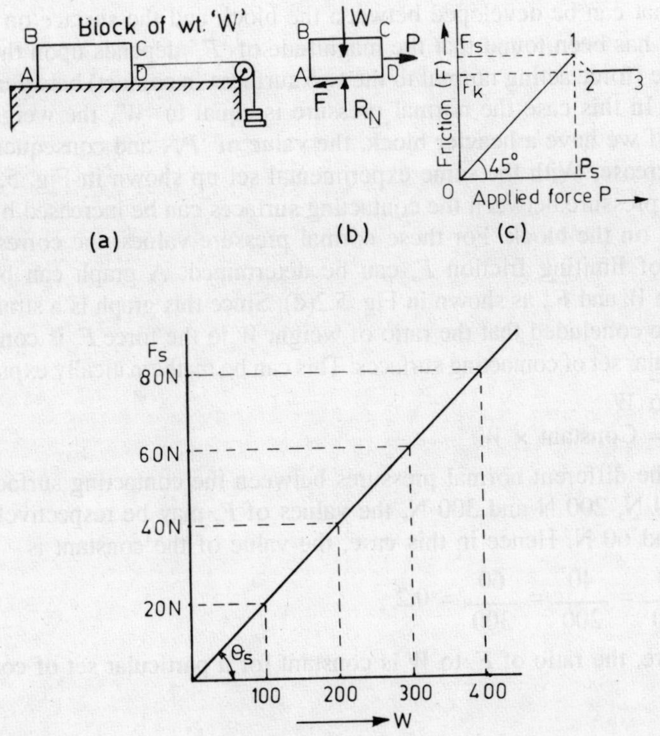

Fig. 5.2 (a, b, c, d)

The results of such an experiment are shown in Fig. 5.2(c). This, of course, is not necessarily very accurate, but helps us to understand the nature of friction.

You will see from the graph that so long as the body is in equilibrium, the frictional force is equal in magnitude (also opposite in direction) to 'P', and continuously increases with 'P'. Hence the line 0-1 in the graph is inclined at an angle of 45°. When the body starts moving, the frictional force slightly decreases and the body continues to move. The horizontal portion '2-3' of the graph represents this case. Between the points 1 and 2, the curve is uncertain and is shown by a dotted line. The straight portion 0-1 of the graph is the zone of static friction because the body is not moving and the portion beyond point 1 is known as the zone of kinetic friction (because the body is in motion).

From the graph shown in Fig. 5.2(c), we can see that during the range of static friction, the frictional force 'F' is equal to the applied force 'P' and can be determined from the conditions of static equilibrium. However, there is a maximum value of 'P' beyond which the block is no longer in equilibrium. At this value of P, the body starts moving and this limiting value of P is shown as 'P_s' in the graph.

5.4.1 *Static Friction*

The frictional force 'F_s' corresponding to 'P_s' is the maximum frictional force that can be developed between the block and the surface on which it rests. It has been found that the magnitude of 'F_s' depends upon the normal pressure (force acting normal to the two surfaces in contact) between the two bodies. In this case the normal pressure is equal to 'W', the weight of the block. If we have a heavier block, the value of 'P_s', and consequently 'F_s', will increase. With the same experimental set up shown in Fig. 5.2(a), the normal pressure between the contacting surfaces can be increased by adding weights on the block. For these normal pressure values, the corresponding values of limiting friction F_s can be determined. A graph can be drawn between W and F_s, as shown in Fig. 5.2(d). Since this graph is a straight line, it can be concluded that the ratio of weight W to the force F_s is constant for a particular set of contacting surfaces. This can be mathematically expressed as,

$$F_s \propto W$$

or $F_s = \text{Constant} \times W$

If the different normal pressures between the contacting surfaces were say, 100 N, 200 N and 300 N, the values of F_s may be respectively 20 N, 40 N and 60 N. Hence in this case, the value of the constant is

$$\frac{20}{100} = \frac{40}{200} = \frac{60}{300} = 0.2$$

Therefore, the ratio of F_s to W is constant for a particular set of contacting surfaces.

i.e. $\dfrac{F_s}{W} = \text{constant}$

or $F_s = \mu_s W$

where μ_s represents the constant and is called the *coefficient of static friction*.

5.4.2 *Kinetic Friction*

When the block starts moving, we find that the frictional force F_k remains fairly constant. It has been found that this force 'F_k' developed during the motion of the block also depends upon the normal pressure between the two contacting surfaces (which in Fig. 5.2 is equal to 'W'). If the block is heavier a larger value of 'P_k' is required to keep it moving. The frictional force 'F_k' also would be larger; but still it is found to be proportional to 'W'. A relationship similar to that of static friction is applicable in this case.

Thus,

$$F_k \propto W$$

or $$\frac{F_k}{W} = \text{Constant } (\mu_k)$$

or $$F_k = \mu_k \cdot W$$

where μ_k is called the *coefficient of Kinetic friction.*

The value of μ_k is found to be less than μ_s for the same pair of contact surfaces. In general, we can say that the maximum available frictional force is proportional to W. The coefficient of friction to be used (μ_k or μ_s) depends upon the condition of the *Contacting Surfaces* (that is, whether these are in relative motion, or at rest).

5.4.3 *Cone of Friction*

Let us analyse the block in the experimental set up of Fig. 5.2(a) from a different point of view.

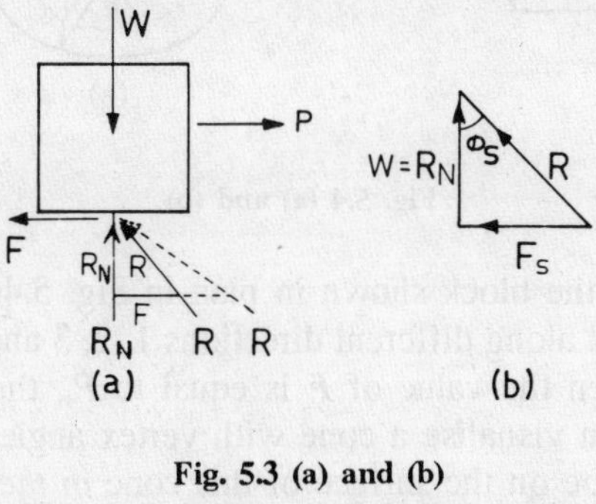

Fig. 5.3 (a) and (b)

In Fig. 5.3(a) is shown the freebody diagram of the block. We find that the reaction of the horizontal surface on the block is composed of two parts, namely, the horizontal frictional force 'F' and the normal reaction 'W'. These two can be combined by the parallelogram law, and the total reaction (R) can be obtained. This is shown in Fig. 5.3(a).

As the value of 'P' is increased, the value of 'F' increases while the value of '*W*' remains constant. Thus, it can be seen that the total reaction '*R*', and its inclination to the vertical axis increases due to an increase in the value of '*F*'. This is shown in Fig. 5.3(a) by the triangle shown in the dotted lines.

However, *F* cannot increase indefinitely. The maximum, value of '*F*' is 'F_s' corresponding to the value of 'P_s', and is given by $\mu_s \cdot W$. This condition is shown in Fig. 5.3(b). At this stage the frictional force is equal to F_s, and the angle between the total reaction '*R*' and '*W*' is equal to 'ϕ_s'. ϕ_s is known as the *angle of static friction*. Any increase in the value of '*P*' beyond 'P_s' will cause the motion of the block. The block at this stage is said to be in a state of *impending motion.*

The value of angle of static friction 'ϕ_s' can be calculated as,

$$\tan \phi_s = \frac{F_s}{R_n} = \frac{F_s}{W} = \mu_s.$$

That is, the angle of static friction is such that its tangent has a value equal to the coefficient of friction. Note carefully that the relationship $\mu_s = \tan \phi_s = \frac{F_s}{W}$, applies *only in the case of impending motion.* When the value of frictional force is less than 'F_s', the angle ϕ and the frictional force must be *calculated from* the conditions of quilibrium.

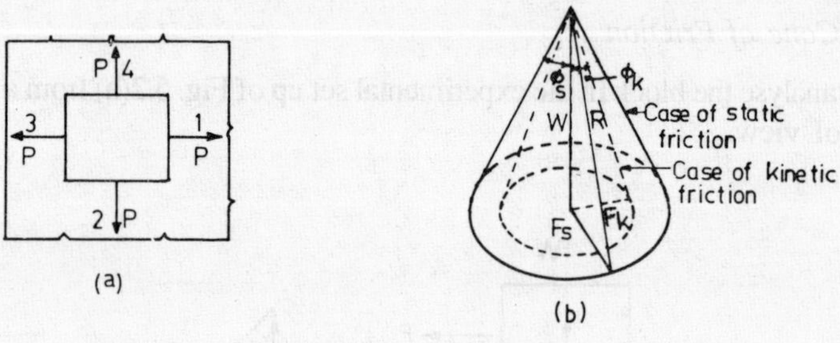

Fig. 5.4 (a) and (b)

Now, consider the block shown in plan in Fig. 5.4(a). Here, the force '*P*' is shown applied along different directions 1, 2, 3 and 4 alternatively. In all these cases, when the value of *F* is equal to P_s, the frictional force is equal to F_s. One can visualise a cone with vertex angle equal to 2 ϕ_s. The total reaction must be on the surface of this cone *in the case of impending motion.* The line of action of the total reaction is the generator of this cone which is obtained by revolving this line about the normal reaction line (of '*W*'). Such a cone is known as the *cone of static friction* and is shown in Fig. 5.4(b).

It can be immediately seen from the above discussion, that the cone of static friction signifies two points. As mentioned before, in the case of impending motion the total reaction lies on the surface of this cone. When the frictional force is less than the maximum available friction 'F_s', the total reaction lies within this cone.

One can similarly visualise a cone of Kinetic friction as shown in dotted lines in Fig. 5.4(b). This is a cone with a vertex angle 2 ϕ_k, where ϕ_k is the *angle of kinetic friction* given by $\tan \phi_k = \mu_k = \frac{F_k}{W}$. Since μ_k is less than μ_s for the same pair of contact surfaces, the cone of Kinetic friction lies within the cone of static friction.

5.5 Laws of Friction

On the basis of a large number of experiments conducted, the following seven laws of friction have been derived. These hold good when the contacting surfaces are dry.

1. **Friction always acts to oppose the motion of the free body (or its tendency to move). Frictional force acts tangential to the surfaces in contact.**

As we have seen before, this is always true. Frictional force comes into play only when the bodies have a tendency to move relatively. When there is no force acting, tending to move the bodies relatively, frictional force is also zero. Since the bodies move tangential to each other and not towards each other, we may say that the frictional force also acts tangential to the surfaces in contact.

2. **If static friction is acting, the value of the frictional force may vary from zero to the maximum available value, depending upon the resultant force tending to cause motion.**

We have seen from the experimental results in art. 5.4 that the frictional force is always equal to the force tending to cause motion and should satisfy the conditions of static equilibrium. However, there is a limit to the amount of frictional force that can be developed between two surfaces in contact, and the maximum value is given by $F_s = \mu_s \cdot W$. It should be carefully noted here that $F_s = \mu_s \cdot W$ is applicable only when motion is impending, and that F_s is the maximum frictional force available between the surfaces in contact. In cases where motion is not impending, frictional force must be equal to the force or the component of the force tending to cause motion.

3. **The maximum value of static friction (i.e. the limiting friction when motion is about to take place) is equal to $\mu_s \cdot W$, where μ_s is the coefficient of static friction and W is the normal pressure between the surfaces.**

This law is quite important because it relates the frictional force F_s to the normal pressure. As the normal pressure is increased, the maximum value of static friction also increases. It is a common experience that when you have to move a heavy body, it is advantageous to apply a pull at a slight inclination to the horizontal direction. (Fig. 5.5). The effective horizontal force tending to cause motion is, of course, decreased but at the same time the normal pressure between the surfaces is also decreased due to the vertical component of the applied force. The reader can try this by trying to move a heavy object by applying a force horizontally, and also in a direction which is slightly inclined upwards.

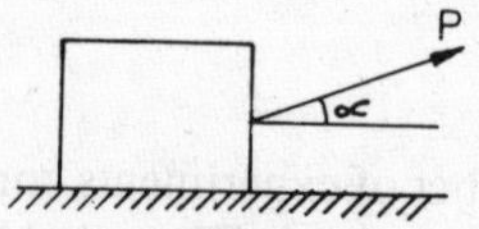

Fig. 5.5

4. The maximum available friction force F_s is independent of the area of surfaces in contact.

This law states that the maximum friction force developed does not depend upon the area of the surfaces is in contact. This can be explained as follows.

Strictly speaking no surface is actually smooth. The many imperfections of a surface can perhaps be not identified by the naked eye, but can be easily seen using a high magnification microscope. What appears to be a highly polished surface to the naked eye, will have a number of 'hills and valleys' as shown in Fig. 5.6. When two such surfaces are in contact, actual contact area is much less than the apparent surface area. The contact area is actually only that which is due to the meshing of the 'hills and valleys' of one surface with those of the other. The pressures at these points of contact are quite high, and at these points there is a fusion of the particles of one surface with those of the other. Such fusion is technically known as a 'cold weld'. When the normal pressures between the surfaces increase, more 'hills and valleys' mesh together and more 'cold welds' are formed.

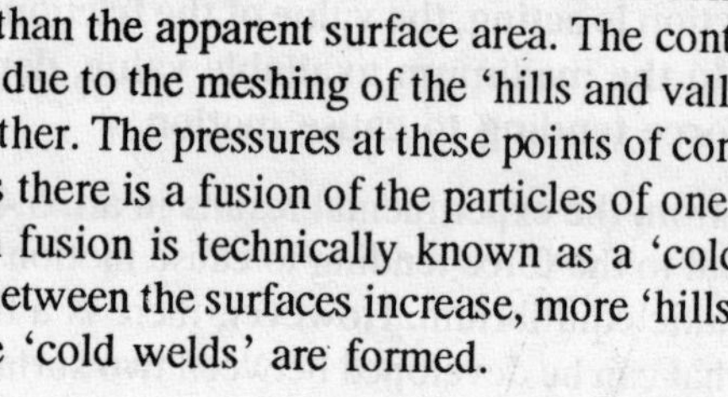

Fig. 5.6

When one body tends to move over another, such 'welds' formed due to high contact pressures have to be broken before motion is possible. The frictional force developed is only at these contact points, and has to be overcome for motion to take place. This also explains the dependence of the maximum frictional force on the normal pressure between the surfaces. The more the normal pressure, the more are the 'welds' or points of contact between the surfaces, and the more is the amount of frictional force developed.

Now, whatever is the *apparent* surface area, the actual contact area depends only on the normal pressure. Therefore, even if the surface areas of two sets of bodies are different, keeping all other characteristics the same, the actual contact area will be the same and consequently the frictional force developed will be the same. Therefore, the frictional force developed is not dependent upon the surface area.

5. If kinetic friction is acting, the frictional force is constant at its limiting value.

Once the motion of one body over another has started, as we have seen from Fig. 5.2(c), the frictional force slightly decreases, but remains at a constant value. While the body is continuously moving, the meshing of 'hills and valleys' of one surface with those of the other is not as effective as they are when at rest. The frictional force consequently decreases.

Actually, when the body starts moving, at low velocities, the kinetic friction is nearer to the static friction value and at high velocities it decreases rapidly to a somewhat constant value. The same is the case (but reversed in order) when the body comes to rest from its motion.

6. The kinetic friction is equal to $\mu_k \cdot W$, where μ_k is the coefficient of kinetic friction and W is the normal pressure.

As we have already explained in previous laws, the kinetic frictional force is slightly less than the maximum static friction. μ_k, therefore, is slightly less than μ_s.

7. The angle between the total reaction and its normal component (normal pressure) is called the angle of friction. The tangent of this angle is known as the coefficient of friction.

This has already been discussed in art. 5.4.3. The relationship is expressed as $\tan \phi = \mu$. It should be carefully noted that the above relationship holds good only at the maximum static friction value, i.e., when motion is impending.

The above 7 laws form the basis of the principles of dry friction. These laws can be applied so long as the normal pressure is not too great. For very high pressures surface characteristics are changed and the frictional coefficient changes.

5.6 Coefficient of Friction

As has already been stated, the coefficient of friction is the ratio of the amount of limiting frictional force developed to the normal pressure between the contacting surfaces. There are two coefficients of friction, namely μ_s and μ_k. The coefficient of kinetic friction, μ_k, is slightly less than the coefficient of static friction, μ_s.

It is also easily understood that if two surfaces in contact are rough, the coefficient of friction is consequently high. Thus the friction coefficient measures the roughness between a pair of contacting surfaces and incorporates the geometric property of the mating contours. Therefore, it is meaningless to speak of a friction coefficient for a single surface.

Table 5.1 gives a set of values of static friction coefficients for pairs of contacting surfaces. These values are quite approximate and should be cautiously used in practical circumstances. However, for the purpose of discussion in this chapter, these values may be taken to be valid.

Table 5.1 *Values of coefficients of Static Friction*

Surfaces	*Coefficient of Friction*
1. Dry masonry on Brick work	0.6–0.7
2. Masonry on dry clay	0.5
3. Masonry on wet clay	0.3
4. Earth on Earth	0.25–1.0
5. Timber on Stone	0.4
6. Timber on Timber	0.2
7. Timber on metals	0.2–0.6
8. Iron on stone	0.3–0.7
9. Metals on Metals (Dry)	0.15–0.25

5.7 Solution of Problems Involving Frictional Forces

Problems involving frictional forces generally fall into three categories. The method of solution in each case is also different.

In the first types of problems, the body is in equilibrium and motion is not impending. Such problems can be solved by taking into account the frictional force and other external forces and applying the conditions of equilibrium. It is to be noted, however, that the frictional force is less than the limiting static friction, and hence the coefficient of friction is not to be used in calculating the frictional forces.

In the second type of problems, the condition of impending motion is to be investigated. The frictional force is, therefore, equal to the limiting static friction.

In the third type of problems, there exists relative motion between the contacting surfaces, and hence the coefficient of kinetic friction applies.

The following examples illustrate the method of solution.

Example 5.1

A block weighing 10 KN is kept on a rough horizontal surface. The coefficient of static friction between the block and the surface is 0.4. What minimum horizontal force is required to start the block moving?

Solution

The problem is shown diagramatically in Fig. 5.7(a). It is clear from the problem that it is a case of limiting static friction.

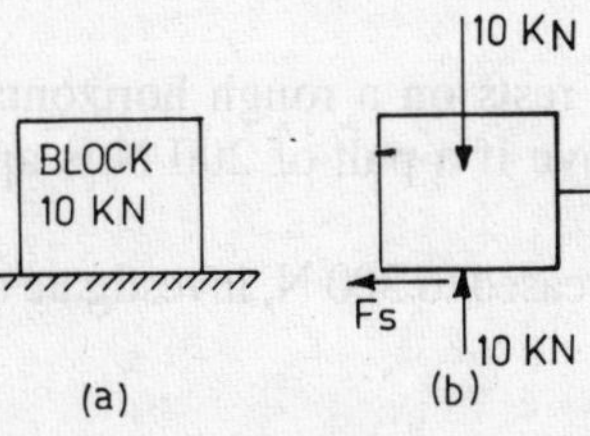

Fig. 5.7

Let P be the horizontal force applied. The normal reaction R_N is equal to 10 KN.

Limiting static friction = $\mu_s \cdot R_N$

= 0.4 × 10 = 4 KN.

From the free body diagram shown in Fig. 5.7(b), it is clear that when motion is impending $P = F_s$.

Therefore, the minimum value of P required to start the block is F_s.

∴ Minimum horizontal force required = **4 KN**

Example 5.2

A block weighing 400 N is resting on a horizontal plane. Determine the force P required to be applied at 30° to the horizontal as shown in the figure to cause motion to impend. μ_s = 0.40

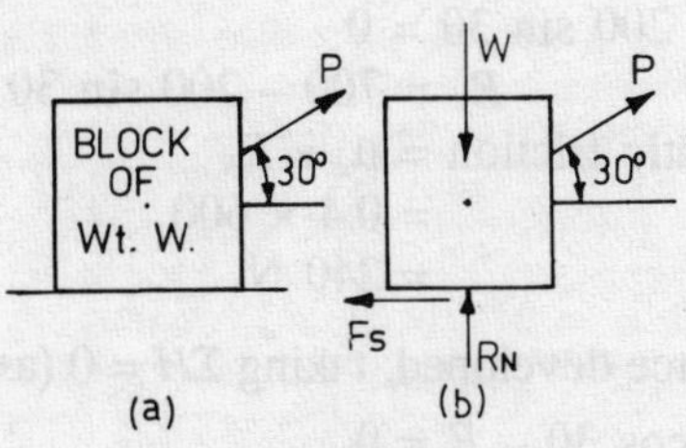

Fig. 5.8

Solution

When motion is impending, the horizontal component of the force P is equal to the limiting static friction. From the freebody diagram,

$$\text{Normal reaction } R_N = 400 - P \sin 30$$

$$\text{Limiting static friction} = 0.4 \times R_N = 0.4(400 - P \sin 30).$$

Now, horizontal component of P = Limiting static friction when motion is impending.

$$\therefore \quad P \cos 30 = 0.4(400 - P \sin 30)$$

$$p = \mathbf{107\ N}$$

Example 5.3

A body weighing 700 N rests on a rough horizontal surface. If $\mu_s = 0.4$,

(a) will the body move if a pull of 200 N is applied to it at 30° to the horizontal?

(b) If this force is increased to 300 N, investigate the condition of the body.

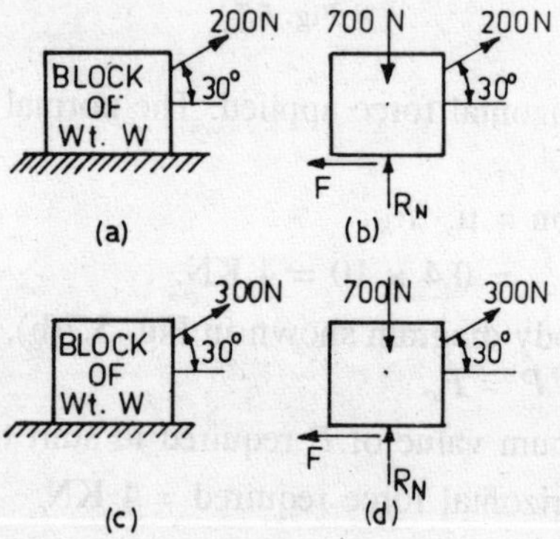

Fig. 5.9

Solution

(a) The condition is as shown in Fig. 5.9(a) and

(b) Since the body does not move vertically $\Sigma V = 0$

$$\therefore \ R_N - 700 + 200 \sin 30 = 0$$

$$R_N = 700 - 200 \sin 30 = 600 \text{ N}$$

$$\text{Limiting static friction} = \mu_s \times R_N$$

$$= 0.4 \times 600$$

$$= 240 \text{ N}$$

If F is the frictional force developed, taking $\Sigma H = 0$ (assuming equilibrium),

$$200 \times \cos 30 - F = 0$$

$$F = 173.2 \text{ N}$$

Since this value of F, (actual frictional force developed) is less than the limiting static friction, the body will be at rest.

(b) The condition is shown in Fig. 5.9(c) and (d)

$$\text{Again taking } V = 0$$

$$R_N - 700 + 300 \sin 300 = 0$$

$$R_N = \mathbf{550\ N}$$

Limiting static friction $= \mu_s \times R_N = 0.4 \times 550$

$= \mathbf{220\ N}$

Now, assuming equilibrium and equating horizontal forces,

$$300 \cos 30 - F = 0$$

$$F = 259.8 \text{ N}$$

Since this force is more than the limiting static friction, the body is not in equilibrium and will move.

5.8 Moment of Frictional Force

Let us consider a block resting on a rough horizontal surface. The free body diagram of the block is shown in Fig. 5.10.

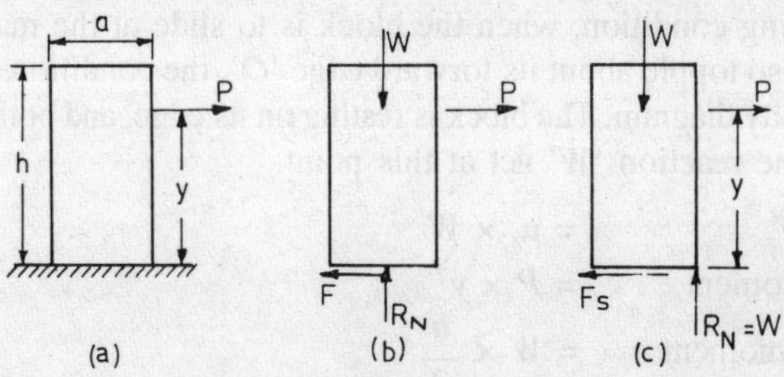

Fig. 5.10

From the free body diagram, it is clear that *P* and *F* are two equal, opposite and non-collinear forces. They, thus, form a couple and tend to rotate the body.

If the distance between *P* and *F* increases, by applying the force *P* at higher levels on the block, the moment of the couple also increases. However, when the block is about to rotate the conditions are as shown in the Figure 5.10(c).

In this case *P* and *F* form a couple that topples the body, and *W* and R_N (= *W*) form a couple that prevents this rotation. For a given block and surface conditions, there is a maximum height at which the force *P* can be applied without toppling the body. If the height '*y*' is more than this, the body topples and if it is less, the block slides. This condition is investigated in the problems below.

Example 5.4

A block of height '*h*' and weight *W* is resting on a horizontal surface on its square base of side '*a*'. A force *P* is applied horizontally to the block. Determine the maximum height at which the force *F* can be applied, if the block is to slide without toppling. Take the coefficient of friction as μ_s.

Solution

From the problem it is clear that the condition of impending motion is to be investigated. The conditions are shown in Fig. 5.11(a).

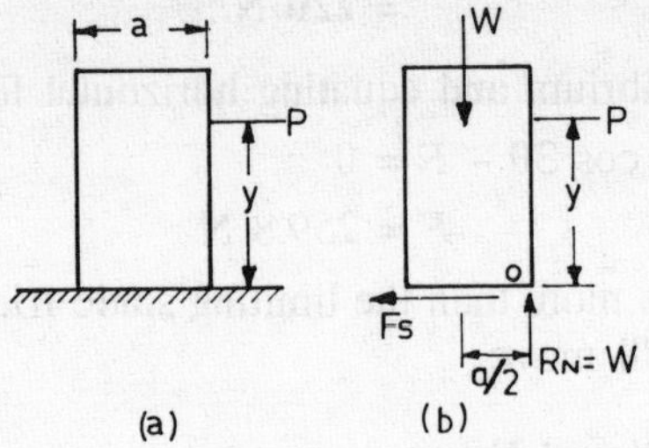

Fig. 5.11 (a) and (b)

Let the force be applied at a height y. It is easy to see that as the height 'y' increases P tends to topple the block, and 'W' tends to prevent this toppling. In the limiting condition, when the block is to slide at the maximum value of 'y' and also topple about its forward edge 'O', the conditions are as shown in the freebody diagram. The block is resting on its edge, and both the frictional force and the reaction 'W' act at this point.

The force F_s $= \mu_s \times W$

Toppling moment $= P \times y$

Stabilising moment $= W \times \dfrac{a}{2}$

For equilibrium in the limiting condition, taking moment about O and setting

$$\Sigma M = 0,$$

$$P \times y - W \times \frac{a}{2} = 0$$

$$y = \frac{W}{P} \cdot \frac{a}{2} = \frac{W}{\mu_s \cdot W} \frac{a}{2} \quad (\because P = F_s = \mu_s \cdot W)$$

$$y = \frac{a}{2\,\mu_s}$$

This is the maximum height 'y' at which P can be applied to slide the block without toppling.

Note: From $y = \dfrac{a}{2\,\mu_s}$, $\mu_s = \dfrac{a}{2\,y}$. This gives us a method of determining the coefficient of friction. Take a block of sufficient height 'h' as given in the problem above, and apply a horizontal force P at some height. The block will slide or topple depending upon this height. By trials, we can find the height 'y' at which the block slides and also tends to topple. Half the base width in the direction of motion divided by this height gives the coefficient of static friction.

Example 5.5

A block has dimensions of 10 × 20 × 25 cm. It is resting on a 10 × 20 cm face on a rough horizontal suface. The value of $\mu_s = 0.3$. A continuously increasing force is applied along the top edge of the block (a) parallel to the 10 cm edge, and (b) parallel to the 20 cm edge. Determine whether the block will slide or topple eventually in both the cases.

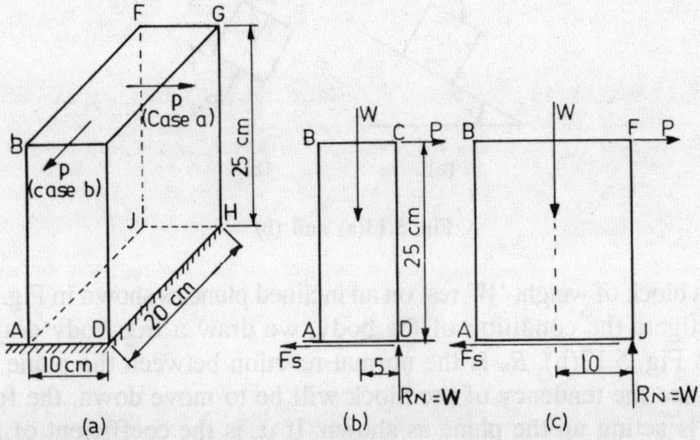

Fig. 5.12

Solution

(a) The condition is shown in Fig. 5.12(b), when the block is about to slide and topple. It is clear from the diagram that

The Rotating couple $= P \times 25$

and The Stabilishing couple $= W \times 5$

Since P $= \mu_s \times W = 0.3\ W,$

Rotating couple $= 0.3\ W \times 25 = 7.5\ W.$

Since the rotating couple is more than the stabilising couple, the block will topple.

(b) The conditions are shown in Fig. 5.12(b).

As before, Rotating couple $= P \times 25 = 0.3\ W \times 25 = 7.5\ W$

Stabilising couple $= W \times \dfrac{20}{2} = 10\ W.$

As the stabilising couple is more than the rotating couple, the body, in this case, will slide.

5.9 Rough Inclined Plane

We will now investigate frictional forces on a rough inclined plane. An inclined plane is a plane lying at a certain angle to the horizontal as shown in Fig. 5.13(a). Such planes are quite common, viz., a ramp in a building or a steep hill road. No new principles are involved in the investigation. Since the gravity forces, say the weight of a block, are no longer normal to the plane on which the block is resting, the conditions are different from that of a rough horizontal plane.

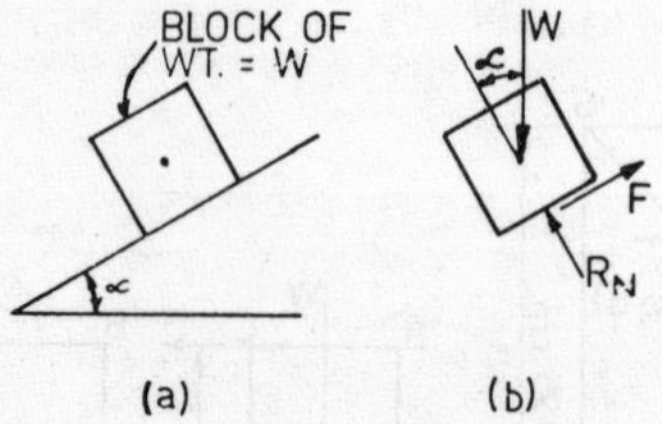

Fig. 5.13(a) and (b)

Let a block of weight '*W*' rest on an inclined plane as shown in Fig. 5.13(a). To investigate the condition of the body, we draw a free body diagram as shown in Fig. 5.13(b). R_N is the normal reaction between the plane and the block. Since the tendency of the block will be to move down, the frictional force '*F*'is acting up the plane as shown. If μ_s is the coefficient of friction, then,

Limiting static friction $= \mu_s \cdot R_N = \mu_s \cdot W \cos \alpha.$

Force acting down the plane tending to cause motion is equal to $W \sin \alpha$. Now it is easy to see that if $W \sin \alpha$ is more than $\mu_s \cdot W. \cos \alpha$, the body will start moving down the plane. In the limiting case, for equilibrium,

$$W \sin \alpha = \mu_s \cdot W \cos \alpha$$

or

$$\tan \alpha = \mu_s \cdot$$

Since $\mu_s = \tan \phi_s$, where ϕ_s is the angle of friction, the above condition gives $\alpha = \phi_s$.

Therefore, a body will be at rest on an inclined plane only when the angle of the inclined plane θ is less than the angle of friction ϕ_s between the body and the plane.

Now let us investigate frictional forces on an inclined plane when forces are applied to cause motion up or down an inclined plane. We will investigate four cases as given below.

(a) Motion up the plane, force '*P*' applied upwards and parallel to the plane

The condition is shown in Fig. 5.14(a). A block of weight '*W*' is resting on an inclined plane inclined at 'θ' to the horizontal. A force *P* is applied to the block, parallel to the plane as shown. The free body diagram of the block is shown in Fig. 5.14(b) when motion is impending. From the figure, selecting X- and Y- axes as shown,

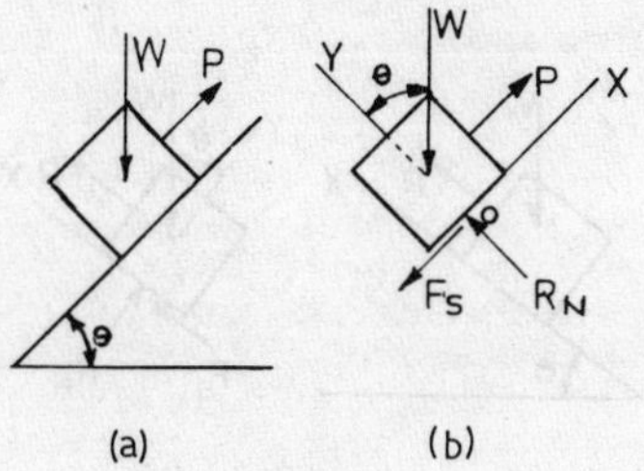

Fig. 5.14

$\Sigma F_y = 0, R_N - W \cos \theta = 0$

$\therefore R_N = W \cos \theta$

Equating forces parallel to the plane, we have

$\Sigma F_x = 0, P - F_s - W \sin \theta = 0$

$P - \mu_s W \cos \theta - W \sin \theta = 0$

$\boldsymbol{P = W (\mu_s \cos \theta + \sin \theta)}$

(b) Motion up the plane, force *P* applied horizontally

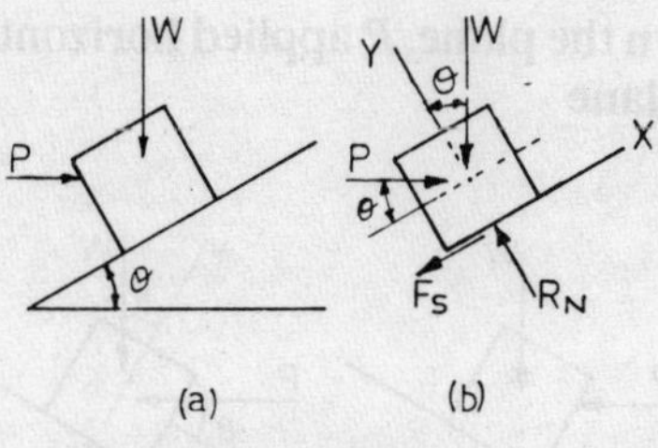

Fig. 5.15

The condition in this case is shown in Fig. 5.15(a). The free body diagram is shown in Fig. 5.15(b), when motion is impending. Selecting X- and Y- axes as shown, and writing the values, we have

$\Sigma F_y = 0$, and $\Sigma F_x = 0$ for equilibrium,

$\Sigma F_y = 0, R_N - W \cos \theta - P \sin \theta = 0$

$R_N = W \cos \theta + P \sin \theta$

$\Sigma F_x = 0, P \cos \theta - W \sin \theta - F_s = 0$

$P \cos \theta - W \sin \theta - \mu_s (W \cos \theta + P \sin \theta) = 0$

$$p = \frac{W(\sin \theta + \mu_s \cos \theta)}{(\sin \theta + \cos \theta)}$$

(c) Motion down the plane, *P* applied parallel to the Plane to prevent motion down the plane

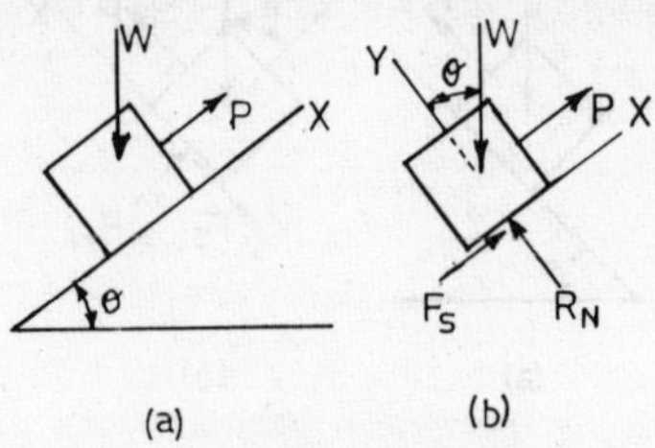

Fig. 5.16(a) and (b)

The condition is shown in Fig. 5.16(a) and the free body diagram when motion is impending in Fig. 5.16(b).

Again,

$$\Sigma F_y = 0, R_N - W \cos \theta = 0$$

$$R_N = W \cos \theta$$

$$\Sigma F_x = 0, F_s + P - W \sin \theta = 0$$

$$\mu_s \cdot W \cos \theta + P - W \sin \theta = 0$$

$$P = W (\sin \theta - \mu_s \cdot \cos \theta)$$

(d) Motion down the plane, *P* applied horizontally to prevent motion down the plane

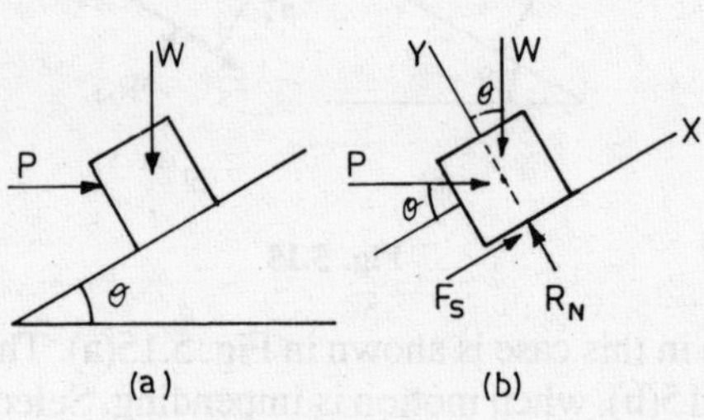

Fig. 5.17 (a) and (b)

The condition is shown in Fig. 5.17(a) the free body diagram in Fig. 5.17(b).

$$\Sigma F_y = 0, R_N = W \cos \theta + P \sin \theta$$

$$\Sigma F_x = 0, P \cos \theta - W \sin \theta + F_s = 0$$

$$P \cos \theta - W \sin \theta + \mu_s (W \cos \theta + P \sin \theta) = 0$$

$$P = \frac{W (\sin \theta - \mu_s \cos \theta)}{(\sin \theta + \cos \theta)}$$

The four cases solved above are given for illustrative purpose only. It is not necessary, it is even undesirable, to remember these formulae for application in specific cases. It is far more elegant to draw the free body diagrams in any specific case, and try to obtain the solution from first principles.

Graphical Construction

We have already seen that $\mu_s = \tan \phi_s$, where ϕ_s is the angle of static friction. ϕ_s is the angle between the total reaction R(vector sum of R_N and F_s) and the normal reaction R_N. We observe in the above four cases that when motion is impending the equilibrium of the body is governed by three forces, P, W and R, and hence these forces should form a closed triangle. The force triangle can be drawn since W and the angles between the forces are known. The force P can be scaled off from the force triangle.

The force triangles for all the four cases are shown in Figs. 5.18 to 5.21. It is again emphasized that in all the four cases we are considering a condition of limiting friction, and consequently the angle between W and R is either $(\theta + \phi)$ or $(\theta - \phi)$ depending upon the direction of the frictional force, i.e., acting down or up.

Example 5.6

A block weighing 200 N is in contact with an inclined plane. Will the block move down under its own weight? Determine the minimum magnitude of a force applied parallel to the plane to prevent the motion down the plane. What force P is required to just cause motion up the plane? $\mu_s = 0.2$ and the angle of the inclined plane is 30°.

Solution

$\mu_s = \tan \phi_s = 0.2$

It has been shown in art 5.9 that the inclination of the inclined plane should be less than the angle of friction for the equilibrium of the body

$\tan \phi_s = 0.2$

$\phi_s = \tan^{-1}(0.2) = 11.30°$.

Since the inclination of the plane (= 30°) is more than ϕ_s, the body will start moving down the plane.

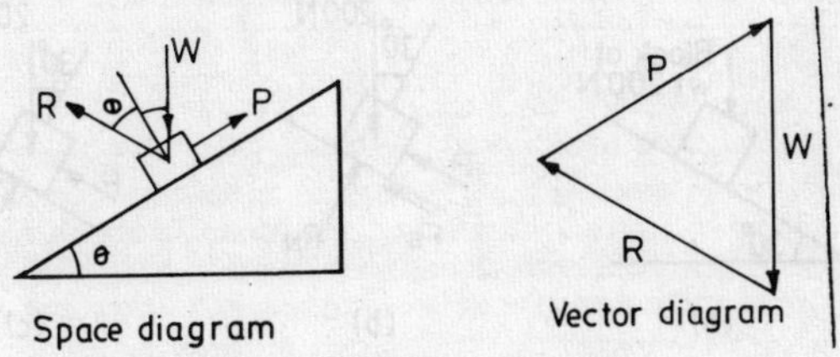

Fig. 5.18

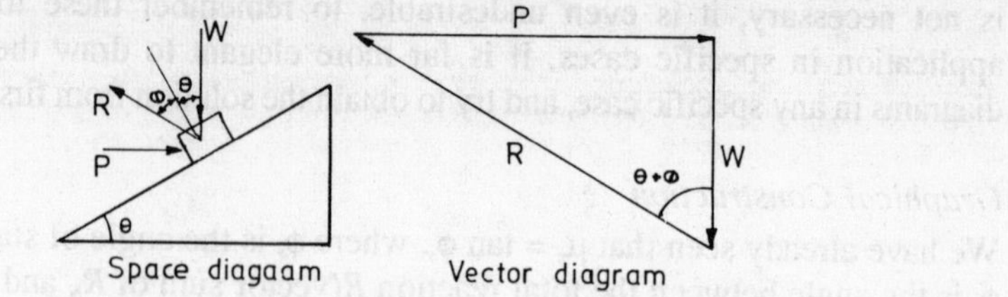

Fig. 5.19

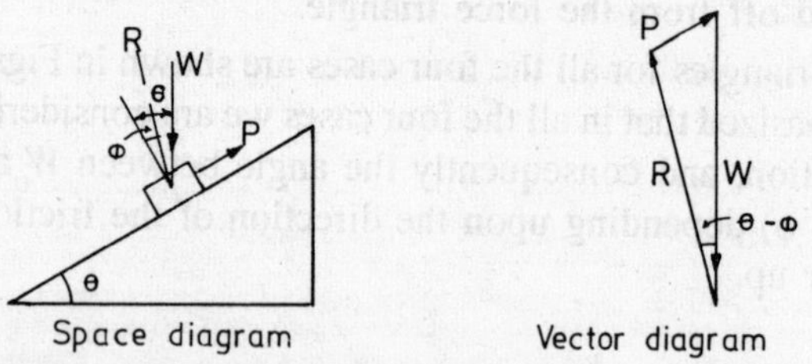

Fig. 5.20

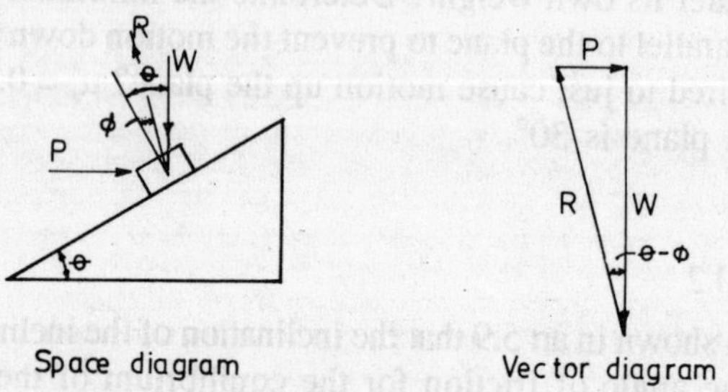

Fig. 5.21

In Fig. 5.22(b), the case with impending motion down the plane is shown. From the free body diagram,

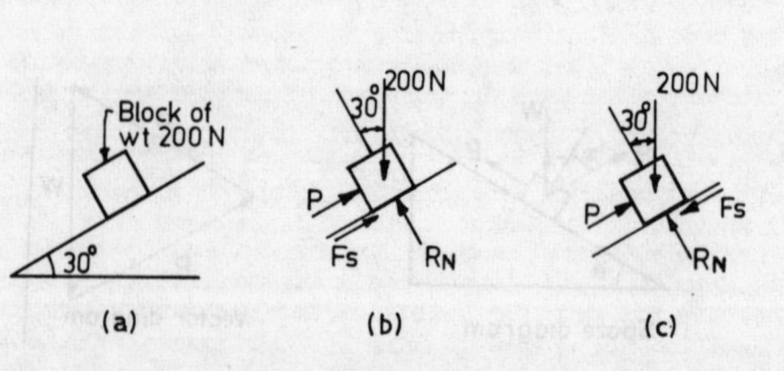

Fig. 5.22

$$R_N = W \cos 30 = \frac{\sqrt{3}}{2} W$$

$$F_s = \mu_s R_N = 0.2 \times \frac{\sqrt{3}}{2} W$$

Equating the forces parallel to the plane, we get

$$P - W \sin 30 + F_s = 0$$

$$P - W \sin 30 + 0.2 \frac{\sqrt{3}}{2} W = 0$$

$$P = -0.2 \frac{\sqrt{3}}{2} W + W \sin 30 = \mathbf{65.36\ N}$$

Case (c) Fig. 5.22(c) shows the case of impending motion up the plane. Notice that the frictional force acts down the plane. From the free body diagram,

$$R_N = W \cos 30 = 200 \frac{\sqrt{3}}{2} = \mathbf{173.2\ N}$$

$$F_s = 0.2 \times 173.2 = \mathbf{34.64\ N}$$

Equating forces parallel to the plane, we have

$$P - F_s - W \sin 30 = 0$$

$$P - 34.64 - 200 \times \frac{1}{2} = 0$$

$$P \quad = \mathbf{134.64\ N}$$

Example 5.7

Determine the minimum value and the direction of a force P required to cause motion of a 200 N block to impend up a 30° inclined plane. The coefficient of friction is 0.20.

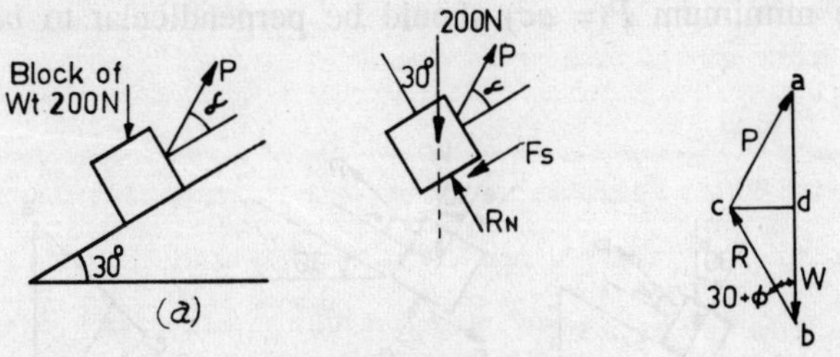

Fig. 5.23 (a) and (b)

Solution

Fig. 5.23(a) shows the body on the plane. Let the minimum force be 'P'

inclined at an angle α to the horizontal as shown. Frictional force acts down the plane. From the free body diagram in Fig. 5.23(b), we have

$$R_N = (200 \cos 30 - P \sin \alpha)$$

Limiting friction

$$F_s = 0.2\,(200 \cos 30 - P \sin \alpha)$$

Equating forces parallel to the plane, we have

$$P \cos \alpha - 200 \sin 30 - 0.2(200 \cos 30 - P \sin \alpha) = 0$$

$$P \cos \alpha + 0.2\,P \sin \alpha = 200 \times \frac{1}{2} + 0.2\,\frac{\sqrt{3}}{2} \times 200$$

$$P\,(\cos \alpha + 0.2 \sin \alpha) = 134.64$$

For P to be minimum, $(\cos \alpha + 0.2 \sin \alpha)$ should be maximum; differentiating and equating to zero, we get

$$-\sin \alpha + 0.2 \cos \alpha = 0$$

or $\tan \alpha = 0.2$ or $\alpha = \phi_s = 11.28°$

$$P\,(\cos \alpha + 0.2 \sin \alpha) = 134.64$$

$$P \cos \alpha\,(1 + 0.2 \tan \alpha) = 134.64$$

$$P \cdot \cos \alpha\,(1.04) = 134.64$$

$$P = \frac{134.64}{1.04 \times 0.9806}$$

$$= 132.0 \text{ N}$$

Alternative Solution

From the free body diagram, F_s and R_N constitute the total reaction R. It must be inclined at an angle ϕ_s to the direction of R_N. The three forces which keep the body in equilibrium are P, W and total reaction R. We can draw the triangle of forces under these conditions for minimum value of P. In Fig. 5.24, we first draw the line ab indicating the weight of block W to some scale. The line bc is then drawn at angle of $(30 + \phi_s)$ to the vertical. For the value of P to be mimimum $P(= ac)$ should be perpendicular to bc as shown.

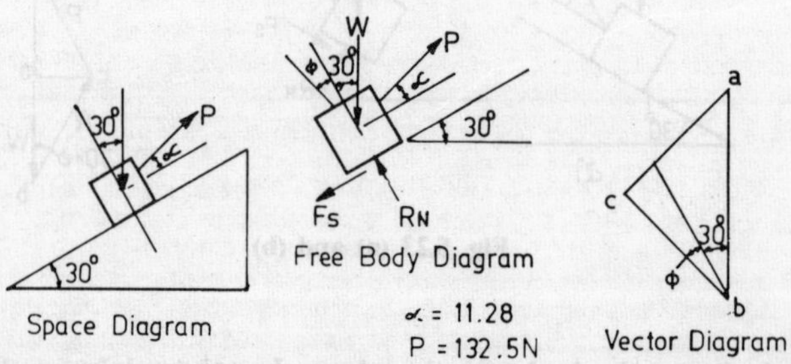

Fig. 5.24

It is easy to see from the triangle of forces that angle *acd* = (30 + ϕ_s) which means that the minimum force *P* is inclined at an angle ϕ_s to the plane.

This problem has also been solved graphically in Fig. 5.25. When the force diagram is drawn to scale as mentioned above, the value of *P* and the angle α can be measured directly from the diagram.

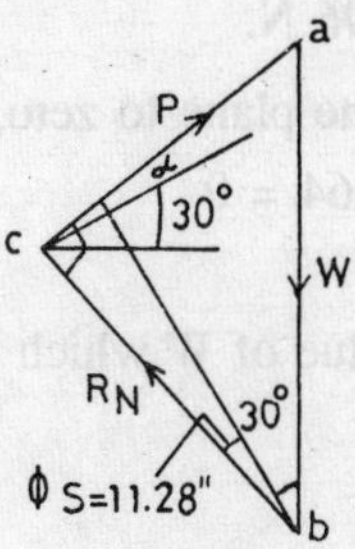

Fig. 5.25

Example 5.8

Determine the range of values of weight *W* such that the 200 N block shown in the Fig. 5.26(a) will neither move up the plane nor slip down the plane. The coefficient of static friction for the contact surfaces is 0.30.

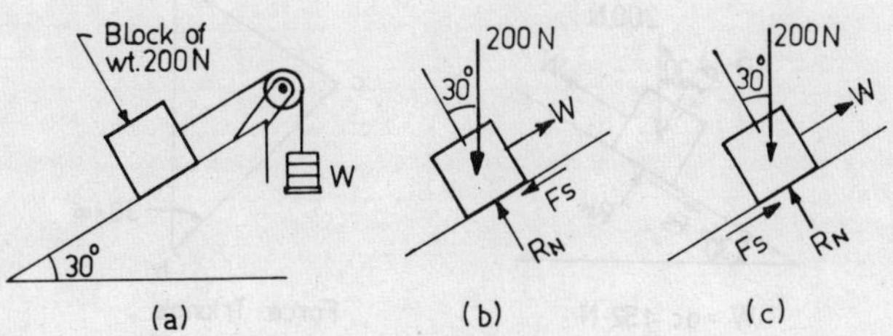

Fig. 5.26 (a), (b) and (c)

Solution

The 200 N body by itself will slip down the plane as μ_s = tan ϕ_s = 16.7° < 30°.

(a) *Motion up the Plane*

The free body diagram is as shown in Fig. 5.26(b). The friction force *F* is acting down the plane to prevent the motion up the plane. In this case,

R_N = 200 cos 30 = **173.2 N**

F_s = 0.3 R_N = 0.3 × 173.2 = 51.96 N.

The tension in the cord is *W* as shown. Equating forces parallel to the plane,

W – 200 sin 30 – 51.96

W = **151.96 N**

This is the maximum value of *W*, so that the body does not move up the plane.

(b) *Motion Down the Plane*

The free body diagram in this case is as shown in Fig. 5.26(c).

The frictional force now acts upward to oppose the motion downward. As before,

$$R_N = 200 \cos 30 = 200 \frac{\sqrt{3}}{2} = 173.2 \text{ N}$$

$$F_s = 0.3 \times 173.2 = \mathbf{51.96\ N.}$$

Equating forces parallel to the plane to zero, we have

$$W - 200 \times \sin 30 + 34.64 = 0$$

$$W = \mathbf{48.04\ N}$$

This is the minimum value of W which will prevent the motion down the plane.

Graphical Solution

Fig. 5.27 gives the graphical solution to this problem. The three forces that keep the body in equilibrium are W, 200 N, and the total reaction which is the vector sum of R_N and F_s. The magnitude of F_s remains the same, but its direction is different in case (a) from that in case (b). The force triangle is drawn by first drawing the 200 N force to scale and drawing a line inclined

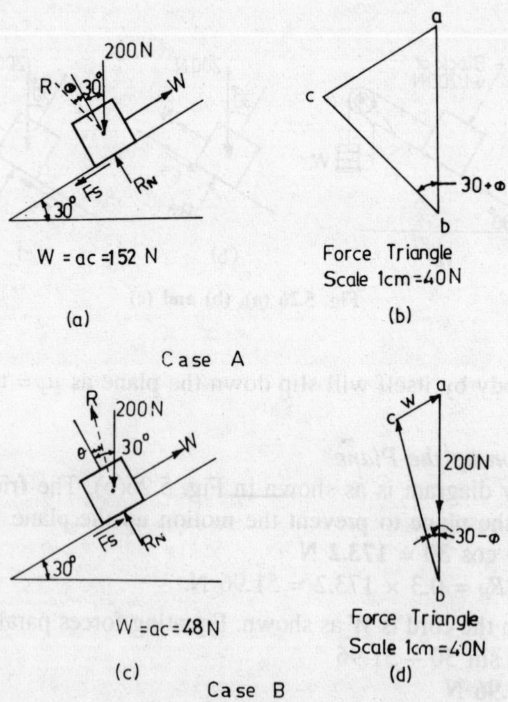

Fig. 5.27

at (30° + $\phi_s°$) for case (a) and (30° – $\phi_s°$) for case (b) from the bottom end of the 200 N vector. A line drawn parallel to the line of action of *W* from the top end of the 200 N vector completes the triangle. The magnitude of *P* can now be scaled off from the diagram.

Example 5.9

Determine the friction force in magnitude and direction acting on the 500 N block if the horizontal force *P* applied as shown in Fig. 5.28 is (a) 250 N and (b) 50 N.

$\mu_s = 0.30$ and $\mu_k = 0.20$

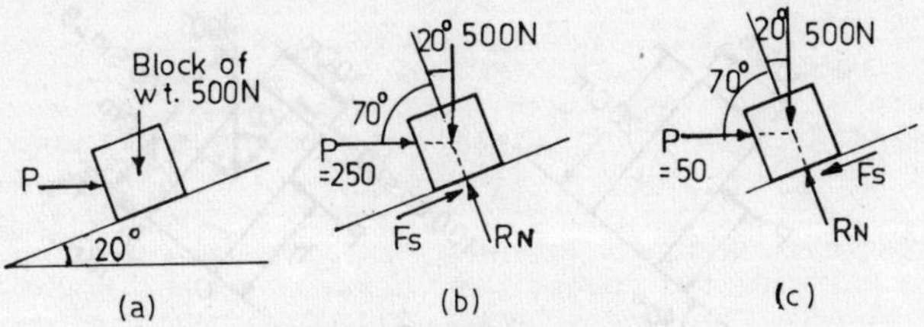

Fig. 5.28 (a), (b) and (c)

Solution

(a) The free body diagram is shown in Fig. 5.28(b). Since it is not known whether the body is in equilibrium or not, we start with the assumption that the body is in equilibrium. If the friction force is F_s, then

Limiting friction $F_s = \mu_s \cdot R_N \cdot$

$$= 0.3 \times (500 \cos 20 + 250 \sin 20)$$
$$= 0.3 \times (500 \times 0.9375 + 250 \times 0.3410)$$
$$= \mathbf{166.6\ N}$$

For equilibrium parallel to the plane,

$$250. \cos 20 - 500 \sin 20 + F_s = 0$$
$$F_s = 500. \sin 20 - 250. \cos 20$$
$$= 171.0 - 235.0 = \mathbf{-64\ N}$$

Since *F* is negative, the force tends to move the body up and F_s acts down For this condition,

$$250. \cos 20 - 500. \sin 20 - F_s = F_s = 0$$
$$F_s = \mathbf{63.90\ N}$$

Since the maximum frictional force available is 166.6 N, the body will be in equilibrium.

(b) The free body diagram is shown in Fig. 5.28(c).

For equilibrium parallel to the plane,

$$50 \cos 20 - 500. \sin 20 - F_s = 0$$
$$F_s = 500. \sin 20 - 50 \cos 20 = \mathbf{-124\ N}\ \text{(acting up)}$$

Limiting friction $= \mu_s . R_N$

$$= 0.3\ [500 \cos 20 + \sin 20]$$
$$= \mathbf{146\ N}$$

Since the limiting friction is more, the body will be in equilibrium.

Example 5.10

A homogeneous block weighing 100 N rests upon an inclined plane as shown in Fig. 5.29. If the coefficient of friction is 0.30, determine the greatest height at which a force *P* parallel to the plane may be applied so that the block will slide up the inclined plane without slipping. Find also the magnitude of *P* for this condition.

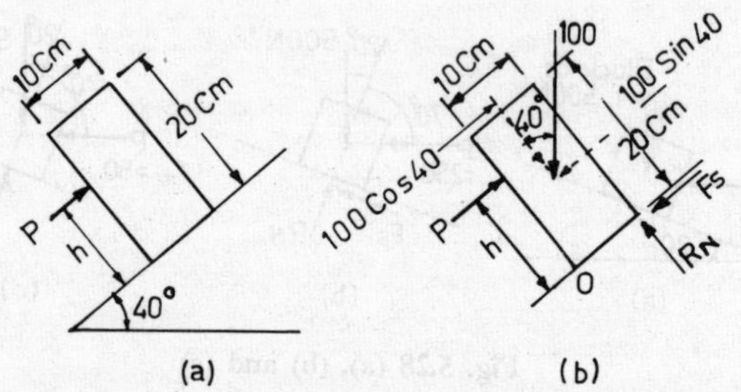

Fig. 5.29

Solution

In the limiting case, the conditions are as shown in Fig. 5.29(b). The whole normal reaction is concentrated at the right corner as shown in the figure.

Applying the conditions of equilibrium to the free body, we have

$$\Sigma V = 0;\ R_N - 100 \times \cos 40 = 0$$
$$R_N = 100 \times \cos 40 = \mathbf{76.6\ N.}$$
$$\Sigma M = 0;\ \text{taking moments about point 'O'},$$
$$(P \times h - 100 \times \sin 40 \times 10) - 100 \times \cos 40 \times 5 = 0$$
$$P \times h = 642.8 + 383.0 = 1025.8\ \text{N}$$

To find the value of *P*, since it is a case of limiting friction,

$$\Sigma H = 0;\ P - 100 \times \sin 40 - F_s = 0$$
$$F_s = \mu_s \times 100 \times \cos 40 = 0.3 \times 100 \times \cos 40 = 22.98\ \text{N}$$
$$P = 64.28 + 22.98 = 87.26\ \text{N}$$
$$\therefore\ h = \frac{1025.8}{87.26} = \mathbf{11.75\ cm.}$$

5.10 Other Problems in Friction

We now take up a few more examples involving frictional forces. The basic principles in the solution of these problems remain the same. The following examples are given to illustrate the approach to the solution of problems involving friction. The importance of drawing free body diagram (s) and assuming the correct direction for the frictional force are emphasized in the examples. Both graphical and analytical solutions are possible. The graphical solutions are just another way of performing the analytical calculations.

Example 5.11

Fig. 5.30(a) shows a ladder placed against a wall. The coefficient of friction is μ_s both at the floor and wall with the ladder. Determine the height upto which the resultant of W_1 and W_2 (= W) can move up (as a man climbs up the ladder).

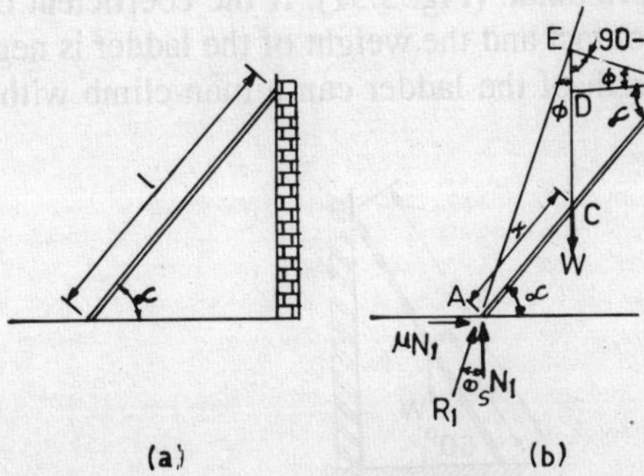

Fig. 5.30

The conditions shown in the free body diagram in Fig. 5.30(b) are when motion is impending. W is the total gravity weight including the weight of the ladder and the weight of say, a man standing on the ladder. The total reactions at the floor and wall will be directed along BE and AE, each making an angle ϕ_s with the normal reactions N_1 and N_2 as shown. The total gravity force W now should pass through E as it is the third force of the three forces (W, R_1 and R_2) which keep the ladder in equilibrium. The distance x can be calculated now from the figure.

$$x = (l - \text{BC})$$

$$BC = \text{BD} \sec \alpha$$

In ΔABE, BE $= l \cos (\alpha + \phi_s)$

In ΔBDE, BD $=$ BE $\cos \phi_s = l \cos (\alpha + \phi_s) \cos \phi_s$

$$\text{BC} = \text{BD} \cdot \sec \alpha = \cos (\alpha + \phi_s) \cos \phi \cdot \sec \cdot \alpha \cdot$$

$$x = l - \text{BC}$$

$$\therefore\ x = 1 - l \cos (\alpha + \phi_s) \cdot \cos \phi_s \sec \alpha \cdot$$

$$= l(1 - \cos (\alpha + \phi_s) \cdot \cos \phi_s \sec \alpha \cdot$$

Some important conclusions can be arrived at from this result.

The motion of the ladder is such that end B slips down and end A slips to the left. The frictional forces are consequently directed upwards at B and to the right at A. However, ϕ_s is the maximum angle of the total reactions at A and B. Therefore, for a given angle α and length l of the ladder, x is the maximum distance from the bottom at which the resultant gravity force can act.

Further it can be seen from the equation for 'x' that $x = l$, when $\alpha = 90 - \phi_s$.

When α is more than $(90 - \phi_s)$, x is more than l. Therefore, when the angle α is $(90 - \phi_s)$ we may conclude that the total gravity wt. can go upto the top of the ladder without slipping.

Example 5.12

A ladder is resting on a horizontal floor and leans against a vertical wall making angle of 60° with the horizontal. (Fig. 5.31). If the coefficient of friction is 0.25 for all surfaces in contact and the weight of the ladder is neglected, find what per cent of the length of the ladder can a man climb without causing the ladder to slip.

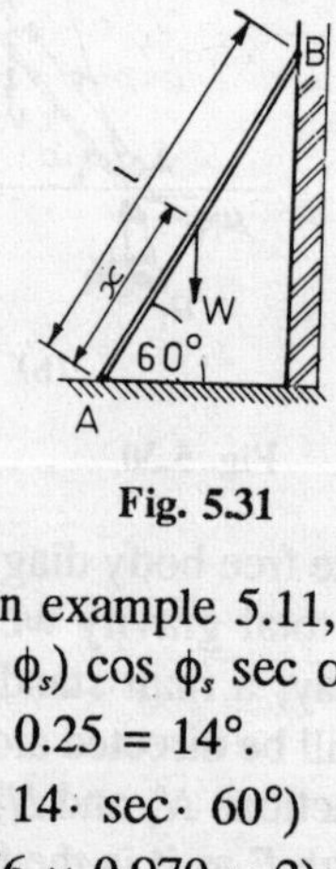

Fig. 5.31

Using the formula derived in example 5.11,

$$x = l[1 - \cos(\alpha + \phi_s)\cos\phi_s \sec\alpha]$$

here $\alpha = 60°$, $\phi_s = \tan^{-1} 0.25 = 14°$

$$x = 100\,(1 - \cos 74 \cdot \cos 14 \cdot \sec \cdot 60°)$$

$$= 100\,(1 - 0.2756 \times 0.970 \times 2)$$

$$= \mathbf{0.465}\ l$$

The percent of length of ladder =46.5%

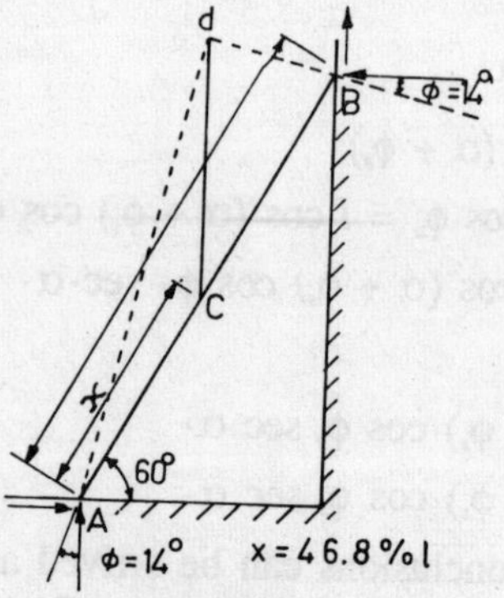

Fig. 5.32

Alternatively the problems can be solved graphically as shown in Fig. 5.32. The two total reaction lines are drawn at A and B inclined at 14° to the normal to meet at *d*. We drop a vertical line from point *d* to meet the ladder line at point '*c*'. AC gives the distance upto which the man can climb.

Example 5.13

A ladder is 7 m long and weighs 250 N. The CG of the ladder is 2.75 m from the bottom end as shown in Fig. 5.33. If the coefficient of friction between the ladder and the wall is 0.3 and between the ladder and the floor is 0.4, when it is kept as shown, determine the safe height upto which a man weighing 750 N can climb without making the ladder slip.

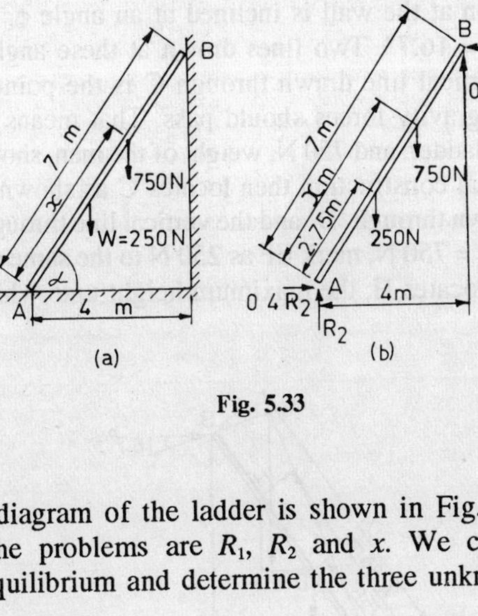

Fig. 5.33

Solution

The free body diagram of the ladder is shown in Fig. 5.33(b). The three unknowns in the problems are R_1, R_2 and x. We can apply the three conditions of equilibrium and determine the three unknowns.

$\alpha = \cos^{-1}(4/7) = 55.1°$

(i) $\Sigma H = 0$

$0.4\ R_2 = R_1$

(ii) $\Sigma V = 0$

$R_2 + 0.3\ R_1 - 250 - 750 = 0$

$R_2 + 0.3\ (0.4\ R_2) - 1000 = 0$, substituting for R_1 from (i)

$$1.12\ R_2 = 1000$$

$$R_2 = \frac{1000}{1.12} = \mathbf{893\ N}$$

$$R_1 = 0.4 \times 893 = \mathbf{357.2\ N}$$

Taking moments about A,

$\Sigma M_A = 0.$

$$250 \times 2.75 \times \cos\alpha + 750 \times x \times \cos\alpha - 0.3\ R_1\ 7\cos\alpha - R_1 \times 7\sin\alpha = 0$$

$$250 \times 2.75 \times 0.5714 + 750 \times x \times 0.5714 - 0.3 \times 357.2 \times 7 \times 0.5714 - 357.2 \times 0.8206 \times 7 = 0$$

$$393 + 429\ x - 429 - 2052 = 0$$

$$x = \frac{2088}{429} = \mathbf{4.867\ m}$$

Graphical Solution

The graphical solution for this problem is shown in Fig. 5.34. The total reaction at the floor is inclined at an angle ϕ_2 to R_2 such that $\tan\phi_2 = 0.4$ or $\phi = 21.8°$, and the total reaction at the wall is inclined at an angle ϕ_1 to R_1, such that $\tan\phi_1 = 0.3$, i.e. $\phi_1 = 16.7°$. Two lines drawn at these angles to R_1 and R_2 intersect at C. A vertical line drawn through C is the point through which the resultant of the gravity forces should pass. That means the resultant of 250 N, weight of the ladder, and 750 N, weight of the man, should pass through G. A simple graphical construction then locates C as shown in Fig. 5.34. A horizontal line is drawn through 'D' and the vertical line through 'C' intersects this line at 'E'. If DE = 750 N, mark EF as 250 N to the same scale. A vertical line through F then locates H, the maximum height upto which the man can climb.

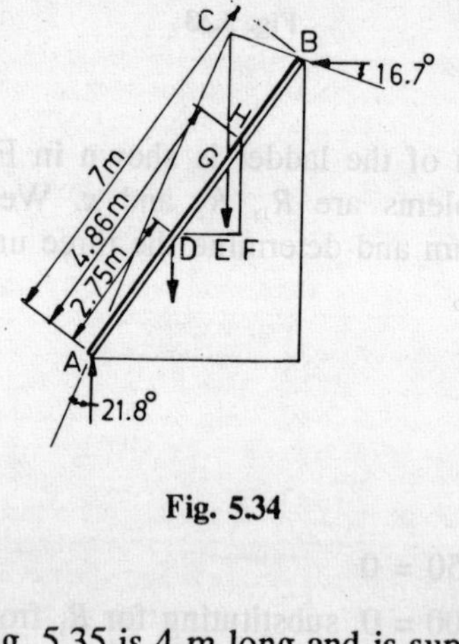

Fig. 5.34

Example 5.14

The ladder shown in Fig. 5.35 is 4 m long and is supported by a horizontal floor and a vertical wall. The value of the coefficient of friction at the wall is 0.2 and at the floor 0.4. The weight of the ladder is 180 N and is acting at its centre.

The ladder carries the weight of a man at 3 m from the bottom along its length. If the man weighs 750 N, determine the reaction components at the floor and wall when the ladder is about to slide. Compute also the minimum angle at this position of the ladder without slipping to the left.

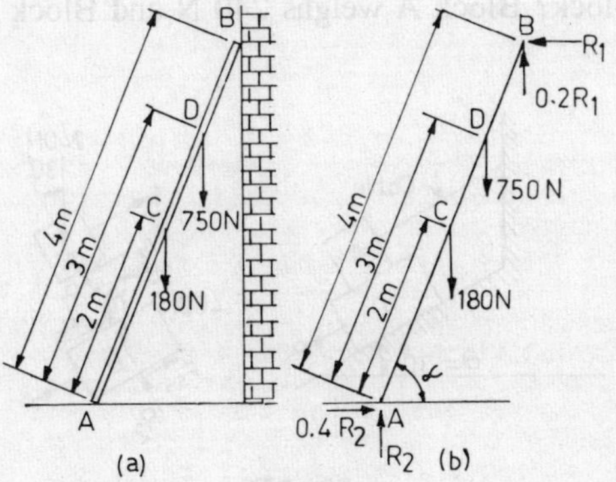

Fig. 5.35

Solution

The free body diagram of the ladder is shown in Fig. 5.35(b). R_1 and R_2 are the normal pressures at the wall and floor. When the ladder is on the verge of slipping, the vertical frictional force at the wall is 0.2 R_1 and the horizontal frictional force at the floor is 0.4 R_2. The directions of the frictional forces are shown at the instant when the ladder is about to slip down on the wall and to the left at the floor.

We can apply the conditions of equilibrium to determine R_1 and R_2.

(i) $\Sigma H = 0$

$0.4\ R_2 = R_1$

(ii) $\Sigma V = 0$

$R_2 + 0.2.\ R_1 = 180 + 750 = 930$ N

Substituting for R_1 from (i),

$R_2 + 0.2\ (0.4\ R_2) = 930$

$$R_2 = \frac{930}{1.08} = \mathbf{861.\ N}$$

$$R_1 = 0.4\ R_2 = 0.4 \times 861 = \mathbf{334.4\ N}$$

To determine the minimum angle α at which the ladder may be kept, we take moments about point A and equate them to zero.

$0.2\ R_1 \times 4 \cos\alpha + R_1 \times 4 \sin\alpha - 180 \times 2 \cos\alpha - 750 \times 3 \times \cos\alpha = 0.$

$0.2 \times 344.4 \times 4 \cos\alpha + 344.4 \times 4 \sin\alpha - 360 \cos\alpha - 2250 \cos\alpha = 0.$

$1377.6 \sin\alpha - 2334.5 \cos\alpha = 0$

$$\frac{\sin\alpha}{\cos\alpha} = \frac{2334.5}{1373.6}$$

$\tan\alpha = 1.695$

$\alpha = \mathbf{59.5°}$

Example 5.15

In Fig. 5.36, if the coefficient of friction is 0.60 and angle $\theta = 30°$, what force P applied downward to B and parallel to the inclined surface will start the motion of the block? Block A weighs 240 N and Block B weighs 400 N.

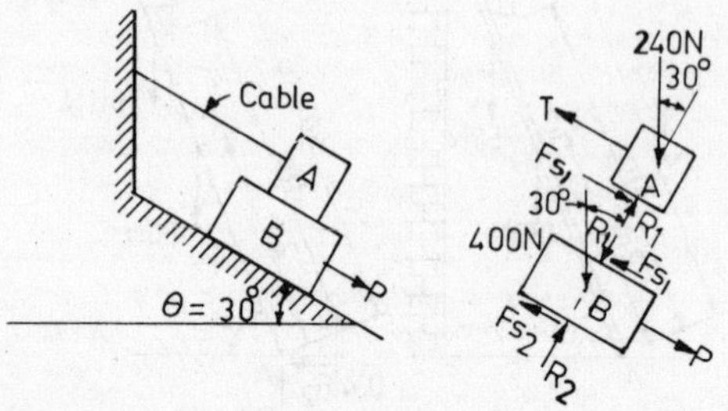

Fig. 5.36

Solution

We first draw the free body diagram of the block A and B as shown in Fig. 5.36(b). Note carefully the direction of the frictional forces. Since B has impending motion down the plane, the frictional forces act upwards the plane on both faces of B, and act downwards on the face of A in contact with B. We consider equilibrium along (x) and normal (y) to the inclined plane, and write the equations of equilibrium for the block A as

$$\Sigma Y = 0,$$

$$R_1 - 240 \times \cos 30 = 0$$

$$R_1 = 207.8 \text{ N}$$

$$F_{s1} = \mu_s \cdot R_1 = 0.6 \times 207.8 = \mathbf{124.7\ N}$$

The tension in the cable can also be found if necessary.

$$\Sigma X = 0,$$

$$T = F_s + 240 \times \sin 30 = \mathbf{244.7\ N}$$

Writing the equilibrium equations for the free body diagram of B,

$$\Sigma Y = 0,$$

$$R_{N2} = 400 \cos 30 + R_{N1} = 400 \cos 30 + 207.8 = \mathbf{554.2\ N}$$

$$\Sigma X = 0,$$

$$P = 124.7 + 0.6 \times 554.2 - 400 \times \sin 30° = \mathbf{257.2\ N}$$

Summary

No surface is perfectly smooth. The irregularities of a surface are perhaps not visible to the naked eye, but can be easily observed under a microscope.

Where two surfaces are in contact under a normal pressure, the actual area of contact is much less than the apparent surface areas. The 'hills and valleys' of one surface mesh with those of the other and where such meshing takes place, the pressure or force of contact is quite high and results in what is known as 'cold welding'.

When two surfaces in contact tend to move relatively, a force is set up tangentialy to the surfaces which resists the motion. This force is known as 'frictional force'. Frictional force comes into play only when the surfaces move or tend to move relative to each other.

Friction can be of two types: (i) dry friction or Coulomb friction occurs between surfaces which are dry, and (ii) fluid friction occurs between surfaces where they are separated by a layer of fluid. Only dry driction is dealt with in this book.

When there is only a tendency for the surfaces to move relatively, the frictional force developed is known as static friction. When one body actually slides over another, the frictional force developed are known as sliding friction or kinetic friction.

It is found that the maximum static friction is always more than the kinetic friction.

There are 7 laws which govern the phenomenon of dry friction as stated below:

(i) Frictional force always opposes the motion (or tendency of motion), of one body over another. The frictional force always acts tangential to the surface.

(ii) If static friction is acting, the value of frictional force may vary from zero to the maximum depending upon the force causing the tendency to move.

(iii) The maximum value of the static friction (or the limiting friction when motion is approaching) is $\mu_s \cdot W$., where μ_s is the coefficient of static friction and W is the normal pressure between the surfaces.

(iv) The maximum static friction is independent of the area of the surfaces in contact.

(v) If kinetic friction is acting, the frictional force is constant at its limiting value.

(vi) The value of kinetic friction is given by $\mu_k \cdot W$, where μ_k is the coefficeint of kinetic friction and W is the normal pressure between the surfaces.

(vii) The angle between the total reaction and the normal component of the total reaction is known as the angle of friction ϕ_s. Further $\tan \phi = \mu$, where μ is the coefficient of friction. (In the case of impending motion)

Review Questions

1 Define the term 'friction'.

2 Distinguish between static and kinetic friction.

3 Explain with illustrative sketches an experimental set up and procedure to determine the coefficient of friction between two materials.
4 Explain what is wrong with the following statement: 'Coefficient of static friction of steel is 0.3'.
5 State the seven laws of friction.
6 Give reasons to show that the amount of frictional force developed does not depend upon the area of contact between the forces.
7 A body rests on an inclined plane and no external forces are applied to it. Will there be any frictional force acting on it? Draw a sketch of the body on the inclined plane and show the forces acting on the body.
8 Is the frictional force always equal to $\mu_s \times R_N$? Explain your answer.
9 What is angle of friction?
10 Is the angle between the normal reaction and total reaction always equal to the angle of friction? Explain your answer.
11 If the normal pressure between two contacting surfaces is increased, will the coefficient of friction increase? Explain your answer.
12 What is cone of friction? Explain the significance of the cone of friction.
13 Discuss the conditions under which a body slides or topples on the application of a force to cause motion.
14 Explain with illustrative sketches the moment of the frictional force and its use in determining the coefficient of friction between two materials.

Problems

5.1 A block weighing 200 N rests on horizontal surface. It requires a horizontal force of 120 N to start moving and a horizontal force of 90 N to slide it at uniform speed. Determine the coefficients of static and kinetic friction.
(Ans. $\mu_s = 0.6$, $\mu_k = 0.45$)

5.2 If the coefficient of friction of wood over stone is 0.60, how heavy a wooden box can be pulled over a stone floor with a horizontal force of 180 N?
(Ans. 300 N)

5.3 If it requires a force of 75 N to pull with uniform speed a box weighing 125 N over a cement floor, what is the coefficient of kinetic friction?
(Ans. $\mu_k = 0.6$)

5.4 A block of stone is pulled over a horizontal surface by a pull of 1550 N at an angle of 10° to the horizontal. If the coefficient of friction for the surfaces is 0.55, what is the weight of the stone?
(Ans. $W = 3044$ N)

5.5 A machine weighing 3 kN is resting on a horizontal floor. A force of 750 N applied by means of a rope inclined at an angle of 20° above the horizontal is able to move the machine. Calculate the value of the coefficient of friction.
(Ans. $\mu_s = 0.26$)

5.6 A metal box weighing 10 kN is pulled along a level surface, at uniform speed, by applying a horizontal force of 3500 N. If another box weighing 6 kN is now placed on top of this box, what force must now be applied to move the box at uniform speed?
(Ans. 5600 N)

5.7 It requires a force of 300 N to move a body weighing 400 N uniformly up an incline of 30°. Find the coefficient of friction for the surfaces.
(Ans. $\mu_k = 0.29$)

5.8 In a post office, parcels move down an inclined sheet metal chute. If the coefficient of friction is 0.15, find the minimum angle of inclination of the chute.
(Ans. 8.53°)

5.9 Determine the force required to start a body weighing 800 N down an inclined plane. The force is applied along the plane which is inclined at 10° to the horizontal. $\mu_s = 0.3$.
(Ans. 97.43 N)

5.10 A body weighing 100 N rests on a plane making angle of 10°. A horizontal force of 30 N is just sufficient to cause motion impend up the plane. Determine the value of the coefficient of friction.

(Ans. μ_s = 0.117)

5.11 A body weighing 600 N rests on an inclined plane. A horizontal force of 200 N is just sufficient to cause motion up the plane. If the coefficient of friction is 0.15, what angle does the plane make with the horizontal?

(Ans. 9.9°)

5.12 A block weighing 1000 N rests on an inclined plane 4 : 3 (4 horizontal to 3 vertical). The coefficient of friction is 0.4. Determine the value of the force applied parallel to the plane to cuase motion to impend (i) up the plane (ii) down the plane. Compute also the frictional force when the force *P* is (a) 600 N (b) 900 N.

Ans. (i) 920 N (ii) 280 N (applied upwards to hold the block) (a) Zero (b) 300 N

5.13 In the figure shown, determine the horizontal force required to cause the motion of the block (a) to impend down the plane and (b) to impend up the plane. The coefficient of friction is 0.55.

(Ans. (a) 20.8 N (b) 1651.4 N)

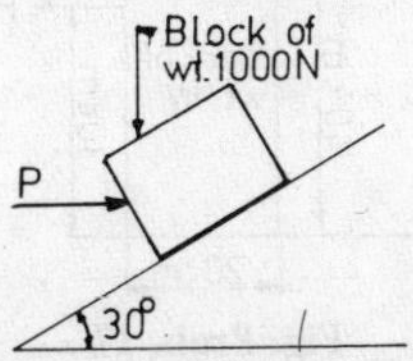

Fig. Prob. 5.13

5.14 In the figure shown, the block weighing 200 N starts moving up the plane when the weight on the hanger is 90 N. Determine with what maximum weight on the hanger wil! the body still start moving down the plane. If the surface is now polished, and it is found that the body moves down the plane when the maximum weight on the hanger is 30 N, calculate the value of the coefficient of friction. Weight of hunger is 10 N.

(Ans. 26.8 N; μ_s = 0.15)

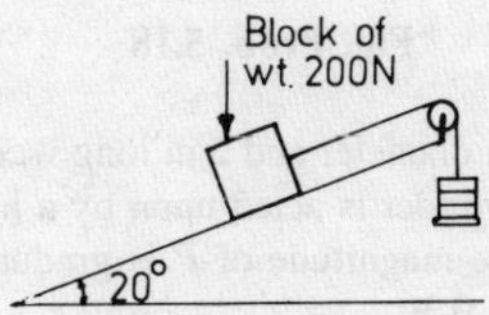

Fig. Prob. 5.14

5.15 A block weighing 600 N is placed on an inclined plane inclined at 30° to the horizontal. Compute the value of a horizontal applied force necessary (a) to just start the block up the plane (b) just prevent motion down the plane. If the horizontal force is equal to 200 N, determine the value of the frictional force μ_s = 0.40.

(Ans. (a) 251.8 N (b) 67.81 N (c) 126.8 N)

5.16 For the system of blocks shown in the figure, find the least value of the force required to cause the blocks move to the left and the corresponding angle α, if the coefficient of friction is 0.20.

(Ans. *P* = 249.6 N; α = 11.3°)

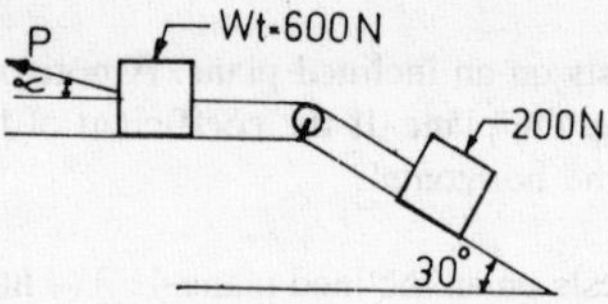

Fig. Prob. 5.16

5.17 The rectangular block shown in the figure has a force '*P*' applied at the height shown. If the value of the force '*P*' is increased uniformly, will the block slide or topple? $\mu_s = 0.3$.

(Ans. The block will slide)

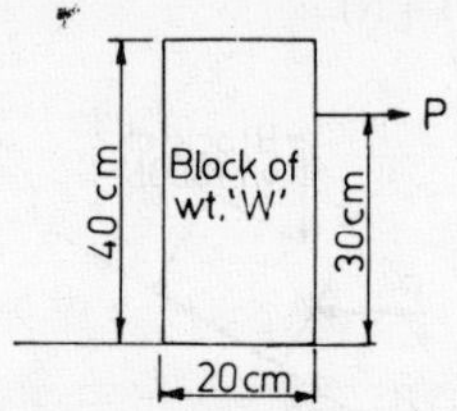

Fig. Prob. 17

5.18 The rectangular block shown in the figure just topples when the height of the applied force is 8 cm. Determine the value of the coefficient of friction.

(Ans. $\mu_s = 0.62$)

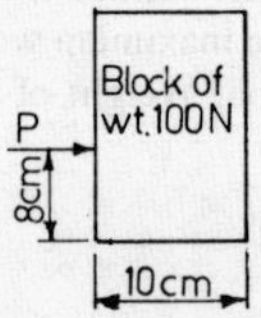

Fig. Prob. 5.18

5.19 A homogeneous cylinder 1 m diameter and 2 m long weighs 2000 N and rests on an end on a horizontal plane. The cylinder is acted upon by a horizontal force *P* at the level of the CG of the cylinder. If the magnitude of *P* is gradually increased, will the cylinder start to slide or topple? $\mu_s = 0.30$.

(Ans. Cylinder will slide)

5.20 If the cylinder of problem 5.19 is resting on an inclined plane inclined at an angle of 30° to the horizontal, at what maximum height can a force *P* be applied to the cylinder so that the cylinder slides up to inclined plane without toppling? $\mu_s = 0.30$.

(Ans. 1.12 m)

5.21 The bar shown in the figure weighs 1000 N. If $\mu_s = 0.10$ for all contacting surfaces and $P = 500$ N, investigate the condition of the body.

(Ans. The bar slides down at A)

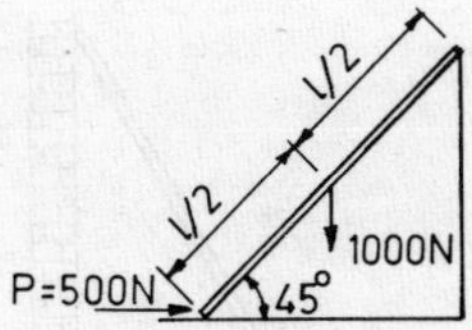

Fig. Prob. 5.21

5.22 A block weighing 1500 N rests on an inclined plane making an angle of 20° with the horizontal. If $\mu_s = 0.20$ and $\mu_k = 0.15$, determine (i) whether the block is in equilibrium and (ii) the frictional force acting on the block.

(Ans. (i) Slides down (ii) 211.4 N)

5.23 An inclined plane makes an angle of 15° with the horizontal. If a block weighing 150 N is released from the top of the plane with an initial velocity, determine whether the block will move down or stop. Also calculate the frictional force acting on the block. $\mu_s = 0.30$, $\mu_k = 0.15$.

(Ans. The block will stop; 21.7 N)

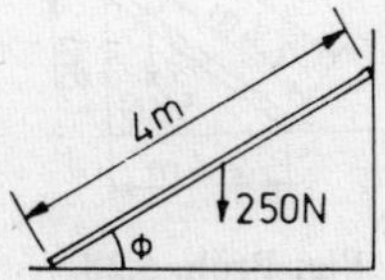

Fig. Prob. 5.24

5.24 Determine the minimum angle θ at which the uniform plank weighing 250 N may be kept as shown in the figure. μ_s for all contacting sufaces = 0.20.

(Ans. θ = 67.6°)

5.25 A cement concrete electric pole 6 m long is lying on a step as shown. If $\mu_s = 0.40$ for all contacting surfaces and the pole weighs 3 KN, calculate the value of the pull required in the cable to move it.

(Ans. 395.4 N)

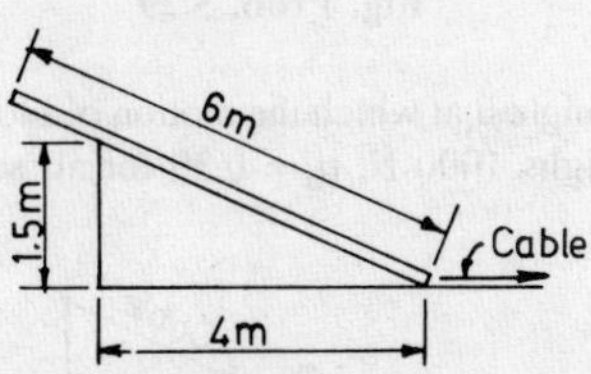

Fig. Prob. 5.25

5.26 A ladder is kept as shown in Fig. If $\mu_s = 0.40$ on the floor and 0.20 on the wall, find how far up the ladder can a man weighing 600 N climb without the ladder slipping? Neglect the weight of the ladder.

(Ans. 0.72 *l*)

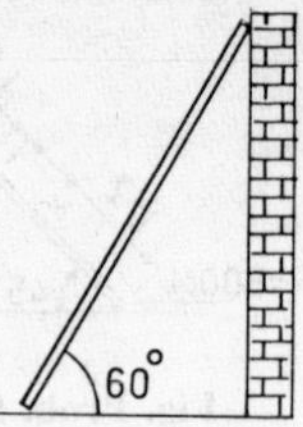

Fig. Prob. 5.26

5.27 In problem 5.26, if the ladder weighs 400 N, how far up the ladder can the man climb?

(Ans. 0.86 *l*)

5.28 Determine the force *P* required to move the 5 KN plank to the left from its rest position. μ_s = 0.50 for all contact surfaces.

(Ans. *P* = 200 N)

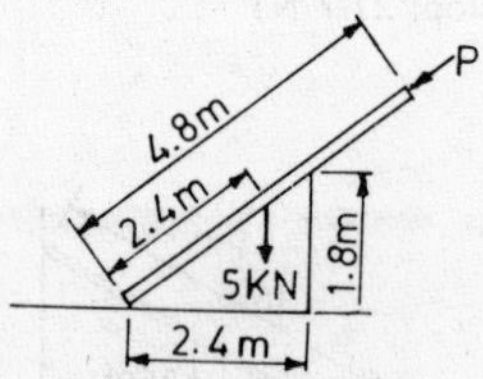

Fig. Prob. 5.28

5.29 Determine the value of *P* required to move the block B to the right μ_s = 0.30.

(Ans. *P* = 120 N)

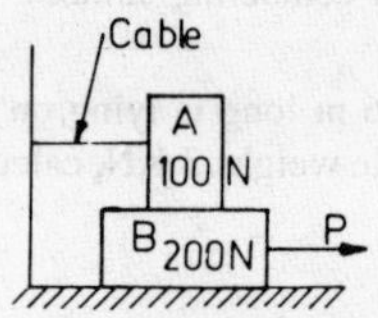

Fig. Prob. 5.29

5.30 Determine the value of angle θ at which the motion of block B impends. Block A weighs 600 N and block B weighs 1000 N. μ_s = 0.30 for all surfaces in contact.

(Ans. θ = 33.4°)

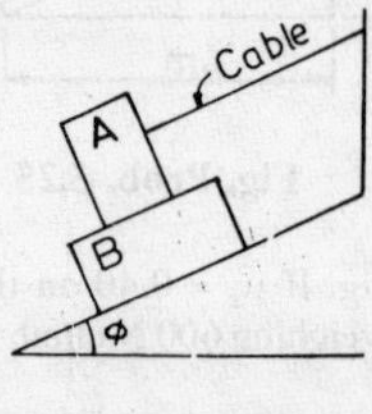

Fig. Prob. 5.30.

6. KINEMATICS

After having mastered the content of this chapter, you will be able to:

- Define the term '*Kinematics*'.
- Define the basic terms related to motion of bodies.
- Derive relationship between parameters of rectilinear motion of a particle including motion under gravity.
- Solve problems of calculating parameters of rectilinear motion of a particle including motion under gravity.
- Interpret the graphs/curves of time *vs* motion parameters of a particle to solve problems of rectilinear motion.
- Explain the rectilinear relative motion.
- Derive relationships between motion parameters for rectilinear relative motion of particles.
- Solve for motion parameters of rectilinear relative motion of particles.
- Explain with examples the curvilinear motion of a particle.
- Define motion parameters of circular motion of particles.
- Derive relationship of motion parameters of angular motion of particles including radial & tangential accelerations.
- Derive relationship of rectilinear and angular motion parameters of particles.
- Define simple harmonic motion (SHM).
- Derive relationship of motion parameters of S.H.M.
- Explain curvilinear motion in terms of two directional rectilinear motion (rectangular coordinate axes).
- Derive expressions of motion parameters for projectiles.
- Solve problems of motion parameters of non rectilinear motion as in case of (a) circular (b) projectiles.
- Explain Kinematics of Rigid bodies.
- Define instantaneous centre in plane motion of rigid bodies.
- Solve simple problems of motion parameters in motion of rigid bodies.

CHAPTER 6

Kinematics

6.1 Introduction

Kinematics is a part of dynamics which deals with the problems of finding motion parameters of bodies irrespective of the forces causing the motion. Alternatively, this subject deals with the determination of the position of a body, its velocity, and acceleration at any instant of time. Some of the practical problems in kinematics are: design of the trajectories of air crafts, missiles, etc, design of gears, cams, linkages for achieving certain motions.

6.2 Motion

We already know that everything in this universe is in motion e.g. a car moving on the earth's surface is in motion with respect to a fixed reference on the earth. The earth itself is in motion as it is rotating about its axis and at the same time moving around the sun. The sun and its planets are also moving relative to stars which in turn are moving in space. The position of a body in space is determined relative to some reference system. The basic reference frame for Mechanics is the primary inertial system or astronomical frame of reference which is an imaginary set of rectangular axes attached to the mean position of the so called fixed stars. From the measurements carried out relative to this reference system, it is evident that the laws of Newtonian Mechanics are valid as long as the velocities of bodies involved are negligible compared with the velocity of light. Measurements made with respect to this reference are said to be absolute. This reference system is considered as fixed in space. In case the calculations of trajectories for rockets, missiles, etc. are carried out with respect to a reference frame attached to the earth's surface, corrections for the absolute motion of the earth have to be applied. For most of the engineering problems these corrections are very small, and hence may be neglected. On this basis, the measurements may therefore be made with respect to a fixed reference frame attached to the earth's surface, and measurements thus made in most of the engineering problems will be absolute.

In developing the subject of Kinematics, we shall first be considering the motion of particles and then the motion of rigid bodies.

6.3 Motion of Particles

Motion of particles does not mean that we are actually dealing with motion of small capsules. On the other hand we are dealing with rigid bodies of finite size, may be an aircraft, a missile, or an automobile. This rigid body is assumed as a particle, thereby ignoring its rotation about its own axis. This assumption can be made when the size of the body as compared to the path it traverses is very small. The amount of error caused in such cases is small and thus neglected. (Consider a case of a car moving on a curved path. The distance travelled by the outer wheel will be greater than the one travelled by the inner wheel. However, for a curve with a larger radius, the distance travelled by the outer and the inner wheels will not be appreciably different from each other or the distance moved by the CG of the car).

6.3.1 *Types of Motion of a Particle*

The type of the motion undergone by a particle is defined by the path traversed by it. There are thus two basic types of motion of a particle—Rectilinear and Curvilinear.

Rectilinear Motion

A particle is said to possess rectilinear motion when it moves in a straight line (Fig. 6.1).

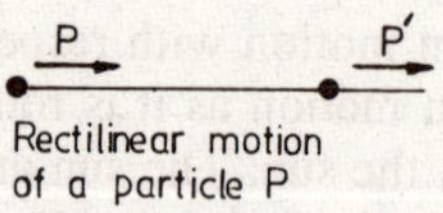

Fig. 6.1

Curvilinear Motion

When a particle moves on a curved path, it is said to possess a curvilinear motion (Fig. 6.2).

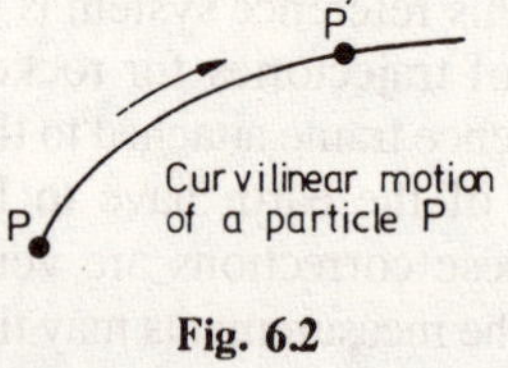

Fig. 6.2

Absolute and Relative Motion

Motion of a particle is also determined by making measurements of the motion with respect to the reference frame. If the measurements are made with respect to a fixed reference frame, the motion is said to be *absolute*. In case the measurements of the motion of a particle are made with respect to a moving reference frame, the motion is said to be *relative*.

Plane and Spatial Motion

If the path of motion i.e. trajectory of a particle lies in a plane, it is called a *plane motion*, e.g. the motion of a freely falling body. If the trajectory of a particle is not contained in a plane, it is called a *Spatial motion*. e.g. motion of an aeroplane flying at different heights and in different directions.

6.3.2 *Basic Concepts*

Displacement

The *change of position* of a particle or a body with reference to certain fixed frame of axes is termed as displacement.

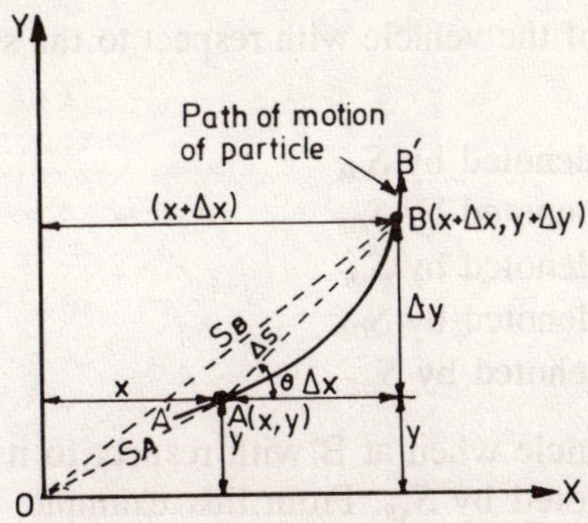

Fig. 6.3

Let us consider a particle moving along a path shown by A′ ABB′ in Fig. 6.3. The path A′ ABB′ lies in a plane X-Y and position of any point can be defined by its coordinates (x, y). At any instant of time (t) let the particle be at a position $A(x, y)$ at a distance S_A, from origion O. After a small time interval Δt, i.e. at a time $(t + \Delta t)$ the particle moves to the position B at a distance S_B from origin O along the path of motion. The position B is defined by coordinates $(x + \Delta x, y + \Delta y)$.

This change of position of particle Δs from A to B is called its displacement AB. This displacement Δs can be defined by its two components, Δx along X-axis, and Δy along Y-axis. The magnitude of the displacement Δs is given as

$$s = \sqrt{(\Delta x)^2 + (\Delta y)^2} \qquad \text{... 6.1}$$

The direction of the displacement can be obtained as

$$\theta = \tan^{-1}\left(\frac{\Delta y}{\Delta x}\right) \text{ with the direction of X-axis} \qquad \text{... 6.2}$$

Thus the displacement of the particle from A to B is represented by vector *AB* in magnitude, direction and sense. As such, it is evident that the displacement *AB* is not the same as the displacement *BA*, since the displacement *BA* is opposite in sense to the displacement *AB*. Consider a vehicle moving from a starting point O along a path ABCDE as shown in Fig. 6.4.

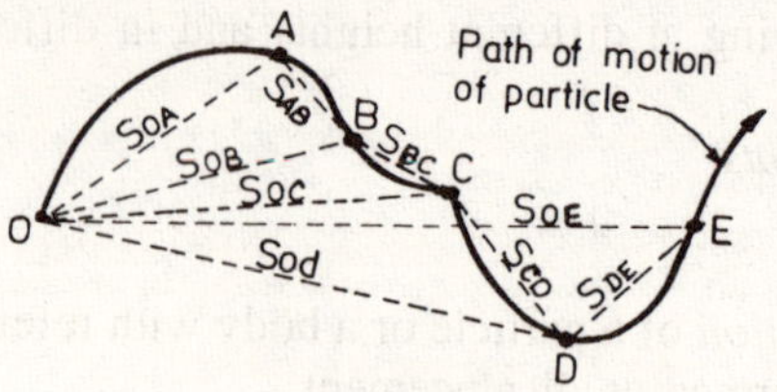

Fig. 6.4

The displacements of the vehicle with respect to the starting point O, when it reaches at

A will be OA denoted by S_{OA}
B will be OB denoted by S_{OB}
C will be OC denoted by S_{OC}
D will be OD denoted by S_{OD}
E will be OE denoted by S_{OE}

Displacement of vehicle when at B with respect to its previous position *A* will be *AB* and denoted by S_{AB}. From this example, it is evident that the distance travelled by the vehicle and the displacement covered are not the same. *The displacement at any instant is the shortest distance* between the origin and the position of the vehicle at that instant.

From the above it is clear that to define displacement completely, the beginning and the end points must also be specified in addition to the magnitude, direction, and sense of the displacement. Since displacement is a vector, it can be added, subtracted or resolved like other vectors. The distance, on the other hand, is a scalar quantity and has only magnitude, and hence is added or subtracted only arithmetically.

In this and the following articles, to define the magnitude of the displacement, the distance unit metre (m) shall be used. In case of small displacements mm or cm, and in case of large displacements km shall be used to represent the magnitude of displacements.

Velocity

The rate of change of position of a body with respect to time is called velocity, which, in other words, is position change per unit time. And like displacement, velocity is also a vector quantity.

Consider a particle *P* moving from a position *A* to a position *B* in time '*t*' along a certain path of motion as shown in Fig. 6.5. Let it further move from B to B′ in a small interval of time $\Delta\, t$. The change of position from *A* to *B* is denoted by the displacement '*S*' of the particle *P*. After a small interval

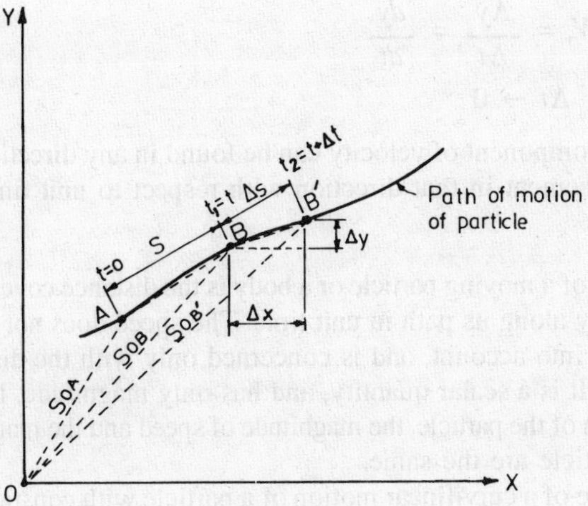

Fig. 6.5

of time Δt, the displacement of the particle when it is at position B' is $(S + \Delta S)$. The change in position of the particle from B to B' during a small interval of time Δt is equal to ΔS. Thus the average velocity V_{av} during this small duration of time Δt is given as

$$V_{av} = \frac{\text{Change in position}}{\text{Time in which the change occurs}} = \frac{\Delta S}{\Delta t}$$

When this time interval Δt is very small, and tends to zero, the velocity becomes instantaneous velocity at the instant of time 't' after the start of motion.

$$\text{i.e. Instantaneous velocity } V = \frac{\Delta S}{\Delta t} = \frac{dS}{dt} \qquad \text{... 6.3}$$

$$\text{as } \Delta t \rightarrow 0$$

Normally instantaneous velocity at any time 't' is referred to as velocity V at time 't'. The direction of velocity is the same as that of the displacement at the instant of measurement. Like other vectors, the velocity vectors can also be added, substracted or resolved. The displacement Δs can be resolved in to the components Δx and Δy respectively along X- and Y- directions, and hence the velocity V can also be resolved in to two components V_x and V_y along the same directions X and Y respectively.

The component of velocity along X direction is as under

$$V_x = \frac{\Delta x}{\Delta t} = \frac{dx}{dt} \qquad \text{... 6.3(a)}$$

$$\text{as } \Delta t \rightarrow 0$$

The component of velocity along Y direction is

$$V_y = \frac{\Delta y}{\Delta t} = \frac{dy}{dt} \qquad \text{... 6.3(b)}$$

as $\Delta t \rightarrow 0$

Thus the component of velocity can be found in any direction by considering the displacement in that direction with respect to unit time.

Speed

The speed of a moving particle or a body is the distance covered by the particle or the body along its path in unit time. The speed does not take the direction of motion into account, and is concerned only with the distance covered in unit time. It is a scalar quantity, and has only magnitude. In case of straight line motion of the particle, the magnitude of speed and the magnitude of velocity of the particle are the same.

In case of a curvilinear motion of a particle with constant speed, velocity is a variable quantity due to changing direction of the motion. Thus speed is constant if the magnitude of rate of covering distance per unit time is constant, while velocity remains constant only if the magnitude, as well as the direction, of the distance covered per unit of time is constant.

The units of velocity or speed for defining their magnitude shall be taken as metre per second (ms^{-1}) or km per hour (km h^{-1}).

Acceleration

Consider a particle moving with a variable velocity 'V' at any instant of time 't', which becomes ($V + \Delta V$) after a small interval of time 'Δt'. The average acceleration 'a_{av}' is then defined as the mathematically average rate of change of velocity with respect to time (see Fig. 6.6).

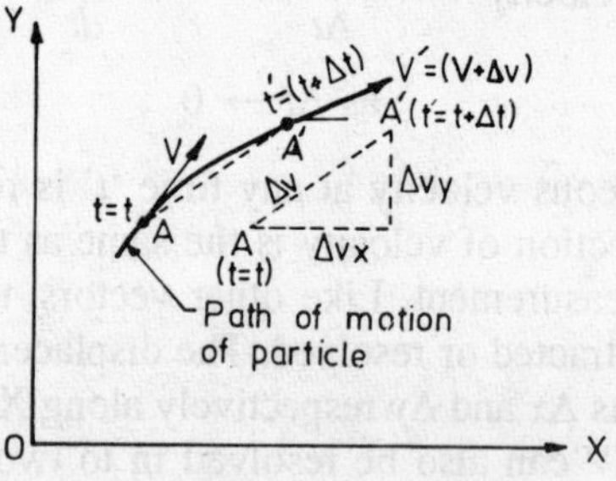

Fig. 6.6

$$a_{av} = \frac{\text{Change in velocity}}{\text{Time interval}} = \frac{\Delta V}{\Delta t}$$

As the time interval 'Δt' reduces, and tends to become zero, the average acceleration becomes instantaneous acceleration '*a*' at an instant of time '*t*'. Thus mathematically

$$a = \frac{\Delta V}{\Delta t} = \frac{dV}{dt} \quad \text{... 6.4}$$

as $\Delta t \to 0$

We know from equation 6.3 that $V = \frac{dS}{dt}$, and substituting this in equation 6.4, we have

$$a = \frac{dV}{dt} = \frac{d}{dt}\left(\frac{dS}{dt}\right) = \frac{d^2S}{dt^2} \quad \text{... 6.4(a)}$$

Since velocity and displacement are vector quantities, the acceleration also is a vector quantity. As the velocity 'ΔV' has components along X and Y directions, acceleration also has components along X and Y directions.

Thus, the component of accleration along *x* direction is given by

$$a_x \frac{\Delta V_x}{\Delta t} = \frac{dV_x}{dt} = \frac{d}{dt}\cdot\frac{dx}{dt} = \frac{d^2x}{dt^2} \quad \text{... 6.4(b)}$$

Similarly, the component of acceleration along Y direction is as

$$a_y = \frac{dV_y}{dt} = \frac{d^2y}{dt^2} \quad \text{... 6.4(c)}$$

Like other vector quanties, acceleration can be added, substracted or resolved.

The common units for defining the magnitude of acceleration are metre/sec/sec(ms^{-2}), or km/hour/hour(km h^{-2}).

6.4 Rectilinear Motion of Particles

The motion of a particle is said to be rectilinear if it moves in a straight path. If a particle moves along different straight paths at different intervals of time as shown in Fig. 6.7, even then the motion of the particle will be called rectilinear, but the computation of motion parameters will be done separately for each portion (*AB*, *BC*, *CD*, *DE*...).

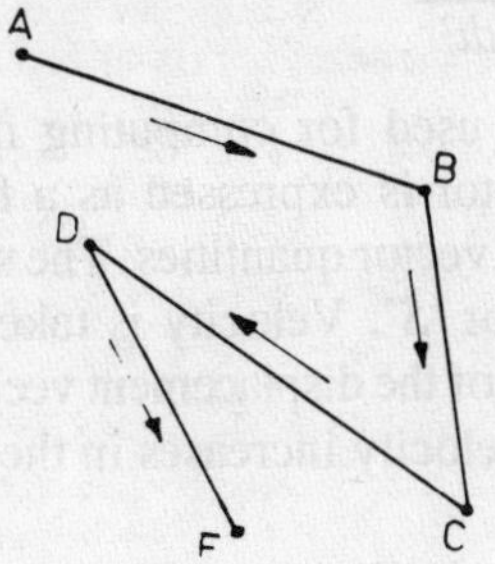

Fig. 6.7

Rectilinear motion can be of uniform or variable velocity. In case of variable velocity, again the acceleration may be uniform or varying. In rectilinear motion, since the displacements and distances will both be measured along the same straight path the numerical values of *velocity* and *speed* will be the same at a particular *instant* of time.

Consider a particle moving 40 m due east in 5 seconds, and then moving 30 m due north in another 5 seconds. The total distance travelled is $(40 + 30) = 70$ m, but the displacement from the starting point is $\sqrt{40^2 + 30^2} = 50$ m, due θ° north of east. This is shown in Fig. 6.8.

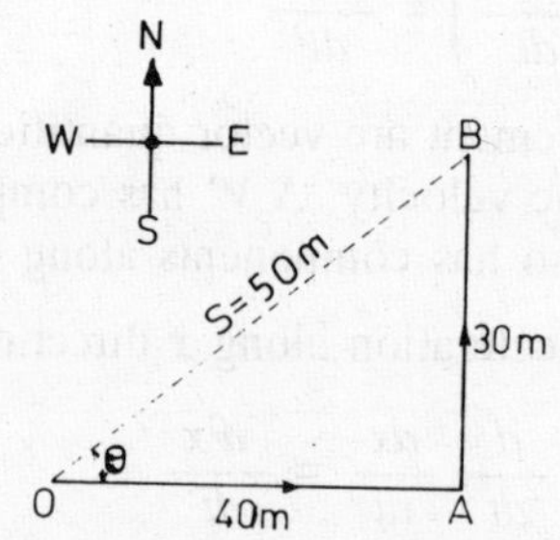

Fig. 6.8

The speed of the particle when between O to A is the same as the magnitude of the velocity between O and A. Similarly the speed between A and B is the same as the magnitude of the velocity between A and B. But the average speed between O and B is not the same as the velocity between O and B, since the displacement OB is not equal to the arithmetic sum of OA and OB. Thus, it should be remembered that the total distance travelled is not the same as the net displacement, and that the speed is equal to the magnitude of the velocity only in a straight path.

The general equations of motion also hold good for the rectilinear motion. The equations are

$$V = \frac{dS}{dt} \quad \text{... 6.3}$$

$$a = \frac{dV}{dt} = \frac{d^2S}{dt^2} \quad \text{... 6.4}$$

These equations can be used for computing motion parameters if the displacement or position vector is expressed as a function of time '*t*'. The quantities '*S*', '*V*', and '*a*', are vector quantities. The signs '*V*', and '*a*', depend on the sign of the basic vector '*S*'. Velocity is taken positive if the particle moves in the positive direction of the displacement vector. Similarly acceleration is considered positive if the velocity increases in the positive direction of the displacement.

In subsequent paragraphs rectilinear motion equations for different cases shall be studied for calculating motion parameters.

6.4.1 *Rectilinear Motion of a Particle having Zero Acceleration*

There are many examples in every day life, as well as in engineering, where one has to determine the motion of bodies moving with constant velocity or zero acceleration. A vehicle moving along a straight path with constant speed or zero acceleration is one such example. This happens when the vehicle in motion is under dynamic equilibrium of forces i.e. the accelerating and retarding forces are just equal and opposite. In other words the forces causing motion and forces opposing motion are balanced.

Fig. 6.9

In order to determine the motion parameters in such a case, let us consider a particle moving with a uniform velocity along a straight path *AP* as shown in Fig. 6.9. Since the velocity is constant, the displacement in equal intervals of time will be the same throughout the motion. Hence the instantaneous velocity of the particle at any instant will be equal to its average velocity at all the positions. Mathematically Instantaneous velocity at any time

$$V = V_{av} = \frac{\text{displacement '}S\text{' during the interval}}{\text{Time interval '}t\text{',}}$$

or $V = \dfrac{S}{t}$

or $S = V \cdot t$ ... 6.5

If the initial displacement of the particle is S_o from the origin at the time $t = 0$, the total displacement S of the particle at any time 't' with reference to the origin will be

$S = S_o + V \cdot t$ ... 6.5(a).

This result can also be obtained by considering a small displacement Δs in a small interval of time Δt.

i.e. $V = V_{av}$ (Constant) $= \dfrac{\Delta S}{\Delta t} = \dfrac{dS}{dt}$

as $\Delta t \to 0$

Or $V = \dfrac{dS}{dt}$

Or $dS = V \cdot dt$

$\therefore \quad S = \Sigma\, dS = \int dS$

$= \int V \cdot dt = = V \int dt$ (As V = constant)

$= V \cdot (t) +$ Constant S_o, where S_o is the initial displacement at time $t = 0$ with respect to the origin.

$$\therefore \quad S = S_o + V \cdot t \qquad \text{... 6.5(a)}$$

Generally the origin is assumed at the starting point of motion at $t = 0$, and in such cases, Equation 6.5 will be applicable. When the origin is not at the starting point of motion, then Equation 6.5(a) is used.

Example 6.1

The motion of a particle is defined by the displacement time relation $S = t^3 - 9t^2 + 15t + 5$, in which 'S' and 't' are expressed in metres and seconds respectively. Determine

(a) when the velocity will be zero

(b) the position and the total distance travelled when the velocity becomes zero the second time.

(c) the position and the total distance travelled when the acceleration becomes zero.

(d) total distance travelled when the velocity becomes $+15$ ms^{-1} the second time.

Solution

$$S = t^3 - 9t^2 + 15t + 5 \qquad \text{.... (1)}$$

$$V = \frac{dS}{dt} = 3t^2 - 18t + 15 \qquad \text{.... (2)}$$

$$a = \frac{dV}{dt} = 6t - 18 \qquad \text{.... (3)}$$

(a) $V = 0 = 3t^2 - 18t + 15$

$\therefore \quad t = 1$ or 5 sec.

Thus velocity first becomes zero at $t = 1$ sec, and then again at $t = 5$ secs.

(b) at $t = 0$, $V_0 = 0 - 0 + 15 = +15$ ms^{-1}

at $t = 1$, $V_1 = 0$

at $t = 2$, $V_2 = 3(2)^2 - 18(2) + 15 = -9$ms^{-1}

at $t = 3$, $V_3 = 3(3)^2 - 18(3) + 15 = -12$ms^{-1}

at $t = 4$, $V_4 = 3(4)^2 - 18(4) + 15 = -9$ms^{-1}

at $t = 5$, $V_5 = 3(5)^2 - 18(5) + 15 = 0$

The velocity first reduces to zero at $t = 1$ sec., and then becomes negative and again becomes zero at $t = 5$ secs. Thus the distance travelled upto $t = 5$ secs is the arithmetic sum of displacements from $t = 0$ to $t = 1$, and $t = 1$ to $t = 5$ seconds.

$$S_0 = 0 - 0 + 0 + 5 = +5 \text{ m}$$
(at $t = 0$)
$$S_1 = 1^3 - 9(1)^2 + 15(1) + 5 = +12 \text{ m}$$
(at $t = 1$ sec)
$$S_5 = 5^3 - 9(5)^2 + 15(5) + 5 = -20 \text{ m}$$
(at $t = 5$ sec)

$\therefore$ Total distance travelled $= |S_1 - S_0| + |S_5 - S_1| = AB + BC$ (see Fig. 6.10)

$= |12 - 5| + |-20 - 12| = 39$ m

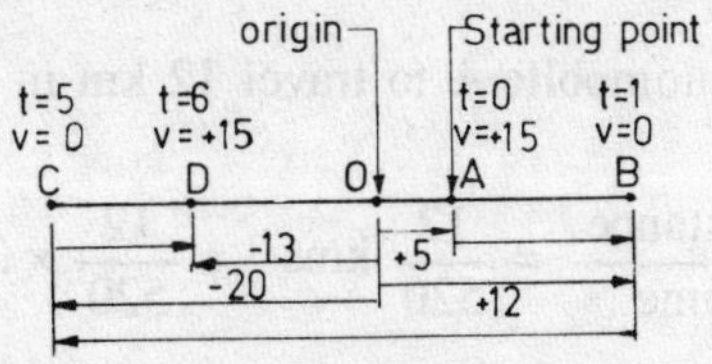

Fig. 6.10

(c) $a = 0 = 6t - 18$

$t = 3$ sec.

$$S_3 = 3^3 - 9(3)^2 + 15(3) + 5 = -4 \text{ m}$$
(at $t = 3$ sec)

$\therefore$ Distance travelled $= |S_1 - S_0| + |S_3 - S_1|$

$= |12 - 5| + |-4 - 12| =$ **23 m.**

(d) $V = +15 = 3t^2 - 18t + 15$

Or $3t^2 - 18t = 0 \quad \therefore t = 0$ or 6

at $t = 6$ secs, velocity becomes $+ 15\text{ms}^{-1}$ the second time

$\therefore S_6 = 6^3 - 9(6)^2 + 15(6) + 5 = -13$ m

(at $t = 6$ sec)

$\therefore$ Distance travelled $= |S_1 - S_0| + |S_5 - S_1| + |S_6 - S_5|$

$= AB + BC + CD$

$= 7 + 32 + 7 =$ **46 m.**

The motion of the particle occurs from A to B, B to C and C to D as shown in Fig. 6.10.

Example 6.2

An automobile '*A*' moving with a uniform velocity crosses a service station on a straight road. Another automobile B following the first one crosses the same service station 40 seconds later. The automobile B overtakes A at a distance of 12 km from the service station. If the automobile B has a uniform velocity of 90 km/hour, find the uniform velocity of the automobile A.

Solution

Both the automobiles are having, uniform rectilinear motion.

$$S = 12 \text{ km}, \; t_A = t, \; t_B = (t_A - 40) \text{ secs}$$

$$V_A = ? \qquad V_B = 90 \text{ km/hour} = \frac{90}{3600} = \frac{1}{40} \text{ kms}^{-1}$$

Time required by the second automobile B to travel 12 km, $t_B = \dfrac{12 \times 3600}{90}$

$$= 480 \text{ sec}$$

Thus the Ist automobile A to travel 12 km in time $t_A = (t_B + 40)$

$$= 520 \text{ seconds}$$

$$\therefore \text{ Velocity } V_A = \frac{\text{Distance}}{\text{time}} = \frac{12}{520} \text{ kms}^{-1} = \frac{12}{520} \times 3600 = 83.08 \text{ km/hour}$$

Example 6.3

A workpiece is being machined on a planer. The cutting stroke of the machine is 1200 mm. The total time of cutting and return stroke for the machine is 10 seconds. If the cutting speed is half the return speed, calculate the cutting stroke speed and cutting stroke time.

Solution

Stroke length S = 1200 mm

$$t_c + t_r = 10 \text{ sec.}$$

Let the cutting stroke speed, and the return stroke speed be V_c and V_r respectively,

$$\therefore \qquad 2\,V_c = V_r$$

Time required for cutting stroke, $t_c = \dfrac{1200}{V_c}$

Time required for return stroke, $t_r = \dfrac{1200}{V_r} = \dfrac{1200}{2V_c} = \dfrac{600}{V_c}$

$$\therefore \text{ Total time of the cycle } t = \frac{1200}{V_c} + \frac{600}{V_c} \qquad \text{... (1)}$$

$$10 = \frac{1200}{V_c} + \frac{600}{V_c}$$

$$V_c = \frac{1800}{10} = 180 \text{ mms}^{-1}$$

$$\therefore t_c = \frac{1200}{180} = 6.67 \text{ sec.}$$

6.4.2 *Rectilinear Motion with Uniform Acceleration*

One of the most common cases of rectilinear motion is that of constant acceleration. This happens when a particle or a body is acted upon by a force of constant magnitude and direction. For example, a body falling freely under gravity is acted by a constant gravitational force and the body is under constant acceleration; a train acted upon by a constant draw bar pull will be under constant acceleration.

With the help of general differential equations of motion, the relations of displacement(S), velocity(V), acceleration(a), and time (t) for rectilinear motion with constant acceleration can be derived. Recalling differential equation 6.3 and 6.4, we have

$$V = \frac{dS}{dt},\ a = \frac{dV}{dt}$$

Rewriting the second equation we have $dV = a \cdot dt$ ... (i)

Consider a rectilinear motion of a particle from A to B and then to B′ with a constant acceleration as shown in Fig. 6.11.

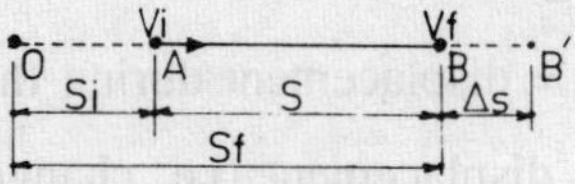

Fig. 6.11

In above equation (i), since the acceleration 'a' is constant, on integrating the equation $dV = adt$, we have

$$\int_{V_i}^{V_f} dV = \int_0^t a \cdot dt. = a \int_0^t dt.$$

or $\quad (V)_{V_i}^{V_f} = a(t)_0^t$

or $\quad (V_f - V_i) = a \cdot t - 0$

or $\quad V_f = V_i + a \cdot t$ ... 6.6

This is the relation of final velocity 'V_f' (instantaneous) at the end of time 't' in terms of initial velocity (V_i) at the beginning of the rectilinear motion with constant acceleration 'a'.

Rewriting differential equation for final velocity V_f at B we have

$$V_f = \frac{dS}{dt}$$

or $$dS = V_f\, dt \qquad \text{... (ii)}$$

The total displacement during the period from the start of motion to the end of time 't' can be obtained by integrating equation (ii) within limits. Thus

$$\int_{S_i}^{S_f} dS = \int_0^t V_f \cdot dt$$

where S_i is position vector at the start of the motion, and S_f is the final position vector at the end of time t.

or $$(S)_{S_i}^{S_f} = \int_0^t (V_i + at)\, dt,$$

or $$(S_f - S_i) = \left[(V_i \cdot t + \frac{at^2}{2}) \right]_0^t,$$

V_i is initial velocity, and is independent of time t, a is the constant acceleration.

or $$(S_f - S_i) = V_i \cdot t + \frac{1}{2}\, a \cdot t^2$$

or $$S = V_i \cdot t + \frac{1}{2}\, a \cdot t^2 \qquad \text{... 6.7}$$

where $S = (S_f - S_i)$ = displacement during the time t.

This equation represents displacement (i.e. change of position $(\overline{S_f - S_i})$ in terms of initial velocity V_i, constant acceleration 'a' and time interval 't'. Rewriting the differential equations, we have

$$V = \frac{dS}{dt}$$

$$a = \frac{dV}{dt}$$

Dividing, we get $\frac{V}{a} = \frac{dS}{dV}$ or $VdV = adS$ ... (iii)

Integrating the above equation (iii) within limits, we have

$$\int_{V_i}^{V_f} VdV = \int_{S_i}^{S_f} adS$$

or $$\left[\left(\frac{V^2}{2} \right) \right]_{V_i}^{V_f} = \left[\left(a \cdot S \right) \right]_{S_i}^{S_f}$$

or $$\frac{1}{2}\,(V_f^2 - V_i^2) = a\,(S_f - S_i)$$

Putting $(S_f - S_i)$ = displacement 'S'

$$\frac{1}{2}(V_f^2 - V_i^2) = a \cdot S$$

or $$V_f^2 = V_i^2 + 2\,a \cdot S \qquad \text{... 6.8}$$

This equation relates final velocity V_f at a certain time, initial velocity V_i, constant acceleration 'a' and the total displacement 'S' during the interval. The above Equations 6.6, 6.7 and 6.8 are basic Equations of rectilinear motion with constant acceleration, and are used to find the magnitude of unknown quantities from the magnitudes of the given quantities. The direction of these vector quantities is directly linked to the direction of the position vector or displacement.

Alternative Approach

The above relations 6.6 to 6.8 can also be found by simple arithmetic considerations as follows.

Consider a particle moving with constant acceleration 'a' and initial velocity V_i at A, and velocity V_f at B as shown in Fig. 6.11.

Change in velocity from A to B = $(V_f - V_i)$

Rate of change of velocity = a

∴ Total change of velocity in time 't' in motion from A to B = $a \cdot t$

Thus $$V_f - V_i = a \cdot t$$

or $$V_f = V_i + a \cdot t \qquad \text{... 6.6}$$

Since the acceleration is constant, the average change in velocity is constant between A and B, and hence

$$\text{Average velocity from A to B} = \frac{(V_i + V_f)}{2}$$

∴ Displacement S from A to B = av · vel × time

or $$S = \frac{V_i + V_f}{2} \cdot t$$

or $$S = \frac{1}{2}(V_i + V_i + a \cdot t)\,t$$

or $$S = V_i \cdot t + \frac{1}{2}\,a \cdot t^2 \qquad \text{... 6.7}$$

Using the relation $V_f - V_i = at$, $S = \left(\dfrac{V_i + V_f}{2}\right) \cdot t$

Eliminating 't' from these equations, we get

$$\frac{S}{(V_f - V_i)} = \frac{(V_i + V_f)}{2\,a}$$

or $$V_f^2 - V_i^2 = 2\,a \cdot s$$

or $$V_f^2 = V_i^2 + 2\,a \cdot s \qquad \text{... 6.8}$$

The above relations can now be summarised as below

$$V_f = V_i + a \cdot t \quad \text{... 6.6}$$

$$S = V_i + \frac{1}{2} a \cdot t^2 \quad \text{... 6.7}$$

$$V_f^2 = V_i^2 + 2\, a \cdot S \quad \text{... 6.8}$$

It may be noted that the displacement (S) and time (t) in these equations are measured from the starting point of motion at which the initial instantaneous velocity is V_i; V_f is the final instantaneous velocity at the end of time 't'.

It may further be noted that these equations give the magnitudes of different quantities while the directions are based on the initial direction of motion. All quantities are considered positive if they are increasing in the positive direction of the initial motion, and are considered negative in the direction opposite to the initial direction of the motion.

In case of rectilinear motion caused by two agencies in two different directions the resultant rectilinear motion can be determined by the law of parallelogram of vectors.

Example 6.4

A vehicle having rectilinear motion is moving with a velocity of 30 kmh^{-1} reaches a down grade and starts accelerating at a uniform rate of 0.10 ms^{-2}. Determine the velocity of the vehicle and the distance travelled after 2 minutes from the time it reaches the down grade.

Solution

$$V_i = 30 \text{ kmh}^{-1} = 30 \times \frac{10}{36} = 8.33 \text{ ms}^{-1}$$

$$a = 0.10 \text{ ms}^{-2};\ t = 2 \times 60 = 120 \text{ sec.}$$

$$V_f = V_i + a \cdot t$$

or $V_f = 8.33 + 0.1 \times 120 = 20.33 \text{ ms}^{-1}$ (= 73.2 kmh^{-1})

Displacement

$$S = V_i \cdot t + \frac{1}{2} a \cdot t^2$$

$$= 8.33 \times 120 + \frac{1}{2} \times (0.10)\,(120)^2$$

$$= 1000 + 720 = 1720 \text{ m}$$

Example 6.5

A vehicle having rectilinear motion is moving with a velocity of 60 kmh^{-1}. The vehicle is brought to a halt in a distance of 110 m by the application of brakes. Determine the uniform rate of deceleration and the time for which the brakes were applied.

Solution

$V_i = 60 \text{ kmh}^{-1} = 60 \times \dfrac{10}{36} = 16.66 \text{ ms}^{-1}$

$S = 110$ m; $a = ?$ $t = ?$ $V_f = 0$

$V_f^2 = V_i^2 + 2\, a \cdot S$

$0^2 = (16.66)^2 + 2\, a(110)$, $a = -1.26 \text{ ms}^{-2}$

$V_f = V_i + a \cdot t$, $0 = 16.66 + (-1.26)\, t$

$t = 13.2$ secs

Example 6.6

A body in rectilinear motion is found to travel 24 metres in 3 seconds and 80 metres in the next 5 seconds. What distance will it travel in 10 seconds from the starting point, if its acceleration is constant throughout?

Solution

S_3 ($t = 3$ sec) = 24 m, S_8 ($t = 3 + 5$ sec) = 24 + 80 = 104 m

$S = V_i\, t + \dfrac{1}{2}\, a \cdot t^2$

Putting the values we have

$24 = V_i\,(3) + \dfrac{1}{2}\, a \cdot (3)^2 = 3V_i + 4.5\, a$ ... (1)

Also $104 = 8V_i + 32\, a$ ... (2)

Solving (1) and (2)

$V_i = 5 \text{ ms}^{-1}$, $a = 2 \text{ ms}^{-2}$

$\therefore S_{10}$ (in 10 seconds) $= 5(10) + \dfrac{1}{2}\,(2)\,(10)^2 =$ **150 m**

Example 6.7

A train moving on a straight path starts from a station *A* and moves with an acceleration of 0.20 ms^{-2} for an interval of 1.5 minutes. Then it moves with uniform velocity for 10 minutes. Thereafter the train slows down by decelerating at 0.5 ms^{-2} and halts at a station *B*. Determine the distance travelled by the train between the stations *A* and *B*.

Solution

$V_{i_1} = 0$; $a_1 = +0.2 \text{ ms}^{-2}$; $t_1 = 1.50$ min = 90 secs

$V_{f_1} = V_2 = ?$; $a_2 = 0$; $t_2 = 10$ min = 600 secs

$V_{i_3} = V_2$, $V_{f_3} = 0$; $a_3 = -0.5 \text{ ms}^{-2}$; $t_3 = ?$

Consider the journey of the train in 3 portions, viz. accelerating, uniform and decelerating.

Accelerating journey

$V_{f_1} = V_i + a_i t = 0 + 0.2(90) = 18 \text{ ms}^{-1}$

$S_1 = V_i \cdot t + \dfrac{1}{2} a_1 (t_1)^2 = 0 + \dfrac{1}{2} (0.2)(90)^2 = 810 \text{ m}$

Uniform velocity journey

$V_{i_2} = V_{f_1} = 18 \text{ ms}^{-1}; a_2 = 0, t = 600 \text{ secs}$

$S_2 = 18(600) + 0 = 10800 \text{ m}$

Decelerating journey

$V_{i_3} = V_{i_2} = 18 \text{ ms}^{-1}; a_3 = -0.5 \text{ ms}^{-2}$

$V_{f_3} = 0$

$V_{f_3}^2 = V_{i_3}^2 + 2\, a_3 \cdot S_3 = (18)^2 + 2(-0.5)S_3$

Or $S_3 = \dfrac{18 \times 18}{2 \times 0.5} = 324 \text{ m.}$

$\therefore$ Total distance $S = S_1 + S_2 + S_3$

$= 810 + 10800 + 324 = \mathbf{11934 \text{ m}}$

Example 6.8

A motorist is driving his car at 90 kmh^{-1} when he observes that a traffic light 240 m ahead turn red. The traffic light is timed to remain red for 15 seconds. If the motorist wishes to pass the light without stopping just as it turns green, determine

(a) the required uniform deceleration of the car

(b) the speed of the car as it passes the traffic light.

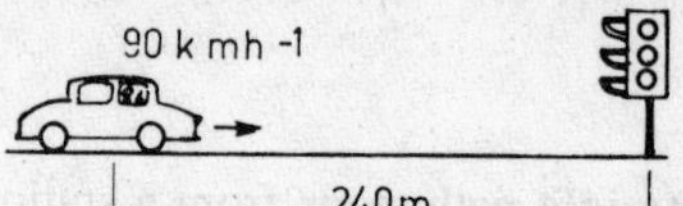

Fig. 6.12

Solution

$V_i = 90 \text{ kmh}^{-1} = 25 \text{ ms}^{-1}$

$S = 240 \text{ m}, t = 15 \text{ sec}, a = ?\ V_f = ?$

$S = V_i\, t + \dfrac{1}{2}\, a \cdot t^2$, or $240 = 25 \times 15 + \dfrac{a}{2} (15)^2$

or $a = (-25 \times 15 + 240) \dfrac{2}{225} = -1.2 \text{ ms}^{-2}$

$V_f = V_i + a \cdot t = 25 + (-1.2)(15) = 7.0 \text{ ms}^{-1}$

Example 6.9

A car starts from origin, with a certain initial velocity in a straight line, with an acceleration of $a = -4.50$ ms^{-2}. The velocity after 6 seconds becomes 9 ms^{-1} in the opposite direction. Determine

(a) The initial velocity V_i,

(b) The net displacement, and

(c) The total distance travelled.

Solution

V_i = ?, $a = -4.5$ ms^{-2}; $V_f = -9$ ms^{-1}, $t = 6$ secs

(a) $V_f = V_i + at$

$-9.0 = V_i - 4.5(6)$, $\therefore$ $V_1 = 18$ ms^{-1}

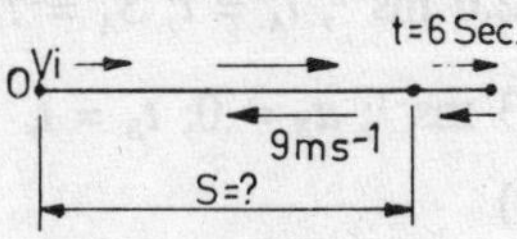

Fig. 6.13

(b) $S = V_i \cdot t + \frac{1}{2} a \cdot t^2 = 18 \times 6 - \frac{1}{2} \times 4.5(6)^2 = 27$ m

(c) Since the velocity changes its sign during its motion, the distances travelled will be in opposite directions, and hence the total distance travelled will be found by taking arithmetic sum of two distances in opposite directions without considering their signs.

Let the velocity becomes zero in t_1 time. Then

$0 = 18 - 4.5\, t_1$, $\therefore$ $t_1 = 4$ secs

Distance in 4 secs = Displacement in 4 secs

$$S_1 = 18(4) - \frac{4.5}{2}(4)^2 = 36 \text{ m}$$

Distance travelled in next 2 secs (t = 4 secs to t = 6 secs)

$S_2 = 0 - \frac{4.5}{2}(2)^2 = -9$ m (After 4 secs, the velocity becomes zero; thus $V_i = 0$ for next motion and t = 2 secs)

Thus total distance = $|36| + |-9| = 45$ m

Example 6.10

Two cars A and B have a velocity of 120 kmh^{-1} in the same direction. car A is 200 m ahead of car B when the brakes are applied to car A, causing it to decelerate at a constant rate of $2ms^{-2}$. In what time will car *B* overtake *A*, and how far will each car have travelled? Find the velocity of Car *A* at the time of overtaking.

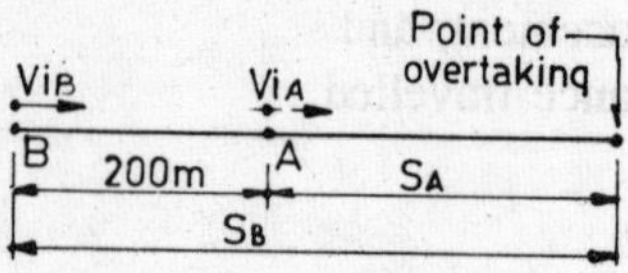

Fig. 6.14

Solution

Car A: $V_{i_A} = 120 \text{ kmh}^{-1} = 120 \times \dfrac{10}{36} = 33.3 \text{ ms}^{-1}$,

$V_{f_A} = ?;\ a_A = -2.0 \text{ ms}^{-2},\ t_A = t;\ S_A = ?$

Car B: $V_{i_B} = V_{f_B} = 33.3 \text{ ms}^{-1};\ a_B = 0;\ t_B = t,$

$S_B = (S_A + 200)$

$S = V_i\, t + \dfrac{1}{2}\, at^2$

$S_A = (33.3)\, t + \dfrac{1}{2}(-2.0)\, t^2 = (33.3\, t - t^2)$ ... (1)

$S_B = 33.3t + 0 = 33.3t$ ... (2)

But $S_B = S_A + 200$

$\therefore \quad 33.3t = (33.3t - t^2) + 200$

or $\quad t = 10\sqrt{2} = \mathbf{14.14 \text{ secs}}$

$\therefore$ Substituting, we get

$S_A = 33.3 \times 10\sqrt{2} - 200 = \mathbf{271.33 \text{ m}}$

$S_B = 33.3 \times 10\sqrt{2} = \mathbf{471.33 \text{ m}}$

V_f (at the time of overtaking) $= V_i + at$

$V_{f_A} = 33.33 - 2.0\,(10\sqrt{2}) = \mathbf{5.05 \text{ ms}^{-1}}$

Example 6.11

A small boat has a velocity of 3 ms^{-1} across the direction of flow of water in the river which is flowing at 4 ms^{-1}. Find the resultant velocity of the boat and determine the displacement in 5 secs after starting the motion of the boat from the left bank.

Solution

$V_x = 3 \text{ ms}^{-1};\ V_y = 4 \text{ ms}^{-1}$

Resultant velocity $V = \sqrt{V_x^2 + V_y^2} = \sqrt{3^2 + 4^2} = 5 \text{ ms}^{-1}$

Direction of the velocity θ with the direction of flow

$$\theta_y = tan^{-1}\left(\frac{V_x}{V_y}\right) = \tan^{-1}\left(\frac{3}{4}\right)$$

Example 6.12

A ship starts from rest with an acceleration of 1 ms^{-2} due east. From north there is strong wind which causes the ship to shift southward at a constant velocity of 5 ms^{-1}

(a) Find the resultant velocity and position of the ship with respect to starting point after 5 secs.

(b) Find the above after 10 secs.

Solution

V_y = 5 ms^{-1} (constant),

V_{x_i} = 0; a_x = 1 ms^{-2}, t = 5 secs

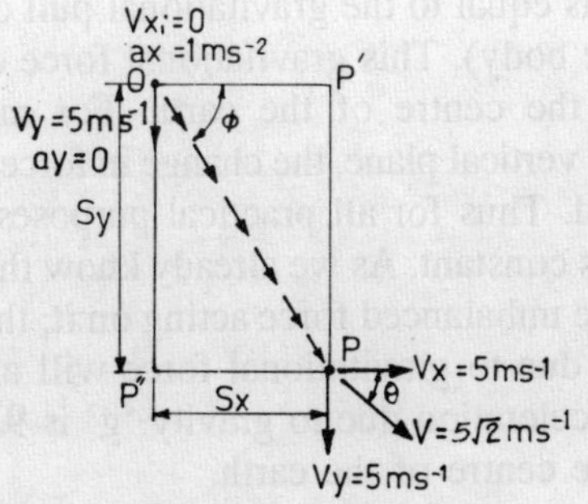

Fig. 6.15

Let the initial position of the ship be at O, and let it come to position P after 5 secs (See Fig. 6.15). After 5 secs, let the velocity of ship in eastern direction = V_x. Then V_x = 0 + 5 × 1 = 5 ms^{-1} (Eastward)

Also due to wind V_y = 5 ms^{-1} (Southward) constant

Resultant velocity $V = \sqrt{(5)^2 + (5)^2} = 5\sqrt{2}$ ms^{-1} = 7.07 ms^{-1}

$$\theta = \tan^{-1}\left(\frac{5}{5}\right) = 45° \text{ South of East}$$

Displacement

$$\text{Eastward } S_x = 0(5) + \frac{1}{2}(1)(5)^2 = 12.5 \text{ m}$$

$$\text{Southward } S_y = 5(5) = 25 \text{ m}$$

Thus, the position after 5 secs will be

$$S = \sqrt{S_x^2 + S_y^2} = \sqrt{12.5^2 + 25^2} = 27.95 \text{ m}$$

$$\theta = \tan^{-1}\left(\frac{25}{12.5}\right) = \tan^{-1}(2) = 63° \ 26' \text{ South of east.}$$

(b) $V_x = 0 + 1 \times 10 = 10 \text{ ms}^{-1}$

$V_y = 5 \text{ ms}^{-1}$

$\therefore \ V_r = \sqrt{5^2 + 10^2} = 5\sqrt{5} = 11.180 \text{ ms}^{-1}, \ \theta = 63° \ 26'$ East of South.

$$S_x = 0\,(10) + \frac{1}{2}\,(1)\,(10)^2 = 50 \text{ m}$$

$$S_y = 5(10) + \frac{1}{2}\,(0)\,(10)^2 = 50 \text{ m}$$

$$\therefore \quad S = \sqrt{50^2 + 50^2} = 50\sqrt{2} = 70.7 \text{ m}$$

along direction $\phi = 45°$ East of South.

6.4.3 *Rectilinear Motion Under Gravity*

For a body falling freely or projected vertically upwards, the unbalanced force acting on the body is equal to the gravitational pull of the earth (also known as the weight of the body). This gravitational force depends on the distance of the body from the centre of the earth. For motions involving small displacements in the vertical plane, the change in force of gravity is very small, and hence neglected. Thus for all practical purposes the gravitational force acting on the body is constant. As we already know the acceleration of a body is proportional to the unbalanced force acting on it; therefore the acceleration acting on the body due to gravitational force will always be constant. The average value of acceleration due to gravity 'g' is **9.80 ms^{-2}**, and is always directed towards the centre of the earth.

Reference Axes

In solving problems of freely falling bodies in upward direction, with respect to the earth's surface as origin, the axis is assumed positive. Thus all the displacements, velocities and accelerations acting in the *upward* direction with respect to the origin at the earth's surface are considered *positive,* and those occuring *downward* are considered *negative.* Since the acceleration due to gravity is directed towards the centre of the earth, it is always taken as -9.80 ms^{-2} with respect to the upward positive motion. The signs of various quantities of motion are explained in Fig. 6.16 for a particle in a vertical direction.

Consider the motion of a body thrown vertically up from the origin O at the earth's surface. The body first goes up till it reaches the highest point *A*, and then it starts falling from *A* and again passes through the origin O, and continues to fall downwards below the reference ground level to a point *B* in a well or a shaft. The various quantities of motion have different signs during the particle's or body's motion in different zones. These are given below for clarification.

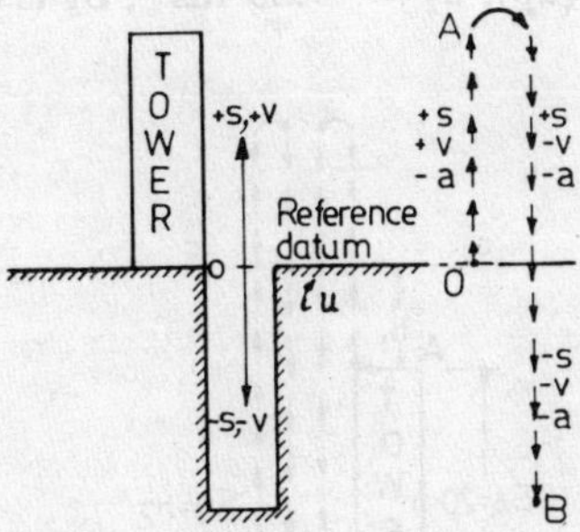

Fig. 6.16

O to A (Upward Journey)

(i) Displacement S (or position vector) + ve
(ii) Velocity V + ve
(iii) Acceleration due to gravity 'a' – ve

A to O (Downward Journey above the Origin O)

(i) Displacement S (or position vector) + ve
(ii) Velocity V – ve
(iii) Acceleration 'a' due to gravity. – ve

O to B (Downward Journey below the Origin O)

(i) Displacement S – ve
(ii) Velocity V – ve
(iii) Acceleration 'a' due to gravity. – ve

Thus, in the above case the displacements and velocities measured upwards from the origin are taken as positive, and those measured downward are taken as negative. In case the initial motion is downwards, the +ve velocities and +ve acceleration is considered downwards with reference to the origin at the starting point.

Example 6.13
A stone is thrown vertically up with a velocity of 15 ms^{-1} from the top A of a tower 20 m high. At the same instant a second stone is thrown vertically up with a velocity of 20 ms^{-1} from the base B of the tower. Find when and where the two stones will be at the same height?

Solution

$$V_{i_1} = 15 \text{ ms}^{-1} \text{ (up)}; \ a_1 = -9.80 \text{ ms}^{-2};$$

S_A(Position vector) = 20 m

V_{i_2} = 20 ms^{-1} (up); $a_2 = -9.80$ ms^{-2}, S_B (Position vector) = 0

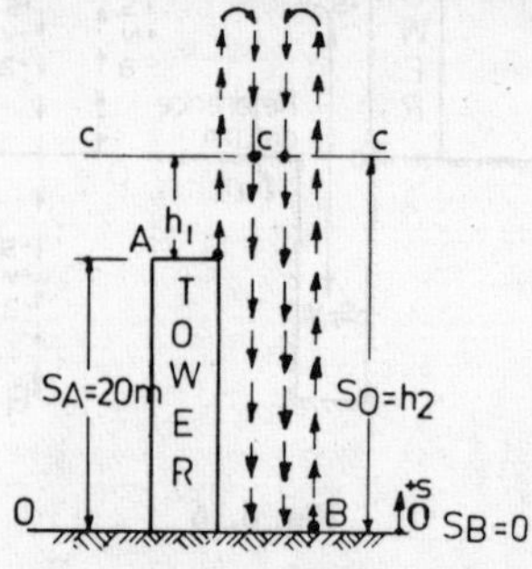

Fig. 6.17

Let the two stones are at a common level CC at a height h_1 above the top A of the tower, and h_2 above the base B of the tower. (See Fig. 6.17)

Displacement of stone No. 1 = (Position C) – (Position A) = $(S_C - S_A)$

Using the equation of rectilinear motion, we have

$$(S_C - S_A) = V_{i_1}\, t + \frac{a_1}{2}(t)^2,$$ where t is time after which they are at common level.

or $$(S_C - 20) = 15\,t - \frac{9.80}{2}(t)^2 = 15\,t - 4.9\,t^2 \quad \text{... (1)}$$

Considering the second stone thrown from B, we have

$$(S_C - S_B) = V_{i_2}\cdot t + \frac{a_2}{2}\cdot t^2$$

$$S_C - 0 = 20t - \frac{9.80}{2}t^2 = 20t - 4.9t^2 \quad \text{... (2)}$$

Substituting the value of S_C from Eqn (2) in Eqn (1) above, we have

$(20t - 4.9t^2) - 20 = 15t - 4.9t^2$, or $t = 4$ secs

$$\therefore \quad S_C = 20(4) - 4.9(4)^2 = +\,1.60 \text{ m}$$

Thus the two stones are at the same level 4 seconds after they are thrown and the common level is 1.60 m above the base of tower taken as reference datum.

Example 6.14

A stone is dropped down a well and the sound of splash is heard after $4\frac{8}{35}$ seconds. If the velocity of sound is taken as 343 ms^{-1}, find the depth of the well up to the water level.

Solution

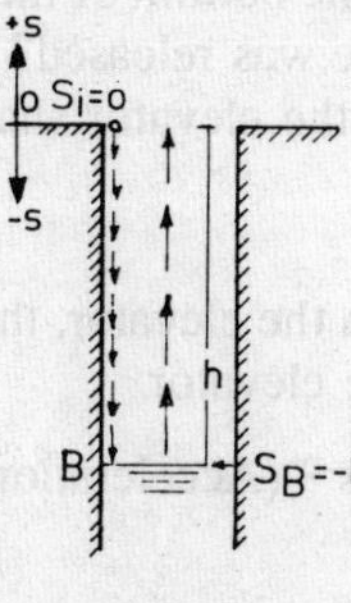

Fig. 6.18

$$V_{i_1} = 0;\ t = 4\frac{8}{35} \text{ secs} = (t_1 + t_2);\ S_i = 0$$

$$a_1 = -9.80 \text{ ms}^{-2};\ V_{i_2} = 343 \text{ ms}^{-1};\ S_B = -h = ?;\ a_2 = 0$$

Let the depth of the well be h metres, and the stone takes t_1 secs to reach the water level 'B' in the well. After the stone strikes water, the sound of splash travels up to reach the observer at the ground level (See Fig. 6.18).

Motion of Stone

$$\text{Displacement} = (S_B - S_i) = V_{i_1} \cdot t + \frac{1}{2}\ a_1\ t_1^2$$

or $\quad (-h - 0) = 0 - 4.9t_1^2$ or $h = 4.9t_1^2$ ... (1)

(As S_B position vector w.r.t ground $= -h$)

Motion of Sound of Splash

$$(0 - S_B) = V_{i_2} \cdot t_2 + \frac{1}{2}\ a_2 \cdot t_2^2$$

$$(0 - (-h)) = 343\ t_2 + 0, \text{ or } h = 343\ t_2 \qquad \text{... (2)}$$

$$\text{But } t_1 + t_2 = \text{total time} = 4\frac{8}{35} = \frac{148}{35} \text{ secs}$$

$$\therefore\ t_2 = \left(\frac{148}{35} - t_1\right) \qquad \text{... (3)}$$

Substitution t_2 in equation (2) above, we have

$$h = 343\left(\frac{148}{35} - t_1\right) \qquad \text{... (2a)}$$

$$\text{Thus } h = 4.9t_1^2 = 343\left(\frac{148}{35} - t_1\right)$$

or $t_1^2 + 70t_1 - 296 = 0, \quad \therefore\ t_1 = 4$ secs or -74 secs

Hence $h = 4.9t_1^2 = 4.9(4)^2 =$ **78.4 m**

Example 6.15

A stone is released from an elevator moving up at a speed of 6 ms^{-1}. The stone takes 4 seconds to reach the bottom of the elevator shaft. (a) How high was the elevator when the stone was released? (b) With what velocity does the stone strike the bottom of the elevator shaft?

Solution

Since the stone is released from the elevator, the initial velocity of the stone will be the same as that of the elevator.

i.e. $V_i = 6.0$ ms^{-1}; $a = -9.8$ ms^{-2} (Acceleration due to gravity)

$t = 4$ secs, $h = ?$

Using the motion equation Displacement $S = V_i\, t + \dfrac{a}{2}\, t^2$, we have

$$h = +6.0(4) - \frac{9.8}{2}(4)^2 = -54.4 \text{ m.}$$ (–ve sign indicates that the stone will be below the level at which it was released)

i.e., the point of release of stone will be 54.4 m above the bottom of the elevator shaft.

(b) $V_f = V_i + at = 6.0 - 9.8(4) = -33.2$ ms^{-1} (–ve sign shows that the velocity of the stone will be downward)

Example 6.16

Drops of water are observed to drip from a faucet at uniform interval of time. As any drop *B* begins to fall freely, the preceding drop *A* has already fallen 0.40 m. Determine the distance by which the drop *A* has fallen when the distance between *A* and *B* has becomes 1.24 m.

Solution

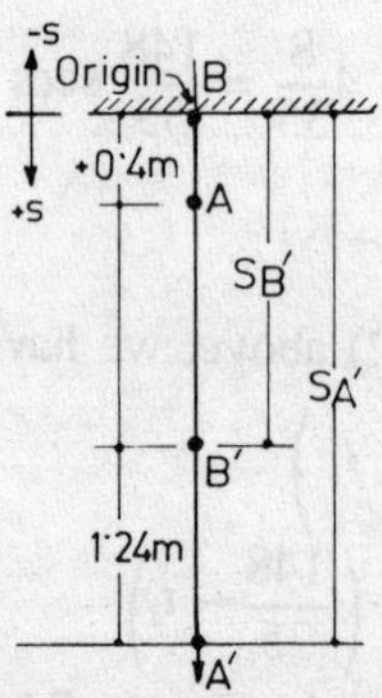

Fig. 6.19

Let the two consecutive drops *A* and *B* be as shown in Fig. 6.19. Let *A* travels from its position *A* to *A′*, when the drop *B* travels from *B* to *B′* in time interval '*t*'. Assume the initial position of *B* as the origin. Consider the downward direction as positive, and thus

$S_A = +0.4$ m, $S_B = 0$; $a_A = a_B = +9.8$ ms^{-2}

Using the equation of motion we have

Displacement of B = $(S_{B'} - S_B) = V_{i_t} \cdot t + \frac{1}{2}\, a \cdot t^2$

or $\qquad S_{B'} = +4.9t^2 \qquad$ ... (1)

Displacement of A = $(S_{A'} - S_A) = V_i\, t + \frac{a}{2}\, t^2$

or $\qquad (S_{A'} - 0.4) = V_i \cdot t + 4.9t^2 \qquad$ ... (2)

Substituting $\quad S_{A'} = (S_{B'} + 1.24)$, and $S_{B'} = +4.9t^2$, we have

$(S_{B'} + 1.24) - 0.4 = V_i\, t + 4.9t^2$

or $\qquad 4.9t^2 + 0.84 = V_i\, t + 4.9t^2$, or $V_i\, t = +0.84 \qquad$ ... (3)

V_i is the initial velocity of the drop A when it had already fallen by a distance of 0.40 m. Hence

$$V_i^2 = 0^2 + 2 \times 9.8 \times (0.40),$$

$\therefore \qquad V_i = \pm\sqrt{0.80 \times 9.8} = \pm\, 2.8$ ms^{-1} $\qquad$ ... (4)

Substituting this value in equation (3) above, we get

$\pm(\pm\, 2.80)\, t = +0.84$, $\quad \therefore\ t = \pm\, 0.30$ secs (Only positive value will be considered)

From equation (2) above

$$(S_{A'} - 0.40) = +2.8(0.3) + 4.9(0.3)^2 = 1.281 \text{ m}$$

$\therefore \qquad S_{A'} = +\, 1.681$ m

6.4.4 *Graphical Solutions of Problems Involving Time Motion Curves*

Let us consider a body falling freely under the influence of the gravitational pull of the earth. The body, as a result of this gravitational pull is subjected to a constant acceleration '*g*' at all instants of time. Representing acceleration of the body with respect to time '*t*' we get a straight line parallel to the time axis as shown in Fig. 6.20.

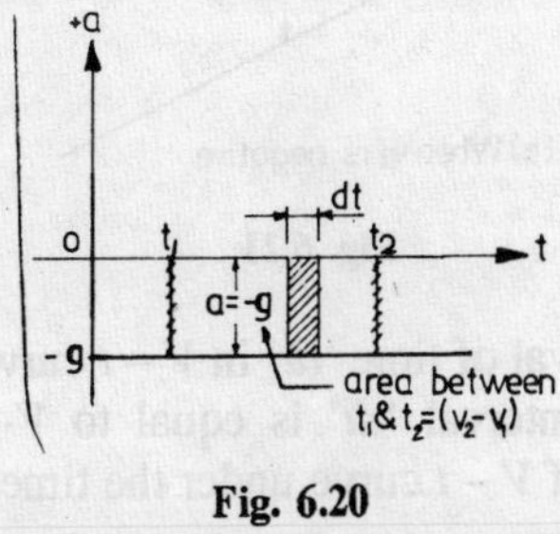

Fig. 6.20

From Fig. 6.20, we find that the line representing the acceleration due to gravity is below the time axis, and is negative. This is so because the acceleration due to gravity is directed towards the centre of the earth, and hence, taking the downward direction as negative, it is negative.

Consider a small interval of time dt, Area of the $a - t$ diagram under this time interval 'dt' is equal to $a \cdot dt = -gdt$. Since we know that $a \cdot dt = dv$ i.e. the area '$a \cdot dt$' represents the change in velocity 'dv' during the small time interval 'dt'. From this it can be concluded that *the area under '$a - t$' curve between any two instants of time 't_1', and 't_2' gives the change in velocity ($V_2 - V_1$) during the interval between t_1 and t_2.*
Knowing the initial velocity V_i of the body at time $t = 0$, we can find the velocity V at any instant of time t. In a particular case, if at $t = 0$;

$$V = V_i;$$

$$t = t_1,\ V = V_1,$$

$$t = t_2;\ V = V_2$$

Then we have $V_1 = V_i$ + Area of $a - t$ diagram from $t = 0$ to $t = t_1$

Or $\quad V_1 = V_i - g\,t_1 \qquad$ ... (1)

Similarly $\quad V_2 = V_i - gt_2 \qquad$ ... (2)

It is therefore possible to draw the $V - t$ curve from $a - t$ diagram as shown in Fig. 6.21.

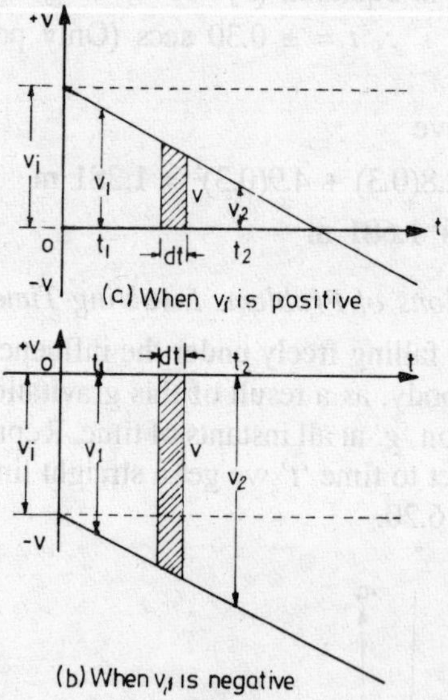

Fig. 6.21

Consider a small interval of time 'dt' in $V - t$ curve. The area of the $V - t$ diagram under this time interval 'dt' is equal to $V \cdot dt$. But we know that $V \cdot dt = dS$. Thus, the area of $V - t$ curve under the time interval 'dt' represents the change in the position of the particle or displacement of the particle, during the above time interval. From this it can be concluded that *the area under $V - t$ curve between any two instants of time t_1 and t_2 give the change in position $(S_2 - S_1)$ during this interval of time $(t_2 - t_1)$* if the initial position of the particle at time $t = 0$ is S_i, we have positions of the particle at any time t_1 and t_2 as given below

$(S_1 - S_i)$ = area of $V - t$ diagram from $t = 0$ to $t = t_1$

and $(S_2 - S_i)$ = area of $V - t$ diagram from $t = 0$ to $t = t_2$

Thus $S_1 = S_i$ + Area of $V - t$ diagram from $t = 0$ to $t = t_1$

$S_2 = S_i$ + Area of $V - t$ diagram from $t = 0$ to $t = t_2$

The area under curve is taken with its proper sign i.e. positive or negative depending upon whether the area is above or below the time axis.

Graphical Representation and Interpretation of Equations of Geometry of Motion

Recalling the general equations of motion, and examining the curves of motion parameters with respect to time shown in Figs. 6.22 to 6.24, we have

$$V = \frac{dS}{dt}, \text{ or } V \cdot dt = dS \qquad \text{... (1)}$$

$$a = \frac{dV}{dt}, \text{ or } a \cdot dt = dV \qquad \text{... (2)}$$

From equation (1) it is evident that the total displacement during the period from $t = 0$ to $t = t$ will be the summation of small displacements i.e.

$$\int_{S_i}^{S} dS = (S - S_i) = \int_{t=0}^{t=t} V \cdot dt$$

or $(S - S_i)$ = Area of $V - t$ curve from $t = 0$ to $t = t$

i.e. Displacement during certain time interval = Change of position during the time interval = Area of V–t curve during the time interval considered ... (6.9)

From equation (2), it can be similarly derived that

$$(V - V_i) = \int_{t=0}^{t=t} a \cdot dt$$

i.e. Change of Velocity during certain time interval = Area of $a - t$ curve during the time interval considered. ... (6.10)

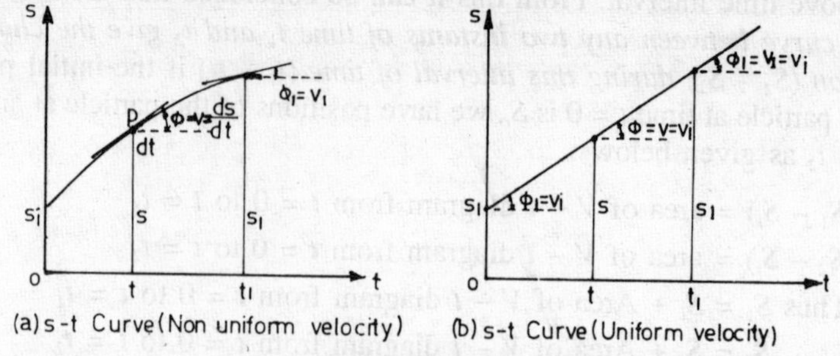

Fig. 6.22

Consider the displacement time curve (or position-time curve) shown in Fig. 6.22 a and b. The ordinate at any instant of time 't' represents the position of the particle at that instant, and if initial position of the particle is zero then the ordinate at any instant denotes directly the displacement of the particle during time interval 't'. The slope of the tangent drawn to $S - t$ curve at any point represents velocity of the particle at that instant of time 't'. In case the particle has a constant velocity, the slope $\frac{dS}{dt}$ will also be constant, and the $S - t$ curve in that case will be a straight line.

Now consider the velocity–time curve as shown in Fig. 6.23 a, b and c. The ordinate at any instant of time 't' represents the velocity of the particle at that instant of time. The slope of the tangent drawn to the $V - t$ curve at any point represents the acceleration at that instant. In case the acceleration is constant, the slope $\frac{dV}{dt}$ will be uniform, and the $V - t$ curve will be oblique straight line (shown in Fig. 6.23b), while if the acceleration is zero then the slope of $V - t$ curve will be zero, and the $V - t$ curve will be parallel to the time axis (shown in Fig. 6.23c). In all these $V - t$ curves, the area of

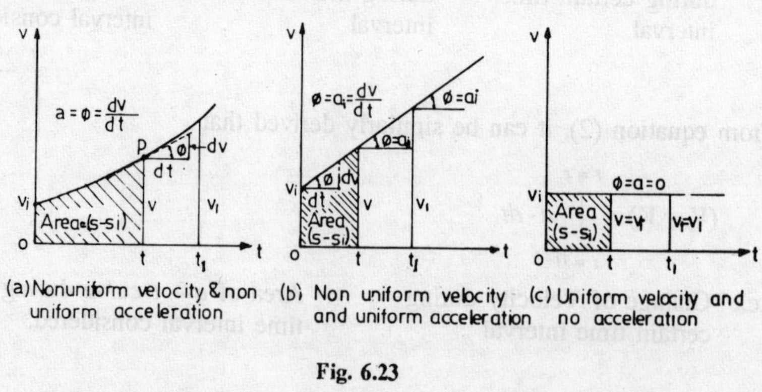

Fig. 6.23

the curve during certain time interval represents the total *change in position* of the particle during the interval considered. If the initial position of the particle is zero then the area under the curve for certain time interval represents the displacement directly.

Consider the acceleration–time curve as shown in Fig. 6.24 a, b and c. The ordinate at any instant of time t represents the acceleration at that instant of time. The slope of the tangent drawn to the $a - t$ curve at any point represents the rate of change of acceleration. In case the slope of $a - t$ curve is variable, the acceleration variation is non-uniform as shown in Fig. 6.24a, while if the slope is uniform, the rate of variation of acceleration is uniform as shown in Fig. 6.24b. If the slope is zero, it represents the case of uniform acceleration as shown in Fig. 6.24c. The area of $a - t$ curve during certain interval of time represents the change in velocity during that interval of time as shown in Figs 6.24a, b, c.

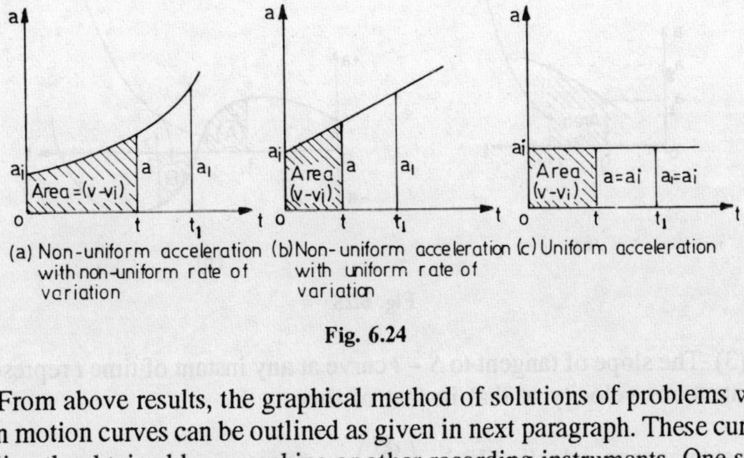

(a) Non-uniform acceleration with non-uniform rate of variation (b) Non-uniform acceleration with uniform rate of variation (c) Uniform acceleration

Fig. 6.24

From above results, the graphical method of solutions of problems with given motion curves can be outlined as given in next paragraph. These curves are directly obtained by a machine or other recording instruments. One such example is that of earth quake measuring instruments which record $S - t$, $V - t$ or $a - t$ curves with reference to certain axes.

Procedure for Graphical Solutions of Motion Parameters

(1) Area under the $V - t$ curve measures the *change in position* (Fig. 6.25a)

i.e. $(S_2 - S_1)$ = Shaded area of $V - t$ curve during time interval t_1 to t_2

Positive area above t-axis represents increase in position vector while area below t-axis indicates decrease in position vector (Fig. 6.25b).

(2) Area under the $a - t$ curve measures the *change in velocity* during the interval considered while it does not give the velocity (Fig. 6.25a)

i.e. $(V_2 - V_1)$ = Shaded area of $a - t$ curve during the time interval t_1 to t_2.

Area above the t-axis indicates increase in velocity, while area below t axis indicates decrease in velocity (Fig. 6.25b).

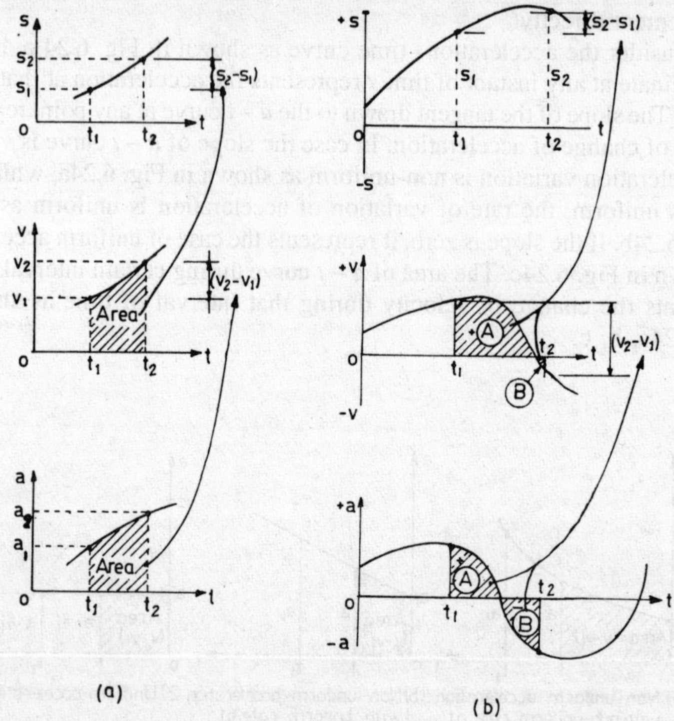

Fig. 6.25

(3) The slope of tangent to $S - t$ curve at any instant of time t represents instantaneous velocity at that instant of time

i.e. instantaneous velocity $(V)_p = \left(\dfrac{dS}{dt}\right)_p = (\theta)_p$

(See Fig. 6.26a)

or $V_1 = \theta_1$

In case the slope of the $S - t$ curve increases the velocity also increases and vice-versa. If the slope at any instant is positive the instantaneous velocity will be positive, while if the slope of $S - t$ curve is negative the instantaneous velocity will be negative (Fig. 6.26a and b).

(4) The slope of tangent to $V - t$ curve at any instant of time 't' represents the instantaneous acceleration at the instant of time t

i.e. $(a)_p = \left(\dfrac{dV}{dt}\right)_p = \phi_p$ or $a_1 = \phi_1$ (See Fig. 6.26b and c)

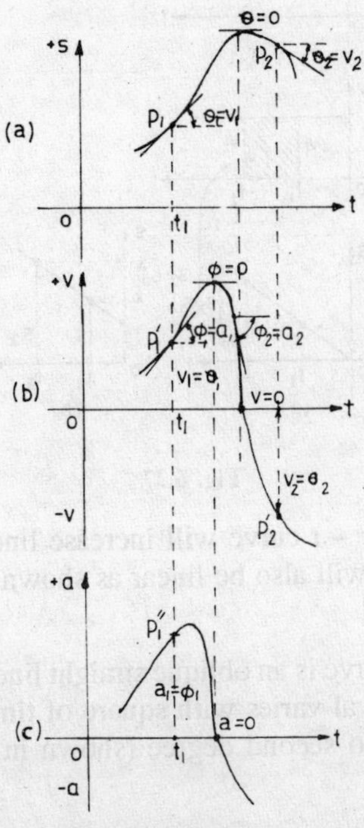

Fig. 6.26

If the slope ϕ of the tangent to $V - t$ curve increases, the acceleration also increases and vice versa, while if the slope at any instant is negative, the instantaneous acceleration will also be negative (See Fig. 6.26b and c).

From above results, the following conclusions can be drawn

(1) If the velocity is uniform, $V - t$ graph will be a straight line parallel to time–axis, and thus

(a) the $V - t$ curve will have zero slope of the tangent at any point, i.e. *acceleration will be zero* (shown in Fig. 6.27a) and

(b) the area in the curve will increase linearly with increase in t, and hence the change in position vector ($S_2 - S_1$) will increase linearly and is represented by the oblique straight line (shown in Fig. 6.27 a and b).

(2) If the acceleration is constant and different from zero, the $a - t$ curve will be a straight line parallel to time axis as shown in Fig. 6.28 a and b. Thus

(a) the slope of tangent at any point is zero, and hence there is no variation in acceleration,

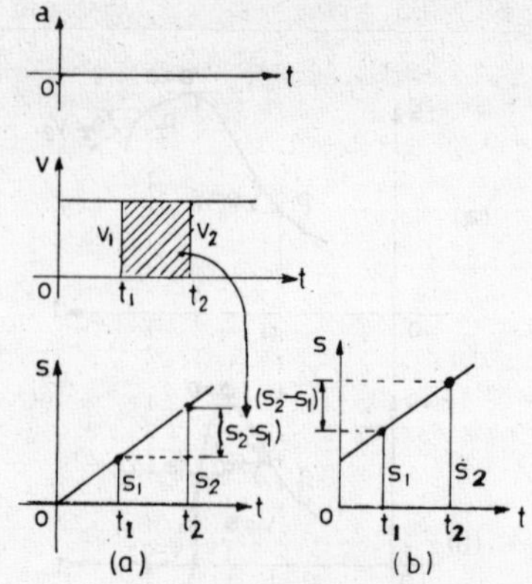

Fig. 6.27

(b) the area under $a-t$ curve will increase linearly with t, and hence the variation of velocity will also be linear as shown by the oblique straight line (Fig. 6.28 a, b), and

(c) since the $V-t$ curve is an oblique straight line, the area of $V-t$ curve during certain time interval varies with square of time t, and hence the $S-t$ curve will be parabolic to second degree (shown in Fig. 6.28 a, b).

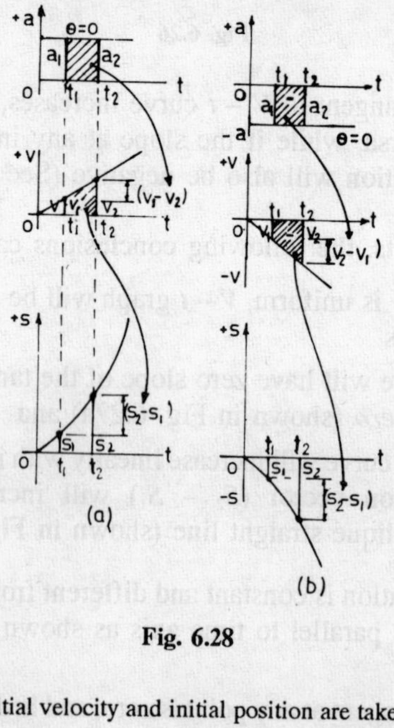

Fig. 6.28

Note: In above cases the initial velocity and initial position are taken as zero but other values can also be taken.

(3) If the acceleration varies linearly with time, the $a - t$ curve will be an oblique straight line (shown in Fig. 6.29 a, b), and thus

(a) the tangent at any instant has constant slope representing uniform rate of variation.

(b) the shaded area of $a - t$ curve will vary with square of time interval, representing square variation (parabolic) of velocity time curve (shown in Fig. 6.29 a, b) and

(c) the shaded area of $V - t$ curve will vary as cube of time, and hence $S - t$ curve will be a cubic parabola (shown in Fig. 6.29 a, b).

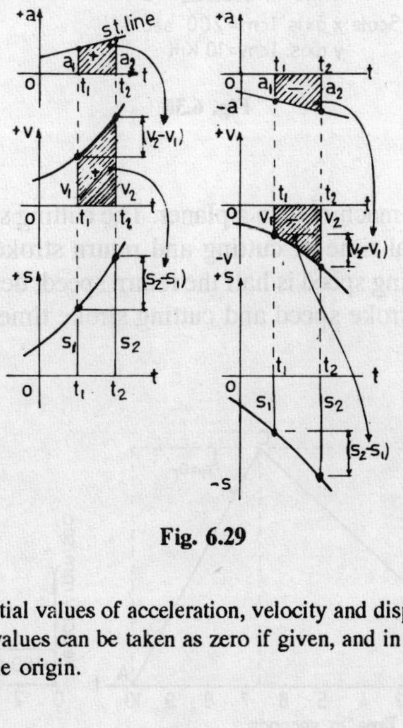

Fig. 6.29

Note: In above cases, initial values of acceleration, velocity and displacements are not shown as zero, but these values can be taken as zero if given, and in that case the curves shall be starting from the origin.

Example 6.17

A car moves with a uniform speed along a straight path and covers a distance of 2 km in 1 min 20 seconds. Determine the time required to cover a distance of 60 km using graphical method.

Procedure

Draw a displacement *v/s* time curve taking $S = 2$ km at $t = 80$ seconds, and $S = 0$, at $t = 0$. This curve will be a straight line because the velocity is uniform. From the curve so drawn, measure the time corresponding to a displacement $S = 60$ km(OB′). This time from the Fig. 6.30 will be $t = OB'$ = 2400 seconds.

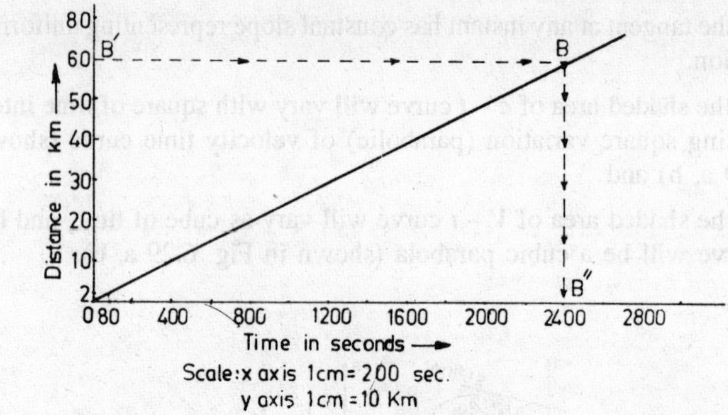

Fig. 6.30

Example 6.18

A work piece is being machined on a planer. The cutting stroke of the machine is 1200 mm. The total time of cutting and return stroke for the machine is 10 seconds. If the cutting speed is half the return speed, determine by graphical method, the cutting stroke speed and cutting stroke time. Show the graph of *V – t*.

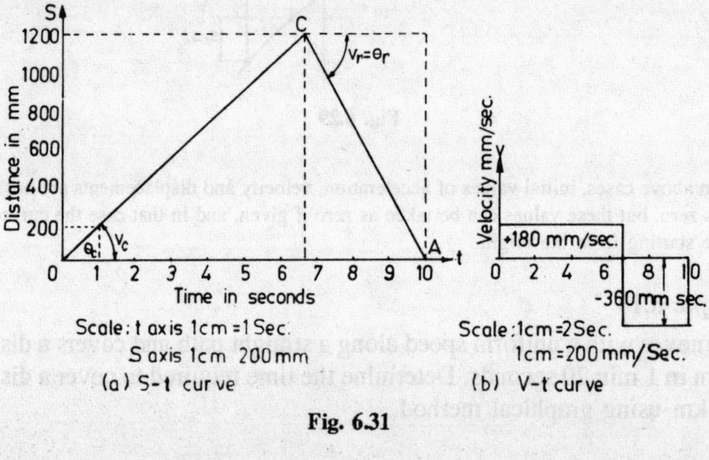

Fig. 6.31

Procedure

Since the return stroke velocity is double the cutting stroke velocity, the time of return stroke shall be half the cutting stroke time. Therefore, the time of cutting stroke and return stroke will respectively be 2/3(10) secs, and 1/3(10) sec. At the end of cutting stroke, and at the beginning of return stroke, the displacement shall be 1200 mm. Thus drawing the displacement–time curve as shown in Fig. 6.31 the slopes of tangents measured as

$$\theta_c = V_c = \frac{1200}{\frac{2}{3}(10)} = 180 \text{ mm/sec.}$$

$$\theta_r = V_r = \frac{-1200}{\frac{1}{3}(10)} = -360 \text{ mm/sec}$$ (–ve sign of velocity indicates opposite direction)

The time of cutting stroke $t_c = \frac{2}{3}(10) = 6.67$ secs.

Example 6.19

A train starts from station A and gains speed on a straight track at the rate of 1 ms^{-2} for the first 5 seconds, thereafter, at the rate of 2 ms^{-2} for next 10 seconds. It then moves at uniform speed for 1 min 20 seconds. Brakes are then applied which bring the train to stop in 8 secs. Draw $a - t$, $V - t$, and $S - t$ curves for the motion of the train. From the curves, find the distance between the stations and the maximum speed attained.

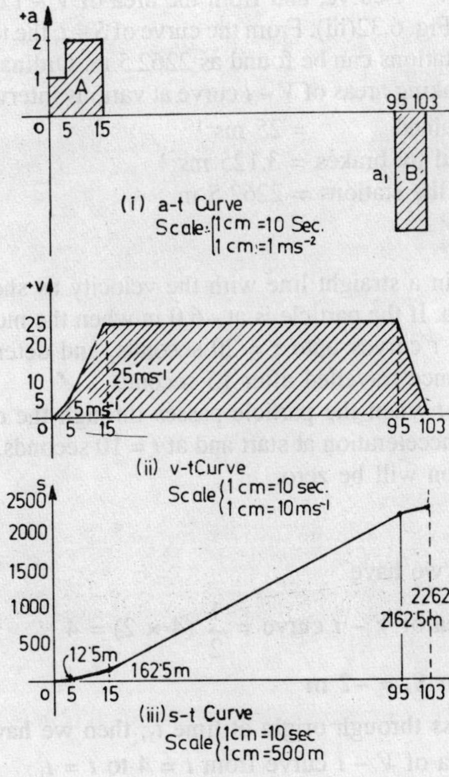

Fig. 6.32

Solution

Fig. 6.32(i) shows the $a - t$ curve in which the ordinate is $a = 1\ ms^{-2}$ from $t = 0$ to $t = 5$ seconds.

Area of $a - t$ diagram from $t = 0$ to $t = 5$ secs $= 1\ ms^{-2} \times 5$ secs $= 5\ ms^{-1}$

$\therefore\ V_5 - V_0 = 5\ ms^{-1}$ or $V_5 =$ **5 ms⁻¹** $(V_0 = 0)$

From $t = 5$ to $t = 15$ secs

Oridinate of $a - t$ curve $= 2\ ms^{-2}$

$\therefore$ Area of $a - t$ curve between $t = 5$ to $t = 15$ secs will be $= 2 \times 10 = 20\ ms^{-1}$

$\therefore\ V_{15} - V_5 = 20\ ms^{-1}$, or $V_{15} = 20 + 5 = \mathbf{25\ ms^{-1}}$

From $t = 15$ to $t = 95$ seconds, acceleration $a = 0$,

$\therefore\ V_{95} = V_{15} = \mathbf{25\ ms^{-1}}$

Since the velocity at $t = 103$ seconds is zero, we have $0 - V_{95}$ = area of $a - t$ curve from $t = 95$ to $t = 103$ seconds, or $(0 - 25)$ = area $a_1 \times (8)$

$\therefore\ a_1 = \dfrac{25}{8} = -3.125\ ms^{-2}$ (–ve sign shows retardation)

We complete the $V - t$ curve, and from the area of $V - t$ curve, draw $S - t$ curve as shown in Fig. 6.32(iii). From the curve of $S - t$, the total displacement between the two stations can be found as 2262.5 m. Ordinates of $S - t$ curve are computed by taking areas of $V - t$ curve at various intervals. From graphs

Max. Velocity attained $= 25\ ms^{-1}$

Retardation offered by brakes $= 3.125\ ms^{-2}$

Distance between the stations $= 2262.5$ m.

Example 6.20

A particle moves in a straight line with the velocity as shown by its $V - t$ curve in Fig. 6.33a. If the particle is at –6.0 m when the motion starts, draw the $S - t$ and $a - t$ curves upto $t = 20$ seconds, and determine

(i) the total distance travelled after 15 seconds

(ii) the values of t when the particle passes through the origin, and

(iii) the values of acceleration at start and at $t = 10$ seconds. Also find when the acceleration will be zero.

Solution

From $V - t$ curve, we have

$$S_4 - S_0 = \text{Area of } V - t \text{ curve} = \frac{1}{2}(4 \times 2) = 4$$

or $S_4 - (-6) = 4$, or $\mathbf{S_4 = -2\ m}$

Let the particle pass through origin at time t_1; then we have

$$0 - (-2) = \text{Area of } V - t \text{ curve from } t = 4 \text{ to } t = t_1$$

or $$2 = (t_1 - 4) \times 2$$

$\therefore$ $$t_1 = 5 \text{ secs (i.e. when } S_5 = 0)$$

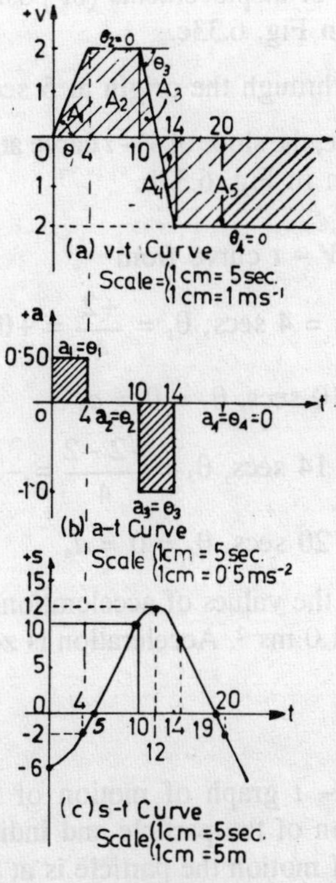

Fig. 6.33

From $V - t$ diagram, it is evident that $V = 0$ at $t = 12$ secs

Thus the position vector will be maximum when velocity = 0

$$\therefore\ S_{max.} - S_5 = 2(10 - 5) + \frac{1}{2}\ (2)\ (2) = 10 + 2$$

$S_{max.}$ (at t = 12 sec) = **12 m.** (Since $S_5 = 0$)

$$S_{15} - S_5 = 2\ (10 - 5) + \frac{1}{2}\ (2)\ (2) - \frac{1}{2}\ (2)\ (2) - 2(15 - 14)$$

or $$S_{15} = 10 + 2 - 2 - 2 = \mathbf{+8.0\ m}$$

Let S become zero again at $t = t_2$ say, then we have

$$S - S_{15} = -2(t_2 - 15)$$

$$0 - 8 = -2\ (t_2 - 15),\ t_2 = \mathbf{19\ seconds}$$

The total distance travelled at t = 15 secs will be the arithmetic sum of distances travelled up to t = 15 secs, i.e.

$$D_{15} = |0 - (-6)| + |12 - 0| + |8 - 12| = \mathbf{22\ m}$$

From various values of displacements (or position vectors), the $S - t$ curve is drawn as shown in Fig. 6.33c.

The particle passes through the origin at 5 secs and 19 secs.

To draw the $a - t$ curve, the slope of $V - t$ curve at important points is determined and plotted as shown in Fig. 6.33b.

Slope of tangent to $V - t$ curve from

$$t = 0 \text{ to } t = 4 \text{ secs; } \theta_1 = \frac{+2}{4} = \mathbf{+0.5\ ms^{-2}} = a_1$$

$$t = 4 \text{ to } 10 \text{ secs, } \theta_2 = 0 = a_2$$

$$t = 10 \text{ to } 14 \text{ secs, } \theta_3 = \frac{-2\ -2}{4} = \mathbf{-1.0\ ms^{-2}} = a_3$$

$$t = 14 \text{ to } 20 \text{ secs, } \theta_4 = 0 = a_4$$

From the $a - t$ curve the values of accelerations are at $t = 0$, $a = +0.5$ ms^{-2}; $t = 10$ secs, $a = 0$, $-$ 1.0 ms^{-2}. Acceleration is zero from $t = 4$ to 10 secs and after $t = 14$ secs.

Examples 6.21

Fig. 6.34a shows $a - t$ graph of motion of a particle. Draw $V - t$ and $S - t$ curves of motion of the particle and indicate the important values, if initially at the start of motion the particle is at the origin and has a velocity of -4.0 ms^{-1}.

Solution

From the $a - t$ curve, we find the area at interval of 1 sec, and determine the values of change in velocity during this interval. Since the initial velocity at the start is known, velocities at other times can be determined. $V - t$ curve with important values is shown in Fig. 6.34b.

e.g. $(V_5 - V_0)$ = Area of $a - t$ graph from $t = 0$ to 5 secs.

$$(V_5 - (-4) = \frac{1}{2}(4 \times 4) + \frac{1}{2} \times 1 \times (-2))$$

$$V_5 + 4 = 8 - 1, \text{ or } V_5 = \mathbf{+3\ ms^{-1}}$$

Similarly the $S - t$ curve is drawn by considering the area of $V - t$ curve. $S - t$ curve with important values is shown in Fig. 6.34c.

$(S_2 - S_0)$ = Area of $V - t$ diagram from $t = 0$ to 2 secs.

$$S_2 - 0 = \frac{2}{3}(-4 \times 2) = \mathbf{-5.33\ m}$$ (Area of parabola of 2nd degree = 2/3 × base × height).

From Fig. 6.34 b and c, it is evident that the position vector has maximum value, when the velocity is zero and changes its sign.

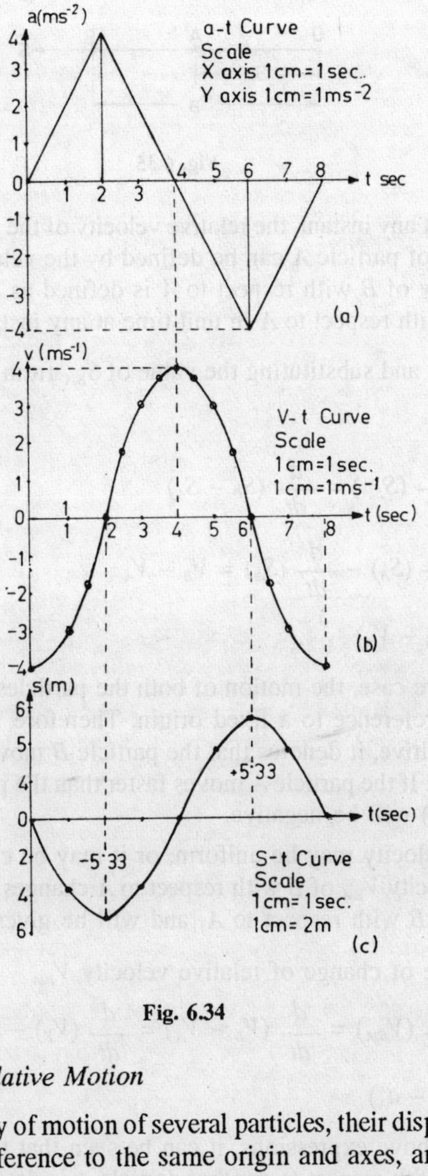

Fig. 6.34

6.4.5 *Rectilinear Relative Motion*

For studying geometry of motion of several particles, their displacements must be measured with reference to the same origin and axes, and their journey times should also be reckoned from the same instant. In such cases the coordinate axes are assumed to be stationary, and the relative position or displacement can easily be found by their vector difference. For example consider particles *A* and *B* moving along a certain direction OS, and having position vectors S_A and S_B respectively at any instant. (shown in Fig. 6.35) Since *A* and *B* are both moving, if at any instant the position of *B* is required in relation to the position of A, it may be defined by a vector

$$S_{B/A} = (S_B - S_A) \quad \text{... (1)}$$

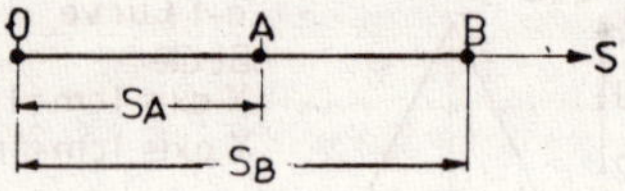

Fig. 6.35

Similarly at any instant, the relative velocity of the particle B with respect to the velocity of particle A can be defined by the velocity vector $V_{B/A}$. This relative velocity of B with respect to A is defined as the change in relative position of B with respect to A in unit time at any instant of time. Using the relation $V = \frac{dS}{dt}$, and substituting the value of $S_{B/A}$ from equation 1 above, we get

$$V_{B/A} = \frac{d}{dt}(S_{B/A}) = \frac{d}{dt}(S_B - S_A)$$

$$= \frac{d}{dt}(S_B) - \frac{d}{dt}(S_A) = V_B - V_A$$

Thus $V_{B/A} = (V_B - V_A)$ ... (2)

In the above case, the motion of both the particles occurs along a fixed direction with reference to a fixed origin. Therefore, if the quantity $V_{B/A} = (V_B - V_A)$ is positive, it denotes that the particle B moves faster with respect to the particle A. If the particle A moves faster than the particle B, the quantity $V_{B/A} = (V_B - V_A)$ will be negative.

Relative velocity may be uniform, or it may be changing with time. If the relative velocity $V_{B/A}$ of B with respect to A changes, there will be relative acceleration of B with respect to A, and will be given by

$a_{B/A}$ = rate of change of relative velocity $V_{B/A}$

Thus $a_{B/A} = \frac{d}{dt}(V_{B/A}) = \frac{d}{dt}(V_B - V_A) = \frac{d}{dt}(V_B) - \frac{d}{dt}(V_A)$

Or $a_{B/A} = (a_B - a_A)$ ... (3)

From the above expressions, it can be seen that the relative motion of any particle B with respect to another particle A is determined by taking the difference of the motion parameter, such as displacement, velocity or acceleration of the particles B and A with respect to the origin. In the above expressions, if the motion vector of A becomes zero, then the relative motion vector of the particle B with respect to the particle A becomes the same as the motion vector of B. Thus, if we superimpose a negative motion vector equal and opposite to that of A on both the particles A and B, we shall get the relative motion of 'B' with respect to A as resultant motion vector of 'B' because the resultant motion vector of A becomes zero.

For example consider two trains *A* and *B* moving on parallel tracks with velocities of 10 ms^{-1} and 40 ms^{-1} respectively in the same direction. Now let us find the relative velocity of *B* with respect to *A*. We superimpose a velocity of –10 ms^{-1} on both the motion vectors.

Apparent resultant velocity V'_A of the train $A = 10 - 10 = 0$,
and apparent resultant velocity V'_B of the train $B = 40 - 10 = 30$ ms^{-1}

Thus, $V_{B/A} = V_B - V_A = V'_B = 30$ ms^{-1}

Similarly if the relative velocity of the train *A* with respect to *B* is required, then we superimpose the velocity equal and opposite to that of the train *B*. Thus,

$$V'_A = 10 - 40 = -30 \text{ ms}^{-1}, \; V'_B = 40 - 40 = 0$$
$$V_{A/B} = V_A - V_B = V'_A = -30 \text{ ms}^{-1}$$

From the above expressions it is clear that a passenger sitting in the train *B* will apparently feel himself to be stationary with train *A* moving backwards at a velocity of 30 ms^{-1}; while a passenger sitting in the train *A* will feel himself stationary and train *B* apparently moving forward at 30 ms^{-1}.

This is a simple case of relative motion on parallel paths, but the same principle of superimposing negative velocity and making one of the motions zero, can be applied to particles moving along non-parallel paths. Consider two particles *A* and *B* having motions along *OA* and *OB* as shown in Fig. 6.36 a, b and c.

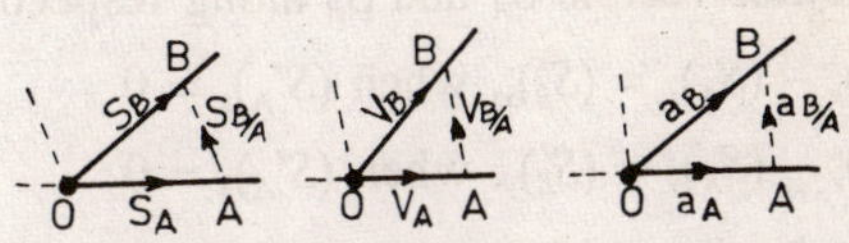

(a) $\vec{S}_{B/A} = \vec{S}_B - \vec{S}_A$ (b) $\vec{V}_{B/A} = \vec{V}_B - \vec{V}_A$ (c) $\vec{a}_{B/A} = \vec{a}_B - \vec{a}_A$

Fig. 6.36

The relative motion of *B* as observed from *A* in relation to displacement, velocity and acceleration will be such that the motion vector of *B* appears to occur *along AB* as shown in respective diagram. It may be stated that the relative motion vector of *A* as observed from *B* will be such that it occurs *along BA'*.

In the above Figure 6.36, the arrow above the quantity represents the corresponding vector quantity and the difference represents the vector difference. For example S_A, V_A and a_A represent respectively the displacement (or position) vector, velocity vector and acceleration vector of *A*. Similarly $(S_B - S_A)$ represents vector difference of vector S_B and vector S_A. These vectors are not along the same line of action, so, their vector sum is obtained by parallelogram or traingle law of vectors. Similar explanation is valid for velocity and acceleration vectors.

Resolution of Rectilinear Relative Motion Vectors

It may be noted that the relative motion vector can also be resolved like other ordinary vectors along any set of axes. The resolved components along any axis are indicated by using the respective subscript with the corresponding relative vector.

Consider two particles A and B, having motion vectors as shown in Fig. 6.37.

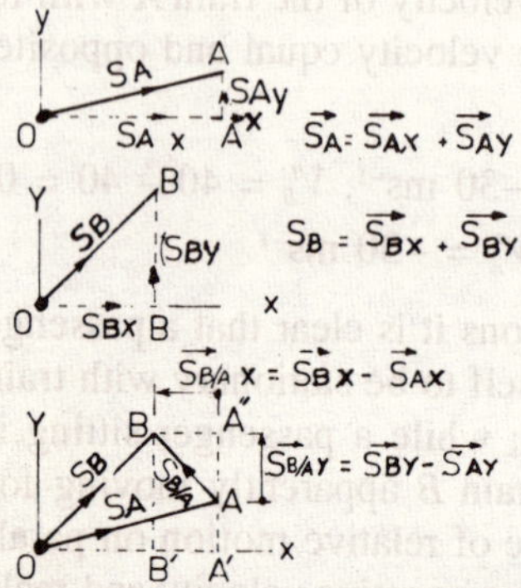

Fig. 6.37

From the figure it can be easily shown that the relative motion vector $S_{B/A}$ has two components $(S_{B/A})_x$ and $(S_{B/A})_y$. These components can be found by considering the initial motion vectors S_A and S_B along respective directions, i.e.

$$(S_{B/A})_x = (S_B)_x - (S_A)_x = (S'_B)_x, \text{ when } (S'_A)_x = 0$$

$$(S_{B/A})_y = (S_B)_y - (S_A)_y = (S'_B)_y, \text{ when } (S'_A)_y = 0$$

These relations can also be written as

$$(S_B)_x = (S_A)_x + (S_{B/A})_x$$

$$(S_B)_y = (S_A)_y + (S_{B/A})_y$$

In these relations the subscripts indicate the component of the respective vector of a particular particle along a particular direction.

With similar notations, valocity and acceleration vectors can also be written as

$$(V_{B/A})_x = (V_B)_x - (V_A)_x = (V'_B)_x, \text{ when } (V'_A)_x = 0$$

$$(V_{B/A})_y = (V_B)_y - (V_A)_y = (V'_B)_y, \text{ when } (V'_A)_y = 0$$

$$(a_{B/A})_x = (a_B)_x - (a_A)_x = (a'_B)_x, \text{ when } (a'_A)_x = 0$$

$$(a_{B/A})_y = (a_B)_y - (a_A)_y = (a'_B)_y, \text{ when } (a'_A)_y = 0$$

Procedure for Solving Problems of Rectilinear Relative Motion

For the solution of rectilinear relative motion problems, the following procedure is adopted:

(i) Draw a sketch showing the motion vectors (such as displacement, velocity or acceleration) of the concerned particles.

(ii) Mark the directions and magnitudes of the motion vectors.

(iii) Apply a negative or opposite motion vector of the magnitude equal to that of the particle with respect to which the relative motion is required (i.e. of the observer) on the motion vector of the particle of which the relative motion is to be observed. The first particle (i.e. observer) will now be apparently considered stationary.

(iv) Determine graphically or analytically the resultant motion vector for the observed particle which will indicate the relative motion vector with respect to the observer.

The above procedure and equations shall now be illustrated by solving examples.

Example 6.22

The ships S_1 and S_2 are 3 km. away from each other at a given instant, S_2 being South-East of S_1. S_1 is travelling at 6 km/hour due East, while S_2 is travelling at 10 km/hour due North. Determine(a) the relative velocity of the Ship S_2 with respect to the Ship S_1, (b) the shortest distance between the two ships, and (c) the time when the two ships are nearest.

Solution

V_1 = 6 km/hr. due East, V_2 = 10 km/hr. due North. Displacement S = 3 km South-East.

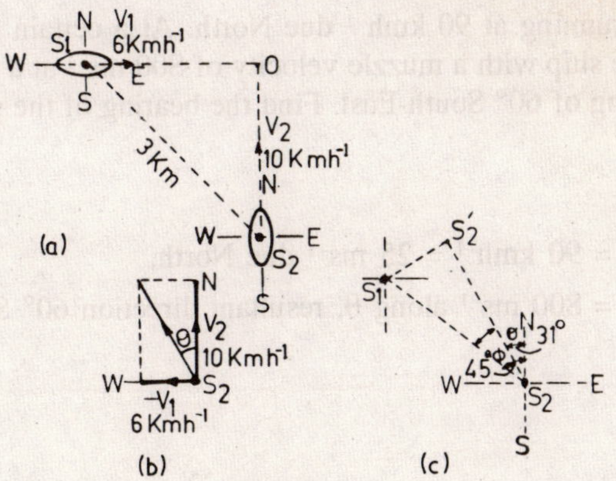

Fig. 6.38

(a) Make the ship S_1 stationary by applying a velocity of 6 kmh^{-1} due West, and find the resultant velocity of S_2 with 10 kmh^{-1} velocity due North and 6 kmh^{-1} velocity due West (–ve of V_1). Thus,

Resultant velocity $V_2' = \sqrt{6^2 + 10^2 + 2\,(-6)10 \cos 90}$

Or $$V_{2/1} = \sqrt{136} = 11.66 \text{ kmh}^{-1}$$

$$\tan\theta = \frac{-6}{10} = -0.60,\ \theta = 31° \text{ West of North.}$$

The ship S_2 will appear to travel at 11.66 kmh^{-1} along a direction 31° West of North to an observer on the Ship S_1.

(b) Now if the Ship S_1 is assumed to be stationary, the ship S_2 apparently moves along a direction 31° West of North, the shortest distance of the ship S_2 from the apparently stationary ship S_1 will be the perpendicular distance on the apparent direction of the ship S_2 from S_1(See Fig. 6.38 c)

$$\angle - S_1S_2S_2' = 45° - 31° = 14°$$

S_1S_2 at the instant = 3 km and is in South-East direction.

$\therefore$ The shortest distance $S_1S_2' = 3 \sin 14° = 0.723$ km.

(c) The apparent distance travelled by the ship S_2 when it reaches the shortest distance point S'_2 is

$$S_2S_2' = S_2S_1 \cos 14° = 2.913 \text{ km}$$

The apparent velocity = 11.66 kmh^{-1}

Thus the time when the two ships are at the shortest distance will be $\frac{2.9113}{11.66} = \frac{1}{4}$ hour (i.e. 15 mins)

Example 6.23

A cruiser is running at 90 kmh^{-1} due North. At a certain instant a shot is fired from the ship with a muzzle velocity of 800 ms^{-1} at a stationary target along a bearing of 60° South-East. Find the bearing of the gun to strike the target.

Solution

$$V_1 = 90 \text{ kmh}^{-1} = 25 \text{ ms}^{-1} \text{ due North,}$$

$$V_2 = 800 \text{ ms}^{-1} \text{ along } \theta, \text{ resultant direction } 60° \text{ SE}$$

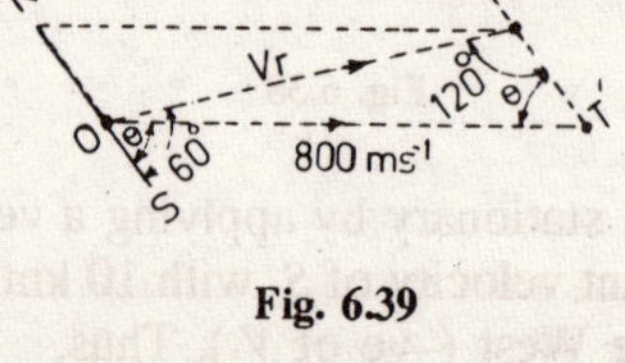

Fig. 6.39

Let the cruiser be at the position O when the shot is fired to hit the target

T at 60°. The shot fired from the cruiser will have two velocities, one of its own 800 ms^{-1} at θ° *SE*(*OT*′), and other of the cruiser 25 ms^{-1} due North(*ON*). The resultant of these velocities should be along the direction *OT* (60° *SE*) to hit the target. From Fig. 6.39

$$\angle TT'O = \angle SOT' = \theta$$

$$\angle T'TO = 180 - 60 = 120°$$

$$\angle T'OT = (60 - \theta)$$

Applying sine rule in Δ T′TO

$$\frac{TT'}{\sin(60-\theta)} = \frac{OT'}{\sin 120} = \frac{OT}{\sin\theta}$$

$$TT' = \text{component of velocity } V_r = 25 \text{ ms}^{-1}$$

$$OT' = \text{component of velocity } V_r = 800 \text{ ms}^{-1}$$

Substituting these values

$$\frac{25}{\sin(60-\theta)} = \frac{800}{\sin 120}$$

Solving, we get θ = 58°27′

Thus the shot should be fired 58°27′ South-East.

Example 6.24

A car moving along a South-East road at 80 kmh^{-1} crosses a railway crossing. At the same time a train also passes the same crossing at 60 kmh^{-1} along the railway track which is due East. Determine (a) the apparent velocity of car as observed by a passenger in the train and (b) the apparent velocity of the train as observed by the driver of the car.

Solution

V_C = 80 kmh^{-1}(South-East), V_T(Train) = 60 kmh^{-1} due East

For finding the apparent velocity of the car ($V_{C/T}$) with respect to the passenger (observer) in the train, the train should be made stationary by superimposing equal and opposite velocity of the train both on the train and the car, and the resultant velocity of the car is found. (see Fig 6.40.)

$$\text{Thus } V_{C/T} = V'_C = \sqrt{(80)^2 + (60)^2 + 2(80)(60)\cos 135°}$$

$$= 56.67 \text{ kmh}^{-1}$$

$$\tan\theta = \frac{80 \sin 135°}{60 + 80 \cos 135°} = \frac{40\sqrt{2}}{3.44} = 16.44$$

or θ = 86.5° with West i.e. 3.5° West of South.

Alternative Approach

Resolving along X

$(V_{C/T})_x = (V_C)_x - (V_T)_x$, **X-direction along East**
Y-direction along North

$$(V_{C/T})_x = 80 \cos(-45°) - 60 \cos(0°) = -3.44$$

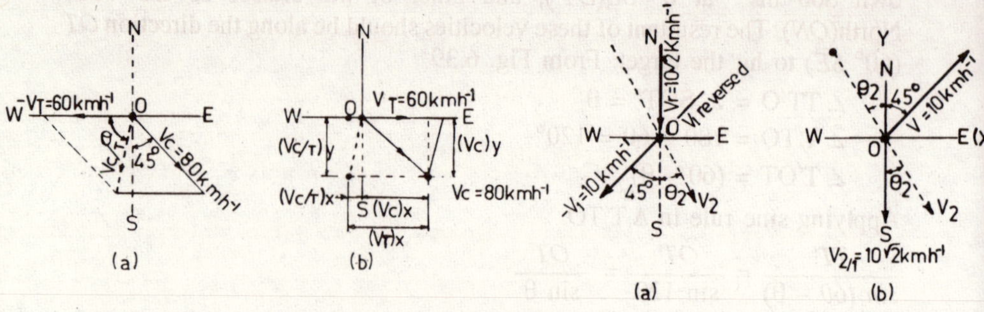

Fig. 6.40

Resolving along Y

$$(V_{C/T})_y = 80 \sin(-45°) - 60 \sin 0° = -40\sqrt{2}$$

$$\therefore \quad (V_{C/T}) = (V_{C/T})_x + (V_{C/T})_y$$

$$= \sqrt{(-3.44)^2 + (-40\sqrt{2})^2}$$

$$= (56.67) \text{ kmh}^{-1} \text{ (i.e. } V_{C/T} \text{ should be in S-W Zone).}$$

$$\tan\theta = \frac{(V_{C/T})_y}{(V_{C/T})_x} = \frac{-40\sqrt{2}}{-3.44} = 16.44$$

$$\theta = 86.5° \text{ with respect to West(i.e } 3.5° \text{ West of South)}$$

The car will appear to move at 56.67 kmh^{-1} along 3.5° West of South to the observer in the train.

The train will appear to move with a velocity of 56.67 kmh^{-1} along 3.5° East of North to the car driver and can be determined in a similar manner as that of the car w.r.t. train.

Example 6.25

A car moves horizontally at a velocity of 20 kmh^{-1} against rain, which appears to fall at $20\sqrt{5}$ kmh^{-1} at an angle of $\tan^{-1}\left(\frac{1}{2}\right)$ with the horizontal.

Find (a) the true velocity of rain and its inclination (b) the apparent velocity of rain if the car speeds upto 40 kmh^{-1} against the train.

Solution

Vel. of the car V_1 = 20 kmh^{-1}, Vel. of the rain V_2 = ?

Relative vel. of rain w.r.t. car $V_{2/1} = -20\sqrt{5}$ kmh^{-1}

$\theta\ 2/1 = \tan^{-1}\left(\frac{1}{2}\right)$ $\sin\theta = 1/\sqrt{5}$, $\cos\theta = 2/\sqrt{5}$

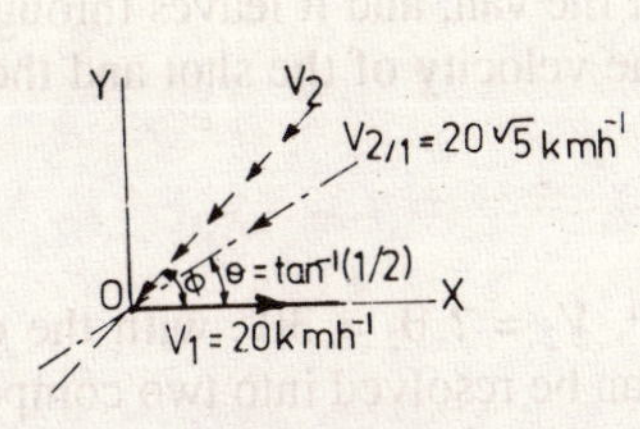

Fig. 6.41

Let the true direction of rain be ø with the horizontal, and its magnitude be V_2. Consider horizontal axis as X-axis, and vertical as Y-axis.

We know that

$$(V_{2/1})_x = (V_2)_x - (V_1)_x, \text{ and}$$
$$(V_{2/1})_y = (V_2)_y - (V_1)_y.$$

Substituting the values in these equations we get

$$(V_{2/1})_x = V_2 \cos \phi - 20 \cos 0$$
$$(-20\sqrt{5}) \cos \theta = V_2 \cos \phi - 20$$
$$V_2 \cos \phi = 20 - 20\sqrt{5} \times 2/\sqrt{5} = -20 \quad \text{... (i)}$$

Similarly,

$$V_2 \sin \phi = -20\sqrt{5} \cdot \frac{1}{\sqrt{5}} = -20 \quad \text{... (ii)}$$

Solving equations (i) and (ii), we get

$$V_2 = -20\sqrt{2} \text{ kmh}^{-1} \text{ and } \phi = 45° \text{ with the horizontal.}$$

(b) Now $V_1 = 40$ kmh^{-1}, $V_2 = -20\sqrt{2}$ kmh^{-1}, $\phi_2 = 45°$

$$(V_{2/1})_x = (V_2)_x - (V_1)_x$$

or $$(V_{2/1})_x = (-20\sqrt{2} \cos 45°) - (40 \cos 0) = -60 \quad \text{... (iii)}$$

Also $$(V_{2/1})_y = (V_2)_y - (V_1)_y$$

or $$(V_{2/1})_y = (-20\sqrt{2} \sin 45°) - (40 \sin 0) = -20 \quad \text{... (iv)}$$

Solving (iii) and (iv)

$$(V_{2/1}) = 20\sqrt{10} \text{ kmh}^{-1}, \text{ at } \theta = \tan^{-1}\left(\frac{1}{3}\right) \text{ with horizontal.}$$

Therefore, when the car speeds to 40 kmh^{-1}, the train apparently appears to fall with a velocity of $20\sqrt{10}$ kmh^{-1} and inclined at an angle of $\theta = \tan^{-1}(1/3)$ with the horizontal.

Example 6.26

A militry van 2.50 m wide and 5.58 m long moving at 90 kmh^{-1} in horizontal direction is hit by a rifle shot in the front right corner at 30° to the direction of motion of the van, and it leaves through diagonally opposite corner of the van. Find the velocity of the shot and the time of travel by the shot in the van.

Solution

$V_1 = 90$ kmh^{-1} = 25 ms^{-1}, V_2 = ? $\theta_2 = 30°$ with the direction of motion. The velocity of the shot can be resolved into two components, i.e. along (x-x' direction) and perpendicular (y-y' direction) to the direction of motion of the van.

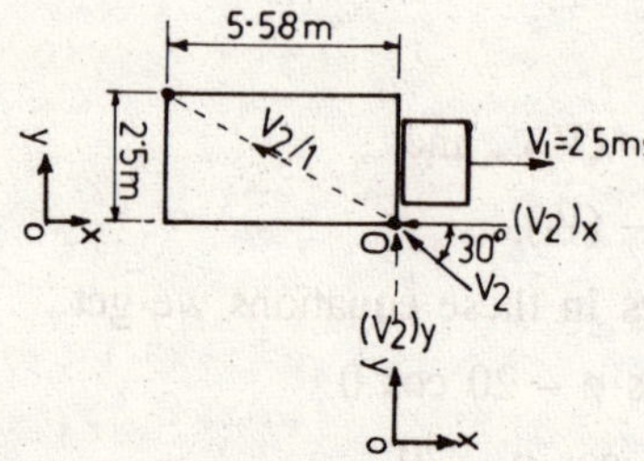

Fig. 6.42

Mathematically

$(V_2)_x = -V_2 \cos 30°$ (–ve because it is against the direction of motion of the van.)

$(V_2)_y = +V_2 \sin 30°$ (towards left)

Since the shot passes through the van diagonally the resultant distances travelled along horizontal and perpendicular directions are

$S_x = -5.58$ m

$S_y = +2.50$ m (towards left)

Resultant relative velocities are

$$(V_{2/1})_x = (V_2)_x - (V_1)_x$$

or $(V_{2/1})_x = -V_2 \cos 30° - 25, \because (V_1)_x = +25$ ms^{-1}

and $(V_{2/1})_y = +V_2 \sin 30°$, uniform

Now the relative distances covered by the shot are $(S_{2/1})_x = -5.58$ m $(S_{2/1})_y = +2.50$ *m. If it happens in time 't'* then

$$(-V_2 \cos 30° - 25)\, t = -5.58 \quad ...(1)$$

$$V_2 \text{ Sin } 30°.\ t = 2.50 \text{ or } V_2 = \frac{+2.50}{t \cdot \sin 30} = \frac{5}{t} \quad ...(2)$$

Solving (1) and (2)

$t = 0.05$ sec, $V_2 = 100$ ms^{-1}

Example 6.27

A car moves horizontally with a velocity of 40 kmh^{-1} when rain appears to fall at 30° from front with the vertical to the driver of the car. When the car speeds up and attains a velocity of 100 kmh^{-1}, the rain appears to the driver to fall at 45° from front with the vertical. Find the true veloocity of the rain fall.

Solution

$$V_1 = +40 \text{ kmh}^{-1},\ V_2 = ?\ \theta_{2/1} = 30°;\ V_{2/1} = V_r$$
$$V'_1 = +100 \text{ kmh}^{-1},\ \theta'_{2/1} = 45°,\ V'_{2/1} = V'_r$$

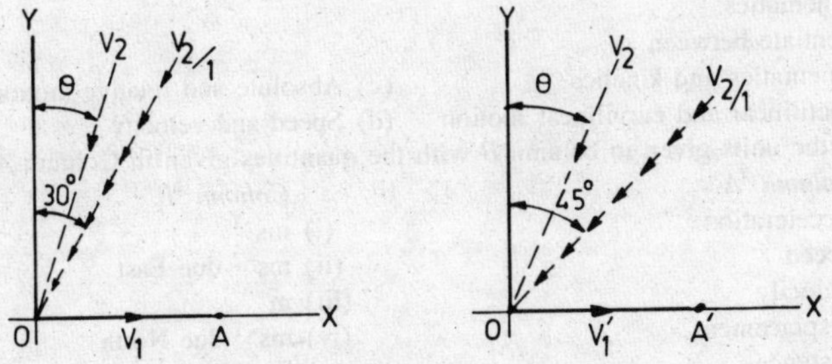

Fig. 6.43

Let the actual rainfall be inclined at an angle of θ with the vertical and have magnitude of velocity V_2 kmh^{-1}.

Using the equations of resolved components for relative velocity along horizontal and vertical directions, we have

Horizontal component $(V_2)_x = (V_1)_x + (V_{2/1})_x$

or $$V_2 \sin\theta = +40 + V_r \sin 30° \quad \text{.... (i)}$$

vertical component $V_2)_y = (V_1)_y + (V_{2/1})_y$

or $$-V_2 \cos\theta = 0 - V_r \cos 30° \quad \text{.... (ii)}$$

Dividing (i) by (ii)

$$\frac{V_2 \sin\theta - 40}{V_2 \cos\theta} = \tan 30° = \frac{1}{\sqrt{3}}$$

or $$V_2 (\cos\theta - \sqrt{3} \sin\theta) = -40\sqrt{3} \quad \text{.... (iii)}$$

Similarly in the second case, when velocity of the car is 100 kmh^{-1}

$$\frac{V_2 \sin\theta - 100}{V_2 \cos\theta} = \tan 45° = 1$$

or $$V_2 (\cos\theta - \sin\theta) = -100 \quad \text{.... (iv)}$$

Dividing (iii) by (iv), we get

$$\frac{(\cos\theta - \sqrt{3}\sin\theta)}{\cos\theta - \sin\theta} = \frac{40\sqrt{3}}{100}$$

or $$\tan\theta = \frac{10 - 4\sqrt{3}}{6\sqrt{3}} = 0.2956$$

$$\theta = 16°\ 28' \text{ with the vertical}$$

Substituting in equation (iv)

$$V_2 = \frac{-100}{0.9590 - 0.2834} = \frac{-100}{0.6756} = -148 \text{ kmh}^{-1}$$

(–ve sign indicates downward direction)

Thus true rain velocity is 148 kmh^{-1} and falling at 16°28′ with the vertical.

Review Questions

1 Define
   (a) Mechanics
   (b) Dynamics
   (c) Statics
   (d) Kinematics
   (e) Velocity
   (f) Acceleration
   (g) Displacement.

2 Differentiate between
   (a) Kinematics and kinetics
   (b) Rectilinear and curvilinear motion
   (c) Absolute and relative motion
   (d) Speed and velocity

3 Match the units given in column *B* with the quantities given in Column *A*

Column 'A'	*Column 'B'*
(i) Acceleration	(i) ms^{-1}
(ii) Speed	(ii) ms^{-2} due East
(iii) Velocity	(iii) m
(iv) Displacement	(iv) ms^{-1} due North
(v) Distance	(v) Radian per second
(vi) Angular speed	(vi) m due West
	(vii) ms
	(viii) ms^2

4 State with reasons whether the following statements are true or false
   (i) If the speed is constant, the acceleration will be zero. T/F
   (ii) If the speed is uniform, the acceleration will be uniform. T/F
   (iii) If the velocity is uniform, the speed will also be uniform. T/F
   (iv) If the speed is uniform, the velocity will also be uniform. T/F
   (v) If the velocity is uniform, the acceleration will be zero. T/F
   (vi) If the acceleration is constant, the velocity will also be constant. T/F
   (vii) If the acceleration is constant, the velocity will be zero. T/F
   (viii) If the velocity is zero, the displacement will also be zero. T/F
   (ix) If the displacement is zero, the velocity will also be zero. T/F
   (x) The speed represents the magnitude of the velocity at any instant. T/F

5 State the type of motion shown in Fig. Q 5(i) and (ii) from A to B, B to C and C to D.

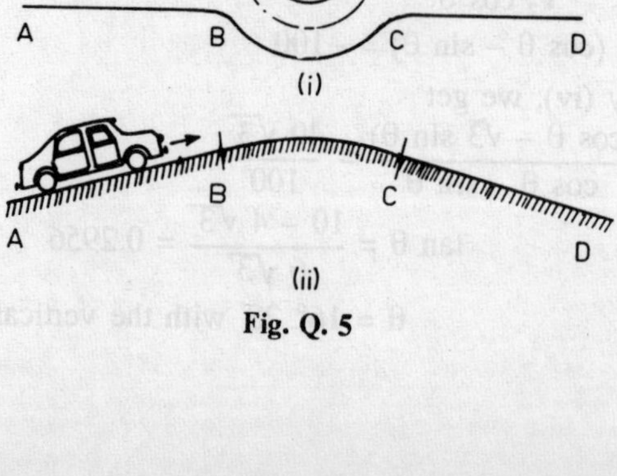

Fig. Q. 5

6 Define (a) Displacement, (b) Velocity, (c) Acceleration and (d) Deceleration

7 Differentiate between

(a) Displacement and distance travelled by a particle

(b) Velocity and speed of a particle

(c) Acceleration and deceleration of a particle

(d) Constant speed and constant velocity of a particle

(e) Average and Instantaneous velocity of a particle

(f) Rectilinear motion with uniform velocity and rectilinear motion with uniform acceleration.

8 Explain the meaning of (a) $\dot{V} = \frac{dS}{dt}$, (b) $a = \frac{dV}{dt}$, (c) $a = \frac{d}{dt}\left(\frac{dS}{dt}\right)$

9 Classify the following cases into rectilinear, non-rectilinear, rectilinear with constant velocity and rectilinear with constant acceleration

(a) A smooth wheel rolls along a smooth sloping plane.

(b) A car moving along a straight path.

(c) A car moving 1 km in each 1 min. along a straight road.

(d) A train moving along a straight railroad comes to halt.

(e) A car travelling along a curved path.

(f) A body falling freely under gravity in still air.

(g) A body thrown in air with a velocity V along a plane at 30° with the horizontal.

10 State whether the following statements are true or false.

(a) In rectilinear motion of a particle, if the particle has negative displacement after a certain time interval, then its acceleration will also be negative. T/F

(b) A particle having constant speed always traces rectilinear motion T/F

(c) In rectilinear motion, the velocity is always constant T/F

(d) In rectilinear motion, the direction of motion is always along a straight line T/F

(e) In a rectilinear motion, the velocity can be constant or variable. T/F

(f) In rectilinear motion, the distance travelled and the displacement occured are one and the same thing. T/F

(g) A particle moving with uniform acceleration describes equal distances in equal intervals of time T/F

11 Compute the net displacement and net distance travelled in the following cases shown in Fig. Q. 11a, b, and c.

(a) A particle moves from A to B and then B to C

(b) A particle move from P to Q and then Q to R

(c) A particle move along MNOPQ.

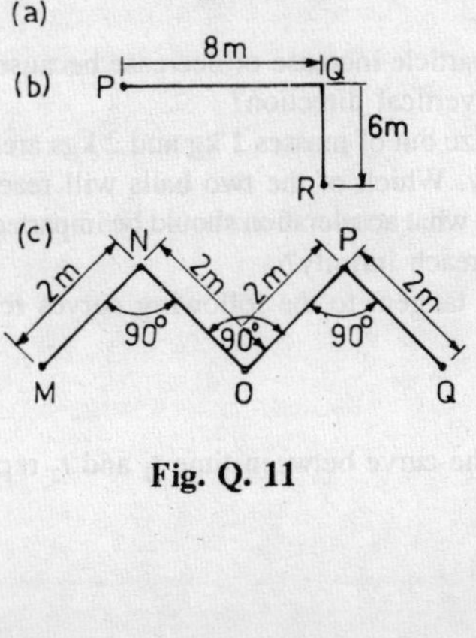

Fig. Q. 11

12 Compute the resultant of the following

(a) A particle having velocities of 3 ms^{-1} due East and 4 ms^{-1} due North.

(b) A particle having accelerations of 5 ms^{-2} due North and 12 ms^{-2} due West.

(c) A boat having velocity of 6 ms^{-1} across the direction of flow in a river having velocity of flow 8 ms^{-1}.

Ans. (a) 5 ms^{-1} along $\tan^{-1}(4/3)$ North East (b) 13 ms^{-1} along $\tan^{-1}(2.4)$ West of North (c) 10 ms^{-1} due $\tan^{-1}(3/4)$ with flow

13 A car moves 20 km along a straight path due East and then returns along the same path after 40 minutes. What is the car's average speed? Is it the same as the average velocity? What is average velocity?

Ans. Average speed = 1 km/min., average vel = 0.

14 A car moving along a straight road passes through a valley along the road as shown in Fig. Q. 14. Name the type of motion the car will perform before, during and after passing the valley.

Fig. Q. 14

15 When the velocity of a particle is constant, does the average velocity over any time interval differ from the instantaneous velocity at the end of the interval?

16 Is it possible that a particle accelerates instantaneously even when it has zero velocity? If so explain with example.

17 Is it possible for a particle to have a constant speed and varying velocity? If so, give an example.

18 Is it possible for a body to have eastward velocity while experiencing a westward acceleration? If so explain with an example.

19 Is it possible for a body to have a constant velocity and varying speed?

Answer the following questions in brief:

20 A ball is thrown vertically up, taking air resistance into account, would you expect the time of rise to be longer or shorter than the time of fall?

21 Can there be motion of a body in two directions having an acceleration in one direction only?

22 A man standing at the top of a tower throws a ball vertically up with a certain velocity V, and then throws another ball vertically downward with the same velocity V. Neglecting the air resistance find which of the two balls will hit the ground with greater velocity.

23 A body is thrown at 45° to the vertical, what is the direction of acceleration due to gravity on the body?

24 Does the velocity of a particle increase or decrease because of acceleration due to gravity when it is thrown in a vertical direction?

25 Two balls of the same size but of masses 1 kg and 2 kgs are thrown vertically upward with the same initial velocity. Which of the two balls will reach the ground first?

26 Neglecting air resistance, what acceleration should be imparted to a rocket projected vertically upwards so that it can reach infinity?

27 What does the slope of tangent to the following curves represent.

(a) $a - t$ curve

(b) $V - t$ curve

(c) $S - t$ curve

28 What does the area of the curve between time t_1 and t_2 represent in the following cases?

(a) $a - t$ curve

(b) $V - t$ curve

(c) $S - t$ curve

29 Correct the following statements, if not correct

(a) Area of $a - t$ curve between two limits of time t_1 and t_2 represents instantaneous velocity during the time interval.

(b) Area of $a - t$ curve between two limits of time t_1 and t_2 represents maximum velocity during the time interval.

(c) Area of $V - t$ curve between two limits of time t_1 and t_2 represents the position vector during the time interval.

(d) Area of $V - t$ curve between two limits of time t_1 and t_2 represents the total change in position vector during the time interval.

30 If $a - t$ curve for the motion of a particle coincides with t–axis, what can be the possible the type of $V - t$ and $S - t$ curves? Can the initial velocity be zero?

31 If $V - t$ curve for a particle's motion is parallel to time axis, what will be possible $S - t$ curve?

32 If the curve of position vector 'S' with respect to time 't' is a straight line parallel to time axis, what does this represent? Explain.

33 If the curve of position vector 'S' with respect to time is an oblique straight line passing through the origin, what does this signify? Show corresponding $V - t$ and $a - t$ curves.

34 What does the point of zero acceleration signify in $a - t$ curve in relation to its corresponding $V - t$ curve?

35 What does the point of zero velocity signify in $V - t$ curve in relation to its corresponding $S - t$ curve?

36 Select the correct acceleration-time graph for a particle thrown vertically up under gravity during its ascending journey.

37 State whether the following statements are true/false.

(a) Two vehicles moving on two straight parallel roads will have more relative velocity with respect to each other when both the vehicles move in the same direction, than their relative velocity when these move in opposite direction. T/F

(b) The relative velocity of a vehicle A with respect to vehicle B will be the same as the resultant of velocity of vehicle A and vehicle B. T/F

(c) The relative velocity of a particle A with respect to the particle B will be equal and opposite to the relative velocity of B with respect to A. T/F

(d) When two trains A and B are moving on straight parallel tracks in opposite directions, the relative velocity of train A with respect to train B will be equal to the numerical difference of velocity of A and B. T/F

(e) When two trains A and B are moving on parallel straight tracks in the same direction, the relative velocity of B with respect to A will be equal to the algebraic difference of velocities of B and A. T/F

(F) two automobiles A and B are moving at right angles, the relative velocity of B with respect to A will be along the diagonal of a parallelogram formed by the velocity of B and opposite velocity of A. T/F

38 Define relative velocity, relative acceleration and relative displacement.

39 You are required to cross a river at right angles to the direction of flow. Explain with sketch what should be the direction of motion of your boat? (No calculations are required).

40 If rain falls vertically and a scooter driver moves due East, what will be the apparent direction of rainfall as observed by the scooter driver? Explain by a simple sketch without calculations.

41 Is it possible to resolve the relative velocity? If so explain with sketch indicating the relations of relative velocity components with respect to the components of the velocities of the two particles under consideration.

6.5 Curvilinear Motion of Particles

A particle is said to possess a curvilinear motion when it traverses a curved path. This curved path may be a simple circular path or any other curve. The motion can be along the part of an open curve or along a complete closed curve and repeated after some interval. Examples of curved path are a car moving along a curved road, a shell dropped from an air craft, a joy ride moving along a circle in horizontal plane or vertical plane, motion of wheel of a jacked car, motion of a fan, etc.

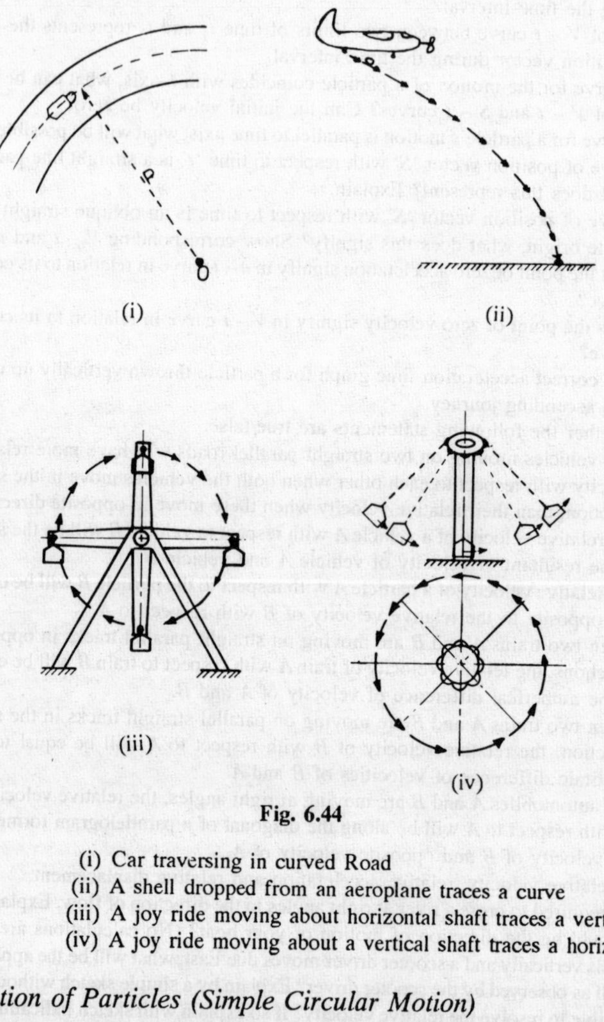

Fig. 6.44

(i) Car traversing in curved Road
(ii) A shell dropped from an aeroplane traces a curve
(iii) A joy ride moving about horizontal shaft traces a vertical circle
(iv) A joy ride moving about a vertical shaft traces a horizontal circle

6.5.1 *Rotation of Particles (Simple Circular Motion)*

We shall first consider simple circular motion in one plane. For example motion of the round stone fixed on a rod hinged at one end in a mortar mill is simple circular motion. The motion of the rod joining roller to the central peg is also a circular one. (Fig. 6.45)

The term angular motion is normally associated with the motion of the line joining the given point (object) and the centre of motion. Thus the angular motion of a particle implies the motion of the line joining the particle and the centre of the motion. Circular motion is generally measured in polar coordinates.

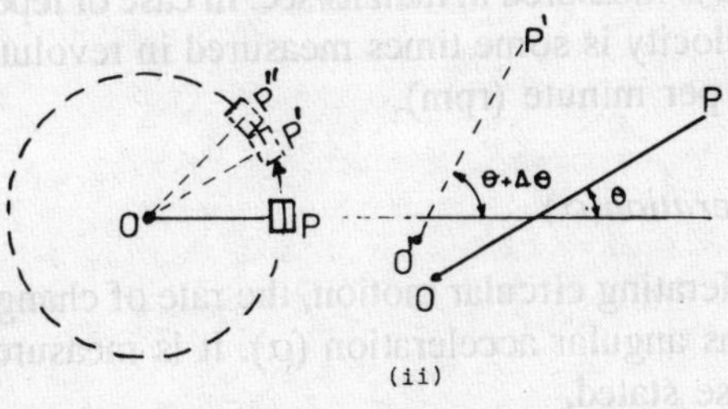

(i) Motion of roller P along circumference of a circle

(ii) Motion of line OP to $O'P'$ in a circular manner

Fig. 6.45

Polar Co-ordinates

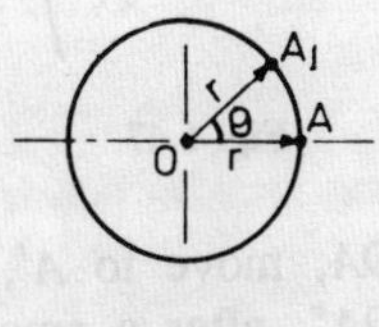

Fig. 6.46

To define the position of any particle in a plane two coordinates are necessary. These are the radius or the distance of the particle from the fixed centre of motion (pole) and angular displacement of the line joining the particle and the centre with respect to the original position of this line. The distance is measured in linear units while the angle is measured in radians unless otherwise specified.

Consider a particle at A along the original line OA, and let it moves to A_1 *(See Fig. 6.46). The polar coordinates to define the position of the particle at* A_1 will be radius $OA_1 = OA = r$, and angle $\theta = \angle A_1OA$. Mathematically these coordinates are represented as (r,θ).

Definitions

Angular Displacement (θ)

The total angle traced by any line joining the particle, and the centre of motion with respect to its original position is termed as angular displacement(θ). It is measured in radians.

Angular Velocity(ω)

The rate of angular displacement(θ) with respect to time is known as angular velocity(ω), and is measured in radians/sec. In case of repetitive circular motion, this angular velocity is some times measured in revolutions per second (rps) or revolutions per minute (rpm).

Angular Acceleration(α)

In case of accelerating circular motion, the rate of change of angular velocity (ω) is termed as angular acceleration (α). It is measured in radians/sec/sec, unless otherwise stated.

6.5.2 *Relationship of Angular Motion Parameters*

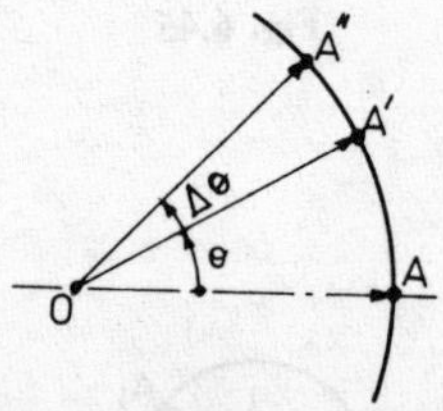

Fig. 6.47

Let a particle at A, along OA, move to A', along OA' in time t, and further move to A'', along OA'', after a small interval of time Δt (See Fig. 6.47).

By definition of angular velocity, the average angular velocity of the particle during small interval of time Δt is given by

$$\omega_{av} = \frac{\angle AOA'' - \angle AOA'}{\text{time } \Delta t} = \frac{(\theta + \Delta\theta) - \theta}{\Delta t}$$

$$= \frac{\Delta \theta}{\Delta t}$$

$$\text{limit } \Delta t \rightarrow 0$$

As the time interval Δt becomes smaller and smaller

$$\omega_{av} \rightarrow \omega$$

Thus instantaneous angular velocity ω at any instant of time t shall be given by

$$\omega = \text{limit } \frac{\Delta\theta}{\Delta t}, \text{ as } \Delta t \rightarrow 0$$

$$\text{or } \omega = \frac{d\theta}{dt} \qquad \text{.... 6.11(a)}$$

By definition, the angular acceleration is the rate of change of the angular velocity, thus

$$\alpha = \frac{d\omega}{dt} = \frac{d}{dt}\left(\frac{d\theta}{dt}\right) = \frac{d^2\theta}{dt^2} \qquad \text{.... 6.11(b)}$$

Eliminating *dt* from equations 6.11(a) and 6.11(b),

$$\alpha\, d\theta = \omega \cdot d\omega \qquad \text{.... 6.11(c)}$$

Signs

In above relations θ, ω, and α are all taken as positive if they are in the same direction, either clockwise or anticlockwise.

Now consider the following different cases:

(a) *Uniform Angular Velocity* (ω)

If the angular velocity is uniform, the angular acceleration is zero, as there is no change in the angular velocity. Mathematically.

$$\omega_f = \omega_i = \omega, \qquad \text{.... 6.12(a)}$$

$$\theta = \omega \cdot t,\ \theta = \theta_i + \omega_i\, t \qquad \text{.... 6.12(b)}$$

where

θ = Total angular displacement in radians,

θ_i = Initial angular displacement in radians (usually zero)

ω_i = Initial angular velocity in radians/sec,

ω_f = Final angular velocity in radians/sec,

α = Angular acceleration in radians/sec/sec,

t = Time in seconds.

(b) *Uniform Angular Acceleration*

Angular acceleration is rate of change of angular velocity and is analogous to linear acceleration. If the angular motion is such that it accelerates uniformly and has initial angular velocity ω_i, the total angular displacement θ, final instantaneous angular velocity ω_f after period '*t*' can be found by the integration method, or otherwise (similar to that of rectilinear motion). These relations with usual notations are as follows.

$\omega_f = \omega_i + \alpha \cdot t.$ .... (a)

$\theta_f = \theta_i + \omega_i \cdot t + \frac{1}{2} \alpha \cdot t^2$ .... (b)

$\omega_f^2 = w_i^2 + 2\,\alpha \cdot \theta$, where $\theta = (\theta_f - \theta_i)$ .... (c)

In the above expressions, the various quantities are considered in radians, while these can also be taken in revolutions or degrees, but necessary conversion constants must be used. These conversion constants are given as under:

1 revolution = 2 π radians = 360°

i.e. Multiply by $\frac{1}{2\pi}$ to convert the radians to revolutions,

Multiply by $\frac{180}{\pi}$ to convert radians to degrees

Multiply by 2π to convert revolutions to radians

Multiply by $\frac{\pi}{180}$ to convert degrees to radians

Note: In all above relations care should be taken to ensure homogeniety of dimensions and consistency of units.

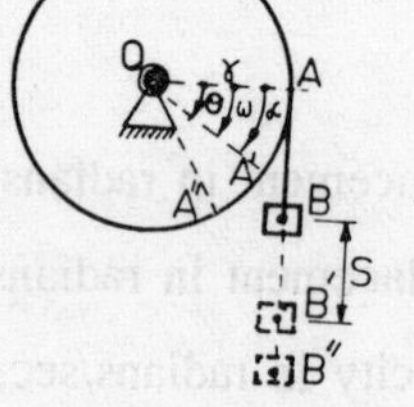

Fig. 6.48

6.5.3 *Relation of Angular and Linear Motion Parameters*

Let a particle A rotate along a fixed circular path and occupy the position A′ as shown in Fig. 6.48. The point *A* on the circumference travels a linear distance *S* = *BB*′, when the line *OA* rotates through an angle θ radians (θ = ∠*AOA*′). The arc distance along the circumference is *AA*′, and the corresponding distance along *AB* is *BB*′. Thus

$S = BB' = \text{Arc } AA' = r \cdot \theta$

Or $S = r \cdot \theta$ .... (1)

where

S = Linear distance = Arc length,

r = radius of circular path,

θ = rotation in radians by which the particle moves.

Differentiating the above equation (1) with time

$$\frac{dS}{dt} = \frac{d\,(r\theta)}{dt} = r\,\frac{d\theta}{dt}, \qquad (\because r \text{ is constant})$$

$$\text{or } V = r \cdot \omega, \qquad \left(\because V = \frac{ds}{dt} \text{ and } \frac{d\theta}{dt} = \omega\right) \qquad \text{.... (2)}$$

Further differentiating equation (2), we have

$$\frac{dV}{dt} = \frac{d}{dt}\,(r\omega) = r \cdot \frac{d\omega}{dt},$$

$$\text{or } a_t = r \cdot \alpha \qquad \left(\because \frac{dV}{dt} = a_t, \ \frac{d\omega}{dt} = \alpha\right) \qquad \text{.... (3)}$$

The linear quantities are measured along tangential directions. The relations of magnitudes of linear and angular parameters as obtained above are summarised as

$$S = r \cdot \theta \qquad \text{.... 6.14(a)}$$

$$V = r \cdot \omega \qquad \text{.... 6.14(b)}$$

$$a_t = r \cdot \propto \qquad \text{.... 6.14(c)}$$

From above equations, it is obvious that the magnitude of linear tangential motion parameters (viz V, a_t) are directly proportional to the radial distances of the particles from the axis of rotation. It may be clarified that the linear motion parameters are along the tangential directions which change from time to time as the particle moves along the circumference i.e. the direction of linear motion parameters of a particle change even though placed at the same radial distance. For example, the direction of linear velocity of a particle at a fixed radial distance varies with the direction of motion from point to point although its magnitude remains constant.

6.5.4 *Radial and Tangential Acceleration of Particle*

Particles moving in a circular path have acceleration not only in tangential direction, but in raidal direction too. The expressions for radial and tangential directions are derived for particles moving with uniform angular motion and with accelerating angular motion.

Case I. *Angular Acceleration Zero (Uniform Angular Velocity.)*

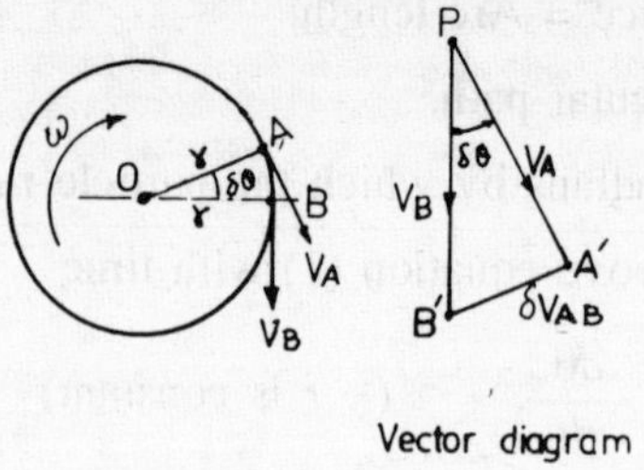

Fig. 6.49

Let a particle be moving with a uniform angular velocity ω along the circular path as shown in Fig. 6.49.

Linear tangential velocity and acceleration of the particle at A will be

$V_A = \omega \cdot r$, and $a_{t_A} = 0$, ... (1)

After a small time interval, let the particle reach B, then

$V_B = \omega \cdot r,\ a_{t_B} = 0$, ... (2)

Though the magnitudes of the linear velocities of the particle are the same at *A* and *B* but their directions are different. This means that there is a change in velocity from *A* to *B*, and hence there must be some acceleration acting on the particle during its motion from *A* to *B*. Let the change in velocity from *A* to *B* = ΔV_{AB}, then by the principle of vectors the change in velocity

$\Delta V_{AB} = V_{AB} = (V_B - V_A)$ .... (3)

We draw a vector diagram as shown in Fig. 6.49

PA′ is parallel and equal to V_A

PB′ is parallel and equal to V_B

As V_A and V_B are numerically equal, the sides PA′ and PB′, they are equal in length.

The change in velocity ΔV_{AB} in small duration Δt when the particle moves from *A* to *B* is thus represented by length A′B′ in the vector diagram.

During the small duration Δt, the direction changes by small angle $\Delta\theta = \omega.\Delta t$. .... (4)

Since the angle $\Delta\theta$ is very small and the vector diagram PA′B′ is an isosceles triangle, the change in velocity ΔV_{AB} = A′B′ can be assumed to occur approximately along the perpendicular to tangential directions, and hence along the radius OA.

From vector diagram

ΔV_{AB} = A′B′ = $V_A \cdot \Delta\theta$

Substituting values of $\Delta\theta$ and V_A, we have

$\Delta V_{AB} = (\omega \cdot r)\ (\omega \cdot \Delta t)$

$\Delta V_{AB} = \omega^2 \cdot r\Delta t$

Since the change in velocity is in radial direction the radial acceleration

$$a_n = \frac{\text{Change in velocity along radius}}{\text{time}}$$

$$\text{i.e. } a_n = \frac{\omega^2 \cdot r\ \Delta t}{\Delta t}$$

$$\text{or } a_n = \omega^2 \cdot r \qquad \text{.... 6.15}$$

This radial acceleration of a particle is of constant magnitude since ω and r are constant, but the direction goes on changing and is always *directed towards* the centre of motion. This radial acceleration of a particle moving with uniform angular velocity is known as '*CENTRIPETAL*' acceleration.

Since angular acceleration '∝' is zero, the tangential accelerations at A and B are zero, i.e. the particle has no tangential acceleration (i.e. $a_t = 0$).

Case II *Constant Angular Acceleration (Accelerating Angular Velocity)*

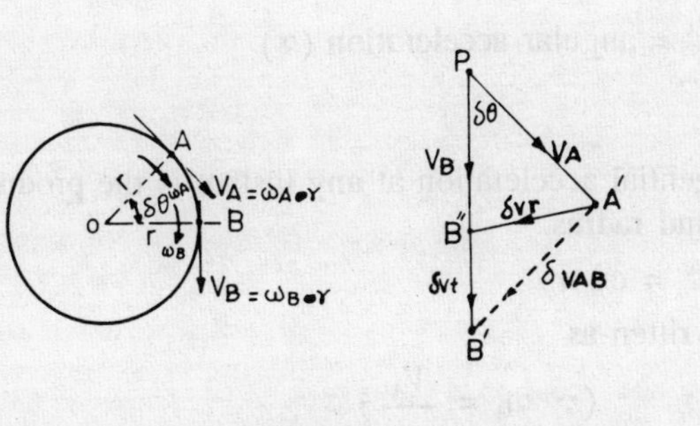

Fig. 6.50

Let a particle have angular motion along a circular path with accelerating angular velocity as shown in Fig. 6.50. Let the angular velocity of the particle be ω_A at A, and let it accelerates to ω_B at B where it reaches in a small interval of time Δt. The instantaneous tangential velocity at A and B are given as

$$V_A = \omega_A \cdot r,\ V_B = \omega_B \cdot r$$

Draw a vector diagram taking PA′ equal and parallel to V_A, and PB′ parallel and equal to V_B. Thus the vector A′B′ represents the resultant change in velocity ΔV_{AB} (see Fig. 6.50). Consider a point B″ in PB′ such that PB″ = PA′ in magnitude. The triangle PA′B″ is an isosceles triangle. Since the time duration Δt is small, the angle $\Delta\theta$ will also be small, and hence the vector A'B' can be approximately considered normal to PA' (i.e. normal to V_A). The total change of velocity ΔV_{AB} can be assumed to consist of normal component ΔV_n and tangential component ΔV_t

$$\Delta V_n = A'B'' = V_A \cdot \Delta\theta, \text{ (As } \Delta\theta \text{ is very small)}$$

$$\Delta V_t = B'B'' = PB' - PB'' = (V_B - V_A)$$

$$\Delta\theta = \Delta t \cdot \omega_A$$

Normal or radial acceleration

$$= \frac{\text{Normal or radial change in velocity}}{\text{Duration of change}}$$

$$\text{i.e. } a_n = \frac{\Delta V_n}{\Delta t} = \frac{(V_A \cdot \Delta\theta)}{\Delta t} = \frac{(\omega_A \cdot r)\,(\omega_A \cdot \Delta t)}{\Delta t}$$

$$\text{or} \quad a_n = \omega_A^2 \cdot r \qquad \text{... 6.16(a)}$$

Thus the radial acceleration (also called normal acceleration) at any instant is equal to the product of square of its instantaneous angular velocity and the radius.

The tangential acceleration is along the tangent at A and is equal to the rate of change of tangential velocity i.e.

$$a_t = \frac{(V_B - V_A)}{\Delta t} = \frac{(\omega_B \cdot r - \omega_A \cdot r)}{\Delta t} = \frac{(\omega_B - \omega_A)\, r}{\Delta t}$$

Since $\dfrac{\omega_B - \omega_A}{\Delta t}$ = angular acceleration ($\propto$)

$$a_t = r \propto \qquad \text{.... 6.16(b)}$$

Thus the tangential acceleration at any instant is the product of its angular acceleration and radius.

The relation $a_n = \omega_A^2 \cdot r$

can also be written as

$$a_n = \frac{V_A^2}{r}; \qquad (\because \omega_A = \frac{V_A}{r})$$

For a straight path the radius $r = \infty$, and hence

$$a_n = \frac{V_A^2}{\infty} = 0$$

Thus it is evident that there will be no radial component in case of straight line motion. The tangential acceleration becomes the same as linear acceleration along the straight line.

6.5.5 *Simple Harmonic Motion (Periodic Motion)*

If a particle moves in such a manner that after a certain fixed interval of time, it acquires the same position and moves in the same direction with the same velocity, it is said to have a periodic motion. The motion of many machine camponents is periodic in nature. This principle of periodic motion is used in operating valves and other mechanisms.

If a body moves in such a manner that its *acceleration* is always *directed towards a fixed point* and is proportional to its *distance from the fixed point*, then the motion of the body is called *Simple Harmonic motion (SHM).*

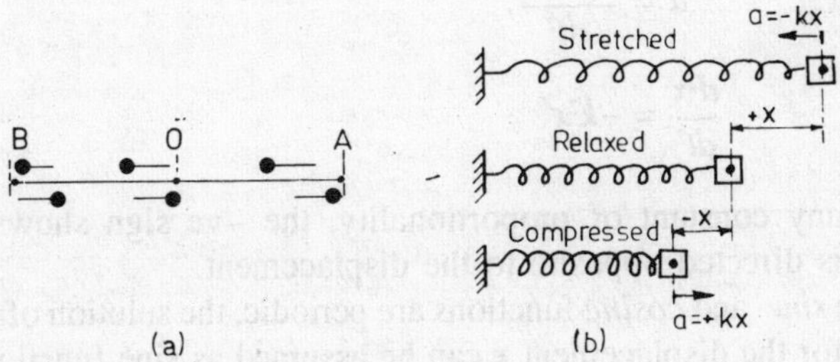

Fig. 6.51

Let a particle move along a straight path BOA and let its acceleration be directed towards O and depend on the distance of the particle from O. The particle first moves from reference position O to A and then A to O, O to B and B to O, and thereafter, it repeats the motion. For example a stretched or compressed spring gives rise to *SHM*.

Definitions

Periodic Time (t_o)

Periodic time is the time taken by the particle for one complete cycle of motion. It is generally measured in seconds.

Frequency (n)

The number of cycles which the particle can make in one second is called frequency (n).

If the period of a cycle is t_o, then the number of cycles in one second (n) is given by

$$n = \frac{1}{t_o} \qquad 6.17$$

Amplitude

The maximum displacement of the particle from its mean position of motion is called the amplitude.

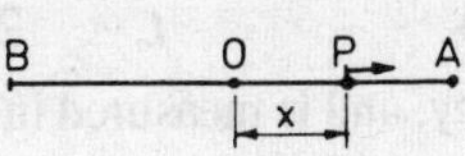

Fig. 6.52

Mathematical Expressions

Consider a particle P at a distance of x from its mean position O. By the definition of *SHM*, the acceleration of the particle at this stage will be proportional to the distance i.e. $a \propto x$, or $a = -k^2 x$. Also if the displacement $S = x$, we have acceleration

$$a = \frac{d^2 x}{dt^2},$$

or
$$\frac{d^2x}{dt^2} = -k^2x \quad .... (1)$$

where k^2 is any constant of proportionality, the –ve sign shows that the acceleration is directed opposite to the displacement.

Since the *sine* and *cosine* functions are periodic, the solution of the above equation (1) for the displacement x can be assumed as sine function of time to start with.

i.e.
$$x = A \sin(\omega t + \sigma) \quad .... (2)$$

Differentiating

$$\frac{dx}{dt} = +\omega A \cos(\omega t + \sigma)$$

Differentiating further, we get

$$\frac{d^2 x}{dt^2} = -\omega^2 A \sin(\omega t + \sigma) \quad .... (3)$$

Putting these values in equation (1) above, we get

$-\omega^2 A \sin(\omega t + \sigma) = -k^2 A \sin(\omega t + \sigma)$

$$\therefore \quad k = \omega$$

Hence the solution will be

$$x = A \sin(\omega t + \sigma)$$

It may be noted that this solution gives rise to a set, or family of solutions. The arbitrary constants A and σ will depend on the starting point, or mean position of the motion. The above function will repeat after one complete cycle after a time $\left(\frac{2\pi}{\omega}\right)$ i.e. if t is replaced by $\left(t + \frac{2\pi}{\omega}\right)$, the motion repeats.

Therefore $\frac{2\pi}{\omega}$ is the time period of one cycle.

$$\therefore \quad t_o = \frac{2\pi}{\omega}, \text{ and } n = \frac{1}{t_o} = \frac{\omega}{2\pi} \quad .... 6.18$$

ω is known as angular frequency, and is measured in radians/sec. The constant A is amplitude, and depends upon the given condition. It is equal to the maximum displacement from the mean position.

$(\omega t + \sigma)$ is called the phase of the motion, and the constant σ is called the phase constant. Constants A and σ are determined from the initial position of displacement and velocity.

Relation of Circular Motion and S.H.M. of a Particle Along the Diameter of the Circle

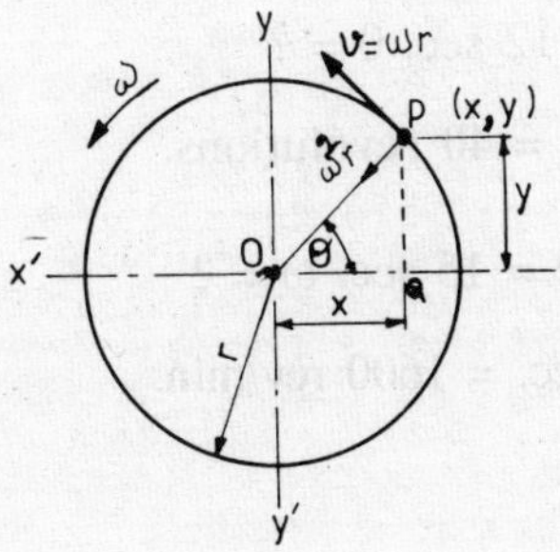

Fig. 6.53

Consider a particle P moving along the circular path with a uniform angular velocity ω. Let Q represent the projection of the particle P on the diameter XOX′. As P moves along the circular path $XYX'Y'X$, the point Q moves along the straight diameter $XOX'OX$.

Consider point P at any instant of time such that OP makes an angle θ with OX, and P has coordinates (x, y). The particle P will have tangential velocity $V = \omega \cdot r$, and normal radial acceleration $a_n = \omega^2 r$

where ω = uniform angular velocity of the particle P in radians/sec. Resolving V and a_n horizontally along X-X' axis, we get

$$V_x = V\cos(90 - \theta) = \omega \cdot r \sin\theta,$$

$$= \omega r \cdot \frac{y}{r} = \omega \cdot y \qquad \left(\because \sin\theta = \frac{y}{r}\right)$$

$$= \omega\left(\sqrt{r^2 - x^2}\right) \qquad (\because r^2 = x^2 + y^2)$$

Also $a_n = \omega^2 \cdot r$,

Resolving horizontally $a_x = a_n \cos\theta$,

or $\quad a_x = \omega^2 \cdot r\cos\theta = \omega^2 x \quad (\because x = r\cos\theta)$

i.e. $\quad a_x = \omega^2 \cdot x$ 6.19

Since ω^2 is constant, it is evident that the horizontal acceleration of the particle is proportional to its horizontal distance x from the mean position O. This shows that the particle P moving along the circle will result in *SHM* of the particle Q, correspondingly moving along the diameter of the circle.

Example 6.26

A fly wheel rotates freely on frictionless bearings at 200 revolutions per minute. Calculate the number of revolutions it has made in 12 seconds after the start. If this wheel rotates by 400 revolutions in 15 seconds, find its angular speed.

Solution

For frictionless bearings, there is no resistance, and hence the wheel will rotate with uniform motion i.e. the acceleration will be zero.

(a) $\omega = 200$ rev/min, $t = 12$ sec. $\theta = ?$

$$\theta = \omega t = 200 \times \frac{12}{60} = \textbf{40 revolutions.}$$

(b) $\theta = 400$ revolutions, $t = 15$ sec. $\omega = ?$

$$\omega = \frac{\theta}{t} = \frac{400}{15} \text{ rev./sec.} = 1600 \text{ rev/min.}$$

Example 6.29

A fly wheel has initial angular speed of 600 rev/min, and it comes to rest in 5 secs due to friction in bearings.

(a) What is its angular retardation? How many revolutions it has made before coming to rest?
(b) If the initial angular speed is 1200 revolutions/min, and the wheel has the same bearings, how long would it take the wheel to come to rest?
(c) What could be the speed after 8 and 12 seconds if the wheel starts with 1000 rev/min.

Solution

(a) $\omega_i = 600$ rpm $= 10$ rps, $\alpha = ?$ $\omega_f = 0$, $t = 5$ sec.

$$\omega_f = \omega_i + \alpha\, t$$

$$0 = 10 + \alpha\,(5), \;\therefore\; \alpha = -2 \text{ rev/sec/sec.}$$

$$\omega_f^2 = \omega_i^2 + 2\,\alpha\,\theta$$

$$\therefore\; \theta = \frac{\omega_i^2}{2\,\alpha} = \frac{+100}{+2 \times 2} = 25 \text{ revolutions.}$$

(b) $\omega_i = 1200$ rpm $= 20$ rps, $\omega_f = 0$, $t = ?$

$$\alpha = -2 \text{ rev/sec}^2$$

$$0 = 20 - 2t, \quad \therefore\; t = 10 \text{ sec(Ans)}$$

(c) $\omega_i = 1000$ rpm $= \frac{100}{6}$ rps, $t_1 = 8$ sec, $t_2 = 12$ sec. $\omega_f = ?$

$$\omega_f = \omega_i + \alpha \cdot t = \frac{100}{6} - 2 \times 8 = \frac{4}{6} \text{ rps} = 40 \text{ rpm}$$

Let the wheel come to rest in t secs

$$0 = \frac{100}{6} - 2\,(t), \;\therefore\; t = \frac{25}{3} \text{ secs}$$

Thus the velocity after 12 seconds will be zero, since there is no positive driving force after the wheel comes to rest due to bearing friction.

Example 6.30

A fly wheel was moving at 1200 rpm when the steam was shut off and makes 400 revolutions before it comes to rest. Find the acceleration and the time taken to come to rest.

Solution

$$\omega_i = \frac{1200}{60\text{ rps}} = 20\text{ rps},\ \theta = 400\text{ rev},\ \alpha = ?\ t = ?$$

$$\omega_f^2 = \omega_i^2 + 2\,\alpha\,\theta$$

$$0 = (20)^2 + 2\alpha(400),\ \therefore\ \alpha = -\frac{400}{800} = -\frac{1}{2}\text{ rev/sec}^2$$

Also $\omega_f = \omega_i + \alpha t$

$$0 = 20 - \frac{1}{2}t,\ \therefore\ t = 40\text{ sec.}$$

Example 6.31

A fly wheel is subjected to a constant turning moment which causes a uniform anticlockwise angular acceleration of 3 rev/sec². If the wheel was turning freely in clockwise direction at 3000 rev/min before the turning moment was applied, find the angular velocity after 20 seconds, and the total number of revolutions made during this period.

Solution

$$\omega_i = \frac{3000}{60} = 50\text{ rps};\ \alpha = -3\text{ rev/sec}^2,\ t = 20\text{ sec},\ \omega_f = ?,\ \theta = ?$$

$$\omega_f = \omega_i + \alpha\,t = 50 - 3\times 20 = \mathbf{-10\text{ rps}/ = 600\text{ rpm}}$$

(600 rpm in anticlockwise direction)

Let the wheel change its direction after t_1 seconds. Then

$$0 = 50 - 3\,t_1,\quad \therefore\ t_1 = \frac{50}{3}\text{ sec}$$

$$t_2 = 20 - \frac{50}{3} = \frac{10}{3}\text{ sec}$$

$$\theta_1 = \omega_i\,t_1 + \frac{1}{2}\alpha\cdot t_1^2 = 50\times\frac{50}{3} - \frac{1}{2}\times 3\times\frac{50}{3}\cdot\frac{50}{3}$$

$$= \mathbf{416\frac{2}{3}\text{ rev.}}$$

$$\theta_2 = 0 - \frac{1}{2}\times 3\left(\frac{10}{3}\right)\left(\frac{10}{3}\right) = \mathbf{-16\frac{2}{3}\text{ rev.}}$$

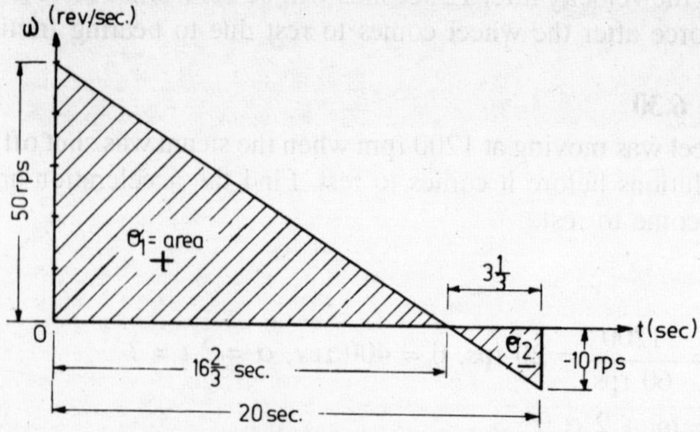

Fig. 6.54

Thus the gross number of revolutions

$$= 416\frac{2}{3} + 16\frac{2}{3} = 433\frac{1}{3} \text{ rev.}$$

Net revolutions will be

$$416\frac{2}{3} - 16\frac{2}{3} = 400 \text{ rev (clockwise)}$$

This can also be found by drawing the $\omega - t$ curve as shown in Fig. 6.54. Since initial displacement $\theta = 0$, we have

Net displacement θ = Area of $\omega - t$ curve

$$= \frac{1}{2}\left(50 \times \frac{50}{3}\right) - \frac{10}{2} \times \frac{10}{3} = 400 \text{ rev}$$

Gross $\theta = \theta_1 + \theta_2$, where θ_1 and θ_2 indicates the numerical values without sign.

$$= \frac{1}{2} \times 50 \times \frac{50}{3} + \frac{1}{2} \times 10 \times \frac{10}{3} = 433\frac{1}{3} \text{ rev}$$

Example 6.32

A fly wheel increases its speed from 300 to 600 rpm in 10 seconds. The diameter of the wheel is 2 m. Find

(a) the angular acceleration,
(b) the number of revolutions during this period of 10 secs, and
(c) the normal and tangential accelerations at the rim of the wheel at the end of 10 seconds.

Solution

$$\omega_f = \frac{600}{60} = 10 \text{ rps.}, \ \omega_i = \frac{300}{60} = 5 \text{ rps}, \ t = 10 \text{ secs}$$

$$r = \frac{2}{2} = 1 \text{ m}, \alpha = ?, \theta = ?, a_n \text{ and } a_t = ?$$

$$\omega_f = \omega_i + \alpha \cdot t$$

or $10 = 5 + \alpha\,(10), \quad \therefore \alpha = +\frac{1}{2} \text{ rev/sec}^2$

Also $\omega_f^2 = \omega_i^2 + 2\,\alpha\,\theta, \quad \therefore \theta = \dfrac{10^2 - 5^2}{2 \times \frac{1}{2}} = 75$ revolutions.

Linear tangential acceleration

$$a_t = \alpha \cdot r,$$

where α is in rad/sec^2 $= \frac{1}{2} \times 2\pi. = \pi$ rad/sec^2

$$\therefore a_t = \pi \cdot 1 = 3.14 \text{ ms}^{-2}$$

$a_n = \omega^2 \cdot r$, where ω is in radians/sec.

At the end of 10 secs

$$\omega = \omega_f = 10 \text{ rps} = (10 \times 2\pi) \text{ rad/sec.}$$

$$\therefore a_n = (20\pi)^2 \cdot 1 = 3947.84 \text{ ms}^{-2}$$

Example 6.33

A roller of a motorised mortar mixer rotates with a uniform angular acceleration for first 8 seconds and describes 100 radians. It then rotates with constant angular velocity for the next 5 seconds, during which time it describes 80 radians. Find the angular acceleration and initial angular velocity.

Solution

$$\theta_1 = 100 \text{ rad.}, t_1 = 8 \text{ secs}, \alpha_1 = ?, \omega_1 = ?$$

$$\theta_2 = 80 \text{ rad.}, \quad t_2 = 5 \text{ secs}, \alpha_2 = 0, \omega_2 = ?$$

During second 5 secs, the roller rotates with uniform angular velocity and describes 80 radians, and hence

$$\omega_2 = \frac{80}{5} = 16 \text{ rad/sec.}$$

This angular velocity is achieved in 8 seconds during which the roller accelerates and describes 100 radians.

Total displacement during first 8 secs = average vel. × time

i.e. $100 = \dfrac{(\omega_i + 16)}{2} \times 8$

or $\quad \omega_i = 9$ rad/sec

Now $\omega_2 = \omega_i + \alpha_1 \cdot t$

or $\quad 16 = 9 + \alpha_1(8), \therefore \alpha_1 = \dfrac{7}{8}$ rad/sec^2 = 0.875 rad s^{-2}.

Example 6.34

A particle is moving with its acceleration directed to and proportional to its distance from a fixed point. It has velocities of 2.0 ms^{-1} and 1.50 ms^{-1} respectively at distances 0.3 m and 0.4 m from its mean position. Find the maximum velocity, amplitude and maximum acceleration of the particle. Establish the relation of the acceleration to its distance from the fixed point.

Solution

Since the acceleration of the particle is directed and proportional to its distance from the fixed point, it is simple harmonic motion.

$$V_{x_1} = 2.0\ ms^{-1},\ x_1 = 0.30\ m,\ V_{x_2} = 1.50\ ms^{-1},\ x_2 = 0.40\ m$$

For SHM

$$V_x = \omega\ \sqrt{(r^2 - x^2)}$$

Substituting the values

$$2.0 = \omega\ \sqrt{(r^2 - 0.3^2)} \quad \text{.... (i)}$$

$$1.50 = \omega\ \sqrt{(r^2 - 0.4^2)} \quad \text{.... (ii)}$$

Dividing (i) by (ii), and squaring

$$\left(\frac{2}{1.5}\right)^2 = \frac{(r^2 - 0.3^2)}{(r^2 - 0.4^2)}$$

or $\quad r^2 = \dfrac{1}{7}\ [16(.16) - 9(.09)] = 0.25, \quad \therefore\ r = 0.50\ m$

From (i)

$$2.0 = \omega\ \sqrt{(0.5^2 - 0.3^2)}, \quad \therefore\ \omega = \frac{2.0}{0.4} = 5\ \text{rad/sec}$$

$$\text{Amplitude } r = 0.5\ m$$

$$\text{max velocity } V_x = \omega \cdot r = 5 \times 0.5 = 2.50\ ms^{-1}$$

$$\because \quad x = 0,\ V_x = \omega\ \sqrt{r^2 - 0^2}$$

Max. acceleration $a_n = \omega^2 \cdot r = 5^2 \times 0.5 = 12.5\ ms^{-2}$

Relation of acceleration and distance x

$$a_x = -\omega^2 \cdot x = -25\ x$$

Example 6.35

In a connecting rod and crank mechanism, the crank is of 20 cm length and rotates with 300 rpm. If the piston is assumed to have SHM, calculate its velocity and acceleration when it is 4 cm from one of the dead ends. Also calculate the maximum values of acceleration and velocity.

Solution

$$\omega = \frac{300 \times 2\pi}{60} = 10\pi\ \text{rad/sec},\ r = 20\ cm = 0.20\ m$$

When piston is 4 cm from dead ends,

$$x = 20 - 4 = 16\ cm = 0.16\ m \text{ from the mean position}$$

For SHM

Acceleration $a_x = -\omega^2 \cdot x = -(10\ \pi)^2 \times 0.16 = -158\ \text{ms}^{-2}$

Velocity $V_x = \omega\ \sqrt{(r^2 - x^2)} = 10\ \pi\ \sqrt{[(0.20)^2 - (0.16)^2)]} = 3.77\ \text{ms}^{-2}$

Max Acceleration $a = -\omega^2 \cdot x$ max $= -198\ \text{ms}^{-2}$

(Acceleration will be maximum when x is maximum i.e. at dead ends)

Max Velocity $V = \omega \cdot r = 10\pi \times 0.20\ \text{ms}^{-1} = 6.28\ \text{ms}^{-1}$

(V_x will be maximum when $x = 0$ i.e. at mean position)

Example 6.36

A reciprocating pump has a piston of stroke length 20 cm, and delivers water at the rate of 5 cycles/sec.

(a) Determine the maximum velocity and acceleration of the piston and RPM of the motor driving the piston through a crank and connecting rod mechanism. Neglect all frictional losses.

(b) Find position, velocity and acceleration of the piston 0.025 second after the start, from the mean position.

Solution

(a) Since the piston is driven by a crank and connecting rod mechanism it is case of SHM of horizontal projection of circular motion.

Radius of the crank $r = \frac{1}{2}$ (Stroke length)

$$= \frac{20}{2} = 10\ \text{cm} = 0.10\ \text{m.}$$

Angular velocity $\omega = 2\pi \times 5$ (As cycle is repeated after 1 rev. of motor)

$= 10\ \pi$ rad/sec.

RPM of Motor = 5 × 60 = 300 rev./min

Max Accl $\omega^2 \cdot r = (10\ \pi)^2\ (0.10) = 98.0\ \text{ms}^{-2}$

Maxvel $\omega \cdot r = 10\pi \times (0.10) = 3.14\ \text{ms}^{-1}$

(b) $t = 0.025$ sec

For SHM

$x = r \cdot \sin \omega t$, $V_x = \omega \cdot r \cos \omega t$, $a_x = -\omega^2 r \sin \omega t$

$\omega t = 10\ \pi\ (.025) = 0.7854\ \text{rad} = 45°$

$\therefore$ $x = 0.10 \sin (45°) = \mathbf{0.0707\ m}$

$V_x = 10\ \pi \times (.10) \cos 45° = \mathbf{2.22\ ms^{-1}}$

$a_x = 98 \sin 45° = \mathbf{69.29\ ms^{-2}}$

6.5.6 *Rectangular Coordinate Axes for Curvilinear Motion*

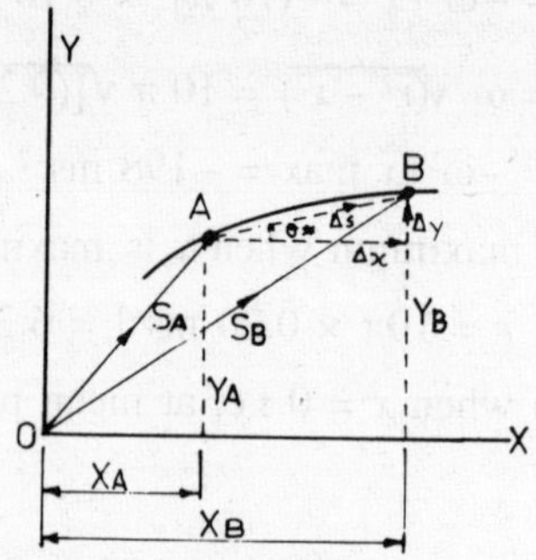

Fig. 6.55

When a particle moves in a plane along a path other then a straight line, the motion is known as curvilinear, and the position of the particle at any instant can be defined by its position vectors, such as S_A, S_B, etc. These position vectors S_A, S_B, are at different angles with respect to certain fixed reference axes. These position vectors can be conveniently defined by two coordinates with respect to two fixed reference axes at mutually perpendicular directions as in Fig. 6.55.

In case of curvilinear motion, the displacement changes both in direction and magnitude. The instantaneous velocity is directed along the tangent, while the acceleration of the particle is not necessarily tangential to the path. Various quantities such as position, velocity and acceleration vectors can be expressed as under:

$$\bar{S} = \bar{x} + \bar{y} \qquad \text{.... (a)}$$

i.e. Vector S_A = Vector x_A + Vector y_A

Similarly $\bar{S}_B = \bar{x}_B + \bar{y}_B$

If the particle moves from A to B in a small time Δt, we have by definition of velocity

$$\text{velocity } V = \frac{\Delta S}{\Delta t} = \frac{dS}{dt}$$

or $$V = \frac{d}{dt}(\bar{x} + \bar{y}) = \frac{dx}{dt} + \frac{dy}{dt}$$

or $$\bar{V} = \bar{V}_x + \bar{V}_y \qquad \text{.... (b)}$$

Differentiating this equation, we get

$$\frac{dV}{dt} = \frac{d}{dt}V_x + \frac{d}{dt}V_y$$

We know that $\frac{dV}{dt}$ = acceleration

$$\therefore \qquad \bar{a} = \bar{a}_x + \bar{a}_y \qquad \text{.... (c)}$$

x, V_x, a_x and y, V_y, a_y are respectively components along X and Y directions for displacement, velocity and acceleration.

Resultant and Components of Displacement, Velocity and Acceleration

In above expressions, it can be easily seen that in rectangular coordinates

$$S = \sqrt{x^2 + y^2}, \ \tan \phi_x = \frac{y}{x}$$

$$V = \sqrt{V_x^2 + V_y^2}, \ \tan \phi_x = \frac{V_y}{V_x} = \frac{dy}{dt} \Big| \frac{dx}{dt} = \frac{dy}{dx}$$

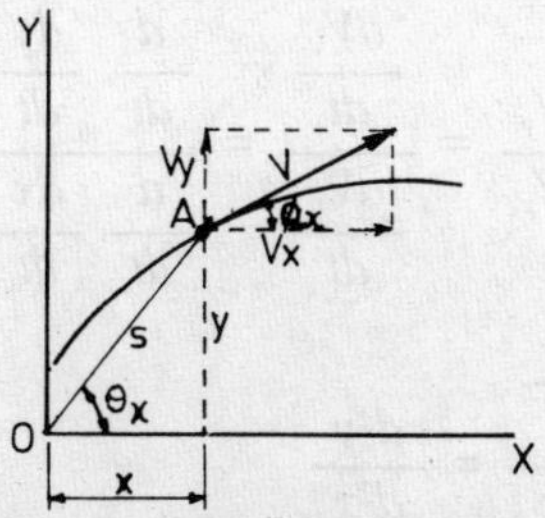

Fig. 6.56

This shows that the resultant velocity V at any instant is tangential to the curve at that instant. By the usual definitions of velocity, acceleration and displacement, it can be shown that

$$V_x = \frac{dS_x}{dt}, \ a_x = \frac{dV_x}{dt}$$

$$V_y = \frac{dS_y}{dt}, \ a_y = \frac{dV_y}{dt}$$

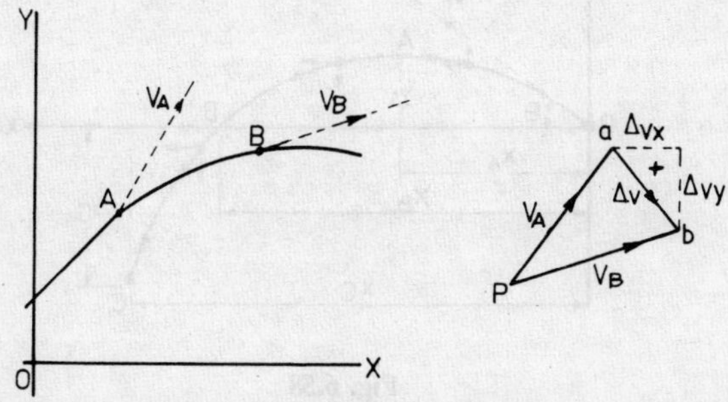

Fig. 6.57

Let V_A be the velocity at point *A*, and V_B be the velocity at *B* after time Δt. Acceleration is the rate of change of velocity, which is determined by drawing a vector diagram.

Let V_A be represented by P_a, and V_B by P_b. The change in velocity ΔV is represented by *ab*.

$$\overrightarrow{\Delta V} = \overrightarrow{\Delta V_x} + \overrightarrow{\Delta V_y}$$

$$\therefore \text{ Acceleration } a = \frac{\Delta V}{\Delta t} = \frac{(\Delta V_x + \Delta V_y)}{\Delta t} = \frac{\Delta V_x}{\Delta t} + \frac{\Delta V_y}{\Delta t}$$

or $$\vec{a} = \vec{a_x} + \vec{a_y}$$

Magnitude $(\Delta V)^2 = (\Delta V_x)^2 + (\Delta V_y)^2$

or $$\left(\frac{\Delta V}{\Delta t}\right)^2 = \left(\frac{\Delta V_x}{\Delta t}\right)^2 + \left(\frac{\Delta V_y}{\Delta t}\right)^2$$

or $$a^2 = a_x^2 + a_y^2$$

Also $$\tan \phi_x = \frac{\Delta V_y}{\Delta V_x} = \frac{\frac{dV_y}{dt}}{\frac{dV_x}{dt}} = \frac{\frac{d}{dt}\frac{dy}{dt}}{\frac{d}{dt}\frac{dx}{dt}}$$

$$= \frac{\frac{d^2y}{dt^2}}{\frac{d^2x}{dt^2}} = \frac{d^2y}{d^2x}$$

Thus the acceleration is not tangential to the path of motion at that instant.

6.5.7 *Projectiles (Air Resistance Neglected)*

To consider the path of projectiles, we shall neglect the effect of air resistance and rotation of earth. The basic principle in solving the curvilinear motion problems of projectiles shall be by resolving motion parameters along the X and Y axes.

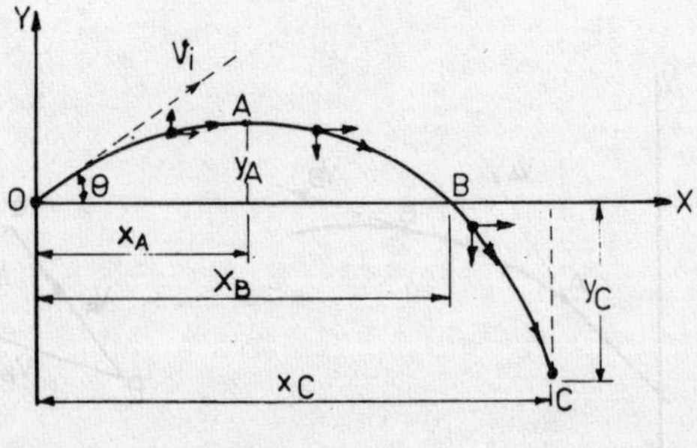

Fig. 6.58

Let a projectile be projected with an initial velocity V_i at an angle θ with X-axis, and let it describe the path OABC as shown in Figure 6.58. The origin of the axes 'O' is considered as the initial point of projection, and the axes are taken as horizontal (X-axis) and vertical (Y-axis). The displacements are taken as positive to the right along X-axis and upward along Y-axis. After the projectile leaves the projection point, it is subjected to initial projecting velocity and acceleration due to gravity in vertical direction (downwards). Thus the motion consists of two rectilinear motions along horizontal and vertical directions. The motion of the projectile at any instant is combination of its two rectilinear component motions at that instant.

Various motion parameters any instant are given as under:

Initial velocity along X, $V_{xi} = V_i \cos\theta$,

Initial velocity along Y, $V_{yi} = V_i \sin\theta$

Resultant Initial velocity $V_i = \sqrt{(V_{xi}^2) + (V_{yi}^2)}$ 6.20

Velocity at any instant after time t,

$$V_t = \sqrt{V_{xt}^2 + V_{yt}^2} \quad \text{.... 6.20(a)}$$

$$V_{xt} = V_{xi} + 0 = V_i \cos\theta$$

$$V_{yt} = (V_{yi} - g \cdot t) = (V_i \sin\theta - g \cdot t)$$

Position at any instant 't'

$$S = \sqrt{x^2 + y^2} \quad \text{.... 6.20(b)}$$

$$x = V_{xi}\, t + 0 = V_i\, t \cos\theta$$

$$y = (V_{yi} \cdot t - \frac{1}{2} g \cdot t^2) = (V_i\, t \sin\theta - \frac{1}{2} g t^2)$$

From the expressions $x = V_i\, t \cos\theta$, $y = (V_i \sin\theta - \frac{gt}{2})\, t$

$$\text{we get } y = (V_i \sin\theta - \frac{g}{2} \cdot \frac{x}{V_i \cos\theta}) \frac{x}{V_i \cos\theta}$$

or

$$y = x \tan\theta - \frac{g \sec^2\theta}{2 V_i^2} x^2 \quad \text{.... 6.21}$$

θ, g, and V_i are constant in the projectile motion,

Eqaution 6.21 of projectile motion is the equation of parabola and hence the path of a projectile is a parabola.

Maximum Height, Range and Time of Flight

Since the horizontal velocity is constant, the horizontal distance travelled is given by constant velocity multiplied by time, while the vertical distance is given by

$y = V_i \cdot t \sin\theta - \frac{1}{2} g t^2$, as it is under acceleration due to gravitity $(-g)$

For y to be maximum $\frac{dy}{dt} = 0 = V_i \sin\theta - \frac{1}{2} g \cdot 2t$

$$\text{i.e. } t \text{ (for max. height)} = \frac{V_i \sin\theta}{g}$$

Putting this value of t, maximum y is

$$y_{max} = V_i \sin\theta.\ \frac{V_i \sin\theta}{g} - \frac{1}{2} g \left(\frac{V_i \sin\theta}{g}\right)^2$$

$$\text{Or } y_{max} = \frac{V_i^2}{2g}.\ \sin^2\theta \qquad \text{.... 6.22}$$

This can also be found by putting vertical velocity = 0, at the highest point, since it changes direction from upwards to downwards.

Putting $V_{yf} = 0$, we get $V_i \sin\theta - gt = 0$

$$\therefore t = \frac{V_i \sin\theta}{g}, \text{ and } V_{max} = \frac{y_i^2}{2g} \cdot \sin^2\theta$$

Let the time of flight when the projectile touches a point on the horizontal line drawn from the point of projection be = T secs. At that instant the vertical distance is zero, and therefore

$$y = 0 = V_i \sin\theta.\ T - \frac{1}{2} g\, T^2,$$

$$\therefore T = 0, \text{ or } \left(\frac{2\, V_i \sin\theta}{g}\right)$$

$$\text{i.e. time of flight } T = \frac{2V_i \sin\theta}{g} \qquad \text{.... 6.23}$$

Since there is no acceleration in the horizontal direction the distance travelled is given by

$$x = V_i \cos\theta \cdot T$$

$$\text{Or} \quad x = V_i \cos\theta \cdot \frac{2V_i \cdot \sin\theta}{g}$$

$$\therefore \quad x = \frac{V_i^2 \sin 2\theta}{g} \qquad \text{.... 6.24}$$

The horizontal distance is also known as horizontal range. This range has different values for given values of V_i and θ. It is maximum for a given value of V_i when the value of $\sin 2\theta$ is maximum, i.e. $\sin 2\theta = 1$

or $\quad \theta = 45°$

$\therefore$ Maximum range $= \dfrac{V_i^2}{g}$, when projected at 45° with the horizontal Eqn.6.25

The above results are summarized below with the usual notations:

(i) Trajectory of the projectile is a parabola,

(ii) Horizonta velocity at any instant = $V_i \cos\theta$,

(iii) Vertical velocity at any instant = $(V_i \sin\theta - gt)$,

(iv) Horizontal distance travelled at any instant = $V_i \cos \theta \cdot t$,

(v) Vertical distance covered at any instant = $(V_i \sin \theta \cdot t - \frac{1}{2} g t^2)$,

(vi) Total time of flight $T = \frac{2 V_i \sin \theta}{g}$,

(vii) Maximum height of flight for a given initial velocity and given angle of projection y_{max} or $(S_y)_{max} = \frac{V_i^2 \cdot \sin^2 \theta}{2 g}$,

(viii) Horizontal range x or $S_x = \frac{V_i^2 \cdot \sin 2 \theta}{g}$,

(ix) Maximum range for a given velocity of projection X_{max} or

$$(S_x)_{max} = \frac{V_i^2}{g}, \text{ when } \theta = 45°$$

where V_i — initial velocity of projection

V_f — final velocity after time t

t — time after projection

θ — Inclination of projection with the horizontal

S_x or x — Horizontal range from the point of projectoin.

S_y or y — Vertical height above the point of projection.

Example 6.37

A shell can be fired with a muzzle velocity of 400 m/sec. What should be th inclination of the gun so that a target at a horizontal distance of 4000 metre can be hit?

Solution

V_i = 400 m/sec, range = 4000 m, $S_y = 0$

$$\text{Range} = \frac{V_i^2 \sin 2\theta}{g} = 4000$$

or
$$\frac{(400)^2 \sin 2\theta}{9.80} = 4000, \; (g = 9.80 \text{ m/sec}^2)$$

or
$$\sin 2\theta = \frac{9.8}{40} = 0.245$$

$\therefore$ $2\theta = 14.2°$ Or $165.8°$

i.e. $\theta = 7.1°$ or $82.9°$ with the horizontal.

Example 6.38

A projectile is fired with a velocity of 400 m/sec and directed at an angle o 30° to the horizontal. Find the range on a horizontal plane through the point of projection. Also find the maximum possible range for this projectile.

Solution

$V_i = 400$ m/sec, $\theta = 30°$, $S_x = ?$ $S_{x_{max}} = ?$

$$S_x = \frac{V_i^2 \sin 2\theta}{g} = \frac{400 \times 400 \times \sin 60°}{9.80} = 14150.0 \text{ metres}$$

$$S_{x_{max}} = \frac{V_i^2}{g} = \frac{400 \times 400}{9.80} = 16350.0 \text{ metres}$$

Example 6.39

A projectile aimed at a target on the horizontal plane falls 850 metres short when fired at an angle of 30° to the horizontal, while it reaches 1350. m ahead of target when fired at 45°. Find the correct distance of the target and the required angle of projection if the velocity remains constant.

Solution

For given initial velocity the range $S_x = \dfrac{V_i^2 \cdot \sin 2\theta}{g}$,

$$\therefore S_{x_1} = \frac{V_i^2 \sin 60°}{g} = 0.866 \frac{V_i^2}{g} \quad \text{.... (1)}$$

$$S_{x_2} = \frac{V_i^2}{g} \sin 90° = \frac{V_i^2}{g} \quad \text{.... (2)}$$

Let the target be at a distance of S_x. Then we have

$$(S_x - 850) = S_{x_1} = 0.866 \frac{V_i^2}{g} \quad \text{.... (1a)}$$

$$(S_x + 1350) = S_{x_2} = \frac{V_i^2}{g} \quad \text{.... (2a)}$$

multiplying equation (2) by 0.866

$$0.866 (S_x + 1350) = 0.866 \frac{V_i^2}{g} \quad \text{.... (2b)}$$

Solving (1a) and (2b)

$$S_x = \frac{(850 + 1169.1)}{0.134} = \mathbf{15067\ m}$$

Putting this value in equation (2a)

$$15067 + 1350 = \frac{V_i^2}{g}$$

$$\therefore V_i = \sqrt{16417 \times 9.80} = 400 \text{ m/sec}$$

For hitting the target, let the angle of projection be α; then $15067 = 16417 \sin 2\alpha$

or $\sin 2\alpha = \dfrac{15067}{16417} = 0.919$

$\therefore 2\alpha = 66.8°, 113.2°$ or $\alpha =$ **33.4°, 56.6°**

Example 6.40

A shell fired from a gun leaves at an angle of 30° to the horizontal. If the velocity of the shell at the time of leaving the gun is 200 ms^{-1}, neglecting air resistance find

(a) the maximum height attained by the shell,

(b) the time taken by the shell to strike the ground,

(c) The horizontal distance covered by the shell before striking the ground,

(d) Velocity at the time of striking the ground.

Solution

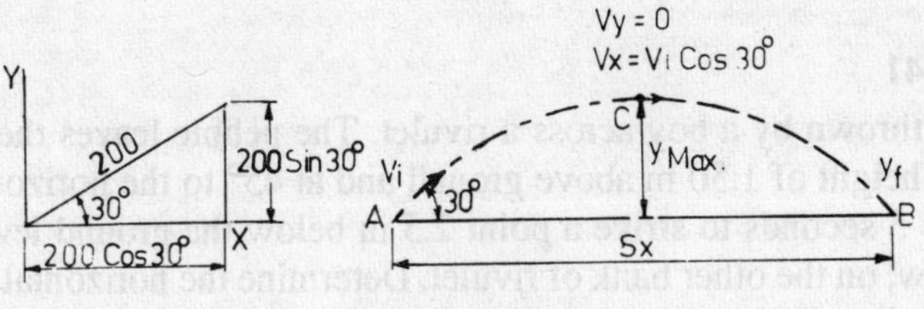

Fig. 6.59

V_i = 200 ms^{-1}, θ = 30°, $S_{y_{max}}$ = ? S_x = ?, t_f = ? V_f = ?

When air resistance is neglected, the shell moves under constant vertical acceleration due to gravity, and zero acceleration in the horizontal direction, Thus

$$a_x = 0 \quad a_y = -g = -9.80 \text{ ms}^{-2}$$

At the starting point A

$$V_{xi} = V_i \cos 30° = \frac{200\sqrt{3}}{2} = 100\sqrt{3} \text{ ms}^{-1}$$

$$V_{yi} = V_i \sin 30° = 200 \times \frac{1}{2} = 100 \text{ ms}^{-}$$

(a) When the shell is at the highest point, its vertical velocity is zero; thus

$$V_{Vf} = 0$$

$$0 = V_{yi} + a_y \cdot t$$

or $0 = 100 - 9.80$ t; $\therefore$ t = **10.204 secs**

$$\therefore S_{y_{max}} = V_{yi} \cdot t - \frac{9.80}{2} t^2 = 100(10.204) - 4.9(10.204)^2 = \mathbf{510 \text{ m}}$$

(b) The time spent in rising to the highest point and in falling back to the ground are the same. Hence the total time of flight

$$t_f = 2 \times 10.204 = 20.408 \text{ seconds.}$$

(c) Since the horizontal velocity remains constant the horizontal distance covered by the shell

$S_x = V_{xi} \cdot t_f = (100\sqrt{3} \times 20.408)\text{m} = \mathbf{3535\ m.}$

(d) Horizontal velocity is constant $(V_{xi}) = 100\sqrt{3}\ \text{ms}^{-1}$. Vertical velocity after 20.408 secs is given by

$$V_{yf} = 100 - 9.8 \times 20.408 = -100\ \text{ms}^{-1}$$

$\therefore$ resultant velocity $V_f = \sqrt{V_{xf}^2 + V_{yf}^2} = \sqrt{(100\sqrt{3})^2 + (-100)^2} = 200$ ms^{-1}

$$\tan\theta_x = \frac{V_{yf}}{V_{xf}} = \frac{-100}{100\sqrt{3}},$$

$\therefore \theta_x = 150°$ with the +ve direction of x.

Example 6.41

A pebble is thrown by a boy across a rivulet. The pebble leaves the hand of the boy at a height of 1.50 m above ground and at 45° to the horizontal. The pebble takes 5 seconds to strike a point 2.5 m below the ground level at the point of throw, on the other bank of rivulet. Determine the horizontal distance travelled by the pebble and the velocity with which it was thrown.

Solution

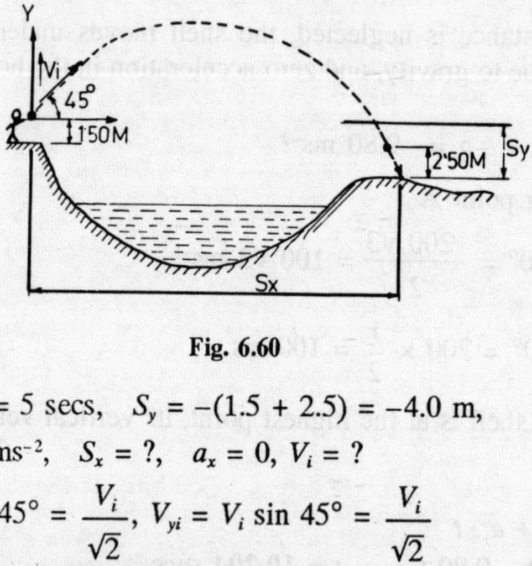

Fig. 6.60

$\theta = 45°$, $t_f = 5$ secs, $S_y = -(1.5 + 2.5) = -4.0$ m,

$a_y = -9.80\ \text{ms}^{-2}$, $S_x = ?$, $a_x = 0$, $V_i = ?$

$$V_{xi} = V_i \cos 45° = \frac{V_i}{\sqrt{2}},\ V_{yi} = V_i \sin 45° = \frac{V_i}{\sqrt{2}}$$

$$S_y = V_{yi} \cdot t + \frac{1}{2} a_y \cdot t^2$$

or $-4.0 = \dfrac{V_i}{\sqrt{2}} \times .5 - \dfrac{1}{2}\ 9.80(5)^2, \quad \therefore V_i = \mathbf{33.5\ ms^{-1}}$

$$S_x = V_{xi} \cdot t = \frac{33.5}{\sqrt{2}} \times = \mathbf{118.5\ m}$$

$(\because a_x = 0)$

Example 6.42

There is a 4.704 m high wall 20 m in front of enemy target. If the gun man is 30 m away from the wall on the opposite side of the target and on the same horizontal plane as the target, find the angle of projection and minimum velocity of projection so that the shell strikes the target.

Solution

There are two conditions to hit the enemy target:

(i) Horizontal range $S_x = 20 + 30 =$ **50 m**

(ii) Height of the projectile at 30 m horizontal distance = **4.704 m**

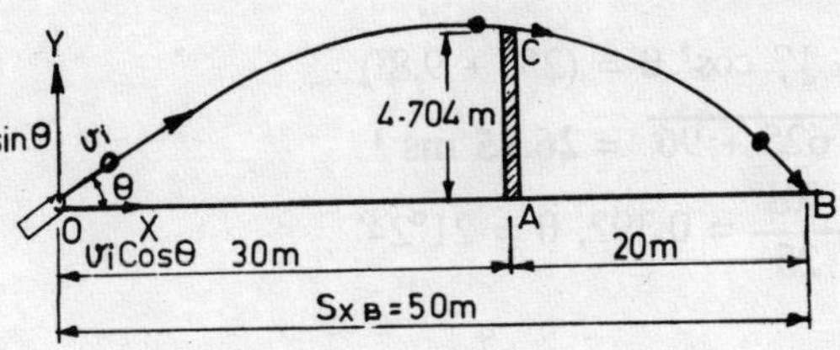

Fig. 6.61

$S_{xA} = 30$ m, $S_{yA} = 4.704$ m, $V_i = ?$, $\theta = ?$

$S_{xB} = 50$ m, $S_{yB} = 0$, $a_x = 0$, $a_y = -9.8$ ms^{-2}

For point B

Vertical distance $S_{yB} = (V_{yi} \cdot t_B + \frac{1}{2} a_y \cdot t_B^2) = 0$

or $V_i \sin\theta \cdot t_B - \frac{9.80}{2} \cdot t_B^2 = 0, \quad \therefore t_B = \frac{V_i \sin\theta}{4.9}$

Horizontal distance OB travelled during t_B

$$S_{xB} = V_i \cos\theta \cdot \frac{V_i \sin\theta}{4.9}, \text{ as } a_x = 0$$

$$50 = \frac{V_i^2 \sin\theta \cos\theta}{4.9} \quad \text{.... (1)}$$

Let the time taken upto the wall be t_A.

Then

$$4.704 = V_i \sin\theta \cdot t_A - \frac{1}{2} \cdot 9.8\, t_A^2 \quad \text{.... (2)}$$

and $30.0 = V_i \cos\theta \cdot t_A$

$$\text{or } V_i \cos\theta = \frac{30}{t_A} \quad \text{.... (3)}$$

Substituting in equation (1)

$$50 = \frac{V_i \sin\theta}{4.9} \cdot \frac{30}{t_A}, \quad \therefore V_i \sin\theta = \frac{50}{30} \times 4.9\, t_A \quad \text{.... (4)}$$

Putting this in equation (2), we get

$$4.704 = \left(\frac{5 \times 4.9}{3} t_A\right) t_A - 4.9\, t_A^2$$

or $$t_A = \sqrt{\frac{4.704 \times 3}{9.8}} = \sqrt{1.44} = 1.20 \text{ secs}$$

From equation (3) and (4)

$$V_i \cos\theta = \frac{30}{1.2} = 25$$

$$V_i \sin\theta = \frac{5}{3} \times 4.9 \times 1.2 = 9.8$$

$$\therefore\ V_i^2 \sin^2\theta + V_i^2 \cos^2\theta = (25^2 + 9.8^2)$$

$$V_i = \sqrt{625 + 96} = \mathbf{26.85\ ms^{-1}}$$

Also $$\tan\theta = \frac{9.8}{25} = 0.392,\ \theta = 21°24'$$

Example 6.43

A ramp inclined at an angle of 30° to the horizontal is made to jump safely over a 6.3 m wide ditch. If the ramp end of ditch is 1.25 m higher than that of other bank of the ditch, find the take off velocity in kmh^{-1} of a jeep for safe crossover.

Solution

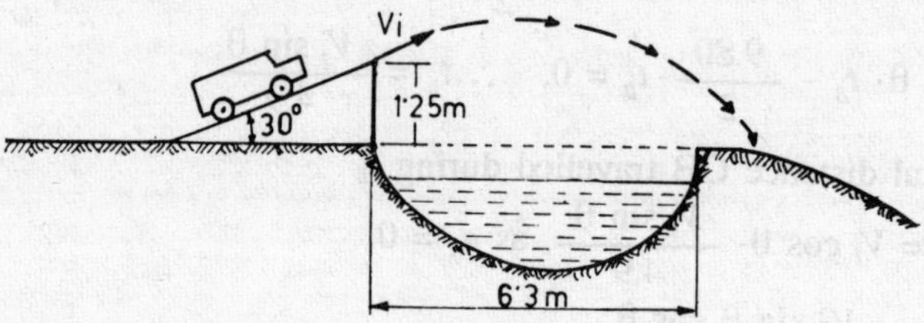

Fig. 6.62

$$V_i = ?,\quad \theta = 30°,\quad S_x = 6.3 \text{ m},\quad S_y = -1.250 \text{ m},$$

$$a_x = 0,\ a_y = -9.8 \text{ ms}^{-2}$$

$$S_x = 6.3 = V_i \cos 30° \cdot t \qquad \text{.... (1)}$$

$$S_y = -1.25 = V_i \sin 30° \cdot t - \frac{1}{2}\, 9.8\, t^2 \qquad \text{.... (2)}$$

Solving (1) and (2), we get

$$(6.3 \sin 30° + 1.25 \cos 30°) = 4.9 \cos 30° \cdot t^2$$

or $$t^2 = \left(\frac{6.3}{4.9} \tan 30° + \frac{1.25}{4.9}\right) = 0.9975 \quad \therefore\ t = 1.0 \text{ sec}$$

From equation (1)

$$V_i = \frac{6.3 \times 2}{\sqrt{3} \times 1.0} = 7.27 \text{ ms}^{-1} = \mathbf{26.2 \text{ kmh}^{-1}}$$

Example 6.44

A soldier fires a bullet at an angle of 30° upwards from his position on a hill to strike a target which is 137.2 m lower than the position of the soldier. The initial velocity of bullet is 117.6 ms⁻¹. Calculate

(a) the maximum height to which the bullet will rise above the point of projection
(b) the total time of flight
(c) the horizontal distance of target from the point of projection, and
(d) the velocity with which it will strike the target.

Solution

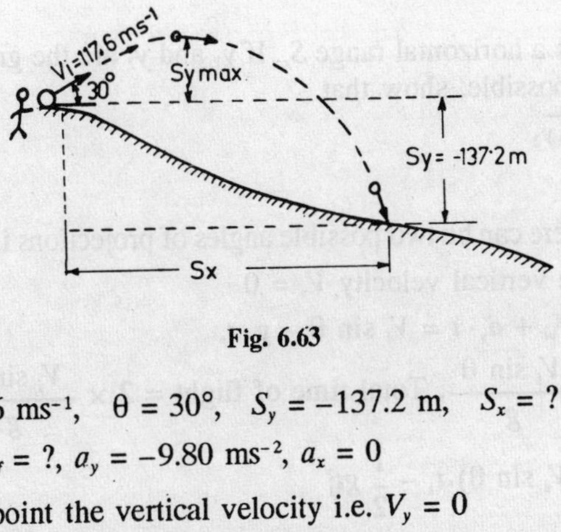

Fig. 6.63

$V_i = 117.6 \text{ ms}^{-1}$, $\theta = 30°$, $S_y = -137.2$ m, $S_x = ?$

$S_{y_{max}} = ?$, $t_f = ?$, $a_y = -9.80 \text{ ms}^{-2}$, $a_x = 0$

At the highest point the vertical velocity i.e. $V_y = 0$

$$0 = V_{yi} + a_y \cdot t_1 = (117.6 \sin 30°) - 9.80\, t_1$$

$$\therefore t_1 = \frac{117.6}{9.8} \times \frac{1}{2} = 6 \text{ secs}$$

$$\text{Max. height } S_{y_{max}} = V_{yi} \cdot t + \frac{1}{2} a_y \cdot t^2$$

$$= 117.6 \sin 30° \times 6 - \frac{1}{2} \cdot 9.8(6)^2$$

$$= \mathbf{176.4 \text{ m}}$$

For target $S_y = -137.2$ m, $S_x = ?$

$$-137.2 = V_{yi} \cdot t + \frac{1}{2} a_y \cdot t^2 = (117.6 \times \sin 30°)\, t_f - \frac{1}{2} \times 9.8\, t_f^2$$

or $\quad t_f^2 - 12\,t_f - 28 = 0,$

$\therefore\ t_f = 6 \pm 8 = \mathbf{14}$ **sec, or −2 (−ve time is not applicable here).**

Thus t_f = **14 seconds**

(c) $\quad S_x = V_{xi} \cdot t_f$, as $a_x = 0$

Or $\quad S_x = (117.6 \cos 30°) \times 14 = \mathbf{1425.8\ m}$

(d) $\quad V_{xf} = V_{xi} = 117.6 \cos 30° = \mathbf{101.85\ ms^{-1}}$

$$V_{yf} = V_{yi} + a_y \cdot t_f$$

$$= 58.8 - 9.8 \times 14 = \mathbf{-78.4\ ms^{-1}}$$

$$\therefore\ V_f = \sqrt{(101.85)^2 + (-78.4)^2} = 128.5\ ms^{-1}$$

$$\tan\theta = \frac{V_{yf}}{V_{xf}} = \frac{-78.4}{101.85} = -0.77$$

$\therefore\ \theta = 37°36'$ **(downward) with the horizontal**

Thus striking velocity is 128.5 ms⁻¹ inclined at 37°36′ with the horizontal in downward direction.

Example 6.45

A projectile has a horizontal range S_x. If y_1 and y_2 are the greatest heights in the two paths possible, show that

$$S_x = 4\sqrt{y_1 y_2}$$

Solution

For the range there can be two possible angles of projections i.e. θ and $(90 - \theta)$

For y(max), the vertical velocity $V_y = 0$

$$0 = V_{yi} + a_y \cdot t = V_i \sin\theta - g \cdot t,$$

$$t_1 = \frac{V_i \sin\theta}{g}, \text{ Total time of flight} = 2 \times \frac{V_i \sin\theta}{g}$$

$$\therefore\ y_1 = (V_i \sin\theta) \cdot t_1 - \frac{1}{2} g t_1^2$$

$$y_1 = \frac{V_i^2 \sin^2\theta}{2g} \qquad \text{.... (1)}$$

$$\text{Similarly } y_2 = \frac{V_i^2 \sin^2(90-\theta)}{2g} = \frac{V_i^2 \cos^2\theta}{2g} \qquad \text{... (2)}$$

$$\therefore\ y_1 y_2 = \frac{V_i^4 \sin^2\theta \cos^2\theta}{4g} = \left(\frac{V_i^2}{2g}\right)^2 (\sin\theta\cos\theta)^2$$

$$\sqrt{y_1 y_2} = \frac{V_i^2}{2g}(\sin\theta\cos\theta) = \frac{V_i \sin\theta . V_i \cos\theta}{2g} \qquad \text{... (3)}$$

$$S_x = V_{xi} \times \text{time of flight} = \frac{V_i \cos\theta . 2V_i \sin\theta}{g} \qquad \text{... (4)}$$

Or $\quad S_x = \dfrac{4\ V_i \cos\theta . V_i \sin\theta}{2g} = 4\sqrt{y_1 y_2}$ which is what was required to be proved.

Example 6.46

A target is '*h*' metres above the plane of projection, and makes an angle β with the horizontal plane through the point of projection. Determine the maximum range on the inclined plane and the angle of projection for the same if the velocity of the projectile fired is V_i. Also find the minimum value of velocity which may be necessary to hit the target.

Solution

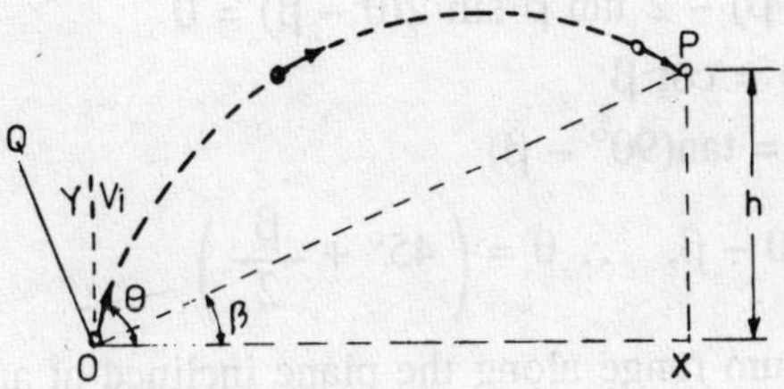

Fig. 6.64

Let the plane along the line of sight for the target be OP, and let the perpendicular axes be OQ. All quantities along OP shall be denoted by subscript P, and those along the perpendicular by Q.

Let the angle of projection with the horizontal be θ and that with OP = (θ – β)

The components of velocity are as under

$$V_{pi} = V_i \cos(\theta - \beta),\ V_{qi} = V_i \sin(\theta - \beta)$$

Components of acceleration due to the gravity are

$$a_p = -g \sin \beta,\ a_q = -g \cos \beta$$

When the target is hit, the displacement along Q is zero, and hence

$$S_q = 0 = V_{qi} \cdot t + a_q \cdot t^2/2$$

$$\text{or}\quad 0 = t \cdot V_i \sin(\theta - \beta) - g \cos \beta \cdot \frac{t^2}{2}$$

$$\text{or}\quad t = \frac{2V_i \sin(\theta - \beta)}{g \cos \beta} \qquad \text{.... (1)}$$

$$S_p = V_{pi} \cdot t + a_p \cdot \frac{t^2}{2}$$

$$= V_i \cos(\theta - \beta)\, \frac{2V_i \sin(\theta - \beta)}{g \cos \beta} - \frac{g}{2} \sin \beta \left(\frac{2V_i \sin(\theta - \beta)}{g \cos \beta} \right)^2$$

$$= \frac{V_i^2}{g \cos \beta}\, [\sin 2(\theta - \beta) - 2 \sin^2(\theta - \beta) \tan \beta]$$

For S_p to be maximum with respect to θ for given values of V_i and β, we have

$$= \frac{d}{d\theta} S_p = 0$$

i.e. $$\frac{V_i^2}{g\cos\beta}\left[\frac{d}{d\theta}\sin 2(\theta-\beta) - \tan\beta\frac{d}{d\theta}2\sin^2(\theta-\beta)\right] = 0$$

$$\therefore \frac{d}{d\theta}\sin 2(\theta-\beta) - \tan\beta\frac{d}{d\theta}(1-\cos 2(\theta-\beta)) = 0$$

or $2\cos 2(\theta-\beta) - 2\tan\beta\sin 2(\theta-\beta) = 0$

or $\tan 2(\theta-\beta) = \cot\beta$

or $\tan 2(\theta-\beta) = \tan(90^\circ-\beta)$

or $2(\theta-\beta) = 90-\beta, \quad \therefore \theta = \left(45^\circ + \frac{\beta}{2}\right)$

Thus for maximum range along the plane inclined at an angle β, the angle of projection with the horizontal must be $\left(45^\circ + \frac{\beta}{2}\right)$

$$S_p\,(\max) = \frac{V_i^2}{g\cos\beta}\left[\sin(90-\beta) - 2\sin^2\left(45^\circ - \frac{\beta}{2}\right)\cdot\tan\beta\right]$$

$$= \frac{V_i^2}{g\cos\beta}[\cos\beta - \tan\beta(1-\sin\beta)]$$

$$= \frac{V_i^2}{g\cos^2\beta}(\cos^2\beta - \sin\beta + \sin^2\beta)$$

$$= \frac{V_i^2}{g\cos^2\beta}(1-\sin\beta) = \frac{V_i^2}{g(1+\sin\beta)}$$

Maximum range along the plane inclined at an angle of β is

$$S_p\,(\max) = \frac{V_i^2}{g(1+\sin\beta)}$$

$S_p = h\,\mathrm{cosec}\,\beta$, where h is height of the target above the horizontal plane.

$$\therefore h\,\mathrm{cosec}\,\beta = \frac{V_i^2}{g(1+\sin\beta)}$$

$$\therefore V_i^2 = gh(\mathrm{cosec}\,\beta + 1)$$

$$V_i = \sqrt{gh(1+\mathrm{cosec}\,\beta)}$$

This is the minimum velocity of projection which is required to hit the target along the plane and at a height 'h' above the point of projection.

Example 6.47

A bomber plane is flying horizontally at a height of 1000 m above a target with a velocity of 400 km/hr. Determine the angle that the line of sight of a target on the ground should make with the horizontal at the time of releasing the bomb. Also determine the position of the bomber plane when the target is hit.

Solution

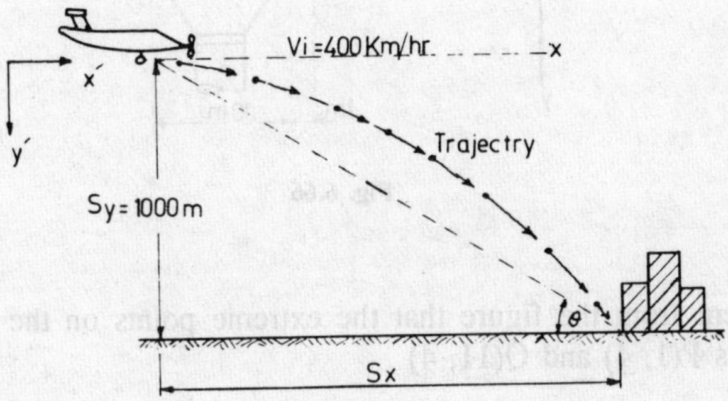

Fig. 6.65

$$V_i = \frac{400 \times 1000}{60 \times 60} = \frac{1000}{9} \text{ m/sec.}$$

Here $a_x = 0,\ V_{xi} = V_i = \left(\dfrac{1000}{9}\right) \text{ ms}^{-1},\ S_x = ?$

$V_{yi} = 0, \quad a_x = 9.80 \text{ ms}^{-2}, \quad S_y = 1000 \text{ m}$

Let the time required to hit the target = t

$$-1000 = 0 \cdot t - \frac{9.8t^2}{2}, \quad \therefore\ t = \sqrt{\frac{2000}{9.8}} = 14.28575.$$

Horizontal distance covered by the Bomb during the period

$$S_x = V_{xi} \cdot t + 0 = \frac{1000}{9} \times 10\sqrt{\frac{20}{9.8}} = 1587.3$$

Hence the angle of line of sight at the time of release of the bomb with the horizontal is given by

$$\tan \theta = \frac{1000}{S_x} = \frac{1000}{1587.3} = 0.633, \quad \therefore\ \theta = 32°.20'$$

Since the horizontal velocity of the bomber plane is same as that of the bomb, the horizontal distance covered by the plane and the bomb shall be the same, so the bomber shall be vertically above the target if the bomber does not change the direction.

Example 6.48

Wheat is discharged at a point 0 from a conveyor belt for collection in a hopper as shown in Figure 6.66. Determine the range of the speed of the belt for which the wheat will fall in the hopper.

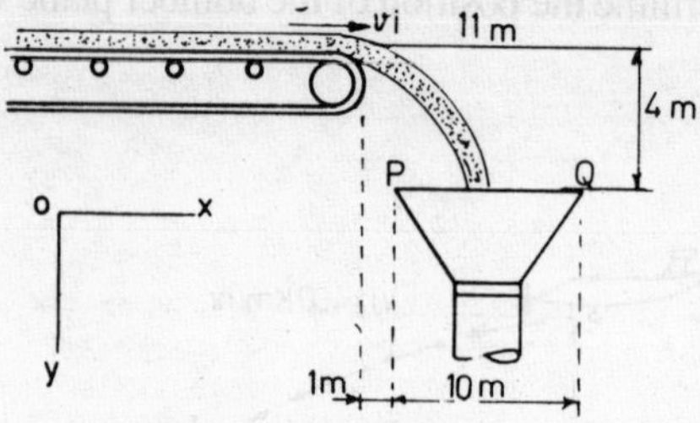

Fig. 6.66

Solution

It is evident from the figure that the extreme points on the hopper have coordinates P(1, 4) and Q(11, 4)

i.e. $S_{xp} = 1, \quad S_{yp} = 4$ and $S_{xq} = 11, \quad S_{yq} = 4$

$$a_{xp} = 0, \quad a_{yp} = 9.80 \text{ ms}^{-2}, \quad a_{xq} = 0, \quad a_{yq} = 9.8 \text{ ms}^{-2}$$

Applying equations $S = V_i \cdot t + \frac{1}{2} a \cdot t^2$

$$4 = V_{yi} \cdot t_1 + \frac{9.8}{2} \cdot t_1^2, \text{ (Since } V_{yi} = V_i \sin\theta = 0, \theta = 0)$$

$$\therefore t = \sqrt{\frac{4 \times 2}{9.8}} = \frac{2\sqrt{10}}{7} \text{ sec}$$

Also

$$1 = V_{xi} \cdot t_1 + \frac{0}{2} \cdot t_1^2, \quad (\because a_x = 0)$$

$$\text{or } V_{xi} = \frac{1}{t_1} = \frac{7}{2\sqrt{10}}$$

$$\text{or } V_i \cos\theta^\circ = \frac{7}{2\sqrt{10}} \quad (\because V_{xi} = V_i \cos\theta)$$

or V_i = **1.1068 ms^{-1}**

For point Q : $S_{xq} = 11$ m, $S_{yq} = 4$ m, $t_2 = t_1$, $(\because S_{yq} = S_{yp})$

$$11 = V_{xi} \cdot t_1 + 0$$

$$\text{or } V_{xi} = \frac{11}{t_1} = \frac{11 \times 7}{2\sqrt{10}} = 12.175 = V_i$$

Thus the range of speed of the belt is 1.1068 ms^{-1} to 12.175 ms^{-1} for which the grains shall fall in the hopper.

Example 6.49

A stone is thrown up at an angle of 30° to the horizontal from a point O with an initial velocity of 30 ms^{-1}. If the net components of acceleration due to gravity and air resistance are $a_x = -1.0\ ms^{-2}$, $a_y = -8.0\ ms^{-2}$, determine the horizontal distance covered by the stone when it reaches a point *A* which is 20 m below the point of projection O.

Solution

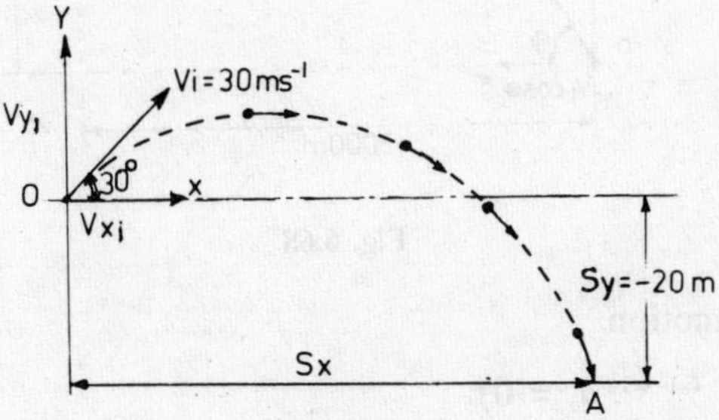

Fig. 6.67

$V_i = 30\ ms^{-1}$, $\theta = 30°$, $a_x = -1.0\ ms^{-2}$, $a_y = -8.0\ ms^{-2}$

$S_y = -20$ m, $S_x = ?$

$V_{xi} = 30 \cos 30° = 25.98\ ms^{-1}$, $V_{yi} = 30 \sin 30° = 15.0\ ms^{-1}$

$$S_y = V_{yi} \cdot t + \frac{1}{2} a_y \cdot t^2, \text{ or } -20 = 15 \cdot t - \frac{8.0}{2} t^2, \quad \text{or } 4t^2 - 15t - 20 = 0$$

$$\therefore t = \frac{15 \pm 23.35}{8} = 4.794, \text{ or } -1.044 \text{ secs}$$

Since negative value is not applicable, $t = 4.794$ secs

$$\therefore S_x = V_{xi} \cdot t + \frac{a_x}{2} \cdot t^2$$

$$= 25.98 \times 4.7940 - \frac{1.0}{2} (4.7940)^2 = \mathbf{113\ m}$$

Example 6.50

A projectile is fired with an initial velocity of 200 ms^{-1} to hit a target P located at 500 m above the level of the gun point O, and at a horizontal distance of 3000 m. Neglecting the air resistance, find the firing angle.

Solution

$V_i = 200\ ms^{-1}$, $S_x = 3000$ m, $S_y = +500$ m,

$a_x = 0$, $a_y = -9.8\ ms^{-2}$

Let the angle of projection be θ.

$V_{xi} = V_i \cos \theta = 200 \cos \theta$

$V_{yi} = V_i \sin \theta = 200 \sin \theta$

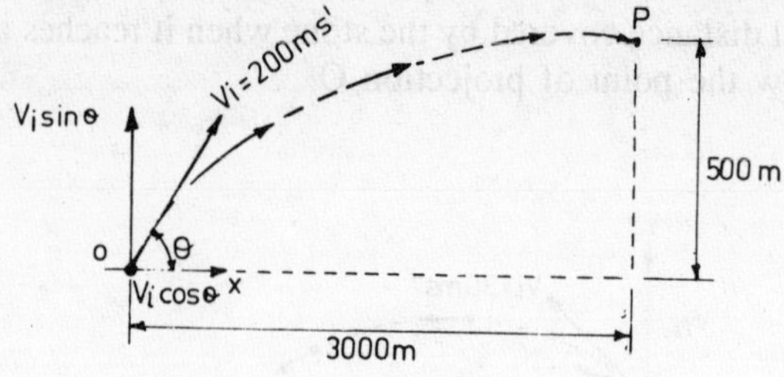

Fig. 6.68

For horizontal motion

$$S_x = V_{xi} \cdot t \quad (\because a_x = 0)$$

$$3000 = 200 \cos\theta \cdot t, \quad \therefore t = \frac{15}{\cos\theta} \qquad \text{.... (1)}$$

For vertical motion

$$S_y = V_{yi} \cdot t + \frac{1}{2} a_y \cdot t^2$$

$$+500 = 200 \sin\theta \cdot \frac{15}{\cos\theta} + \frac{i}{2}(-9.80)\left(\frac{15}{\cos\theta}\right)^2$$

or $\quad 500 = 3000 \tan\theta - 4.9 \times 225 (\sec^2\theta)$

or $\tan^2\theta - \dfrac{3000}{1102.5}\tan\theta + \left(\dfrac{500}{1102.5} + 1\right) = 0$

or $\tan\theta = \dfrac{2.72 \pm \sqrt{7.4 - 5.816}}{2}$

$$= 1.989 \text{ or } 0.731$$

$\therefore \quad \theta = 63°–20' \text{ or } 36°–10'$

Review Questions

42 If the particle moves with a uniform angular velocity, will its linear velocity be uniform?

43 Does a uniform angular velocity result in uniform linear velocity when the particle moves at a constant radial distance?

44 Does the time period of a harmonic periodic motion depend on the amplitude?

45 What type of motion is caused when a simple pendulum displaced by a small angle if the friction and air resistances are neglected?

46 What is the angular speed of the seconds hand of a watch.

Fill in the blanks to complete the statements

47 In SHM of a particle the acceleration is directed to and proportional to its distance from ________________ in its path.

48 (a) In SHM, the ______________ is numerically maximum at the mean position (velocity/acceleration)
(b) In SHM, as the particle moves away from the mean position the acceleration is always (+ve/–ve)
(c) In SHM, the______________is numerically zero at the mean position (velocity/acceleration).
(d) In SHM, as the particle moves away from its mean position its ______________ (velocity/acceleration) numerically decreases.

49 Where is a particle subjected to the maximum acceleration when it moves in SHM?

50 What type of motion will be caused if a spring is compressed and released?

51 Frequency of a periodic motion is 100 cycles/sec. What is its time period?

52 State whether the motion of a body thrown with 20 ms^{-1} velocity upwards at 30° with the horizontal will be
(a) Uniform Rectilinear
(b) Uniform Circular
(c) Accelerating Rectilinear
(d) Curvilinear

53 A stone is dropped vertically from the top of a cliff. State whether the motion of the stone is
(a) Uniformrectilinear
(b) Non-uniform curvilinear
(c) Uniform curvilinear
(d) Non-uniform rectilinear

54 The path of a stone thrown with some velocity in the air without air resistance is
(a) Simple circular
(b) Simple Parabolic
(c) Non-uniform and rectilinear
(d) Uniform and curvilinear

55 When will the range of a body projected with certain velocity be maximum? When the angle of projection with the horizontal is
(a) 0°
(b) 30°
(c) 45°
(d) 60°

56 A body is projected upwards at 30° with the horizontal with initial velocity of 50 ms^{-1}. Which of the following statement(s) are true about the flight of the projectile?
(a) The vertical velocity of projectile is greater than the initial velocity
(b) The vertical velocity remains constant throughout
(c) The horizontal velocity remains constant throughout
(d) The vertical velocity at any instant will be directly proportional to its vertical height from the point of projection.

57 State whether the horizontal distance covered by a projectile projected at an angle θ with the horizontal plane will be directly proportional to
(a) The angle of projection θ
(b) $\tan \theta$
(c) $\sin^2 \theta$
(d) $\sin 2\theta$

58 An aeroplane is flying horizontally with certain velocity and at a certain height, without any air resistance. The engine suddenly fails. Which one of the following things will happen?
(a) The aeroplane will continue to move horizontally at the same height.
(b) The aeroplane will continue to move horizontally with decreasing height.
(c) The aeroplane will fall vertically downwards.
(d) The aeroplane will strike ground at 45°.

59 A shell fired from a gun barrel inclined at 45° with the horizontal fall slightly short of the fixed target. What out of the following alternatives should be done to strike the target?
(a) Increase the height of the gun barrel
(b) Decrease the height of the gun barrel
(c) Increase the inclination of the gun barrel
(d) Decrease the inclination of the gun barrel

60 An arrow is projected at 60° with the horizontal from a bow and string from the tree top to hit a target uniformly moving away from the tree. The arrow falls a little short of the target. Which of the following actions will make it possible to hit the target?
(a) Increase the inclination of projection
(b) Decrease the inclination of projection
(c) Decrease the height of projection plateform
(d) Increase the length of arrow.

61 What will be angle of strike of a pebble with the ground if it is projected at 30° with the horizontal at ground level (neglecting air resistance)
(a) 30°
(b) 45°
(c) 60°
(d) 90°

6.6 Kinematics of Rigid Bodies

When a body is of large dimensions, and it undergoes small displacements, its motion can not be considered as motion of a particle. The motion of such bodies is considered as motion of rigid bodies. The rigid body is assumed to have no deformation of its particles. Only plane motion of rigid bodies shall be considered. The plane motion of rigid bodies consist of

(1) Translation
(2) Rotation and
(3) Translation combined with rotation.

(I) *Translation*

Translation can be rectilinear or curvilinear. Rectilinear translation of a body is the motion in which all particles forming the body move in straight lines (see Fig. 6.69), while in curvilinear motion the particles forming the body move along the congruent curves (see Fig. 6.70).

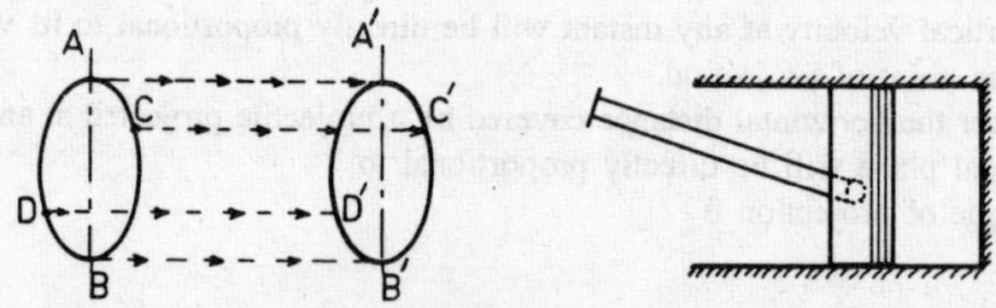

Fig. 6.69

It is evident that the motion of a body can be defined by the motion of any particle on the body, since all the particles on the body move along parallel paths.

(II) *Rotation*

If the motion of a body is such that the particles of the body move along

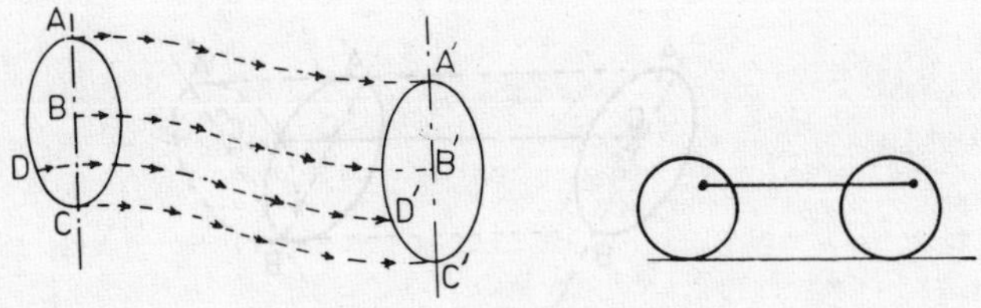

Fig. 6.70

concentric circles whose centre is a point within or outside the body, the motion is called rotation (See Fig. 6.71). The centre of the circles is known as the centre of rotation. This centre of rotation can be a fixed point or a movable point. At one particular instant of time the centre of rotation is a fixed point. Motion of a door about the hinges is a case of rotation with fixed centre, while motion of a connecting rod of an engine is an example of rotation with a moveable centre. It may be noted in case of rotation that any line joining different particles on the body undergoes a uniform angular change while in case of curvilinear translation such line shifts in position without undergoing any angular change with respect to earlier direction.

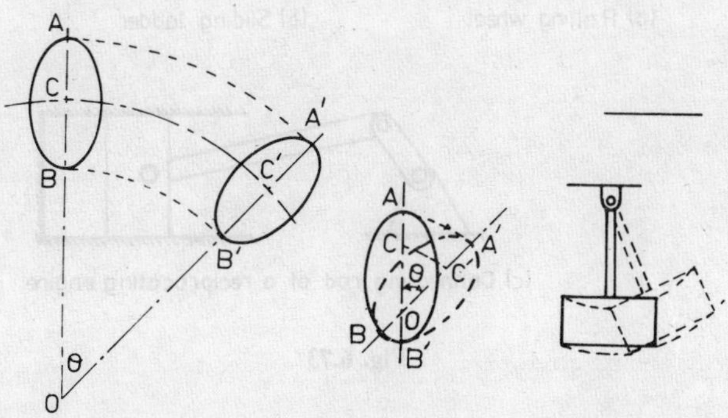

Fig. 6.71

In rotation various particles on the body move along concentric circles.

(III) *Rotation with Translation*

If a body moves in such a way that the particle on the body translate through some distance, and also displace through a certain angle, such a motion is termed as rotation with translation or general plane motion. In this the line joining any two particles does not remain parallel to its original position before displacement, e.g. the rolling wheel or the sliding of ladder (See Fig 6.72, Fig. 6.73).

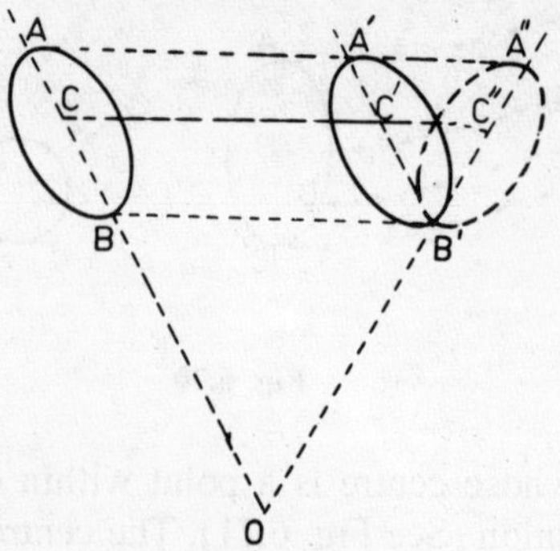

Fig. 6.72

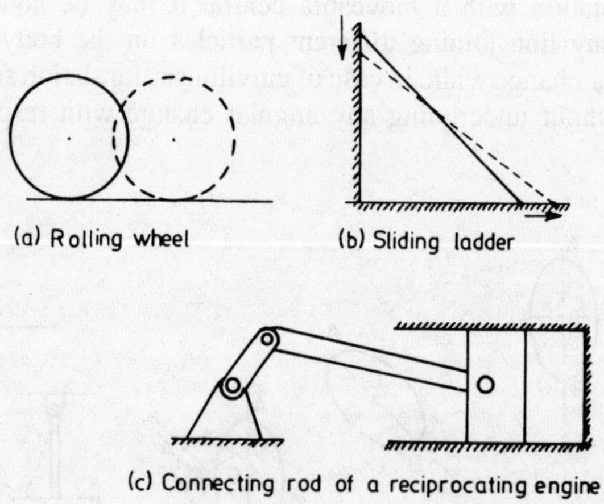

Fig. 6.73

It may be noted that in case of rigid bodies the particles on the body remain at a fixed angular and linear distance apart at all times, i.e. this position does not change with the time (Fig. 6.74).

Distance AB = l_1 = A′B′

CD = l_2 = C′D′

AC = S_1 = A′C′

BD = S_2 = B′D′

The angle between AB and CD = β = the angle between A′B′ and C′D′, where A′B′C′D′ is the displaced position of ABCD (shown in Fig. 6.74).

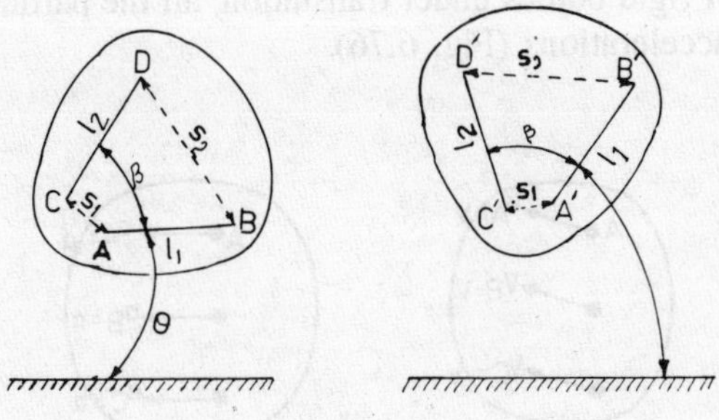

Fig. 6.74

The relations of motion parameters can be established for different type of motions.

6.6.1 *Translation*

Let us assume that a rigid body translates (rectilinearly or curvilinearly) with its particles A and B moving to A' and B' respectively in time Δt as shown in Fig. 6.75.

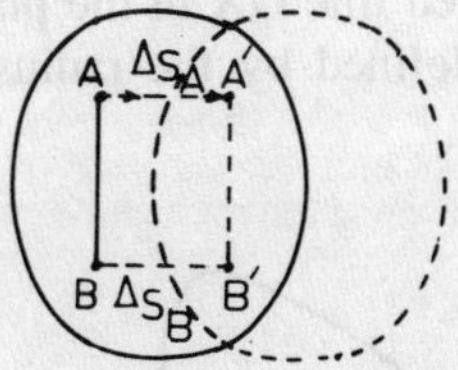

Fig. 6.75

Since the body is rigid, A′B′ = AB; and since the motion is translation, A′B′ is parallel to AB. Hence AA′B′B is a parallelogram. Therefore, vectors AA′ (ΔS_A) and BB′ (ΔS_B), representing the displacements of the particles A and B respectively, are the same in magnitude and direction.

i.e. $$V_A = \underset{\Delta t \to 0}{\text{Limit}} \frac{\Delta S_A}{\Delta t} = \underset{\Delta t \to 0}{\text{Limit}} \frac{\Delta S_B}{\Delta t} = V_B$$

or $V_A = V_B$

Differentiating, we get

$$a_A = \underset{\Delta t \to 0}{\text{Limit}} \frac{\Delta V_A}{\Delta t} = \underset{\Delta t \to 0}{\text{Limit}} \frac{\Delta V_B}{\Delta t} = a_B$$

or $a_A = a_B$

Hence in case of rigid bodies under translation, all the particles have similar velocities and accelerations (Fig. 6.76).

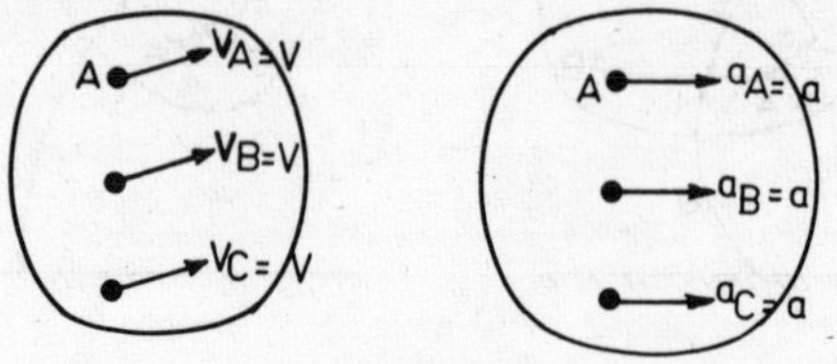

Fig. 6.76

The translatory motion can be with uniform velocity, or it can be with accelerating velocity. The analysis is similar to that of rectilinear motion.

6.6.2 *Rotation*

Consider a rigid body rotating about a fixed axis perpendicular to the plane of motion and intersecting the body in a point O (Fig. 6.77). Let there be any point *P* on the body such that the line OP makes an angle θ at any instant of time with respect to a fixed line OX in the plane of motion. The point *P* in the plane of motion is defined by the radius '*r*' = OP and its angular displacement θ = ∠POX.

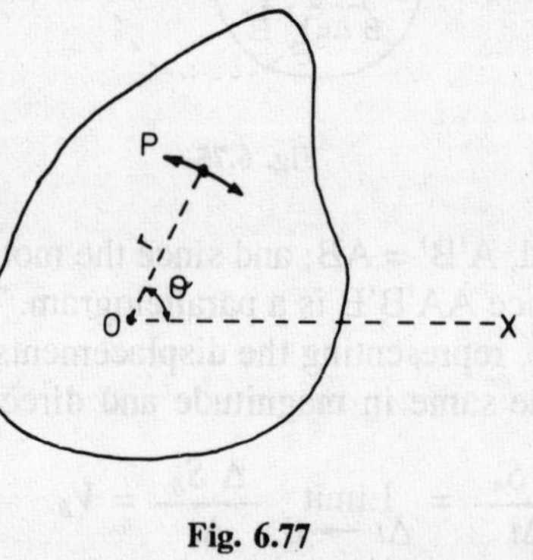

Fig. 6.77

Angular displacement θ can be measured in radians, degrees or revolutions. Various units of angular displacement are connected by the relation:

$$1 \text{ revolution} = 2\pi \text{ radians} = 360 \text{ degrees.}$$

The angular velocity and angular acceleration can be defined by usual procedure of considering respective derivatives of angular displacement with respect to time as under:

$$\text{Angular velocity } \omega = \frac{d\theta}{dt} \quad \text{.... (6.11a)}$$

$$\alpha = \frac{d\omega}{dt} = \frac{d^2\theta}{dt^2} \quad \text{.... (6.11b)}$$

$$\text{Also } \alpha = \omega \frac{d\omega}{d\theta} \quad \text{.... (6.11c)}$$

It may be noted that the displacement θ is generally measured in radians and accordingly angular velocity and acceleration are measured in radians/sec and radians/sec² respectively. The above relations provide magnitudes of various quantities while their directions are determined by clockwise or counter clockwise motion. If θ is measured positive in clockwise direction, then ω and α will also be measured positive in clockwise direction. Rotation can be with uniform angular velocity or with uniformly accelerating motion.

6.6.2(a) *Uniform Rotation*

In case of uniform rotation, the angular velocity at any instant remains constant i.e. $\omega = \omega_i$; where ω_i = initial angular velocity, ω = angular velocity at any instant of time *t*. Angular displacement at any instant is given as

$$\theta = \theta_i + \omega \cdot t \quad \text{or} \quad \omega = \frac{(\theta - \theta_i)}{t} \quad \text{.... (6.12b)}$$

For all particles on the rigid body, the *angular velocity* is the same, but the linear velocity of different particles depends on their distance *r* from the axis of rotation (Fig. 6.78).

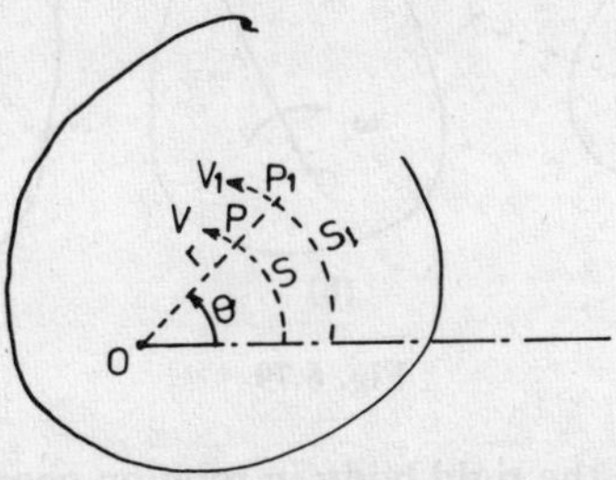

Fig. 6.78

At any instant of time, the linear velocity is given by $V = \omega \cdot r$

6.6.2(b) *Uniformly Accelerated Rotation*

In case of the uniformly accelerating motion the angular acceleration, angular velocity and the angular displacement are related by the following equations, which are similar to linear accelerating motion:

$$\alpha = \frac{(\omega - \omega_i)}{t} = \text{uniform}$$

$$\text{or } \omega = \omega_i + \alpha \cdot t \qquad \text{.... (6.13a)}$$

Total angular displacement

$$\theta = \theta_i + \omega_i t + \frac{1}{2} \alpha \cdot t^2 \qquad \text{.... (6.13b)}$$

$$\omega^2 = \omega_i^2 + 2\alpha (\theta - \theta_i) \qquad \text{.... (6.13c)}$$

where α = angular acceleration in rad/sec^2

ω_i = initial angular velocity in rad/sec

ω = angular velocity at any instant in rad/sec

t = time elapsed after starting in seconds

θ = angular displacement at any instant in radians

θ_i = initial angular displacement at the beginning in radians (usually zero)

If the angular acceleration becomes zero, the motion becomes uniform, and the above equations reduce to those of uniform rotation.

6.6.2(c) *Relations of Linear and Angular Velocities, Accelerations and Displacements*

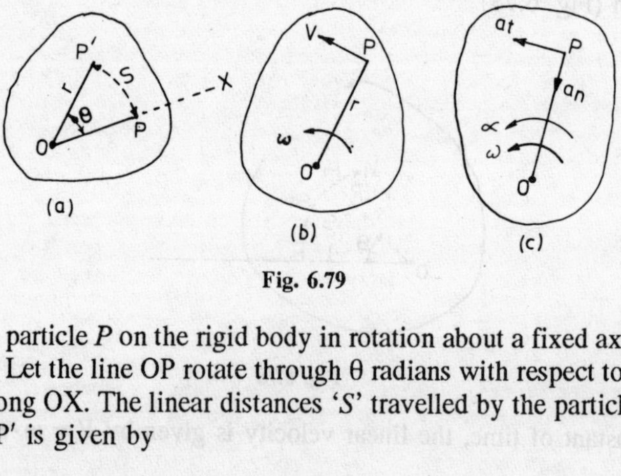

Fig. 6.79

Consider a particle P on the rigid body in rotation about a fixed axis passing through O. Let the line OP rotate through θ radians with respect to its initial position along OX. The linear distances 'S' travelled by the particle P along the path PP′ is given by

$$S = r\theta \qquad \text{.... (6.14a)}$$

where S = linear distance along the path

r = distance of the particle from the centre of the rotation

θ = angular displacement from the original initial axis OX.

By definition of linear velocity

$$V = \frac{dS}{dt} = \frac{d}{dt}(r\ \theta) = \frac{r\ d\theta}{dt} = r \cdot \omega$$

or $$V = r \cdot \omega \qquad \text{.... (6.14b)}$$

It is evident that the linear velocity of the particle depends on its position vector *r* with respect to the axis of rotation. It may be noted that the direction of the linear velocity *V* is always tangential to the angular path of the particle, i.e. it is always at right angles to the radial position vector *r*.

By definition of the linear acceleration, the tangential acceleration a_t is given as

$$a_t = \frac{dV}{dt} = \frac{d}{dt}(\omega r) = r\frac{d\omega}{dt} = r \cdot \alpha$$

or $$a_t = \alpha \cdot r \qquad \text{.... (6.14c)}$$

Recalling the equation for normal or radial acceleration caused by angular motion, we have

$$a_n = \frac{V^2}{r} = \frac{(\omega r)^2}{r} = \omega^2 \cdot r \qquad \text{.... (6.15)}$$

Though the angular acceleration 'α' is the same for all the particles on the rigid body, it is evident from the above equations that the tangential and the normal components of the acceleration of different particles will be different and depends on their position vector '*r*' with respect to the axis of rotation at O. The magnitude of tangential acceleration 'a_t' of a particle at a distance '*r*' remains constant if the angular acceleration 'α' is constant. While the magnitude of normal acceleration a_n of a particle at a fixed distance '*r*' remains constant only if the angular velocity 'ω' is constant. In case of accelerating motion the magnitude of the normal or radial acceleration depends on the square of the angular velocity 'ω' at the instant considered as it changes with time.

6.6.3 *Rotation Combined with Translation*

(General Plane Motion of a Rigid Body)

In case of general plane motion of rigid bodies, the analysis is carried out in two stages by dividing the motion into translation and rotation. Alternatively, such a general plane motion can be converted to pure rotation about an arbitrary point known as '*instantaneous centre*'.

To illustrate this, consider a particle *P* at a position A_1 on a rolling wheel moving in a fixed straight path without any slipping (Fig. 6.80). The wheel

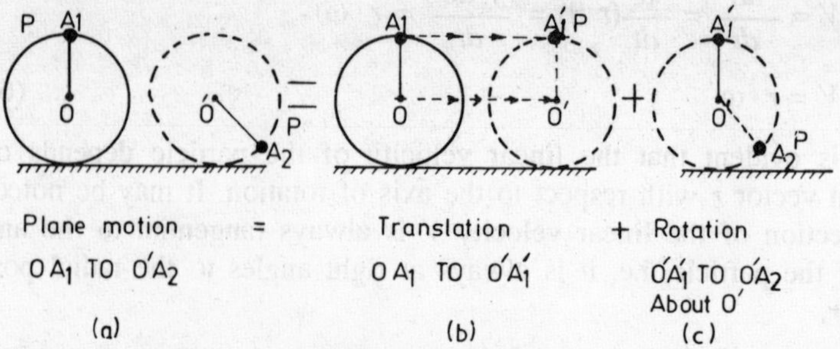

Fig. 6.80

rolls and the particle '*P*' on the wheel at A_1 occupies the position A_2 as shown in Fig. 6.80a. This motion can be assumed to occur by translation of the wheel with the particle *P* moving from A_1 to A_1', and then rotation of the wheel about O' with the particle *P* moving from A_1' to A_2 as shown in Fig. 6.80b and c, respectively. Consider another case of a guided ladder AB moving to A′B′ as shown in Fig. 6.81.

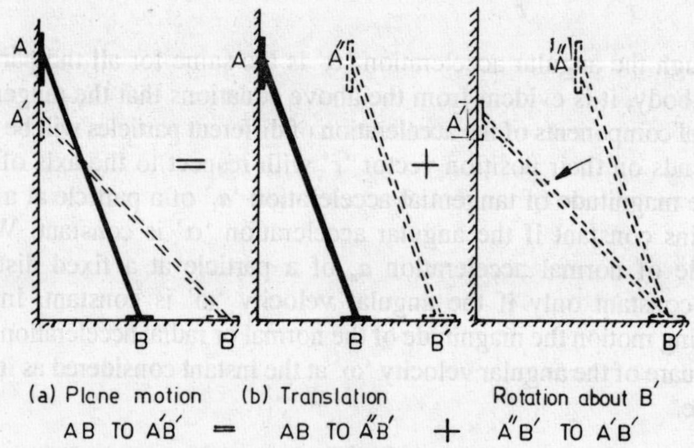

Fig. 6.81

The motion of the ladder AB can be looked upon as translation of the ladder *AB to A″B′* and then the rotation about *B′* from *A″B′ to A′B′* as shown in Fig. 6.81b, c. Thus the total motion of the ladder consists of the translation of B, horizontally and the rotation about B. This can also be considered as translation of A vertically downward and then rotation about A.

Consider the example of a rigid link AB which undergoes a plane motion and occupies a new position A′B′ in the same plane as that of AB (Fig. 6.82a)

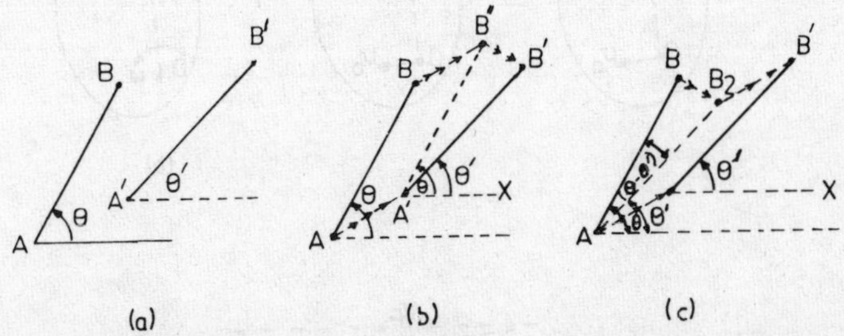

Fig. 6.82

The motion of the link can be looked upon to consist of translation from AB to A′B″, and rotation about the pole A′ from A′B″ to A′B′ (Fig. 6.82 b). It may be noted that the motion of the link can also be looked upon to consist of rotation about the pole *A* from AB to AB_2 and translation from AB_2 to A′B′. The point B could also be considered as pole in above cases.

From above consideration, it is concluded that the plane motion of a rigid body can be completely defined by *motion of translation of certain non-rotating reference point and motion of rotation about such a reference point.*

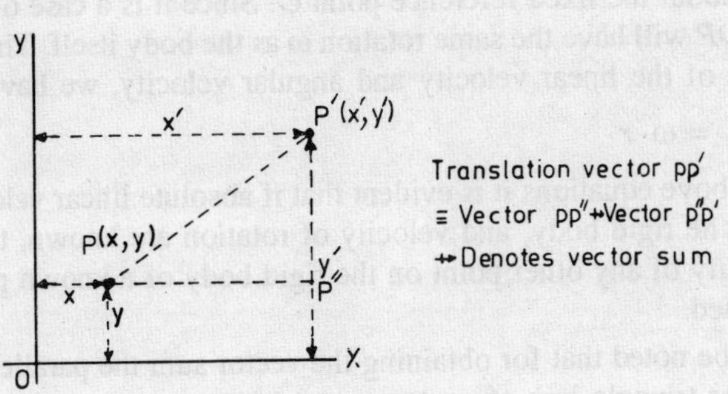

Fig. 6.83

The translation of reference point P(*x*, *y*) can also be defined by its motion along any rectangular coordinate axes X and Y as shown in Fig. 6.83.

6.6.4 *Absolute and Relative Velocity of Particles of Rigid Body in Plane Motion*

Consider a rigid body having a plane motion. Let there be any two points '*P*' and '*O*' on the rigid body having absolute linear velocities of V_p and V_o respectively.

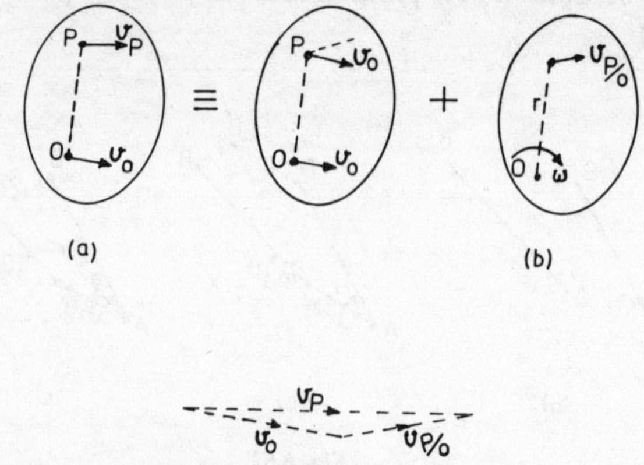

(c) Vector diagram

Fig. 6.84

The velocity V_p of the particle P can be assumed to consist of velocity V_o of any reference point O and a relative velocity $V_{P/O}$ (Fig. 6.84 b, c). This can be represented by vector equation as

$$\bar{V}_p = \bar{V}_o + \bar{V}_{P/O} \text{ (+denotes vector sum)}$$

The relative velocity $V_{P/O}$ can be determined by considering O as a fixed point, in which case the component $V_{P/O}$ is due to rotation ω of the position vector OP about the fixed reference point O. Since it is a case of rigid body the vector OP will have the same rotation ω as the body itself. Thus knowing the relation of the linear velocity and angular velocity, we have

$$V_{P/O} = \omega \cdot r \qquad \text{.... (6.14 b)}$$

From above equations it is evident that if absolute linear velocity of any particle on the rigid body, and velocity of rotation are known, the absolute linear velocity of any other point on the rigid body of a known position can be determined.

It may be noted that for obtaining the vector sum the parallelogram law of vectors or triangle law of vectors are used.

6.6.5 *Instantaneous Centre of Zero Velocity in Plane Motion*

In previous articles the velocity of a particle on a rigid body having plane motion was determined by considering the velocity due to translation and relative velocity due to rotation about some reference point. This can also be done by shifting the reference point to some arbitrary axis whose velocity at the moment is zero and the whole body is assumed to rotate about this arbitrary axis. This axis is called the *instantaneous axis* of zero velocity, and the intersection of this axis with the plane of motion is known as the *instantaneous centre* of zero velocity.

Consider a rigid body in plane motion as shown in Fig. 6.85(a, b, c) and take an arbitrary instantaneous centre of zero velocity at O. By definition of the instantaneous centre the whole body is in pure rotation about this point O. Therefore, the linear velocities of the different points on the rigid body shall be perpendicular to the lines joining the points and the instantaneous centre O. If the direction of linear velocities of at least any two particles are known, the instantaneous centre O can be determined by drawing perpendicular lines to these velocities of the points.

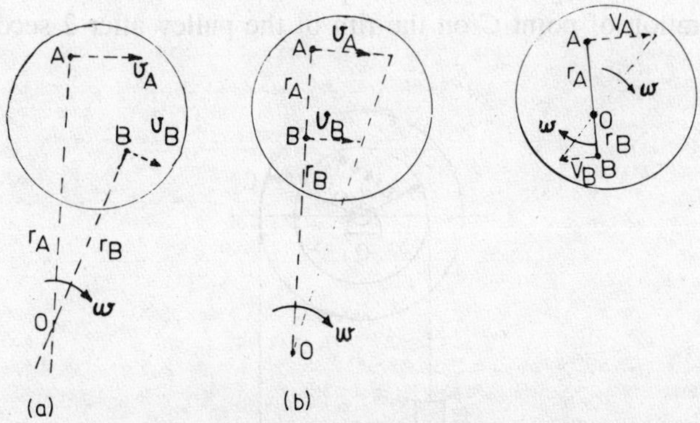

Fig. 6.85

When the velocities at the two points are parallel to each other, we draw a line perpendicular to the velocities, and also draw magnitudes of the two velocities along their directions and draw a line joining these magnitudes and produce to intersect the perpendicular line at O (Fig. 6.85 b). Similarly if velocities of the two points considered are parallel and of opposite nature, the instantaneous centre can be obtained as shown in Fig. 6.85 c.

Since the instantaneous centre has zero velocity and there is pure rotation about this point, the linear velocity of any point on the body can be found by the relation

$$V = \omega \cdot r$$

where, V = linear velocity of any particle on the body,

ω = Angular velocity of body about the instantaneous centre,

r = radial distance of the particle on the body from the instantaneous centre.

The direction of the linear velocity is always perpendicular to its radial vector 'r' and the sense is found by the angular velocity.

It is emphasized that the *instantaneous centre is a fixed point for the instant considered* and it may be *changed after some time*. It may also be noted that the *instantaneous centre of zero velocity is not necessarily the instantaneous centre of zero acceleration.*

The plane motion of rigid bodies can be further explained by illustrative examples.

Example 6.51

A Pulley with two loads connected by inextensible cord is shown in Fig. 6.86. The load A moves downward with an initial velocity of 4.0 ms^{-1} and uniform acceleration of 2.0 ms^{-2}. Find (a) the number of revolutions made by the pulley in 2 seconds (b) velocity and position of load B after 2 seconds (c) the acceleration of point C on the rim of the pulley after 2 seconds.

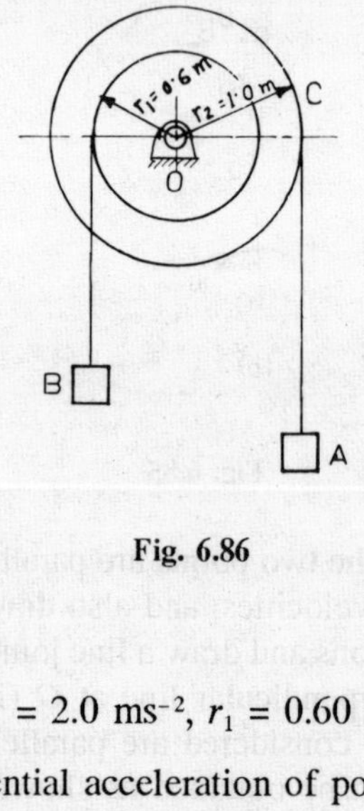

Fig. 6.86

Solution

$$V_{iA} = 4.0 \text{ ms}^{-1},\ a_{iA} = 2.0 \text{ ms}^{-2},\ r_1 = 0.60 \text{ m},\ r_2 = 1.0 \text{ m}$$

Linear velocity and tangential acceleration of point C will be same as that of A, i.e.

$$V_{iC} = V_{iA} = 4.0 \text{ ms}^{-1},\ a_{iC} = a_{iA} = 2.0 \text{ ms}^{-2}$$

Angular velocity and acceleration at C at the beginning will be

$$\omega_i = \frac{V_{iC}}{r_2} = \frac{4.0}{1.0} = 4.0 \text{ rad}\cdot\text{s}^{-1}$$

$$\alpha = \frac{a_{iC}}{r_2} = \frac{2.0}{1.0} = 2.0 \text{ rad}\cdot\text{s}^{-2}$$

(a) $\theta_2 = \omega_i \cdot t + \frac{1}{2}\alpha \cdot t^2 = 4.0 \times 2 + \frac{1}{2} \times 2.0 \times 2^2 = 12 \text{ rad} = 1.91$ revolutions.

(b) $\omega_2 = \omega_i + \alpha \cdot t = 4.0 + 2.0 \times 2 = 8.0 \text{ rad s}^{-1}$

$\therefore\ V_B = \omega \cdot r_1 = 8.0 \times 0.6 = \mathbf{4.8\ ms^{-1}}$

Distance covered by the load B = total rotation × radius

$$= 12 \times 0.6 = \mathbf{7.2\ m}$$

(c) $a_{tC} = 2.0\ \text{ms}^{-2}$ (tangentially)

$$a_{nC} = \omega^2 \cdot r(\text{radially}) = (8.0)^2 \times 1.0 = 64.0\ \text{ms}^{-2}$$

Thus the resultant acceleration will be

$$a_c = \sqrt{(a_{nc})^2 + (a_{tc})^2} = \sqrt{64^2 + 2^2} = \mathbf{64.03\ ms^{-2}}$$

$$\tan\beta = \frac{2}{64} = 0.03125,\ \beta = 1°48° \text{ with the radius which is horizontal.}$$

Example 6.52

A small particle P rests on a horizontal plate which rotates about a fixed vertical axis AB. The plate is initially at rest and is accelerated at a constant rate of 4.0 rad · s^{-2}. If the particle starts sliding when its total acceleration exceeds 5 ms^{-2}, find the time and the angular velocity of the plate when the particle starts sliding. The particle is located at 10 cms from the centre of the axis of the rotation. Also determine the direction of total acceleration when the particle starts sliding.

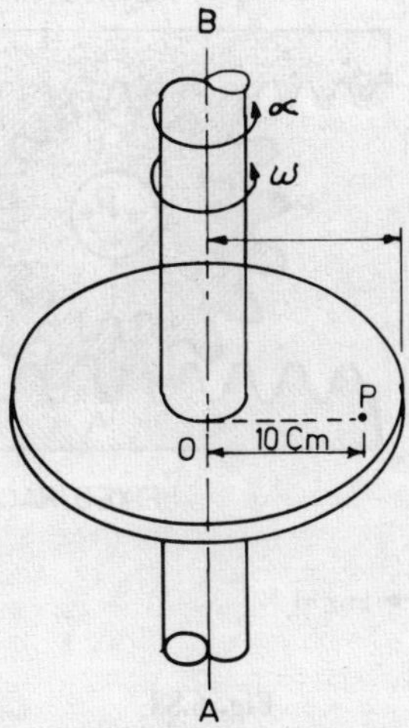

Fig. 6.87

Solution

$\alpha = 4.0\ \text{rad} \cdot \text{s}^{-2}$, $(a_p)_{max} = 5.0\ \text{ms}^{-2}$, $r_p = 0.1$ m,

$\omega_i = 0$, t = ?

$\omega = \omega_i + \alpha \cdot t = 0 + 4.0t = \mathbf{4t\ rad\ s^{-1}}$

$\therefore\ a_{tp} = \alpha \cdot r_p = 4 \times 0.10 = 0.40\ \text{ms}^{-2}$

$a_{np} = \omega^2 \cdot r_p = (4t)^2 \times 0.10 = 1.6t^2$ ms^{-2}

resultant $a_p = \sqrt{(0.4)^2 + (1.6t^2)^2}$

$\therefore \quad (5.0)^2 = (0.4)^2 + 2.56t^4$

or $\quad t^4 = \left(\dfrac{24.84}{2.56}\right)$, $\therefore$ t = **1.765 secs**

$\therefore \quad \omega = 4.0 \times 1.765 =$ **7.06 rad s^{-1}**

The direction of the total acceleration with radius

$$\beta = \tan^{-1} \frac{0.40}{1.6\,(1.765)^2} = \tan^{-1}(0.08) = 4°36' \text{ with the radius}$$

Example 6.53

A double gear shown in Fig. 6.88 rolls in the stationary lower rack R_1, with angular velocity of 5.0 rad s^{-1} in clockwise direction. Determine

(a) the forward velocity of the centre O of the gear

(b) linear velocity of the upper rack R_2 and point P on the outer gear. The diameters of the outer and inner gears are respectively 40 cm and 30 cm.

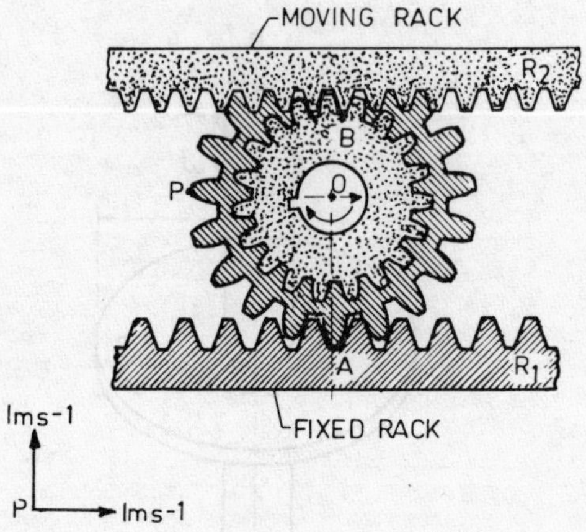

Fig. 6.88

Solution

$r_1 = 0.20$ m, $r_2 = 0.15$ m, $\omega = 5.0$ rad s^{-1}

(a) Linear velocity of the centre O will be the same as the tangential velocity of gear at contact point A.

i.e. $V_o = V_{O/A} = \omega \cdot r_1 = 5 \times 0.2 =$ **1.0 ms^{-1}**

(b) Velocity of upper rack is govered by the gear B, so linear velocity of upper rack = tangential velocity of B + linear velocity of O.

$V_2 = V_o + V_{2/O}$

or $V_2 = V_o + \omega \cdot r_2 = 1.0 + 5 \times 0.15 = \mathbf{1.75\ ms^{-1}}$

The linear velocity of the point P with respect to the centre O is $V_{P/O} = \omega \cdot r_1$ = 1.0 ms^{-1} (Vertically upwards). Thus the resultant absolute velocity of the point P will be V_p = Vector sum of V_o and $V_{p/o}$.

i.e. $\overline{V}_p = \overline{V}_o + \overline{V}_{p/\overline{o}} = \sqrt{(1.0)^2 + (1.0)^2} = \sqrt{2}\ ms^{-1}$ (Both the velocities are at 90° to each other).

$$\beta = \tan^{-1}\left(\frac{1.0}{1.0}\right) = 45°, \text{ with the horizontal radius.}$$

Example 6.54

In a stationary engine the piston, connecting rod, and the crank is shown in Fig. 6.89. The crank OA has a constant clockwise angular velocity of 1800 rpm. When the crank OA is at 45° to the horizontal, determine

(a) the angular velocity of the connecting rod AB,

(b) the velocity of the piston P; the length of the crank OA = 10 cm and that of connecting rod AB = 20 cm.

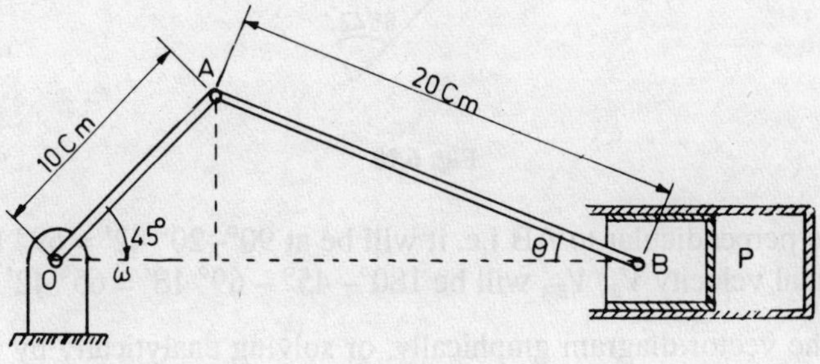

Fig. 6.89

Solution

$$\omega = \frac{1800}{60} \times 2\pi \text{ rad s}^{-1} = 60\pi \text{ rad s}^{-1}$$

$$OA = 0.1 \text{ m}, AB = 0.20 \text{ m}, \angle AOB = 45° = \frac{\pi}{4} \text{ rad}$$

$\angle OBA = \theta$

Velocity of A = $V_{A/O}$, as O is a fixed point.

$V_A = \omega_{OA} \times OA = 60\pi \times 0.10 = 6\pi\ ms^{-1}$

$= 18.84\ ms^{-1}$

The connecting rod AB has a restrain that its end B can only move along horizontal. From ΔOAB.

$$\frac{OA}{\sin\theta} = \frac{AB}{\sin 45°} = \frac{OB}{\sin(180° - \theta - 45°)}$$

$$\text{or } \sin\theta = \frac{OA}{AB}\sin 45° = \frac{10}{20}\frac{1}{\sqrt{2}} = 0.3535$$

$\theta = 20°\ 42'$

Velocity at the point A is known in magnitude and direction while the velocity of end B is known in direction only. Using the equation

$$\overline{V}_B = \overline{V}_A + \overline{V}_{B/A}$$

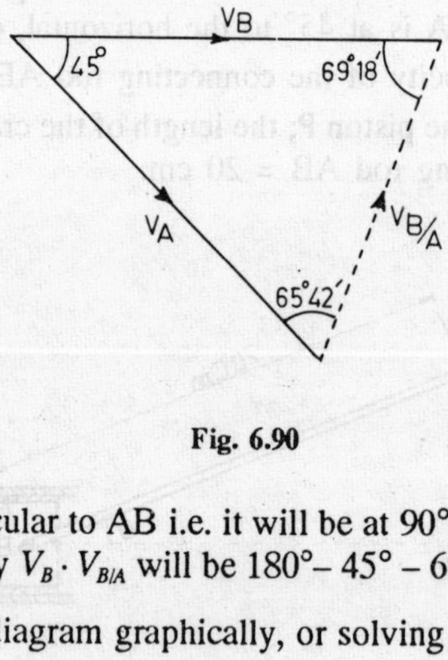

Fig. 6.90

$V_{B/A}$ will be perpendicular to AB i.e. it will be at 90°–20° 42′ = 69° 18′ with the horizontal velocity $V_B \cdot V_{B/A}$ will be 180°– 45° – 69° 18′ = 65° 42′ with V_A

Drawing the vector diagram graphically, or solving analytically by the sine rule we have

$$\frac{V_A}{\sin 69°\ 18'} = \frac{V_B}{\sin 65°\ 42'} = \frac{V_{B/A}}{\sin 45°}$$

$$\therefore\ V_B = \frac{18.84 \sin 65°\ 42'}{\sin 69°\ 18'} = \frac{18.84 \times 0.91}{0.935} = \mathbf{18.3\ ms^{-1}}$$

$$V_{B/A} = \frac{18.84 \sin 45°}{\sin 69°\ 18'} = \frac{18.84 \times 0.707}{0.935} = \mathbf{14.24\ ms^{-1}}$$

$$V_{B/A} = \omega \times AB$$

$$\therefore\ \omega = \frac{14.24}{0.20} = \mathbf{71.2\ rad./sec.}$$

Summary

Kinematics is study of geometry of motion of bodies. In simple problems of kinematics we consider the fixed reference axes attached to the earth's surface. Bodies of finite size, for simplicity, are considered as particles in the study of motion. The motion of the particles may be along a straight line or a curved path.

The motion of particles consists of displacement, velocity, and acceleration, etc. *Displacement* is defined as change of position in a specific direction with respect to certain fixed reference axes. *Velocity* is defined as rate of change of position *with respect to time. Acceleration* is defined as rate of change of velocity with respect to time. If the displacement is denoted by S, then velocity

$$V = \frac{dS}{dt}, \text{ and acceleration } a = \frac{dV}{dt} = \frac{d^2S}{dt^2}$$

In straight line (rectilinear) motion of a particle moving with *Uniform Velocity* (zero acceleration), the relationship of motion parameters is

$$V = \frac{(S_f - S_i)}{t}, \text{ or } S_f = S_i + V \cdot t$$

In straight line motion (rectilinear motion) of particles with *uniform acceleration*, the motion parameters are related as under:

$$V = \frac{dS}{dt}, V_f = V_i + a \cdot t$$

$$(S_f - S_i) = S = V_i\, t + \frac{1}{2} a \cdot t^2, \text{ and } V_f^2 - V_i^2 = 2aS$$

All quantities are to be taken with proper signs according to their direction. If the acceleration is due to gravitational force of earth, the acceleration '*a*' may be replaced by '*g*' (9.8 ms^{-2}) with appropriate sign depending on the direction of motion.

For solving the problems by graphical method, we use the basic equations of motion $V = \frac{dS}{dt}$, or $dS = Vdt$.

i.e. slope of displacement-time curve is velocity at any instant, while the change of displacement is the summation of area of velocity-time curve. Similarly, $a = \frac{dV}{dt}$ or $dV = adt$ i.e. slope of $V - t$ curve is acceleration at that instant of time, while total change of velocity is obtained from the summation of area of $a - t$ curve.

Relative motion study is carried out by making one of the particle stationary by superimposing negative motion vector of the particle with reference to which the motion of the other particle is required to be studied on both the particles, and finding the resultant effect on the particle to be studied. For example, the relative motion of a particle B with respect to another particle A can be shown vectorially as

$$\bar{S}_{B/A} = \bar{S}_B - \bar{S}_A;\ \bar{V}_{B/A} = \bar{V}_B - \bar{V}_A;\ \bar{a}_{B/A} = \bar{a}_B - \bar{a}_A$$

When the particles move along a curved path the motion is termed as *curvilinear*. When the motion is along a circle, it is called simple circular motion. Various motion parameters in polar Coordinate system are interrelated in a way similar to that of rectilinear motion parameters of particles, i.e. angular displacement θ, angular velocity ω and angular acceleration α are related as

$$\omega = \frac{d\theta}{dt};\ \theta_f = \theta_i + \omega \cdot t;\quad \omega_f = \omega_i + \alpha \cdot t$$

$$(\theta_f - \theta_i) = \theta = \omega_i \cdot t + \frac{1}{2}\alpha \cdot t^2;\quad \omega_f^2 - \omega_i^2 = 2\ \alpha\theta$$

The linear and angular motion parameters of a particle at radius r at any instant of time are related as

$$S = \theta \cdot r,\quad V = \omega r;\ a_t = \alpha \cdot r.$$

Circular motion of particles also causes *radial acceleration* due to change of velocity in radial direction, and is directed towards the centre of motion. This is called *Centripetal acceleration* and given by

$$a_n = \omega^2 \cdot r \text{(for uniform magnitude of angular velocity)};\quad a_n = \omega_A^2 \cdot r = \frac{V_A^2}{r}$$

When a particle moves in such a manner that after certain fixed interval of time, it acquires the same position and moves in the same direction with the same velocity, it is said to have *periodic motion*. If a body moves such that its *acceleration is always directed* towards a fixed point and is proportional to its distance from the *fixed point*, the motion is called *Simple Harmonic Motion (SHM)*.

The time taken by a particle for completing one cycle of motion is called its *periodic time*. The *number of cycles* made in one second is called frequency. The *maximum displacement* of the particle from its *mean position* is called *amplitude*.

In Simple Harmonic Motion, $t_o = \dfrac{2\pi}{\omega}$, $n = \dfrac{\omega}{2\pi} = \dfrac{1}{t_o}$

Acceleration of a particle undergoing SHM along the diameter or chord of a circle of radius 'r' is given by: $a_x = \omega^2 \cdot x$

Curvilinear motion of any particle can be looked upon as consisting of combination of rectilinear motion along mutually perpendicular rectangular coordinate axes. All these motion parameters can be resolved vectorially. Along the rectangular *coordinate* axes.

$$\bar{S} = \bar{S}_x + \bar{S}_y; \quad \bar{V} = \bar{V}_x + \bar{V}_y; \quad \bar{a} = \bar{a}_x + \bar{a}_y$$

i.e. $S = \sqrt{S_x^2 + S_y^2}$; $V = \sqrt{V_x^2 + V_y^2}$; $a = \sqrt{a_x^2 + a_y^2}$

Simple *projectiles launched* in air has *acceleration* due to *gravity in vertical direction* while there is *zero acceleration in horizontal direction* (neglecting air resistance).

Resultant velocity of a projectile at any instant $V_r = \sqrt{V_x^2 + V_y^2}$, where V_x and V_y can be found at any instant of time with reference to rectilinear motion equations in X and Y directions.

The vertical position of the projectile at any instant is given by

$$Y = X \tan\theta - \frac{g \sec^2\theta}{2 V_i^2} X^2, \text{ (Parabola)}$$

Max. height achieved by a projectile $Y_{max} = \dfrac{V_i^2}{2\,g} \sin^2\theta$

Flight time $T = \dfrac{2 V_i}{g} \sin\theta$, horizontal distance $X = \dfrac{V_i^2}{g} \sin 2\theta$

Horizontal distance (range) is maximum when the angle of projection θ is $45°$, and is equal to $\dfrac{V_i^2}{g}$.

When bodies of large dimensions undergo small displacements, and their particles do not undergo relative deformations, the motion is called rigid body motion. The motion of these bodies can be plane or spatial, which may be consisting of translation, rotation and combined rotation and translation.

When all the particles constituting the body move along parallel straight lines or congruent curves, it is called respectively rectilinear or curvilinear translation. When the particles of a body move along *concentric circles* whose centre may lie within or outside the body, it is called rotation. This centre is called centre of rotation, and it may be fixed or movable point. Various motion parameters are correlated depending on the type of motion.

The velocity of a particle, on the rigid body undergoing combined translation and rotation, can be determined by establishing an *arbitrary point of zero velocity at that instant*. The body is assumed to rotate about this *instantaneous centre* of zero velocity.

Review Questions

62 Explain linear and angular velocity, and linear and angular acceleration. Give their relations.
63 Give examples and explain plane motion of rigid bodies.
64. Is it possible to convert a plane motion to pure rotation? Explain
65 In plane motion of a rigid body, explain how velocities of two particles are related?
66 If the linear velocities of two particles on a rigid body are known, how would you find the instantaneous centre?
67 Does the instantaneous centre of a motion remain constant at all the times in all types of plane motion?
68 Is it possible to resolve the motion vectors along certain set of rectangular co-ordinate axes?
69 Define instantaneous centre of velocity.
70 Is it true that instantaneous centre of velocity and instantaneous centre of acceleration always coincides? Explain with reasons.
71 If all the particles on a rigid body move with the same angular velocity, will all these particles have the same linear velocity too? Explain with example.

Problems

6.1 The rectilinear motion of a particle is defined by its position vector $S = t^2 - 8t + 15$, where S is in metres and 't' is time in seconds. Determine

(a) at the start of motion, 2 seconds and 5 seconds after the start of motion the following
(i) Position
(ii) Velocity
(iii) Acceleration
(b) the time at which the position of the particle is at the origin.
(c) the time at which the velocity of the particle is zero.

Ans. (a) (i) 15 m, 3 m, 0 (ii) –8 ms^{-1}; –4 ms^{-1}; + 2 ms^{-1} (iii) 2 ms^{-2}; 2 ms^{-2}; 2 ms^{-2}
(b) $t = 3$ secs, 5 secs
(c) $t = 4$ secs.

6.2 The position vector 'S' of a particle is a function of time 't' as $S = t^3 - 12t^2 + 45t - 30$, where S is in metres and t is in seconds. Determine

(a) at the start and 1 second after the start of motion (i) Displacement (ii) Velocity and (iii) Acceleration
(b) the time at which the velocity becomes zero, and
(c) the distance travelled and displacement after 6 seconds after the start of the motion.

Ans. (a) (i) –30 m, + 4 m; (ii) +45 ms^{-1}; + 24 ms^{-1} (iii) –24 ms^{-2}; –18 ms^{-2};
(b) $t = 3$ secs, 5 secs.
(c) Dist. = 62 m
Displ. = +24 m, from the origin.

6.3 A train while passing a station A has a uniform velocity of 60 km/hour. How long will it take to pass another station B at a distance of 15 km? Find the distance of another station C where the train reaches in 40 minutes from A without any stop.

Ans. $t = 15$ mins, $S(A \text{ to } C) = 40$ km.

6.4 A car moving with a constant velocity of 90 km/hour along a straight path crosses a village. After 40 seconds another car moving with a uniform velocity of 108 km/hour crosses the same village. Find when and where the second car overtakes the first.

Ans. $t = 4$ mins after the first car crosses the village at a distance of 6 kms from the village.

6.5 A truck is transporting stone from a quarry to a construction site 20 km away along a straight path. If the speed of the loaded truck is half the speed of the unloaded truck which is 90 km/hour, find the number of trips the truck will make in 6.5 hours. Each loading and unloading takes 15 minutes and 10 minutes respectively.

Ans. 6 trips.

6.6 A truck transporting cement, from a factory to a construction site at a distance of 20 km, makes 8 trips in 8 hours shift including 1 hour for lunch break. The empty truck has twice

the speed of loaded truck and it takes 15 minutes for loading and 12.5 minutes for unloading the truck each time. Find the speed of the empty truck.
Ans. $V = 144 \text{ kmh}^{-1}$

6.7 A straight cable ropeway is used to transport minerals from a quarry site on a hill to a ground base. The trolly can descend at 15 ms^{-1} and ascend at 10 ms^{-1} along the straight ropeway. If the total time in loading and unloading is 10 minutes and each trip including loading and unloading takes 30 minutes, find the length of the ropeway.
Ans. 7.2 km

6.8 A car moving along a straight line at 30 kmh^{-1} speed starts accelerating at a uniform rate and reaches a speed of 66 kmh^{-1} in 1 minute 40 seconds. Find the rate of acceleration and distance travelled during this period.
Ans. a = (0.10 ms^{-2}; S = 1333 m)

6.9 An Automobile moving along a straight line with a velocity of 108 kmh^{-1} is brought to rest in a distance of 150 m by applying brakes. Determine the time for which the brakes were applied.
(Ans. 10 secs.)

6.10 A train travels 50 metres in 5 seconds and 100 metres in the next 5 seconds. Determine whether the train moves with constant velocity or constant acceleration. Compute the initial velocity and the rate of acceleration.
(Ans. V_i = 5 ms^{-1}; a = 2 ms^{-2}).

6.11 A car starting from rest moves in a straight line such that it accelerates for first 10 seconds with a uniform acceleration. It moves with the uniform velocity for next 40 seconds, and then gradually comes to a stop in 5 seconds. If the maximum velocity achieved by the car is 20 ms^{-1}, find the distance covered by the car.
(Ans. S = 950 m)

6.12 A motorist driving his car at a speed of 72 kmh^{-1} finds that the traffic light ahead of him turns red. If he comes to halt at the traffic light in 5 seconds. Find the distance of the car from the traffic lights when the motorist observed the red lights.
(Ans. S = 50 m).

6.13 A train has to cover a distance of 10 km between two stations in 7 minutes. The train can have maximum speed of 90 kmh^{-1}, and the ratardation is at double the rate of acceleration. Find the distance for which the train accelerates, moves with constant speed and decelerates.
(Ans. 333.3 m, 9500 m, and 166.7 m)

6.14 Two automobiles A and B are standing on a straight highway at a traffic signal. As the traffic signal turns green, automobile A accelerates at the constant rate of 1.0 ms^{-2}. Automobile B starts 5 seconds later and accelerates at the constant rate of 1.5 ms^{-2}. Determine
(a) when and where B will overtake A, and
(b) the speed of each automobile at that time.
Ans. (a) 27.24 secs after the automobile A starts. Distance 371 m from the traffic lights.
(b) V_A = 27.2 ms^{-1}, V_B = 33.36 ms^{-1}.

6.15 A particle starts moving in a straight line with the acceleration of 2.0 ms^{-2} for the first 5 seconds, then with acceleration of 1 ms^{-2} for next 3 seconds and finally at –1.5 ms^{-2} for the next 10 seconds. Determine the position of the particle at the end of the motion. Also find the total distance travelled by the particle.
(Ans. S = 114.5 m, Dist. travelled = 117.17 m)

6.16 A scooterist is driving on a straight road at 72 kmh^{-1} when he observes that a traffic signal 200 m ahead of him turns red. The signal is timed at 15 seconds. If the scooterist wishes to pass the traffic light at a speed of 36 kmh^{-1} just when it turns green, find the constant deceleration if the scooter first decelerates, and then moves with constant speed of 36 kmh^{-1}. Also find the time and distance for which the scooter decelerates.
(Ans. 1 ms^{-2}, 10 secs, S = 150 m).

6.17 On a straight highway there are series of traffic light signals which are so arranged that if a vehicle is moving at 54 kmh^{-1} it can reach each signal just as it turns green. A vehicle starts from one traffic signal as it just turns green with a maximum acceleration of 1.5 ms^{-2}. If the distance between two consecutive signals is 750 metres and the maximum speed of the vehicle should be as small as possible, then find the maximum speed, and the time at which it is attained so that the vehicle can pass the next signal just when it turns green

(Ans. V_{max} = 61.3 kmh^{-1}, t_1 = 11.35 seconds).

6.18 A policeman on a motor cycle is escorting a motorcade which is travelling at 54 kmh^{-1}. The policeman decides to take a new position 100 m ahead of his present position. Assuming that the motor cycle can accelerate and decelerate at a rate of 2.5 ms^{-2} and can attain maximum velocity of 90 kmh^{-1}, determine the shortest time in which the policeman can occupy the new position and the distance he travels before taking the new position.

Hint: Assume the distance between the two positions as extra distance to be covered by the motor cycle by its extra speed over 54 kmh^{-1}. Thus total extra distance = 100 m. Thus extra distance is covered by first accelerating to maximum speed of motor cycle, keeping a constant speed at the maximum level, and again retarding to normal speed of 54 kmh^{-1}.
(Ans. Minimum time to occupy the new position = 14 secs. Distance travelled by motor cycle during this speed = 310 m.).

6.19 A fighter plane flying horizontally in a straight line at a net velocity of 300 ms^{-1} is overtaking a bomber plane flying in the same straight line at a net velocity of 200 ms^{-1}. The pilot of the fighter plane fires an air to air missile at the bomber when his plane is 800 m behind the bomber. The missile accelerates at the constant rate of 400 ms^{-2} for 1 sec, and then moves at a constant speed in the same straight line.

(a) how many seconds after firing, the rocket will strike the bomber?

(b) if both the planes continue to move with their respective constant speeds, what will be the distance when the rocket strikes the bomber?

Hint. Initial vel. of the rocket will be the same as that of the fighter plane i.e. 300 ms^{-1}
(Ans. (a) 2 seconds (b) 600 m apart).

6.20 A car and a truck are both travelling at the constant speed of 54.0 kmh^{-1} along a straight path. The car is behind the truck by a clear distance of 12 m. The car driver wish to put his car at a clear distance of 12 m in front of the truck as shown in Fig. Prob. 6.20 and then resume the speed of 54 kmh^{-1}. The maximum deceleration by applying the brakes is 5 ms^{-2}. What is be the shortest time in which the driver of the car can complete the positioning operation if

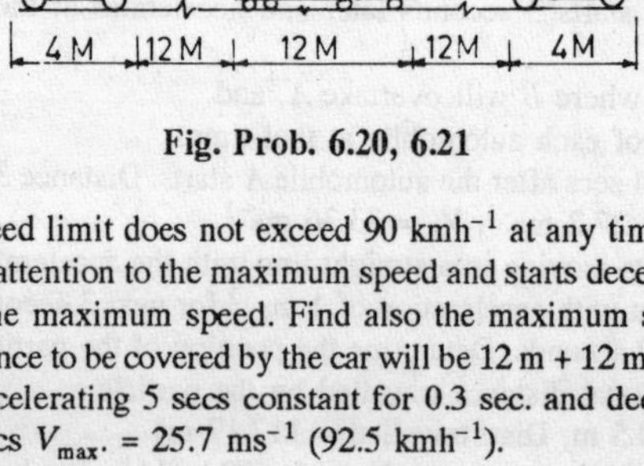

Fig. Prob. 6.20, 6.21

(a) the maximum speed limit does not exceed 90 kmh^{-1} at any time, and (b) the driver does not pay any attention to the maximum speed and starts decelerating immediately after achieving the maximum speed. Find also the maximum speed.

Hint: Total extra distance to be covered by the car will be 12 m + 12 m + 12 m + 4 m = 40 m
Ans. t = 7.5 sec; (accelerating 5 secs constant for 0.3 sec. and decelerating for 2 secs.
(b) t = 7.49 secs $V_{max.}$ = 25.7 ms^{-1} (92.5 kmh^{-1}).

6.21 In the above problem when the car is moving behind the truck at a constant speed of 72 kmh^{-1}, which is the same as the truck speed, the truck driver suddenly applies his brakes and decelerates at a constant rate of 3 ms^{-2}. The car driver applies brakes 2 sec. later to avoid the collision at the rear of the truck. Find the rate of deceleration of the car so that the collision can be just avoided.
(Ans. 3.17 ms^{-2}).

6.22 An express subway train and a local making stops at all stations run on parallel and straight tracks between stations A and E, which are 1600 metres apart. The local train makes stops

of 30 secs duration at each of stations B, C and D between A and E, while the express proceeds to station E without any intermediate stop. Each train accelerates at a constant rate of 1.6 ms^{-2} until it reaches a speed of 16 ms^{-1} and then proceeds at a constant speed. As both the trains approach the next stop, they decelerate by application of brakes at a constant rate of 2 ms^{-2}. Distances between intermediate stations is 400 m each. The express train leaves the station A 3 minutes after the local train has left A, determine (a) which train will reach station E first and (b) when the next train is expected to arrive.
Ans. (a) Local train arrives first at the station E at 3 min 46 secs after it left the station A.
(b) The express train arrives 1 min and 3 secs later than the local train at the station E.

6.23 A police officer observes a car approaching along a straight road at the unlawful speed of 90 kmh^{-1}. He gets on his motor cycle and starts chasing the car just as it passes in front of him. After accelerating for 10 seconds at a constant rate, the officer reaches his top speed of 108 kmh^{-1}. How long does it take the police officer to overtake the car from the time he started? Also find how far it will take him to overtake?
(Ans. t = 30 secs Dist. 750 m).

6.24 A bullet is fired with a velocity of 400 ms^{-1}. It losses half of its velocity in passing 10 cms through a resisting medium of uniform resistance. Find (a) how long it takes in passing through the 10 cms medium? (b) what thickness of this medium is required to bring the bullet to rest? (c) how long will it take to bring the bullet to rest in the above medium if the bullet is fired with initial velocity of 600 ms^{-1}?
Hint: Assume the medium to cause uniform deceleration 'a' when the bullet travels through it.
(Ans. $t_1 = \frac{1}{3000}$ sec. (b) S = 13.333 cm (c) $t_2 = \frac{1}{1000}$ sec.)

6.25 A bullet fired with a velocity of 400 ms^{-1} penetrates a target by 40 cm. If this bullet is fired with the same velocity at the same target but having thickness of 30 cm, find the velocity at which the bullet will emerge from the target. Assume the target to offer the same resistance.
(Ans. V = 200 ms^{-1})

6.26 A burglar's car started from rest with an acceleration of 2 ms^{-2}. A police vigilance party passed the same spot after 5 seconds and followed the burglar's car with a uniform speed of 20 ms^{-1} in the same direction. If the burglar's car can reach a maximum speed of 25 ms^{-1},
(a) Find when and where the police van will overtake the burglar's car? What is the velocity of burglar's car at the time of overtaking?
(b) What happens if the police van passes the spot 6 sec after the burglar's car starts?
(Ans. (a) t = 5 secs from crossing the common spot.
S = 100 m from the spot
V = 20 ms^{-1} for burglar's car
(b) Police van will not be able to overtake the Burglar's car).

6.27 A stone is thrown vertically upwards with a velocity of 10 ms^{-1}. Find how high the stone will rise, and how long it will take to reach the point of throw?
(Ans. h = 5.10 m; t = 2.04 secs).

6.28 A ball 'A' is dropped from Kutab Minar from a height of 30 m above ground; at the same instant another ball 'B' is thrown vertically upwards with a velocity of 15 ms^{-1}. Find when and where the two balls will be at the same height and what will be their velocities at that time.
(Ans. t = 2 secs, h = 10.4 m above the ground, V_A = 19.6 ms^{-1} downward, V_B = 4.6 ms^{-1} downward).

6.29 A ball is thrown vertically upwards and it returns to the ground in 4 seconds. Calculate the initial velocity with which the ball was thrown and the maximum height attained by the ball.
(Ans. V_i = 19.6 ms^{-1} upwards; h = 19.6 m)

6.30 A ball 'A' is projected vertically upward with a velocity of 20 ms^{-1}. Another ball 'B' is projected vertically upwards 1 second later from the same point with the same velocity. Find when and where the two balls will be meeting and what will be their velocities at

the time of meeting.
(Ans. 2.54 sec after the ball A is projected, height 19.1 m, V_A = 4.9 ms^{-1} downwards V_B = 4.90 ms^{-1} upwards).

6.31 A stone is thrown vertically upwards from the top of a tower 39.2 m high with a velocity of 19.6 ms^{-1}. Find the time after which it strikes the ground at the base of the tower. What is the velocity when it strikes the base?
(Ans. t = 5.464 secs, V = 33.95 ms^{-1} downwards).

6.32 A stone is dropped down a well. The splash is heard at the ground 6 secs later. Assuming the velocity of sound to be 320 ms^{-1}, find the depth of the well up to the water level and the time stone will take in reaching water level.
(Ans. depth 150.2 m; t = 5.5314 secs.)

6.33 Two stones were thrown vertically upward at the same moment. The first stone which was thrown with a velocity of 19.6 ms^{-1} reaches the ground 3 secs before the second stone reaches. Determine how high the second stone had gone and what was its velocity of throw.
[Ans. h_2 = 60.025 ms, V_i = 34.3 m^{-1} (upward)].

6.34 A body projected vertically upwards loses half its velocity due to certain obstacle at half the maximum height to which it would have gone. Find the ratio of the velocity with which it strikes the ground to the velocity of projection.
$\left(\text{Ans. } \dfrac{V_f}{V_i} = 0.866\right)$

6.35 A body falling freely from top of a multi-storeyed building under gravity takes 1 second to travel through the lowest storey which is of 5 m height. Find from what height above the base the body has fallen. What is the velocity when it touches the ground.
(Ans. h = (125 + 5) m = 130 m; V = 50.4 ms^{-1}).

6.36 A cage in mine shaft starts to ascend with an acceleration of 1 ms^{-2} from the bottom while a stone is simultaneously dropped from the top of the shaft. If the stone strikes after 10 secs, find the total depth of the mine shaft and the depth at which the stone strikes.
(Ans. h = 540 m; h_1 = 490 m.)

6.37 A cage in mine shaft starting from rest from the surface descends with an acceleration of 2 ms^{-2}. Every 2 secs after the start of the cage a pebble drops in the shaft from the top. Find when and at what depths the 1st, 2nd and 3rd pebble will strike the case with respect to the starting time of the pebbles.
(Ans. t_1 = 0.823 sec; t_2 = 1.646 sec; t_3 = 2.469 sec.
h_1 = 3.32 m; h_2 = 13.28 m; h_3 = 29.90 m.)

6.38 A stone is released from an elevator moving up at a speed of 5.0 ms^{-1}. The stone reaches the bottom of the shaft in 5.0 secs. Find
(a) how high was the elevator when the stone was released.
(b) with what velocity the stone strikes the ground
(Ans. (a) h = 97.5 m; (b) V = 44 ms^{-1} downwards).

6.39 Drops of water fall down a mine shaft at a uniform rate of one drop per second. A mine elevator moving up the shaft at 10 ms^{-1} is struck by a drop of water when it is 100 m below the ground level. When and where will the next drop of water strike the elevator?
(Ans. t = 0.812 sec after the first drop strikes, when the elevator is 91.88 m below the ground level).

6.40 An automobile starts along a straight road, from rest, with an acceleration of 1 ms^{-2} till the speed becomes 60 kms/hour, runs at this speed for some time and then decelerates at a constant rate of 1.5 ms^{-2} before it comes to rest again. If the total distance travelled is 1200 m, find the total time of travel by graphical method.
(Ans. 85.88 secs).

6.41 Draw $a - t$, $V - t$, and $S - t$ curves for a particle which moves with a constant velocity of 20 ms^{-1} for first 5 seconds of its motion, then accelerates with a constant acceleration of 1 ms^{-2} for next 5 seconds and then moves with the constant velocity so attained for another 5 seconds. Indicate maximum velocity and total displacement covered.
(Ans. V_{max} = 25 ms^{-1}
Total displacement = 337.5 m).

6.42 The $a - t$ graph for a particle moving in a straight line is given in the Figure. Draw $V - t$ and $S - t$ graphs indicating important values at $t = 0$, $V = 0$ and $S = +5$ m

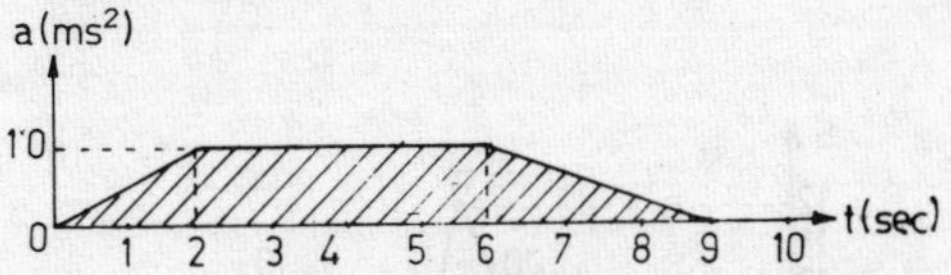

Fig. Prob. 6.42

[Ans. $V_2 = +1.0$ m^{-1}, $S_2 = 5.67$ ms (Position vector)
$V_9 = +6.5$ m^{-1}, $S_9 = 35.67$ m (Position vector).
$V_{10} = +6.5$ ms^{-1}, $S_{10} = 42.17$ ms (Position vector)]

6.43 A particle moves with a constant velocity of 10 ms^{-1} for 10 secs. Draw $a - t$ and $S - t$ curve for the motion of the particle if initially at $t = 0$ the position vector $S = 0$. (Ans. $S_{10} = 100$ m; $a_{10} = 0$)

6.44 Motion of a particle is shown by the $V - t$ graph given below. Answer the following questions:

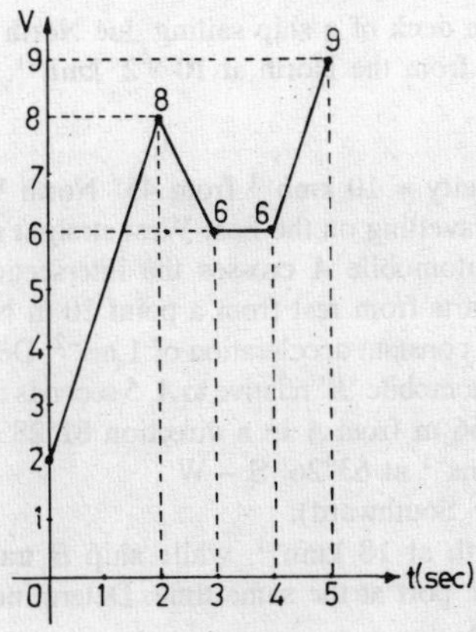

Fig. Prob. 6.44

(i) State the time interval during which the particle is decelerating
(ii) State the time intervals during which the particle is accelerating.
(iii) If initially at $t = 0$, $S = 0$, find the displacement 'S' at 3 secs and 5 secs.
(iv) Draw the $a - t$ graph.

(Ans. (i) $t = 2$ to 3 secs (ii) $t = 0$ to 2 secs and $t = 4$ to 5 secs (iii) $S_3 = 17$ m. $S_5 = 30.5$ m (iv) $a_{0-2} = 3$ ms^{-2}, $a_{2-3} = -2$ ms^{-2}, $a_{3-4} = 0$, $a_{4-5} = +3$ ms^{-2}).

6.45 The displacement-time graph for a particle with certain reference point is shown in the fig. Answer the following:
(i) State the time interval during which the particle accelerated
(ii) State the time interval during which the particle decelerated
(iii) Did the particle stop at any instant? If so, state the time.
(iv) What is the acceleration of the particle between the time 5 secs and 7 secs.
(v) Draw $V - t$ and $a - t$ graphs if at $t = 0$, $V = 0$

(Ans. (i) $t = 0$ to 2 secs)

(ii) $t = 4$ to 5 secs
(iii) $t = 4.5$ secs, $t = 6$ secs.
(iv) $a_{5-7} = 0$).

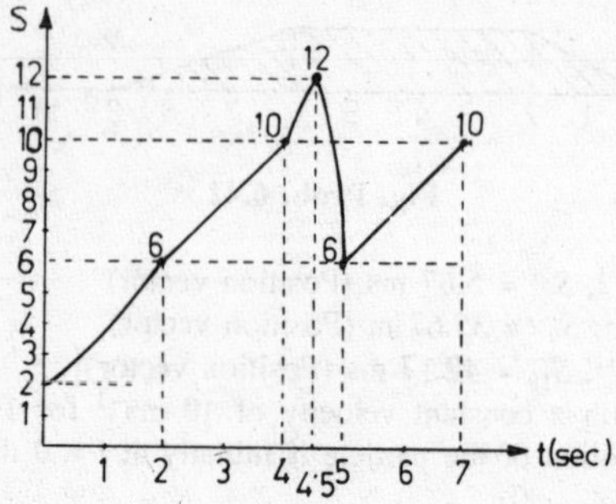

Fig. Prob. 6.45

6.46(*a*) A passenger is on the deck of a ship sailing due North East at 10 kmh^{-1}. The wind appears to him to blow from the North at $10\sqrt{2}$ kmh^{-1}. Find the actual velocity and the direction
of the wind.
(Ans. Actual wind velocity = 10 kmh^{-1} from 45° North West direction).

6.46(*b*) An automobile *A* is travelling on the East-West straight road with a constant speed of 36 kmh^{-1}. When the automobile *A* crosses the intersection of the North-South road, another automobile *B* starts from rest from a point 20 m North of the intersection, and moves Southward with a constant acceleration of 1 ms^{-2}. Determine the position, velocity and acceleration of the automobile '*B*' relative to *A*, 5 seconds after *A* crosses the intersection.
(Ans. (i) *B* will be 50.66 m from *A* in a direction 81°28′ North West.
(ii) $V_{B/A}$ = 11.18 ms^{-1} at 63°26′ S – W
(iii) $a_{B/A}$ = 1 ms^{-2} Southward).

6.47 Ship *A* travels due North at 18 kmh^{-1}, while ship *B* travels towards North-West at 15 kmh^{-1} from the same port at the same time. Determine the relative velocity of
(i) ship B relative to *A*
(ii) Ship *A* relative to *B*, and
(iii) also determine the time when the two ships are 64.65 km apart, if they continue to move along their original courses.
(Ans. (i) 12.93 kmh^{-1} at 55°–7′ S – W
(ii) 12.93 kmh^{-1} at 55°–7′ N – E
(iii) 5 hours).

6.48 To a ship moving due North at 24 kmh^{-1}, the wind appears to blow from the South West at 36 kmh^{-1}. Determine the magnitude and direction of the true wind velocity.
(Ans. 55.62 kmh^{-1} from 27.2° West of South).

6.49 Two aeroplanes *A* and *B* are flying each at a constant altitude of 2000 m vertically above a target. Plane *A* is flying due East at a constant speed of 500 kmh^{-1}, while plane *B* is flying South-West at a constant speed of 750 kmh^{-1}. If Plane *A* passes over the target 10 secs before Plane *B*, determine the relative position and relative velocity of Plane *B* with respect to plane *A* 2 mins after plane *A* passes the target.
(Ans. $S_{B/A}$ = 36.65 along 63° – 45′ SW
$V_{B/A}$ = 1158.8 km/hr along 3°16′ SW).

6.50 Two trains are moving on parallel straight tracks with constant speeds. They take 10 seconds in crossing each other completely. If the lengths of the trains are 100 m and 80 m, and the speed of the longer train is 36 kmh^{-1}, find the speed of the shorter train when
(i) both the train are moving in the same direction.

(ii) both the trains are moving in opposite directions.
(Ans. (i) 100.8 kmh^{-1}
(ii) 28.80 kmh^{-1})

6.51 To a cyclist riding at 12 kmh^{-1} along a horizontal road, rain appears to fall vertically downwards with a velocity of 16 kmh^{-1}. What is the actual velocity of the rain and its direction?
(Ans. 20 kmh^{-1} falling at 36°52′ with the vertical, from behind).

6.52 A cylist finds that rain is falling vertically downwards at 12 kmh^{-1} when he is standing. He then rides his cycle along a horizontal road at 16 kmh^{-1}. Find the relative velocity of rain as it appears to the cyclist.
(Ans. 20 kmh^{-1} falling at 53° – 7′ with the vertical, from the front).

6.53 When a scooter driver is moving with 40 kmh^{-1}. The rain appears falling at 45° with the vertical from the front, while it appears to fall at 30° with the vertical from the front when he slows down his scooter to 24 kmh^{-1}. What is the actual velocity of rain?
(Ans. 38.1 kmh^{-1} at 25°1′ with the vertical).

6.54 A railway track and a national highway are crossing a bridge at right angles. A train moving at 45 kmh^{-1} approaches the bridge when a car moving along the road at 90 kmh^{-1} also approaches the bridge. If the car and the train pass the bridge at the same time, find the position of the car with respect to the train after 1 min.
(Ans. The car will be 750 $\sqrt{5}$ m away from the train along 63°26′ with the rear direction of the train).

6.55 A steamer is travelling due East at 20 kmh^{-1} and the wind is blowing at 48 kmh^{-1} from North to South direction. Find how the smoke from the steamer will appear to a person
(i) Standing in a light house
(ii) Standing on the deck of the steamer itself.
(Ans. (i) To a person in light house the smoke will appear to blow at 52 kmh^{-1} towards 22° – 37′ SE
(ii) To a person on the deck of steamer the smoke appears to blow at 48 kmh^{-1} towards South).

6.56 A person cycling Eastwards at 8 kmh^{-1} finds that the wind seems to blow directly from the North. On doubling his speed, it appears to come from North-East. Find the magnitude and direction of wind velocity.
(Ans. V = 11.312 kmh^{-1}, θ = 45°).

6.57 Two ships, A and B are, at a given instant, 10 km away from each other, A being SW of B. A is travelling at 24 km/hr due North, and B is travelling at 10 km/hr due East. Find
(a) the velocity of A as observed from the ship B
(b) the shortest distance between them, and the time when they are at shortest distance.
[Ans. (a) 26 km/hr 22°37′ West of North.
(b) 9.246 km., Time 0.14646 hrs. (8.79 min)].

6.58 At noon a ship A is observed to be 20 km due East of another ship B, and is travelling at 28 km/hr in a direction 30° West of North. The ship B is travelling North East at 20 km/hr. Determine the velocity of A relative to B, and find the least distance apart and the time when this occurs.
(Ans. $V_{A/B}$ = 30 km/hr, least distance = 18.8 km, Time = 37.6 min.)

6.59 A submarine A is travelling due North at 40 km/hr. Another submarine B at a distance of 25 km in North East direction of A, decides to intercept submarine A in minimum possible time. If the maximum speed of B is 30 km/hr, find the time and point of interception.
(Ans. 30 min, 20 km.)

6.60 A submarine travelling on a course of bearing 80° East of North with a speed of 21 kms/hr reports the presence of an enemy ship on a course of bearing 75° West of South with a speed of 15 km/hr with respect to the submarine. Asuming that the submarine and the enemy ship are at the same place at the time of the observation.
(a) Find the actual direction and the speed of the enemy ship.
(b) If a war ship stationed 150 km South of the submarine immediately starts at 18 km/hr to intercept the enemy ship, what bearing should it take. Find the time required to intercept.

(Ans. (a) Velocity of enemy ship = 6.16 km/hr.
87°48′ South-East.
(b) Bearing of the war ship 20°
East of North. Time required = 8 hrs 45 mins.)

6.61 Find the angle of projection of a projectile so that its horizontal range is maximum. Find the maximum range also.

$\left(\text{Ans. } \alpha = 45°, \text{ Range } S_x = \frac{V_i^2}{g}\right).$

6.62 A stone is projected into the air at a given direction α with the horizontal with a given initial velocity V_i. Show that the path of the stone will be a parabola and find its equation (neglect air resistance).

[Ans. $X^2 = -\frac{2V_i^2}{g}\cos^2\alpha\,(X\tan\alpha - Y)$].

6.63 A particle projected at 30° with the horizontal reaches a point $176\sqrt{3}$ m away from the point of projection. If it is projected with the same velocity but at 45°, find the distance at which the particle strikes the ground. Find the initial velocity of projection.
Ans. $V_i = 58.8\text{ ms}^{-1}$; $S_x = 352.8$ m.

6.64 If S_x is the range of a projectile on a horizontal ground, and S_y its maximum height in the vertical plane when projected with certain initial velocity V_i and inclination α with the horizontal, find the maximum possible horizontal range with the same initial velocity.

$\left[\text{Ans. } (S_x)_{max} = \left(2S_y + \frac{S_x^2}{8S_y}\right)\right].$

6.65 A shell is fired with a velocity of 200 ms^{-1} at an angle of 60° with the horizontal. Find
(a) equation of the trajectory of the shell,
(b) maximum height attained by the shell,
(c) total time of flight,
(d) the horizontal distance travelled by the shell,
(e) the time when the shell is at a horizontal distance of 200 m from the point of projection, and
(f) the time when the shell is 980 m high.
(Ans. (a) $Y = \sqrt{3}X - 4.9 \times 10^{-4}X^2$
(b) $Y_{max} = 1530.6$ m
(c) $T = 35.348$ Sec.
(d) $X = 3534.8$ m
(e) $t = 19.8$ sec
(f) $t = 7.07$ sec or 28.28 sec.)

6.66 A stone is thrown with a velocity 39.2 ms^{-1} along a direction 30° with the horizontal. Find the velocity of the stone at 1 sec, 2 sec, 3 sec and 4 sec. after projection. Also find the maximum height, horizontal range, and maximum possible range with above velocity of projection.
(Ans. $V_1 = 35.33\text{ ms}^{-1}$ upwards at 16°6′ with horizontal,
$V_2 = 33.95\text{ ms}^{-1}$ horizontally
$V_3 = 35.33\text{ ms}^{-1}$ downwards at 16°6′ with horizontal,
$V_4 = 39.2\text{ ms}^{-1}$ downwards at 30° with horizontal).

6.67 A shot is fired horizontally from the top of vigilance tower of 19.6 m height above the ground. Find the time after which the shot will hit a target on the ground level at a horizontal distance of 1000 m. What should be the initial velocity of the shot and with what velocity it hits the target?
(Ans. $t = 2$ secs; $V_i = 500\text{ ms}^{-1}$
$V_f = 500.4\text{ ms}^{-1}$ at 2°18′ with the horizontal).

6.68 An aeroplane, flying at an altitude of 2000 m with a velocity of 1200 km/hr, drops a bomb to hit a target on the ground in line of motion of the plane. Find the point at which the bomb should be released to a hit the target and the inclination of the target at that time.
(Ans. 6735 m horizontally before the target position, θ = 16°32′ with horizontal)

6.69 A target at a horizontal distance of 14.14 km is to be hit by a shell fired from a gun which can produce a muzzle velocity of 400 ms^{-1} in the shell. Find the inclination of the gun to hit the target in least and maximum possible time after the shell is fired. Also find these times.
(Ans. θ_1 = 30° with horizontal, t_{least} = 40.82 secs
θ_2 = 60° with horizontal, t_{max} = 70.7 secs)

6.70 A solider fires at a target with his gun inclined at 14°20′ with the horizontal and strikes a point about 80 m before the target. In next attempt he fires with his gun inclined at 15°36′ with the horizontal and strikes another point 60 m beyond the target. Find the correct angle of the fire, if in all the cases the muzzle velocity of the gun remains unchanged. Also find the distance of the target from the gun and muzzle velocity of the shot.
(Ans. α = 15°, S_x = 1833.0 m, V_i = 189.2 ms^{-1})

6.71 A player hits a football which just clears a post of 3.775 m height situated at 8.66 m in front of the point of hitting. If the inclination of the ball at the time of hitting the ball is 30°, with the horizontal, find when and where the ball will reach the ground?
(Ans. T = 2.041 secs
S_x = 35.347 ms from the point of hit).

6.72 A car moves along a ramp to cross a small stream 5.66 m wide. The rear wheels are 3 m from the front wheels. Find the take off velocity of the car if the take off point is 2.823 m above the landing point on the other side of the stream and the slope of the ramp is 30°. How long the vehicle will be in the air and at what angle will it touch the ground?
(Ans. V_i = 15 ms^{-1}(54 km/hr), t = 0.667 secs
θ = 4° – 15′)

6.73 A ball is thrown at 60° with the horizontal so as to just clear a fence 3 m high above the ground and 20 m away from the point. If the point of throw is 1.50 m above the ground, find the initial velocity of throw.
(Ans. V_i = 15.38 ms^{-1})

6.74 A ball is thrown with a velocity of 30 ms^{-1} from the top of a tower 30 m high to hit a object on the ground in 4 secs. Find the inclination of ball and the horizontal distance of the target from the base of the tower.
(Ans. α = 23°48′, S_x = 109.80 m)

6.75 A enemy boat is observed to move away with a constant velocity of 10 ms^{-1} in the direction of sight from the top of a hill. The enemy boat is required to be hit by a shell from a gun capable of producing muzzle velocity of 30 ms^{-1} in 5 secs. The boat is at a distance of 100 m at the time of observation. Find the inclination of the gun and the height of the gun above the position of the boat.
(Ans. θ = 0°, S_y = 122.5 m)

6.76 A rocket is fired with a velocity of 100 ms^{-1} and at an angle of 60° with the horizontal. Calculate
(i) time of flight
(ii) total horizontal distance covered
(iii) maximum height reached, and
(iv) inclination of the rocket with the horizontal 3 secs, 9 secs, and 15 secs after launching.
(Ans. t = 17.67 secs, S_x = 883.7 m, S_y = 765 m, θ_3 = 48°50′ upwards, θ_9 = 1°–50′ downwards, θ_{15} = 50°23′ downwards).

6.77 A rocket moves along a curved path with linear velocity of 600 ms^{-1} and it/ is observed that the rocket experiences an acceleration of 10 g in a direction normal to the path. Find the radius of the curved path.
(Ans. r = 3673 m).

6.78 A projectile when projected in air has a velocity of –75.9 ms^{-1} after 3 seconds, and is directed to 48°50′ with the horizontal in upward direction. If the projectile is under the accleration due to gravity, find the component of acceleration normal and tangential to the motion, and hence calculate the radius of curvature of the trajectory at this instant of time.
(Ans. a_t = g sin 48°50′, a_n = g cos 48°50′, r = 893 m).

6.79 Two adjacent corners of the rectangular plate are confined to move in slots at right angles as shown in the figure. If the velocity of the corner *A* is uniform 2 ms^{-1} during the interval of motion, determine the angular velocity and acceleration of edge BC, and the linear velocity and acceleration of the point *B* at the instant when $\theta = 30°$.

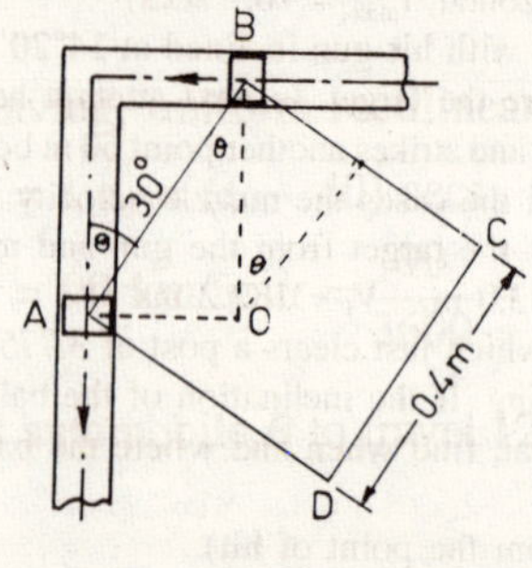

Fig. Prob. 6.79

(Ans. $\omega_{BC} = -10$ rad s^{-1}, $\alpha_{BC} = -100\sqrt{3}$ rad s^{-2}, $V_B = -2\sqrt{3}$ ms^{-1}, $a_B = -80$ ms^{-2})

6.80 A pole AB rests in inclined position on a wall 3 m high as shown. If the end *A* is pulled with a velocity of 2 ms^{-1}, find the angular velocity of the rod AB when the point *A* is 4 m away from the wall.

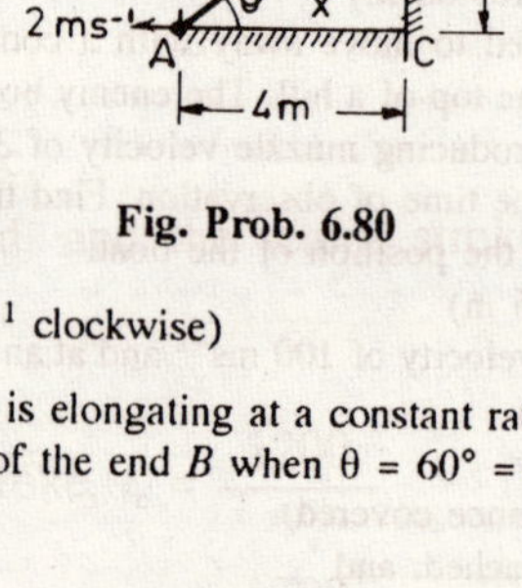

Fig. Prob. 6.80

(Ans. $\omega_{AB} = 0.667$ rad. s^{-1} clockwise)

6.81 The hydraulic linkage PQ is elongating at a constant rate of 0.20 ms^{-1}. Determine the velocity and acceleration of the end *B* when $\theta = 60° = \phi$

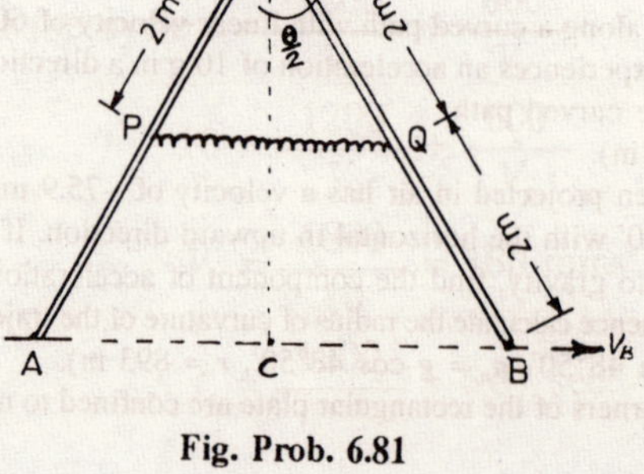

Fig. Prob. 6.81

(Ans. $V_B = 0.4619\ ms^{-1}$, $a_B = \dfrac{8}{300\sqrt{3}}\ ms^{-2}$

$\alpha_B = \dfrac{2}{300\sqrt{3}}$ rad/sec^2).

6.82 A wheel 120 cm in diameter rolls on a horizontal road without slipping at an angular velocity of 4 rad/sec and angular acceleration of 6 rad/sec^2 both clockwise. Determine the absolute acceleration of point *A* shown in figure Prob. 6.82.

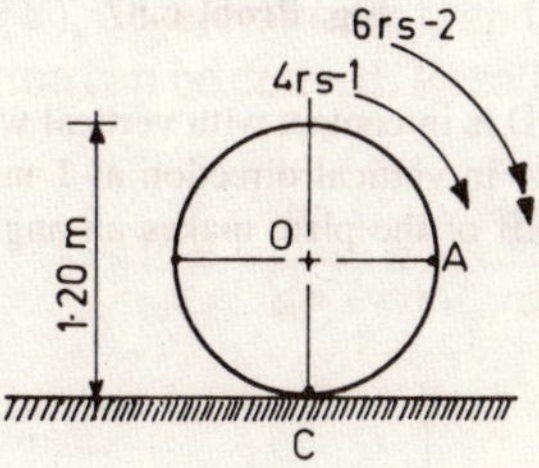

Fig. Prob. 6.82

(Ans. $a_A = 10.252\ ms^{-2}$).

6.83 A ladder AB 4 m long rests with its ends in contact with a horizontal floor and a vertical wall. If at any instant when the inclination of the ladder $\theta = 30°$ with the horizontal the point *B* in contact with the vertical wall moves with a downward linear velocity of $3.464\ ms^{-1}$ and downward linear acceleration of $8.928\ ms^{-2}$, find the linear acceleration and velocity of the point *A*.
Hint $V_B = \omega l \cos\theta$, $a_B = \alpha \cdot l \cos\theta - \omega^2 l \sin\theta$
[Ans. $\omega = -1$ rad/sec., $\alpha = -2$ rad/sec^2
$V_A = 2\ ms^{-1}$ (leftward), $a_A = -9.773\ ms^{-2}$ (right ward)].

6.84 A ladder AB 5 m long rests against a vertical wall at *B* and on a horizontal floor at *A* making angle θ with the horizontal. The end A on the floor is moved with a constant linear velocity of $2\ ms^{-1}$ at the instant when $\theta = 45°$. Find the linear velocity and linear acceleration of the end B along the vertical wall.
Hint $V_A = \dfrac{d}{dt}(l \cos\theta)$, Alternatively Instantaneous centre approach.
(Ans. $V_B = 2\ ms^{-1}$ downward, $a_B = 2.26\ ms^2$, when $\theta = 45°$).

6.85 Three links each of 2 m length are hinged at its ends to form a triangle ABC. At certain instant the whole traingle is moving in such a way that the point *A* moves with a velocity of $4\sqrt{3}\ ms^{-1}$ at 90° to *BC*. Determine the velocity of the point *C*.
(Ans. $V_C = 8\ ms^{-1}$ at 90° to BC).

6.86 The connecting rod AB and crank CB of a reciprocating pump are respectively 120 cm and 30 cm in length. The crank shaft rotates at 900 rpm. Find the velocity of the piston at the instant when the crank is making 30° with the horizontal line from inner dead centre.
Hint Find instantaneous centre with respect to horizontal velocity at *A* and tangential velocity at *B*.
(Ans. $V_A = 17.22\ ms^{-1}$)

6.87 Figure Prob. 6.87 shows a slotted link OP hinged at O and supported by a pin Q sliding freely in the slot. The block carrying the pin at 20 cm height above the hinge slides in the horizontal surface. If the link is at 60° with the vertical and the block moves left ward with a uniform velocity of $2\ ms^{-1}$. Find the angular velocity and angular acceleration of the link.
[Ans. $\omega = 2.5\ rad\ s^{-1}$ anticlockwise,
$\alpha = 12.5\sqrt{3}\ rad\ s^{-2}$ (retardation)]

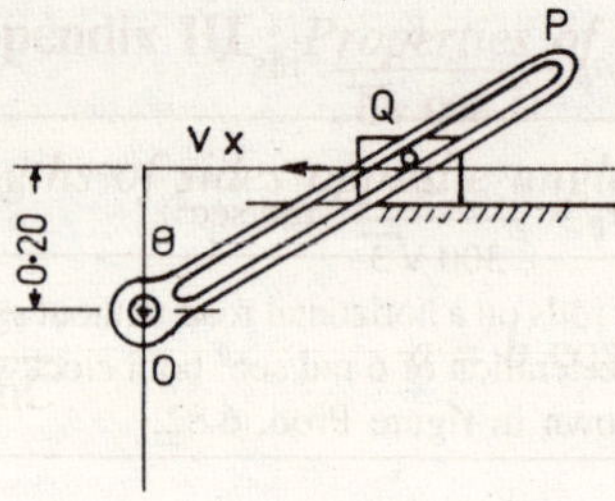

Fig. Prob. 6.87

6.88 A rectangular plate ABCD is in contact with vertical wall and horizontal floor as shown and moves with its end A in vertical direction at 1 ms^{-1}. Find the velocity of other 3 corners when the edge AB of the plate makes an angle of 30° with the horizontal.

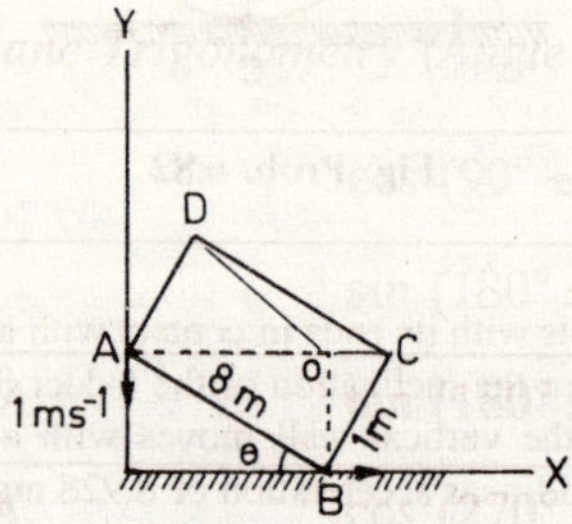

Fig. Prob. 6.88

(Ans. V_B = 0.577 ms^{-1}, V_C = 0.458 ms^{-1}
V_D = 0.936 ms^{-1}).

6.89 A roller of diameter 0.60 m rides between two belts moving in the opposite horizontal directions, as shown in the sketch. Find the instantaneous centre of rotation for the roller assuming no slipping at the points of contact A and B. Find the linear velocity of the actual centre C of the roller. Velocities of the belts at A and B are respectively 4 ms^{-1} and 2 ms^{-1}.

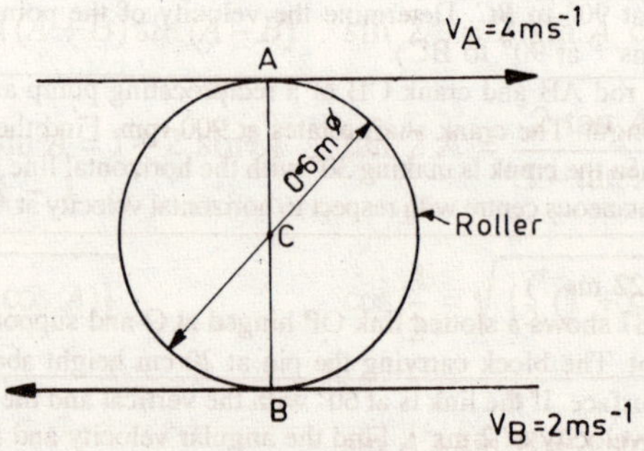

Fig. Prob. 6.89

(Ans. Instantaneous centre O is along AB at 20 cm from B. V_C = 1.0 ms^{-1} right ward).

6.90 A helicopter moves horizontally due East at certain instant at a speed of 90 km/hr while the main blades at the top rotate horizontally at anticlockwise angular velocity of 300 revolution per minute. Determine the instantaneous axis of rotation of the main blades.

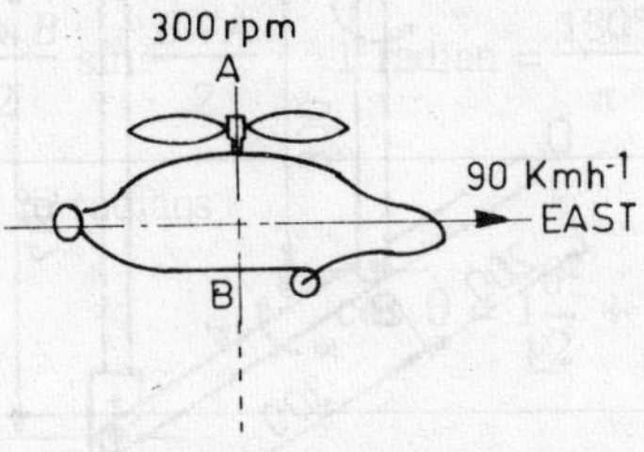

Fig. Prob. 6.90

(Ans. Instantaneous axis will be 0.796 m from the centre towards south).

6.91 A 4 m channel, used to form one edge of an over head slide door, is guided by wheels at *A* and *B* which roll in horizontal and vertical tracks as shown in the figure. If the velocity of the roller B is 1 ms^{-1} vertically upward, determine the angular velocity of the door and the velocity of the end C at the instant when (a) $\theta = 30°$ (b) $\theta = 45°$ and (c) $\theta = 60°$

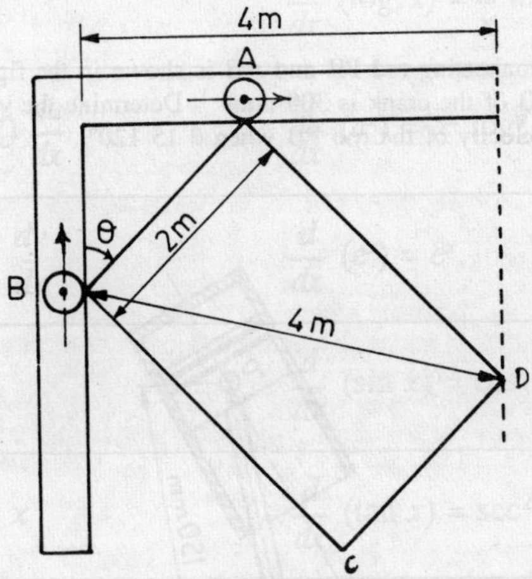

Fig. Prob. 6.91

(Ans. θ = 30°	45°	60°
ω = 1 rad/sec	0.707 rad/sec	0.577 rad/sec
V_C = 2.646 ms^{-1}	1.88 ms^{-1}	1.732 ms^{-1})

6.92 A machine part is shown in the figure. At certain instant the collar *C* moves vertically upwards at 500 mm/sec. Determine (a) the instantaneous centre of the link BC, (b) angular velocities of the link BC and AB (c) Velocity of the Point *B* and (d) Velocity of the mid point of link BC. All dimensions are in mm.

Ans. (a) Instantaneous centre on AB produced at 180 mm from *B* and 240 mm from *C*.

(b) $\omega_{BC} = \dfrac{25}{12}$ rad s^{-1}, Counter clockwise ω_{AB} = 1.25 rad s^{-1}, Clockwise

(c) V_B = 375 ms^{-1} leftward

(d) $V_{C'}$ = 312.5 ms^{-1} upward at an angle $\tan^{-1}\left(\dfrac{3}{4}\right)$ with the vertical).

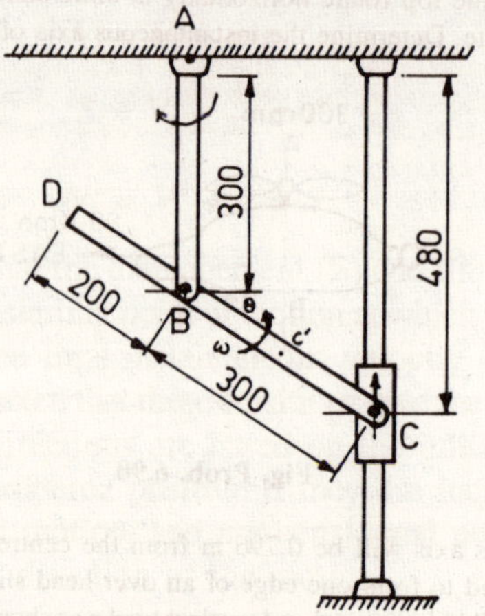

Fig. Prob. 6.92

6.93 A piston P with connecting rod PB and AB is shown in the figure. At the instant the velocity of point B of the crank is 300 mms^{-1}. Determine the velocity of the piston *P*. and the angular velocity of the rod PB when θ 15 120°.

Fig. Prob. 6.93

(Ans. ω_{AB} = 1.045 rad s^{-1}, V_P = 305.0 mms^{-1}).

6.94 Two links AB and BC each of length 130 mm are connected to three collars at *A*, *B* and *C* respectively. Collars at *A* and *C* can move vertically along their respective guides and collar *B* can move horizontally along its guide. At the instant when the collar *B* is at 50 mm from its extreme left end of its run, find the velocity of the collars *B* and *C* if the collar *A* is moving vertically downward at 200 mms^{-1}.

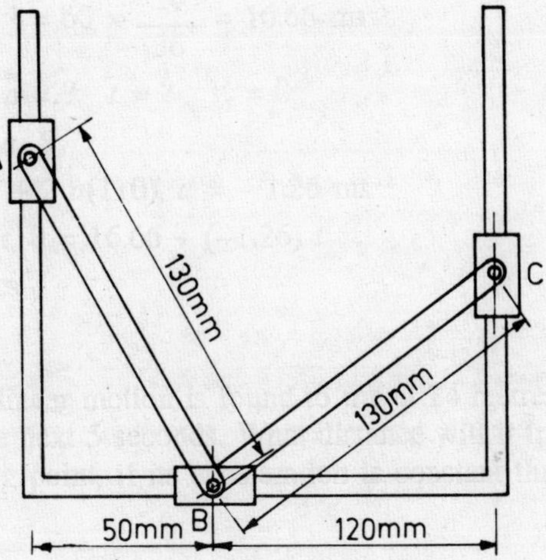

Fig. Prob. 6.94

(Ans. ω_{AB} = 4 rad s^{-1}, Counter clockwise $\quad$ V_B = 480 mms^{-1}, towards right
ω_{BC} = 9.6 rad s^{-1} Counter clockwise, $\quad$ V_C = 1152 mms^{-1}, upward)

7. KINETICS

After having mastered the contents of this Chapter, you will be able to:

- Define the term '*Kinetics*'.
- Explain the scope of kinetics and its inter-relationship with other branches of mechanics.
- Differentiate between state of rest and state of motion with illustrative example.
- State Newton's second law of motion.
- Derive the equation of motion
- Define the terms momentum and impulse
- Explain the principle of conservation of linear momentum.
- Solve problems to determine unbalanced force or motion parameters in case of collision of bodies, thrust of jet etc.
- Explain D'Alembert's principle and use it for solving problems.
- Derive equations for centripetal & centrifugal force in case of circular motion.
- Explain the need for super-elevation of tracks and conditions for preventing overturning of vehicles.
- Solve problems on circular motion, super elevation of tracks and overturning of vehicles.

CHAPTER 7

Kinetics

7.1 Introduction

The discovery of the laws of motion, was a dramatic moment in the history of science. Before Newton's time, the motion of things like the planets was a mystery, but after Newton there was an understanding about these motions. The motion of planets, air crafts, ships, trains, cars, spring mass system, machine elements, etc. could be analysed completely after Newton's laws were enunciated.

Galileo made a great advance in the understanding of motion when he discovered the PRINCIPLE OF INERTIA which states that if an object is left alone or undisturbed, it continues to move with a constant velocity if it was originally moving, or it continues to remain in the state of rest if it was in that state initially. This never appears to be the case in nature, for if we slide a block on a floor, it stops after moving a certain distance as is clear from Figure 7.1. This is because the block is not left to itself, it is being rubbed against the floor.

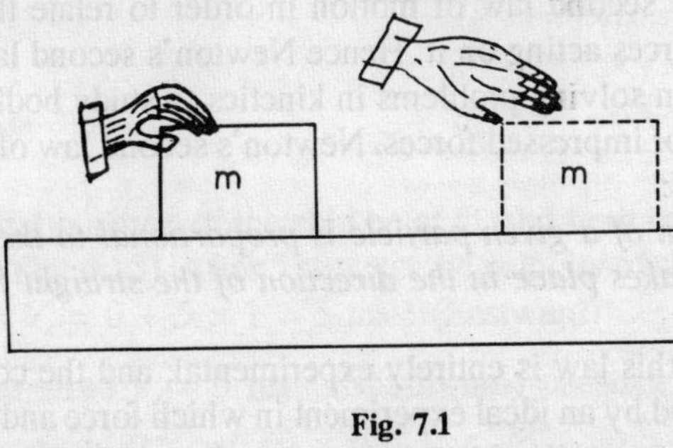

Fig. 7.1

The next thing which was needed was a rule for finding how an object changes its speed or state of rest. This was contributed by Newton. Newton formulated three laws. The first law was mere a restatement of the Galilean Principle of Inertia just described. The second law gave a specific way of determining velocity changes under different influences called forces. The third law described the forces to some extent. For the first two of these laws, we owe much to Galileo and several of Newton's contemporaries, particularly Huyghens, who contributed to their development; but Newton was the first to put them into formal shape in his *Principia* published in 1686. His own contributions were outstanding.

Newton's laws of motion are the basic postulates, or assumptions on which the science of Newtonian dynamics is based, and like other scientific laws, they are justified only in so far as they lead to results in agreement with observations. For example, assuming the laws to be true, the motion of the moon and planets can be calculated. The position of planets, the time of eclipses, etc., are worked out and published several years before they occur. The predicted places and times are found to agree with observations remarkably well, and thus provide excellent confirmation of these fundamental laws, but no proof of these laws can be given.

Kinetics is the branch of mechanics that deals with the problems which require the determination of the effect of forces on the motion of a body, or conversely the forces causing certain motions. Kinetics helps in finding acceleration of a body under the influence of forces. When the acceleration of a body caused by the forces, acting on it, has been determined the principles of kinematics may be applied to determine the geometry of motion, including the displacement and velocity at any instant. In other cases, kinematics may be used to compute acceleration of a body, from which the force required to produce this motion can be determined by applying the principles of kinetics.

7.2 Newton's Second Law of Motion

Newton's first and third laws of motion were used extensively in solving problems in Statics to study bodies at rest. However, when a body is accelerated, that is, when the magnitude or the direction of its velocity changes, it is necessary to use the second law of motion in order to relate the motion of the body with the forces acting on it. Hence Newton's second law of motion is used extensively in solving problems in kinetics to study bodies in motion under the influence of impressed forces. Newton's second law of motion may be stated as follows:

The acceleration of a given particle is proportional to the unbalanced applied force, and takes place in the direction of the straight line in which the force acts.

The validity of this law is entirely experimental, and the concept of the law is best understood by an ideal experiment in which force and acceleration are assumed to be measured without any error. A particle is subjected to a force F_1, the particle is observed to move in a straight line and in the direction of the force F_1. By determining the position of the particle at various instants, its acceleration can be found out. It is observed that its acceleration has a constant magnitude a_1. If the same particle is subjected to forces F_2, F_3, etc. of different magnitude or direction, in turn, as shown in Figure 7.2. We find each time that the particle moves in the direction of the force acting on it and the accelerations produced a_2, a_3, etc. have a constant ratio to the corresponding force. Thus it is observed from this experiment that

$$\frac{F_1}{a_1} = \frac{F_2}{a_2} = \frac{F_3}{a_3} \; .... = \frac{F}{a} = \text{Constant} \qquad .... (7.1)$$

This proves that the unbalanced force on a given particle is proportional to the acceleration induced in the particle in the direction of the unbalanced force.

The constant of proportionality for force and acceleration is a characteristic of the particle under consideration. The constant is denoted by *m* and is known as mass of the particle.

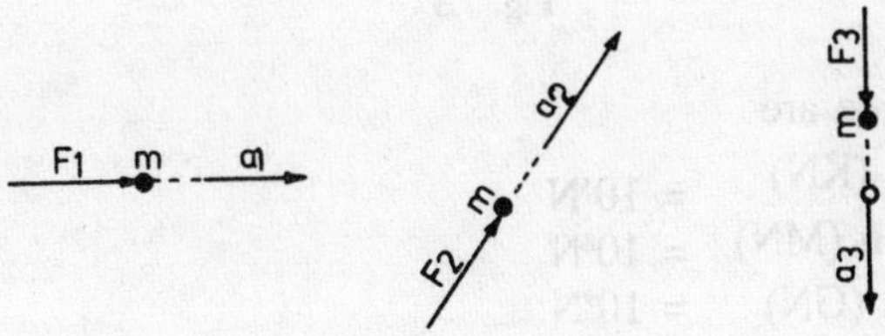

Fig. 7.2

7.3 Equation of Motion

7.3.1. *Derivation of Equation of Motion*

When a force *F* is acting on a particle of mass *m*, the acceleration '*a*' produced must satisfy Newton's second law of motion.

i.e. $F \propto a$ or $F = (\text{Constant})\ a$

or $F = ma$ ....(7.2)

7.3.2. *Units of Mass and Force*

We have seen that the mass of a body is a quantitative measure of inertia, and represents the matter content in it. It is the reluctance of the body to accelerate. Mass is also a property of a body by virtue of which it is always accompanied by a mutual attraction to other bodies. The SI unit of mass is the kilogramme (kg) and the international proto type of this particular quantity of matter is in the custody of the Bureau International des poids et Measures in Paris. Other units of mass are:

1 Tonne (t) $= 10^3$ kg.
1 Gramme (g) $= 10^{-3}$ kg.
1 Milligram (mg) $= 10^{-6}$ kg.

The basic SI unit of force is the Newton (N). It is defined as that force which, when applied to a body of mass one kilogramme, produces an acceleration of one metre/sec/sec in the body. (Ref. Fig. 7.3)

Thus from equation of motion

$$F = ma$$

$$1 \text{ (Newton)} = 1 \text{ (kilogramme)} \times 1 \text{ (metre/sec}^2\text{)}$$

$$1 \text{ N} = 1 \text{ kg} \times 1 \text{ ms}^{-2}$$

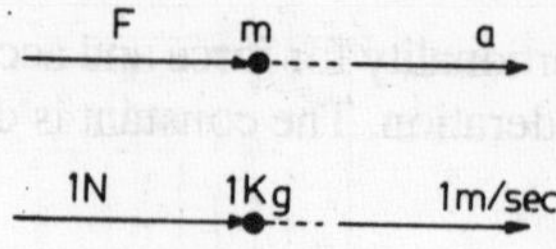

Fig. 7.3

Other units of force are

1 kilonewton (KN) $= 10^3$N
1 meganewton (MN) $= 10^6$N
1 giganewton (GN) $= 10^9$N

Force is also measured in gravitational units expressed as kilogramme force. When acceleration due to gravity acts on a body of mass one kilogramme, the force experienced by the body is one kilogramme force (weight).

$$1 \text{ kgf} = g \text{ Newton} = 9.8 \text{ N}$$

7.3.3. *General Equation of Motion*

Instead of a single force, if the particle is subjected to a number of forces having different magnitudes and direction, then the equation of motion may be written as

$$\Sigma F = m \Sigma a$$

where ΣF represents vector sum of all the forces (i.e. the resultant) acting on the particle having a mass m and the resultant accelerations Σa.

When the forces and accelerations are described by rectangular coordinates x, y, z. Equations of motion may be written as:

$$\left.\begin{aligned} F_x &= m\,a_x \\ F_y &= m\,a_y \\ F_z &= m\,a_z \end{aligned}\right\} \quad \text{....(7.3)}$$

where

$$\Sigma F = \sqrt{(F_x)^2 + (F_y)^2 + (F_z)^2}$$

and $\Sigma a = \sqrt{a_x^2 + a_y^2 + a_z^2}$

For a plane curvilinear motion of a particle, where forces and accelerations are described by normal and tangential components, a_n and a_t, the equations of motion may be written as

$$\left.\begin{aligned} F_n &= m\,a_n = m \times \frac{V^2}{r} \\ F_t &= m\,a_t = m \times \frac{dv}{dt} \end{aligned}\right\} \quad \text{... (7.4)}$$

so,

$$\Sigma F = \sqrt{(F_n)^2 + (F_t)^2}$$

and $\Sigma a = \sqrt{a_n^2 + a_t^2}$

Thus the general form of equation of motion is

$$\Sigma F = m \Sigma a$$

If the resultant force ΣF acting on a particle is zero, then from general equation of motion, we have

$$m \Sigma a = 0$$

since mass m cannot be zero

therefore $\Sigma a = 0$

Thus, if a particle is initially at rest, it will remain at rest; and if it is moving, then it will continue to move with a constant velocity. From this, we see that Newton's first law of motion is a particular case of Newton's second law of motion.

7.4. Linear Momentum of a Particle

Consider a body of mass m acted upon by an average force F for a time t. The average acceleration a is given by

$$a = \frac{v_f - v_i}{t}$$

where v_i and v_f are initial and final velocities respectively

$$\therefore F = ma$$

$$= m \frac{v_f - v_i}{t}$$

$$= \frac{mv_f - mv_i}{t}$$

The product of mass and velocity ($m \times v$) is a measure of the quantity of motion called the *momentum* of the body. It is a vector quantity. The term mv_i represents the momentum of the body at the beginning, and mv_f represents the final momentum at the end of the time t.

Since $F = \dfrac{mv_f - mv_i}{t} = \dfrac{m(v_f - v_i)}{t} = m\dfrac{dv}{dt}$

$$\therefore \quad F = \frac{d\,(mv)}{dt} \quad \text{... (7.5)}$$

Therefore Newtons second law of motion may be restated in the following form

"Change of linear momentum per unit time is proportional to the impressed force, and takes place in the direction of the force."

7.5. Impulse and Momentum

We have seen that the impressed force is equal to the rate of change of linear momentum.

$$\therefore \quad F = \frac{d\,(mv)}{dt} = \frac{(mv_f - mv_i)}{t}$$

$$\text{or } F \times t = mv_f - mv_i \qquad \text{... (7.6)}$$

We now define the impulse of the force F as follows:

"The impulse of a force on a body in a given time is equal to the product of the average force and the time during which it acts."

$$\therefore \quad \text{Impulse} = F \times t$$

$$= mv_f - mv_i$$

The first term mv_f on the right hand side of equation is the final momentum of the body at the end of time t; the second term mv_i is the initial momentum; the difference is the change of momentum. Thus, the Impulse $F \times t$ may be measured by the change in momentum it produces.

i.e. Impulse = Change of Momentum

The idea of impulse is useful in solving two particular type of problems:

(a) When the force F is not known, or cannot be measured easily, e.g. in a sudden blow of hammer striking a nail. In this case the duration of actual impact is very-very small, and hence the force during the impact will be large.

(b) Where the impulsive force F is so large that the body to which it is applied suffers permanent deformation. Some energy is lost in producing this deformation (in the form of heat, etc.) it is not easy to calculate exactly the magnitude of this energy loss, hence, the law of conservation of energy cannot be used in solving these type of problems.

The unit of momentum in S.I. system is kg-m/s. The impulse is equal to the change of momentum, and therefore the unit of impulse is also kg-ms^{-1}. However, impulse $F \times t$, is important where a large force acts for a very short time. Unit of impulse is also given as Newton-seconds (Ns) to emphasise the force and time elements in an impulse. Note that

$$1 \text{ N-s} = 1 \text{ (kg} \times \text{m/s}^2\text{) s}$$

$$= 1 \text{ kg} \times \text{m/s} = 1 \text{ kg-}ms^{-1}$$

$$\therefore 1 \text{ Newton–second} = 1 \text{ kg} - \text{metre/second}$$

7.5.1. *Thrust of a Jet*

The thrust of a jet depends on the rate of change of momentum given to the jet fluid. Let v_1 be the jet fluid velocity relative to the engine at entry, v_2 be the velocity of the jet relative to the engine at exit. Then thrust F = rate of change of momentum.

or $\quad F = m\,(v_2 - v_1)$

where $\quad m$ = mass of fluid flow kg. per second in the jet.

$(v_2 - v_1)$ = Change in the velocity in passing through the engine in ms^{-1}.

F is force in newtons (N)

Directions of thrust and velocities are as indicated in Fig. 7.4.

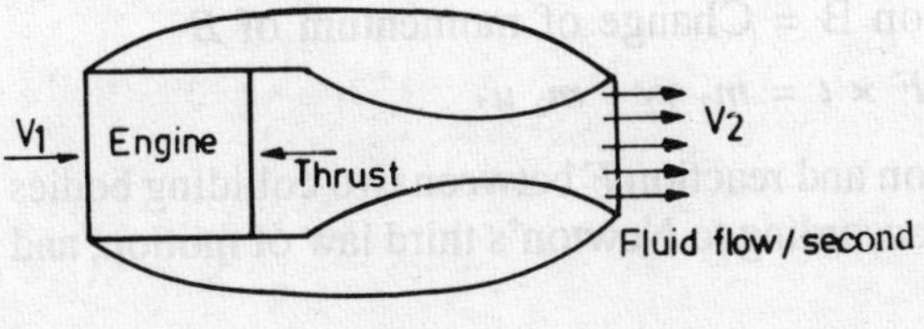

Fig. 7.4

Since we are concerned only with the change in fluid velocity v_1 and v_2, may be either both relative to the engine or both relative to the earth. If v is the velocity of the engine in ms^{-1}, then the power developed at the jet is

Power = Thrust × velocity

$$P = \frac{F \times v}{1000}\ k.W. \qquad \text{... (7.8)}$$

$(1N\ ms^{-1} = 1\ \text{watt})$

7.6. Conservation of Linear Momentum

'If no external force is applied to a body, or a system of bodies, then the total linear momentum of the body or the system remains constant both in magnitude and direction.' This is known as principle of conservation of momentum.

7.6.1. *Collision of Bodies*

Let the bodies A and B have mass m_1 amd m_2 respectively. They are moving with respective velocities of u_1 and u_2 before impact, and v_1 and v_2 after impact as shown in Fig. 7.5.

During the collision, there is an impulse $F \times t$ exerted by A on B. The time t of the impact is very short, and hence the impulsive force is very large.

Therefore the change of momentum due to all other forces external to the system (gravity etc.) may be neglected.

From the law of conservation of momentum, the momentum of the system before collision must be equal to the momentum of the system after collision. This can be proved as follows.

The impulse on the body B is measured by the change of momentum of the body, i.e.

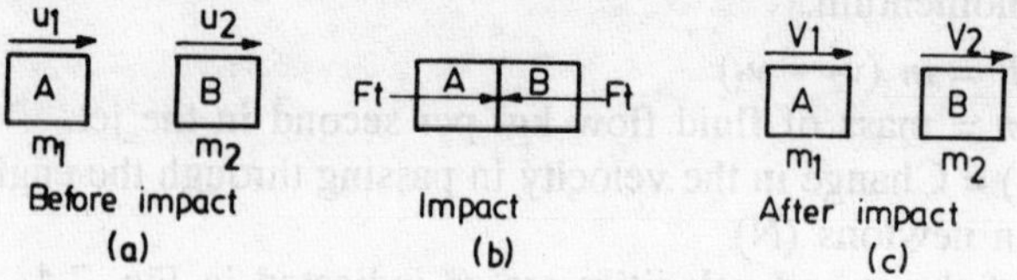

Fig. 7.5

Impulse on B = Change of momentum of *B*

$$F \times t = m_2 v_2 - m_2 u_2$$

Now action and reaction *F* between two colliding bodies A and B is equal in magnitude according to Newton's third law of motion, and acts for the same time *t*.

Impulse on body *A* will also be $F \times t$, but in opposite direction as shown in Fig. 7.5 (b).

Impulse on A = Change of momentum of A

$$-F \times t = m_1 v_1 - m_1 u_1$$

or $$F \times t = -(m_1 v_1 - m_1 u_1)$$

since $$F \times t = -(m_1 v_1 - m_1 u_1) = m_2 v_2 - m_2 u_2$$

$$\therefore \quad m_1u_1 + m_2u_2 = m_1v_1 + m_2v_2 \qquad \text{... (7.9)}$$

Therefore, the total momentum of the system of two bodies taken together is unchanged, although each body seperately suffers a change of momentum of equal magnitude, but opposite sense.

Suppose *A* and *B* represent two railway wagons and initially they are moving with velocities u_1 and u_2 respectively. After collision, suppose they are coupled and they move together in the same direction with a velocity *v*, then according to the law of conservation of momentum

$$m_1u_1 + m_2u_2 = (m_1 + m_2)v \qquad \text{... (7.10)}$$

Although there is no loss of momentum, there is still a loss of kinetic energy, as illustrated in the solved examples later.

Solved Examples

Example 7.1

Find the acceleration produced when a force of 20 N acts on a mass of 8 kg,

Solution

$F = 20$ N

$m = 8$ kg

$a = ?$

$$F = ma$$

$$20 = 8 \times a$$

$$a = \frac{20}{8} = 2.5 \text{ metres/sec}^2$$

$\therefore$ Acceleration produced = 2.5 ms^{-2}

Example 7.2

A block of mass 20 kg is placed on a horizontal table. The coefficient of friction between the block and the table is 0.35. If a force of 200 N is acting on the block as shown in Fig. 7.6 find the acceleration produced in the block.

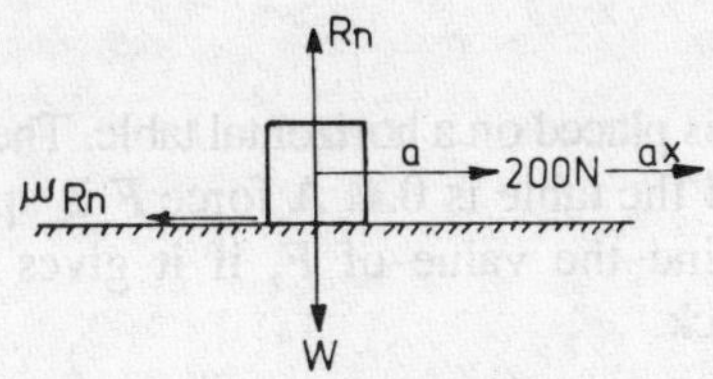

Fig. 7.6

Solution

Unbalanced force $F = 200 - \mu R_n$, where R_n is normal reaction

$$m = 20 \text{ kgs}$$

$$a = ?$$

We know that $R_n = mg = 20 \times 9.8 = 196$ N

$$F = ma$$

$$200 - 0.35 \times 196 = 20 \times a$$

or $$a = \frac{131.4}{20} = 6.57 \text{ m/sec}^2$$

$\therefore$ Acceleration produced = 6.57 metres/sec²

Example 7.3

A motorised boat of mass 20 tonnes drifts in towards a dock at a speed of 2 metres/sec after its engine has been cut off. If the resistance offered by the water is 50 N per tonne, find the distance the boat will travel before coming to rest.

Solution

Resistance offered by the water = 50 × 20 = 1000 N

Mass of the boat = 20 × 1000 kg.

Force = mass × acceleration

$$-1000 = 20 \times 1000 \times a$$

$$a = -\frac{1}{20} \text{ m/sec}^2$$

This acceleration will be in the direction of resistance, i.e. opposite to the direction of motion.

Initial velocity $v_i = 2$ metres/sec

Final velocity $v_f = 0$

Let the distance travelled before coming to rest be 's',

then
$$v_f^2 = v_i^2 + 2\,as$$
$$0 = 2^2 - 2 \times \frac{1}{20} \times s$$
$$s = \frac{4 \times 20}{2} = 40 \text{ metres}$$

∴ The distance travelled = 40 ms.

Example 7.4

A block of mass 50 kg is placed on a horizontal table. The coefficient of friction between the block and the table is 0.4. A force F is applied on the block as shown in Fig. 7.7. Find the value of F, if it gives an acceleration of 2 metres/sec^2 to the block.

Fig. 7.7

Solution

Let R be the normal reaction offered by the table. Free body diagram of the block is shown in Fig. 7.7 (b). Resolving vertically, we get

$$R = 50\text{ g} + F \sin 45 \qquad \text{....(1)}$$

Resultant horizontal force causing acceleration

$$= F \cos 45 - \mu R$$
$$= F \cos 45 - \mu\,(50\text{ g} + F \sin 45)$$

Force = mass × acceleration

$$\therefore \quad F \cos 45 - \mu\,(50\text{ g} + F \sin 45) = 50 \times 2$$

or
$$F \times 0.707\,(1 - 0.4) = 100 + 196$$
$$F = \frac{296}{0.707 \times .6} = 697.8\text{ N}$$

∴ The force F = 697.8 N

Example 7.5

The bob of a pendulum describes an arc of circle in a vertical plane. The distance between the point of suspension and the centre of the bob is 1 metre. If the tension in the cord is 2 times the weight of the bob for the position shown in Fig. 7.8, find the velocity and acceleration of the bob in that position.

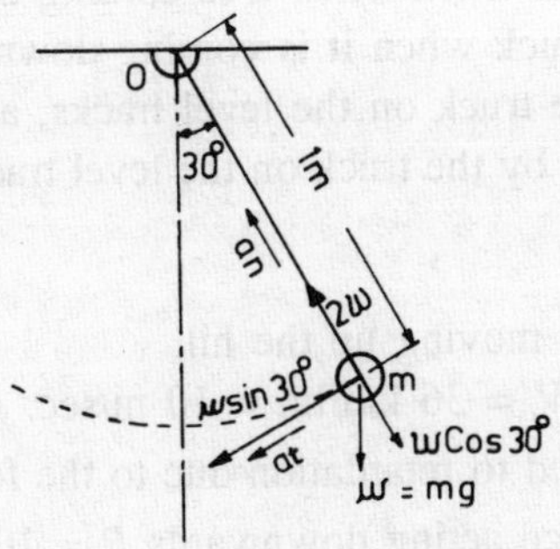

Fig. 7.8

Solution

The mass of the bob is $m = \frac{W}{g}$. Directions of normal and tangential acceleration are shown in Fig. 7.8. Applying Newton's second law of motion, we have

$$F_t = m \times a_t \qquad g = 9.80 \text{ ms}^{-2}$$

$$W \sin 30 = \frac{W}{g} \times a_t$$

$$a_t = g \sin 30 = \frac{9.80}{2} = 4.900 \text{ metres/sec}^2$$

$$F_n = m \times a_n$$

$$2W - W \cos 30 = \frac{W}{g} \times a_n \text{ (as tension } T = 2W \text{ given)}$$

$$a_n = (2 - .866)\, g = 11.11 \text{ metres/sec}^2$$

$$\text{Since } a_n = \frac{v^2}{r}$$

$$\therefore\ 11.11 = \frac{v^2}{1} \text{ or } v = \sqrt{11.11} = 3.33 \text{ metres/sec}$$

(i) Tangential acceleration $= a_t = 4.90$ metres/sec²

(ii) Normal acceleration $= a_n = 11.11$ metres/sec²

(iii) Velocity $= V = 3.33$ metres/sec

Example 7.6

A truck of mass 1.5 tonnes is moving on a level track and then climbs a hill, having an inclination of 30° with the horizontal as shown in Fig. 7.9. At 50 metres up the inclined road, when it is moving at 36 km/h, the engine stops working and brakes are not applied; track resistance is 100 N per tonne. Find the following:

(a) Distance travelled by the truck up the hill after the engine stops,

(b) Acceleration of the truck when it is coming down,
(c) Velocity of the truck when it is coming down,
(d) Retardation of the truck on the level tracks, and
(e) Distance travelled by the truck on the level track before coming to rest.

Solution

(a) When the truck is moving up the hill
velocity of truck V_i = 36 km/hr. = 10 m/sec.

The vehicle is subjected to retardation due to the following resisting forces

(i) Frictional resistance acting downwards $R = 100 \times 1.5 = 150$ N

(ii) Component of the weight along the hill acting downwards

$$= 15 \times 1000 \times 9.8 \sin 30 = 7350 \text{ N}$$

Total resistance against the upward motion of the truck

$$= 150 + 7350 = 7500 \text{ N}$$

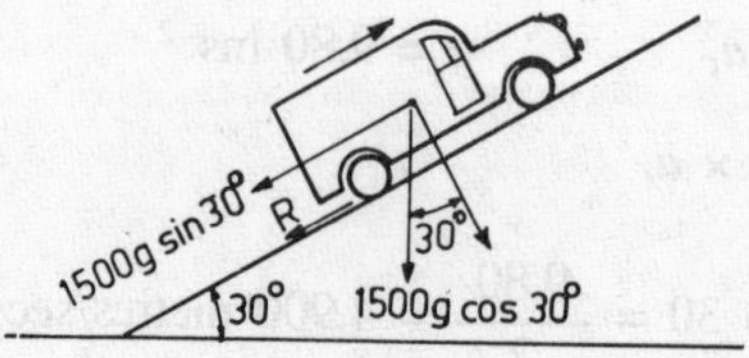

Fig. 7.9

Let the retardation be a ms^{-2}; since Force = mass × acceleration

$$-7500 = 1.5 \times 1000 \times a$$

$$a = -\frac{7500}{1500} = -5 \text{ ms}^{-2}$$

when the truck reaches the highest point along the slope,

final velocity $v_f = 0$

Let the distance travelled by the truck up the hill after the engine stops be s

then $\quad V_f^2 = V_i^2 + 2as$

or $\quad 0 = 10^2 - 2 \times 5 \times s$

$\therefore \quad s = 10$ ms.

(b) When the truck is moving down the hill the component of the weight along the hill will be acting downwards, and frictional resistance will be acting upwards opposite to the direction of motion. Therefore,

Net downward force = $1500 \times 9.8 \times \sin 30 - 100 \times 1.5$

$$= 7350 - 150 = 7200 \text{ N}$$

Let the acceleration of the truck down the hill be a_1 ms^{-2}

$\therefore\ 7200 = 1.5 \times 1000 \times a_1$

$$a_1 = \frac{7200}{1500} = 4.8 \text{ ms}^{-2}$$

(c) Initial velocity = 0, when the truck starts backwards. Let the velocity of the truck when it reaches level track = v_f

Distance travelled $s = 50 + 10 = 60$ metres

$$V_f^2 = V_i^2 + 2\, a_1\, s$$
$$V_f^2 = 0 + 2 \times 4.8 \times 60$$
$$V_f = 24 \text{ metres/sec.}$$

(d) Frictional resistance to the motion = 100 × 1.5 = 150 N. Let the retardation of truck on the level track be a_2 ms^{-2}

Then $\quad -150 = 1500 \times a_2$

$$a_2 = -\frac{150}{1500} = -\frac{1}{10} \text{ m/sec}^2 = 0.1 \text{ ms}^{-2}$$

(e) Initial velocity V_i = 24 metres/sec
Final velocity $V_f = 0$

Retardation $\quad a_2 = -\dfrac{1}{10}$ ms^{-2}

Let the distance travelled by the truck on the level road before coming to rest be s

then $\quad V_f^2 = V_i^2 + 2\, a_2\, s$

$$\text{or } 0 = 24^2 - 2 \times \frac{1}{10} \times s$$

$$\therefore \quad s = 2880 \text{ metres}$$

Example 7.7

Two particles of masses m_1 and m_2 are connected by a light inextensible cord passing over a smooth fixed guide pulley as shown in Fig. 7.10. If m_1 is greater then m_2 find tension T in the string and the acceleration a of the system.

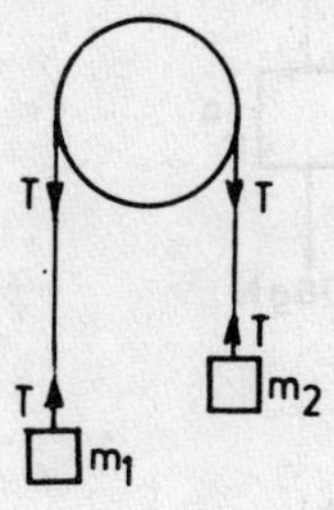

Fig. 7.10

Solution

Since the pulley is smooth, tension T in the string is the same throughout.

Consider the motion of mass m_1

Net downward force on the mass = $(m_1\ g - T)$

$$\therefore \quad (m_1\ g - T) = m_1\ a \qquad \text{....(i)}$$

Now consider the motion of mass m_2

Net upward force on the mass = $(T - m_2\ g)$

$$\therefore \quad (T - m_2\ g) = m_2\ a \qquad \text{....(ii)}$$

Adding equations (i) and (ii), we have

$$(m_1\ g - m_2\ g) = m_1\ a + m_2\ a$$

or
$$a = \frac{(m_1 - m_2)\ g}{(m_1 + m_2)}$$

$$\therefore \text{ Acceleration } a \text{ of the system } = \frac{(m_1 - m_2)\ g}{(m_1 + m_2)}$$

But
$$T - m_2\ g = m_2\ a$$
$$\therefore \quad T = m_2\ (g + a)$$

Substituting for a, we get

$$T = m_2 \left[\frac{m_1\ g + m_2\ g + m_1\ g - m_2\ g}{m_1 + m_2}\right]$$

$$\therefore \text{ Tension in the string } = T = \frac{2\ m_1\ m_2\ g}{(m_1 + m_2)}$$

If
$$m_1 = \frac{W_1}{g}\ \&\ m_2 = \frac{W_2}{g},\ T = \frac{2\ W_1\ W_2}{(W_1 + W_2)}$$

Example 7.8

Two blocks of masses 20 kg and 15 kg respectively are connected by a light inextensible string which passes over a 200 mm diameter pulley of mass 2 kg as shown in Fig. 7.11. The radius of gyration of the pulley may be assumed equal to its radius. Neglecting friction, determine the acceleration of the system, and the tension in the string when the weights are released from rest.

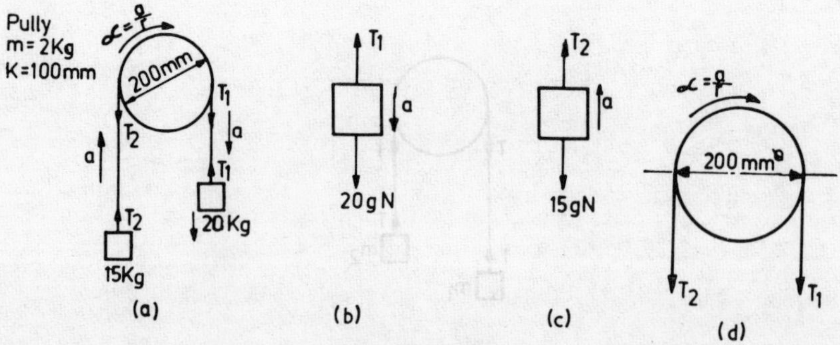

Fig. 7.11

Solution

Let us first consider the 20 kg mass

from Newton's second law

$$F = m\,a$$

$$20\,g - T_1 = 20\,a \qquad \text{....(i)}$$

Now consider the 15 kg mass

$$(T_2 - 15\,g) = 15\,a \qquad \text{....(ii)}$$

Now consider the pulley

Resultant torque on the pulley $= T = T_1\,r - T_2\,r$

$$\therefore\; T = (T_1 - T_2) \times 0.1$$

Moment of inertia I of the pulley $= mk^2$ (where k = radius of gyration)

$= 2 \times (0.1)^2$

$= 0.02$ kgm^2

Angular acceleration of the pulley

$$\propto = \frac{a}{r} = \frac{a}{0.1}$$

from Newton's second law

$$T = I \cdot \propto \qquad \text{(Net torque = } M.I. \times \text{ Angular acc.)}$$

$$(T_1 - T_2)\,0.1 = 0.02 \times \frac{a}{0.1}$$

$$\therefore\; T_1 - T_2 = 2\,a \qquad \text{....(iii)}$$

Form (i) $T_1 = 20\,g - 20\,a$

From (ii) $T_2 = 15\,a + 15\,g$

substituting these values in (iii), we get

$$20g - 20a - (15a + 15g) = 2\,a$$

or $37\,a = 5\,g$

$$\therefore\; a = \frac{5 \times 9.8}{37} = 1.324 \text{ ms}^{-2}$$

substituting value of a in equation (i)

$$20\,g - T_1 = 20 \times 1.324$$

$$\therefore\; T_1 = 169.52 \text{ N}$$

From (iii) $169.52 - T_2 = 2 \times 1.324$

$$\therefore\; T_2 = 166.87 \text{ N}$$

Acceleration of the system $= a = 1.324$ ms/sec^2.

Tension in the string attached to the 20 kg mass = 169.52 N

Tension in the string attached to the 15 kg mass = 166.87 N

Example 7.9

A machine gun fires 200 bullets/min. The muzzle velocity of the bullet is 600 m/sec. The mass of one bullet is 40 gms. Determine the average reaction of the gun against the support if firing is continuous.

Solution

Time taken in firing one bullet $t = \dfrac{60}{200} = 0.3$ sec.

Let F be the average reaction of the gun, in Newtons, against the support. This is also numerically equal to the reaction of gun on the bullet.

The impulse of the bullet $= F \times t$

Change in momentum of the bullet $= m(v_f - v_i)$

$$= \frac{40}{1000} \times (600 - 0)$$

From Newtons second law of motion, the impulse of a force is equal to change in momentum of the body,

$$\therefore \quad F \times t = \frac{40}{1000} \times 600$$

$$\therefore \quad F = 80 \text{ Newtons}$$

Example 7.10

A wagon 'P' of mass 100 tonnes moving at 4 km/h collides with the back of a wagon '*Q*' of mass 40 tonnes, moving in the same direction at 1 km/h. What is the velocity of the wagon '*P*' after the impact and the impulse between them, if the wagon '*Q*' moves off at 6 km/h after the impact?

Solution

Momentum before impact = Momentum after impact

$$100 \times 4 + 40 \times 1 = 100 \times V_p + 40 \times 6$$

$$V_p = \frac{440 - 240}{100} = 2 \text{ Km/h}$$

The impulse on each wagon is equal to the change in momentum. Therefore, for wagon P,

$$\text{Impulse} = mp\ (V_f - V_i)$$

$$= 100 \times 1000\ (2 - 4)\ \frac{1}{3.6}$$

$$(\because 3.6 \text{ kmh}^{-1} = 1 \text{ ms}^{-1})$$

$$= -\ 55555.5 \text{ kgms}^{-1}$$

Exampe 7.11

An elastic pile of mass 1.5 tonnes is being driven in the ground. A hammer of mass 200 kg is dropped freely through 2 metres on to the head of the pile. If the pile is driven 5 cm into the ground in one blow, what is the average resistance of the ground and how much energy is lost in the impact?

Solution

Mass of the pile $M = 1.5$ tonnes $= 1500$ kg

Mass of the hammer $m = 200$ kg

Height through which the hammer is dropped, h = 2 ms

Kinetic energy of hammer just before impact = Potential energy of the hammer

$$\frac{1}{2}\text{mv}^2 = mgh$$

or $$v = \sqrt{2\,gh} = \sqrt{2 \times 9.8 \times 2} = 6.26\ \text{ms}^{-1}$$

Let the velocity of pile with hammer be V after impact

Momentum before impact = Momentum after impact

$$m \times v = (M + m)V$$

or $$200 \times 6.26 = (1500 + 200)\ V$$

$$\therefore \quad V = \frac{200 \times 6.26}{1700} = 0.737\ \text{m/sec}$$

Therefore KE of the system after impact = $\frac{1}{2}(M + m)\ V^2$

or $$\text{KE} = \frac{1}{2}(1500 + 200)\ 0.737^2 = 460\ \text{Joules}$$

Loss of potential energy in descending through a distance x

$$= (M + m)\ gx$$

or $$\text{PE} = (1500 + 200) \times 9.8 \times .05 = 833\ \text{Joules}$$

Work done by hammer and pile = KE Lost + PE Lost

= 460 + 833

=1293 Joules

But work done against the ground resistance R is $R\ x$ 0.05

$\therefore \quad .05\ R = 1293$

$\therefore \quad R = 25860$ N

$\therefore$ Average resistance of the ground = 25860 Newtons

$$\text{Loss of KE at impact} = \frac{1}{2}\text{mv}^2 - \frac{1}{2}(M + m)\ V^2$$

$$= \frac{1}{2} \times 200 \times (6.26)^2 - \frac{1}{2}(1500 + 200) \times .737^2$$

$$= 3458\ \text{Joules}$$

7.7. D' Alembert's Principle

Newton's second law of motion is used to solve the problems in Dynamics by relating external forces acting on a body to the acceleration produced as a result of these forces.

An alternative approach for solution of such problems involving the relationship between forces and accelerations is contained in D' Alembert's Principle which was enunciated in 1743 by D' Alembert. In brief the principle permits the solution of Dynamics problems by the method of Statics. If the basic equation of motion of a particle of mass 'm' is rewritten as $\Sigma F - ma = 0$, it takes the form of an equation of force equilibrium. Then, if an additional force equal to '*ma*' is applied to the rigid body in a direction opposite to the direction of acceleration, the body could then be considered to be in equilibrium under the action of the real forces ΣF, and the additional force '*ma*' added to the system. This additional force '*ma*' is called *inertia force* and it acts in a direction opposite to the direction of acceleration. When inertia force is applied to a body, the body comes in a state of equilibrium, and this artificial state of equilibrium is known as *dynamic equilibrium*. The apparent transformation of a problem in Dynamics to one in Statics by introducing inertia force in a direction opposite to the direction of acceleration is known as D' Alembert's Principle.

7.8. Circular Motion

Let a body of mass '*m*' be attached at the end of string and be rotated about the centre 'O' with angular velocity ω and radius *r*. In chapter of Kinematics we have seen that the acceleration of mass *m* due to its rotation is directed radially inward from A to O and its value is (see Fig. 7.12).

$$a_n = \omega^2 r = \frac{V^2}{r} \qquad \text{....(7.11)}$$

where ω = angular velocity

and V = linear velocity

This acceleration is known as centripetal acceleration and it can take place only if there is a force acting in the direction A to O. This force is causing the body to move in a circular path and is known as the centripetal force. In the present case the force is supplied due to tension in the string acting radially inwards at its connection with the body at A as shown in the Fig. 7.12 (a)

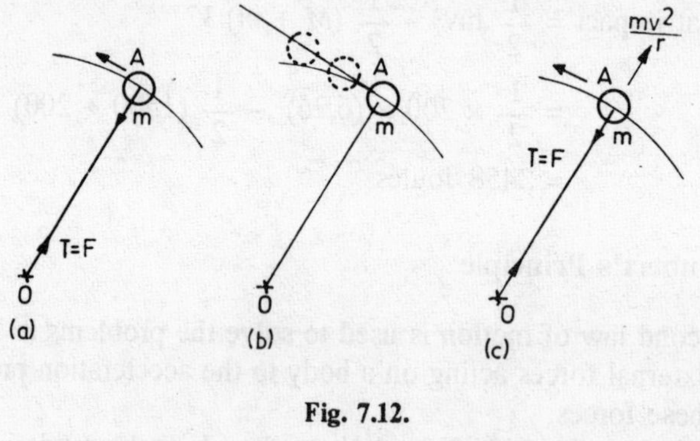

Fig. 7.12.

Magnitude of this centripetal force F is given by

$$F = m \times a_n$$

or

$$F = m \times \omega^2 r = m \times \frac{V^2}{r} \qquad \text{....(7.12)}$$

For the balance of forces acting on the body A, the centripetal force F may be considered as being in equilibrium with an equal and opposite inertia force of magnitude $m\,\dfrac{V^2}{r}$ as shown in the Figure 7.12 (c). This inertia force is acting radially outward, and is known as centrifugal force. This centrifugal force is a reactive force, since it cannot cause the motion by itself. For example, if the string is cut, the active force F due to tension in the string disappears and the body A moves in a straight line tangential to the circular path as shown in Fig. 7.12 (b). It does not fly radially outwards.

7.8.1. *Super Elevation of Tracks*

The super elevation of a railway track is the amount by which the outer rail is raised above the level of the inner rail. The wheels of a train are having flanges, and these flanges are provided in such a way that they remain on the inside of the rail. As the train moves round a curved track, the centripetal force required to give the circular motion is provided by the inward thrust F of the outer rail as shown in Fig. 7.13. To reduce the magnitude of the lateral load, a second rail may sometimes be provided on inside of curve so that the inner wheel flange is contained between two rails. This second rail is provided to take some of the lateral thrust. This side thrust or the lateral load can be eliminated completely at a particular speed by suitable banking of the track. The speed chosen is usually the average speed at which a train may be expected to take the curve. If actual speed of the train is higher than the chosen one, there will be a side thrust on the outer rail; at lower speeds, there will be a side thrust on the inner rail.

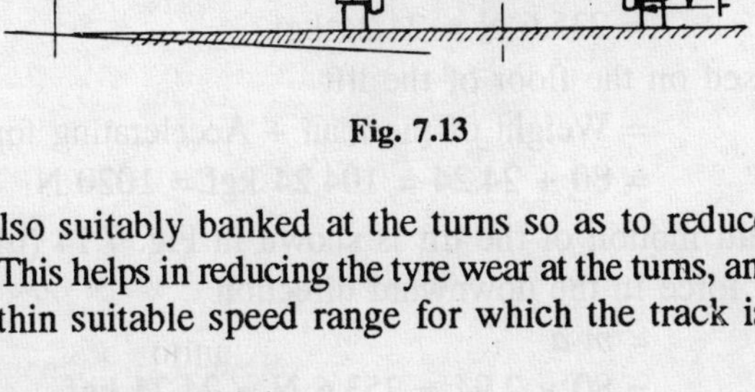

Fig. 7.13

Roads are also suitably banked at the turns so as to reduce the side thrust on the tyres. This helps in reducing the tyre wear at the turns, and it also prevents skidding within suitable speed range for which the track is designed.

7.8.2. *Overturning of Vehicles*

We have seen in the previous article that a train has a tendency to slide outwards when rounding a curve, and it is prevented by providing flanges on the wheels. There is a limiting speed at which sliding just occurs if flanges are not there. There is also a limiting speed at which the vehicle will overturn. As is clear from the Fig. 7.13, the inertia force $m\,\omega^2\,r$ and the side thrust on the rails forms the turning couple, whereas the weight of the train and the ground reaction acting vertically upward forms the restoring couple. As the speed of train (ω increases, value of turning couple ($m\omega^2 r\cdot a$) also increases, and at some limiting speed it will be just equal to the restoring couple. If the speed of the train is increased still further, it will overturn.

Example 7.12

A lift has an upward acceleration of 2.94 metres/sec². What pressure will a man of mass 80 kg exert on the floor of the lift? What pressure would he exert if the lift had an acceleration of 2.94 metres/sec² downwards? What upward acceleration would cause his mass to exert a pressure of 90 kgf on the floor?

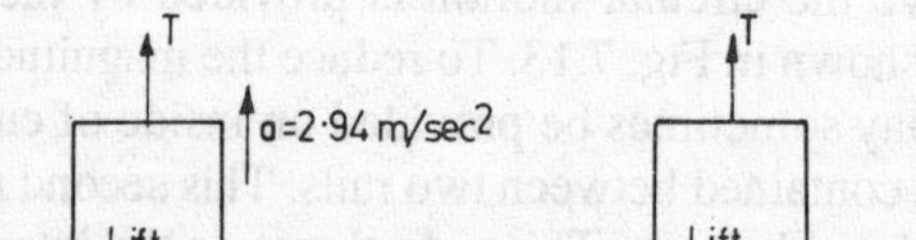

Fig. 7.14

Solution

(a) Upward motion of the lift is shown in Fig. 7.14 (a)

m = 80 kgs

a = 2.94 metres/sec²

Accelerating force = $m\,a$

= 80 × 2.94

= 235.6 N = 24.24 kgf

Pressure caused on the floor of the lift

= Weight of the man + Accelerating force

= 80 + 24.24 = 104.24 kgf = 1020 N

(b) The downward motion of the lift is shown in Fig. 7.14 (b)

Accelerating force in the downward direction

= $m\,a$

= 80 × 2.94 = 253.6 N = 24.24 kgf

Pressure caused on the floor of the lift during

downward motion = Weight of the man – Accelerating force in the downward direction

$$= 80 - 24.24 = 55.76 \text{ kgf}$$
$$= 548 \text{ N}$$

(c) Let the upward acceleration a_1 cause the pressure of 90 kgf on the floor; therefore $90 \times 9.8 = 80 \times 9.8 + 80 \times a_1$

$$\therefore \quad a_1 = \frac{10 \times 9.8}{80} = 1.226 \text{ metres/sec}^2$$

Alternative Approach by D'Alembert's Principle.

The pressure caused on the floor during upward motion by the man is equal to the tension in the cable T due to his weight. As the acceleration acts in the upward direction; by applying the inertia force in the downward direction, the lift may be considered in equilibrium by D'Alembert's Principle. The forces acting on the lift under equilibrium conditions are shown in Fig. 7.15 (a).

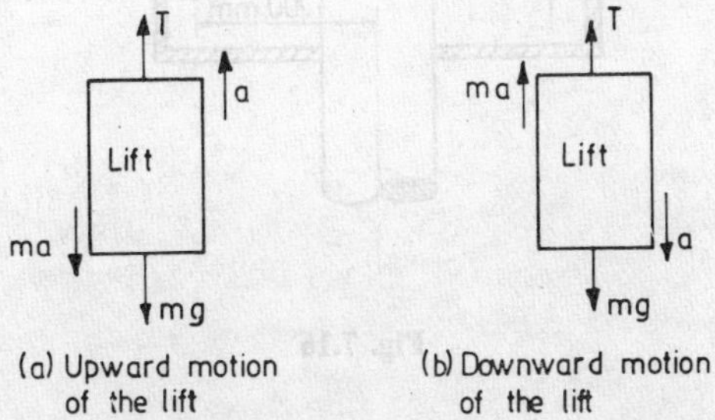

Fig. 7.15

(a) Considering the equilibrium of the forces in vertical direction [see Fig. 7.15 (a)]

$$T = mg + ma$$
$$= 80 \times 9.8 + 80 \times 2.94 = 1020 \text{ N}$$

(b) Considering the equilibrium of the forces in the vertical direction [see Fig. 7.15 (b)]

$$T + ma = mg$$
$$T + 80 \times 2.94 = 80 \times 9.8$$
$$\therefore \quad T = 784 - 236 = 548 \text{ N}$$

$\therefore$ pressure exerted on the floor of the lift during downward motion = 548 N

(c) Let a_1 be the acceleration required to cause a pressure of 90 kgf on the floor, then from Fig. 7.15 (a)

$$T = ma_1 + mg$$
$$90 \times g = 80 \times a_1 + 80 \times 9.8$$

$$\therefore \qquad a_1 = \frac{9.8\,(90 - 80)}{80} = 1.226 \text{ metres/sec}^2$$

∴ Desired acceleration is 1.226 ms^{-2}.

Example 7.13

A centrifugal clutch, in rest position, is shown in Fig. 7.16. The rotating weights are each of mass 1 kg and spring stiffness is 15 kN/m. The centre of mass of each weight is at 200 mm radius in the rest position. Calculate the radial force on the clutch face at 800 rpm.

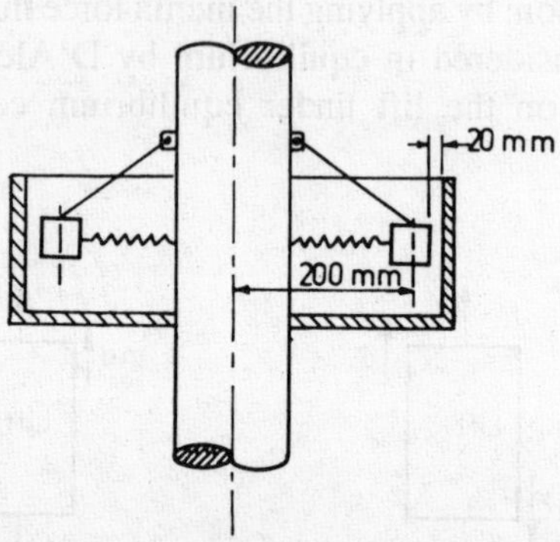

Fig. 7.16

Solution

As the speed of the shaft increases, weights, fly out, until they engage the inside face of clutch drum. Force on each rotating weight is then

(i) Spring force F acting radially inwards

(ii) Inertia force $m\omega^2 r$ acting radially outwards

(iii) Reaction R of the clutch drum on the wt. acting radially inwards.

For equilibrium of forces

$$F + R = m\omega^2 r$$

At the engagement the spring force is

$$F = \text{Stiffness} \times \text{extension}$$

$$= 15 \times 1000 \times 0.020 = 300 \text{ N}$$

Radius of rotation at engagement = 220 mm = 0.220 m

$$\text{Inertia force} = m\omega^2 r = 1 \left(\frac{2\pi \times 800}{60} \right)^2 \times 0.220 = 1560 \text{ N}$$

$\therefore \quad 300 + R = 1560$

or $\qquad R = 1260$ N

Radial force on the clutch face is equal to 1260 N.

Example 7.14

A car of mass 800 kgs traverses an unbanked curve of 150 m radius to 60 kms/h. The wheel base width is 2 metres and the centre of gravity of the car is 500 mm above the road. Calculate the normal reaction at each wheel.

Solution

The forces acting on the car are

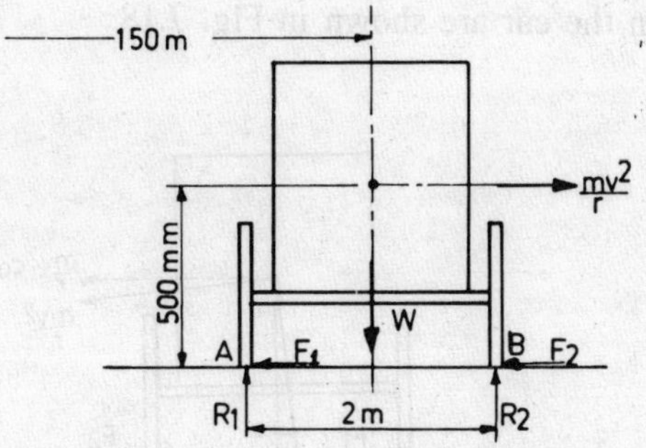

Fig. 7.17

(a) The normal reactions R_1 and R_2 at inner and outer pair of wheels.

(b) The weight $w = mg$, acting vertically downward through the centre of gravity.

(c) The inertia force $\dfrac{mV^2}{r}$, acting radially outward at the centre of gravity.

(d) Frictional forces F_1 and F_2 at each wheel as shown in Fig. 7.17. Taking moment about inner wheel track A

$$R_2 \times 2 = \frac{mV^2}{r} \times .5 + w \times 1$$

$$V = 60 \text{ km/h} = \frac{60 \times 1000}{60 \times 60} = \frac{50}{3} \text{ metres/sec}$$

$$R_2 \times 2 = \frac{800\left(\dfrac{50}{3}\right)^2}{150} \times 0.5 + (800 \times 9.8) \times 1$$

$$\therefore \quad R_2 = 370.3 + 3920 = 4290.3 \text{ N}$$

Hence normal reaction at each outer wheel $= \dfrac{4290.3}{2} = 2145.1$ N

Resolving the forces vertically

$$R_1 + R_2 = W$$

$$R_1 + 4290 = 800 \times 9.8$$

$$\therefore \quad R_1 = 7840 - 4290 = 3550 \text{ N}$$

Hence normal reaction at each inner wheel $= \dfrac{3550}{2} = 1775$ N

Example 7.15
Calculate the angle of banking on a bend of 80 m radius so that car can travel round the bend at 40 km/h without side thrust on the tyres.

Solution
Let the angle of banking be θ
Forces acting on the car are shown in Fig. 7.18

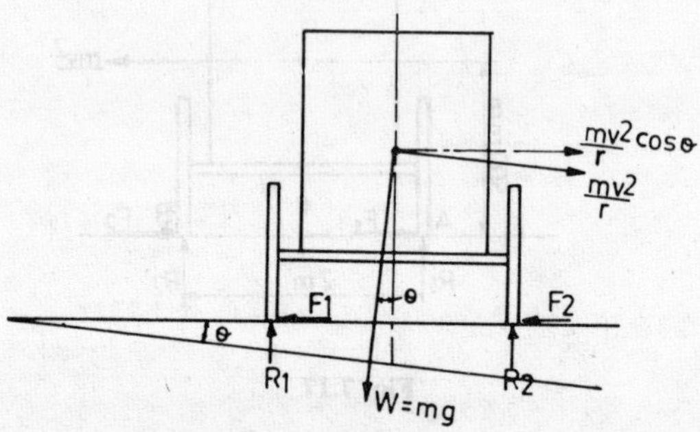

Fig. 7.18

Resolving the forces parallel to the banked plane

$$F_2 + F_1 + W \sin\theta = \frac{mV^2}{r}\cos\theta$$

If there is to be no side thrust, $F_1 = F_2 = 0$

$$\therefore\ mg\sin\theta = \frac{mV^2}{r}\cos\theta$$

or $\quad \tan\theta = \dfrac{V^2}{gr}$

Since $\quad v = 40 \text{ kms/hr} = \dfrac{40\times1000}{60\times60} = \dfrac{100}{9}$ metres/sec

$$\therefore\quad \text{Tan}\,\theta = \frac{\left(\frac{100}{9}\right)^2}{9.8\times80} = 0.158.$$

or $\quad \theta = 9°$

Therefore angle of banking required is 9°.

Summary

Kinetics

Kinetics is the branch of mechanics that deals with the problems which require the determination of the effect of forces on the motion of a body, or conversely the forces causing such motion.

Newton's Second Law of Motion

Newton's Second Law of Motion states that the acceleration of a given particle is proportional to the unbalanced applied force and takes place in the direction of the straight line in which the force acts.

Equation of Motion

$$F = m\,a$$

where F = resultant force on a particle

m = mass of particle

a = resultant acceleration of the particle.

Momentum

The product of mass and velocity ($m \times V$) is a measure of the quantity of motion called the momentum of the body.

Impulse

The impulse of a Force on a body in a given time is equal to the product of the average force and the time during which it acts.

$$\therefore \quad \text{Impulse} = F \times t = mV_f - mv_i$$

Conservation of Linear Momentum

If no external force is applied to a body or a system of bodies then the total linear momentum of the body or the system remains constant both in magnitude and direction. This is known as principle of conservation of momentum.

D Alembert's Principle

The apparant transformation of a problem in Dynamics to one in Statics by introducing inertia force '*ma*' in a direction opposite to the direction of acceleration '*a*', is known as D'Alembert's Principle.

Super Elevation

Super elevation is the raising of outer rail or the far end of the pavement by '*h*' from the inner rail or the near end of the pavement to avoid side thrust on the rails or tyres.

Review Questions

1. State Newton's Second Law of Motion.
2. Define Impulse and Momentum.
3. Explain how D'Alembert's Principle can be used in solving problems in dynamics?

4. Fill in the following blanks:
   (a) Momentum of a body is the product of its mass and ....
   (b) Impulse is the product of force and ....
   (c) In a curved track the amount by which the outer rail is raised above the inner rail is called ....
   (d) If on a given mass m, forces F_1, F_2, F_3 produces acceleration a_1, a_2, a_3, then the ratios $\frac{F_1}{a_1}, \frac{F_2}{a_2}, \frac{F_3}{a_3}$ will always be equal to...........
   (e) A force of 10 N acting on a body of mass 2.5 kgs produces an acceleration of ...., if same force acts on an another body of mass 5 kgs will produce an acceleration of............
5. A ball attached at the end of a string is revolved in a horizontal plane. If the string suddenly breaks, the ball will move.
   (a) towards the centre along radius
   (b) away from the centre along radius
   (c) tangentially
   (d) along a straight line inclined 45° to the tangent
   Tick mark the correct response
6. If a trolley moving with uniform velocity comes under the action of two equal and opposite forces having the same line of action, the trolley will,
   (a) Continue to move with same uniform velocity
   (b) Continue to move with acceleration
   (c) Continue to move with retardation
   (d) Not move at all.
   Tick mark the correct response.
7. Match the units in coloumn *B* with appropriate item under column *A*. One unit from B can go for more than one item under *A*.

Column A	*Column B*
(a) Force ( )	(a) kg
(b) Mass ( )	(b) N
(c) Momentum ( )	(c) m
(d) Impulse ( )	(d) $kgms^{-1}$
(e) Impulsive force ( )	(e) Ns
(f) Weight ( )	(f) Nm
(g) Tractive effort ( )	(g) Nms^{-1}
(h) Tension ( )	

8. Two 10 N weight are supported as shown in Fig. 7.19. The spring balance will indicate
   (i) 10 N
   (ii) 20 N
   (iii) 0 N
9. Equation of motion is given by
   (i) $\Sigma F = W \Sigma a$
   (ii) $\Sigma F = mg \Sigma a$
   (iii) $\Sigma F = m \Sigma a$
   (iv) $\Sigma F = \frac{m}{g} \Sigma a$

   where F = force
   W = weight of body
   m = mass of body
   a = acceleration of the body

   Tick mark the correct response.
10. The Centrifugal force is given by
   (i) $F = \frac{mV}{r}$
   (ii) $F = \frac{mV}{r^2}$

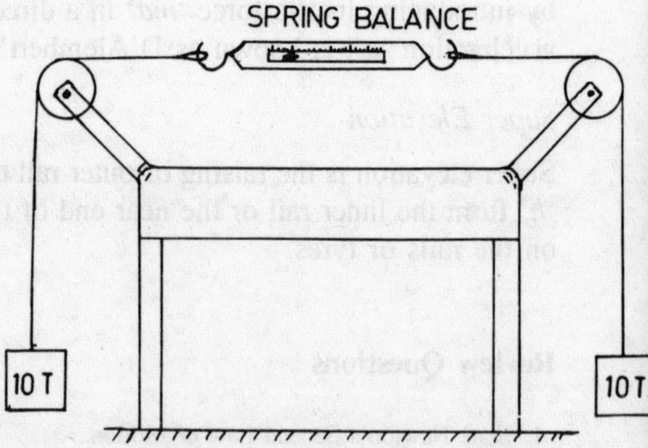

Fig. 7.19

(iii) $F = \dfrac{mV^2}{r}$

(iv) $F = \dfrac{mV^2}{r^2}$

Tick mark the correct response.

11. The law of conservation of linear momentum can be put in the symbolic form as

(i) $m_1u_1 + m_2V_2 = m_1V_1 + m_2u_2$

(ii) $m_1u_1 + m_2u_2 = m_1V_1 + m_2V_2$

(iii) $m_1u_1 + m_1V_1 = m_2u_2 + m_2V_2$

(iv) $m_1V_2 + m_2V_1 = m_1u_2 + m_2u_1$

Tick mark the correct response.

12. Impulse is given by

$$\text{Impulse} = mV_f + mV_i$$
$$\text{Impulse} = F \times t + mV_f + mV_i$$
$$\text{Impulse} = F \times t \times mV_i$$
$$\text{Impulse} = mV_f - mV_i$$

where F = force

t = time

m = mass

V_f = final velocity

V_i = initial velocity

Tick mark the correct response

Problems

7.1 A block of mass 5 kgs is placed on a smooth horizontal plane and is acted upon by a force of 15 N in the horizontal direction. Find the distance travelied by the block in 10 seconds
(Ans. 150 metres).

7.2 A block of mass 10 kgs is placed on a horizontal table. The coefficient of friction between the block and the table is 0.2. If a force of 50 N is acting on the block as shown in Fig. 7.2, find the acceleration produced in the block.

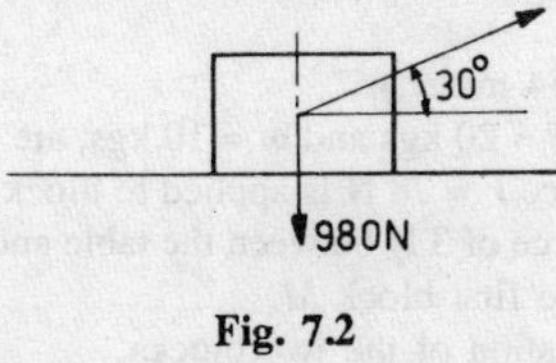

Fig. 7.2

(Ans. 2.37 m/sec²).

7.3 A mass of 50 kgs is acted upon by a constant force which in 5 seconds produces a velocity of 12 m/sec. Find the force if the mass was initially at rest.
(Ans. 120 N).

7.4 A motorised boat of mass 10 tonnes drifts towards a dock at a speed of 0.4 m/sec after its engine has been shut off. A man on the dock pushes it with a force of 500 N to stop it in minimum time. Find the distance the boat will travel before coming to rest.
(Ans. 1.6 metres).

7.5 A motor cycle with its rider has mass of 1/2 tonne. when it is moving at a speed of 36 km/h its engine is cut off. If the ground offers resistance at a rate of 200 N/tonne, find the distance travelled by the motor cycle before coming to rest.
(Ans. 250 metres).

7.6 An observer first measures the velocity of an approaching object to be 10^{-2} m/sec, and then 1 sec later, to be 3×10^{-2} m/sec. If the object has a mass of 5 gm, what conclusion can the observer draw about

(a) the amount of force that has been acting and

(b) the work done by the force.

(Ans. 1×10^{-4} N; 2×10^{-6} Joules).

7.7 A train has a mass of 200 tonnes. If the frictional resistance amounts to 50 N per tonne, what steady pull must the locomotive exert in order to increase the speed on level track from 30 km/hr to 60 km/hr in a period of 90 secs.

(Ans 28.52 kN).

7.8 An electric locomotive together with its train has a mass of 150 tonnes. The average tractive resistance to motion is 80 N per tonne while starting from rest up a gradient of 1 in 100. Calculate the time taken to travel the first 1 km if the average tractive effort exerted by the locomotive is 50 kN.

(Ans. 1 min 53.4s)

7.9 A diesel locomotive of mass 50 tonnes haul a train of wagons of mass 250 tonnes up an incline of 1 in 400. The frictional resistance is 100 N/tonne mass. Calculate the maximum retardation (a) if the brakes are applied at the locomotive only; (b) if the brakes are applied at every wagon and the locomotive. The coefficient of friction is 0.6.

(Ans. 1.0878 m/sec²; 5.9045 m/sec²).

7.10 A car of mass 1000 kgs is travelling at 72 kms/hr along a horizontal straight road when the driver suddenly sees a fallen tree blocking the road 100 m ahead. The driver applies the brakes as soon as his reaction time (0.75 sec) allows and comes to rest 5 m short of the tree.

(a) Assuming constant deceleration caused by the brakes, what is the decelerating force? What fraction is it of the weight of the car? (g = 9.8 m/sec²)

(b) If the car had been on a downward grade of $\sin^{-1}\frac{1}{20}$ with the brakes supplying the same decelerating force as before, with which speed would the car have hit the tree?

(Ans. 2.5×10^3 N; $\frac{25}{98}$; 7.707 ms⁻¹).

7.11 A piece of string of length 1 m, which can support a maximum tension of 20 Newtons, is used to whirl a particle of mass 1/2 kg in a circular path. What is the maximum linear speed with which the particle may be whirled if the circle is in

(a) Horizontal plane

(b) Vertical plane.

(Ans. 6.131 m/sec; 5.4954 m/sec.).

7.12 Two blocks, of masses M = 20 kgs and m = 10 kgs, are in contact on a horizontal table. A constant horizontal force F = 78 N is applied to block M as shown in Fig. 7.12. There is a constant frictional force of 3 N between the table and block m but no frictional force between the table and the first block M.

(a) Calculate the acceleration of the two blocks.

(b) Calculate the force of contact between the blocks.

(Ans. 2.5 m/sec²; 28 N).

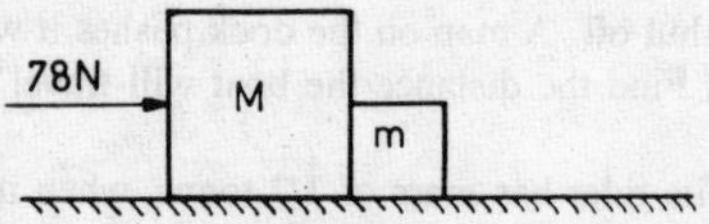

Fig. 7.12

7.13 A prisoner in jail decides to escape by sliding down a rope provided by an accomplice. He attaches the top end of the rope to a hook outside his window. The rope has a mass of 5 kgs, and the prisoner has a mass of 75 kgs. The hook can stand a pull of 700 N without giving way. If the prisoner's window is 25 metres above the ground, what is the least velocity with which he can reach the ground, starting from rest at the top end of the rope. (Ans. 7.48 m/sec.)

7.14 A rider with a bicycle has a total mass of 100 kgs. Find the angle which the bicycle must make with the vertical in travelling round a curve of radius 10 m at a speed of 10 km/hr. Find also, the frictional resistance offered by the ground, if no skidding of wheels occurs at this speed. (Ans. 4°30′; 77.16 N).

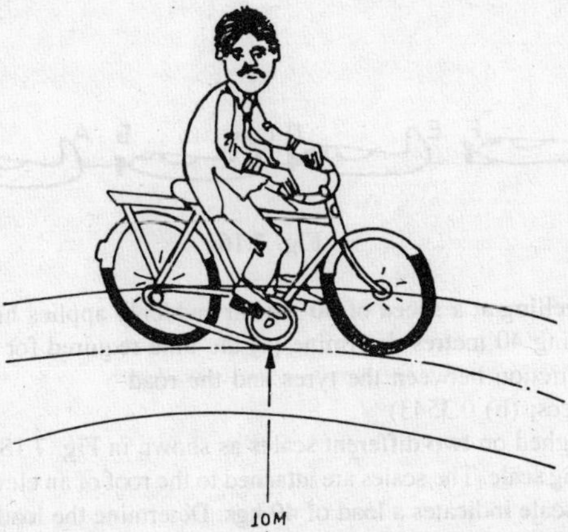

Fig. 7.14

7.15 A trick motor cyclist rides his motor bike around a 'well of death' in the form of a vertical cylinder as shown in Fig. 7.15. The maximum frictional force parallel to the surface of the cylinder is equal to a fraction μ of the normal force exerted on the motor bike by the wall,

(a) At what speed must the motor bike rider go to avoid slipping down?

(b) At what angle to the horizontal must he be inclined?

(c) If $\mu = 0.5$ for rubber tyres on dry wall, and the radius of the cylinder is 10 m, at what minimum speed must the motor bike rider start riding the vertical wall and what angle should he make with the horizontal?

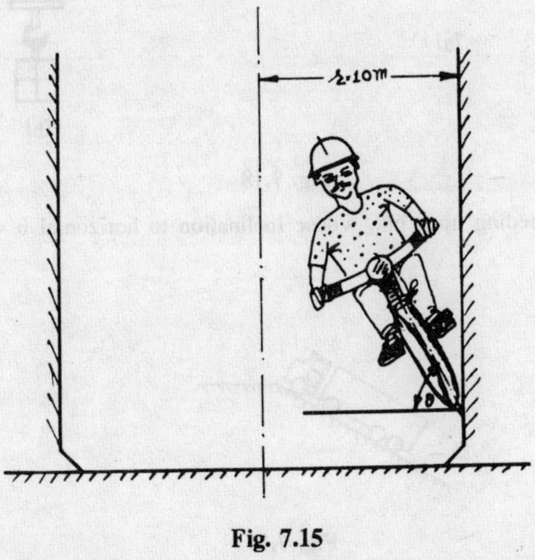

Fig. 7.15

(Ans. (a) $V_{min} = \sqrt{\frac{gr}{\mu}}$; (b) $\theta = \tan^{-1} \mu$
(c) 14 m/sec = 50.4 km/hr.; 26° 35′)

7.16 Three identical gliders each of mass M are being towed through the air in tandem, as shown in Fig. 7.16. Initially, they are travelling at a constant speed and the tension in the tow rope AB is T. The tow plane then begins to accelerate with an acceleration a, what are the tensions in ropes, AB, CD, and EF immediately after the acceleration begins?
(Ans. To + 3 Ma; 2/3 To + 2 Ma; 1/3 To + Ma)

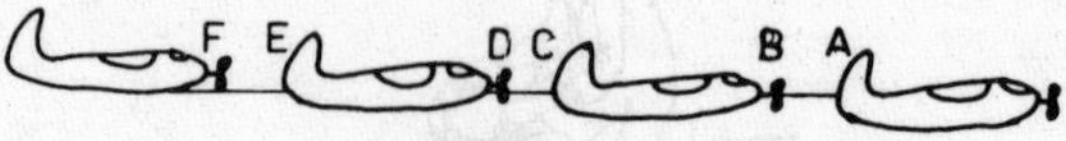

Fig. 7.16

7.17 A motorist travelling at a speed of 60 kms/hr suddenly applies his brakes and comes to rest after skidding 40 metres determine (a) the time required for the car to stop (b) the coefficient of friction between the tyres and the road.
(Ans. (a) 4.8 secs; (b) 0.3543)

7.18 One box is weighed on two different scales as shown in Fig. 7.18. Scale a is lever type; scale b is a spring scale. The scales are attached to the roof of an elevator when the elevator is at rest; each scale indicates a load of 49 kgs. Determine the load which each scale will indicate when the elevator is accelerated upward at a rate of 2 m/sec^2.
(Ans. 49 kgs., 59 kgs.).

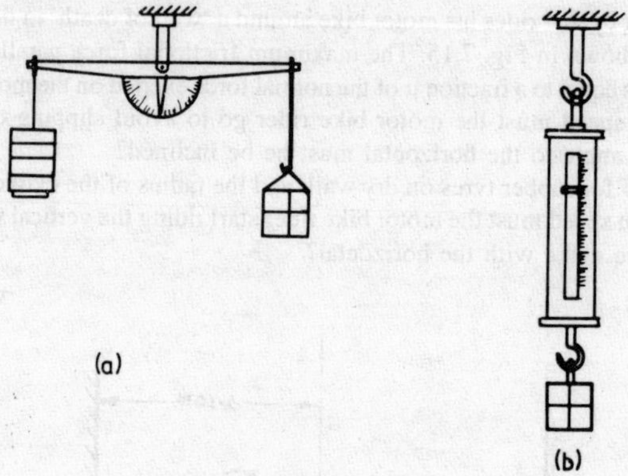

Fig. 7.18

7.19 A truck is proceeding up a hill, whose inclination to horizontal $\theta = \sin^{-1}\frac{2}{100}$, at a

Fig. 7.19

constant speed of 50 Kms/hr. If the driver does not change the setting of his throttle or shift gears, what will be the acceleration of the truck as it starts moving on the level section of the road?
(Ans. a = 0.196 m/sec^2).

7.20 The 10 kg collar was moving down the rod with a velocity of 5 metres/sec when a force F was applied to the horizontal cable. Assuming a coefficient of friction of 0.20 between the collar and the rod, determine the magnitude of the force F if the collar stopped after moving 2 m more down the rod. (Fig. 7.20)
(Ans. 97.85 N).

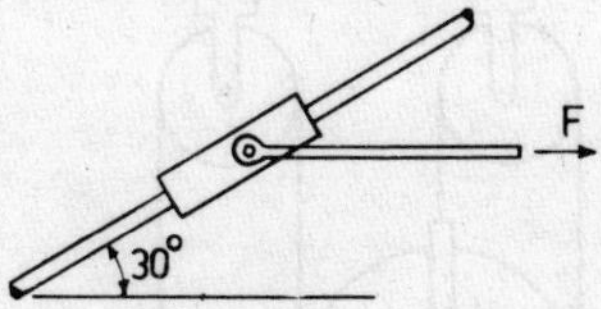

Fig. 7.20

7.21 Two blocks weighing 250 kgs and 100 kgs respectively are attached to the ends of a flexible but inextensible cord overhanging a pulley as shown in Fig. 7.21. Assuming the pulley to be weightless and neglecting axle friction, determine (a) the acceleration of Block A (b) the velocity of block A after 5 secs and the (c) velocity of block A after it has moved through 10 metres.
(Ans. 4.2 m/sec^2; 21 m/sec; 9.165 m/sec.).

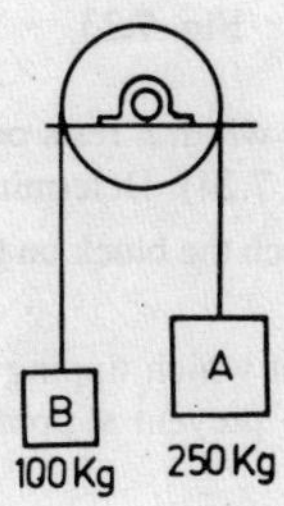

Fig. 7.21

7.22 Two weights A and B are connected by the arrangement shown in Fig. 7.22. Neglecting

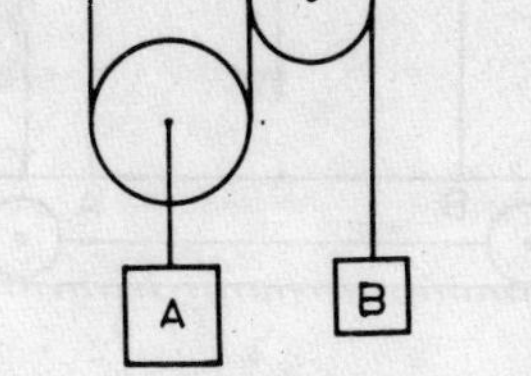

Fig. 7.22

friction and the inertia of the pulley and cord, find the acceleration of the weight *B*. Take mass of *A* as 40 kgs and mass of *B* as 30 kgs.
(Ans. 2.45 m/sec^2).

7.23 The 150 kg block A is connected to a counter weight B by the cable arrangement shown in Fig. 7.23. If the system is released from rest and the block A is observed to move with an acceleration of 2 m/sec^2 directed upward, find (a) the mass of block B, and (b) the corresponding tension in the cable.
(Ans. 155.3 kg; 590 N).

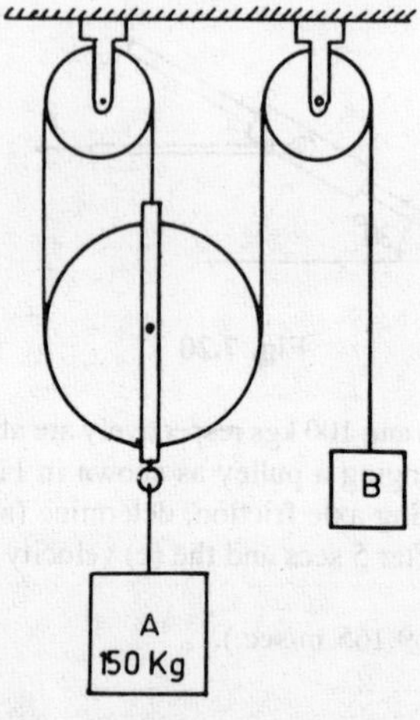

Fig. 7.23

7.24 A block of weight *W* height *h* and width *b* rests on a Flat car which moves horizontally with constant acceleration *a* (Fig. 7.24). Determine

(a) the value of acceleration at which the block on the car starts slipping, if the coefficient of friction is μ, and

(b) the value of the acceleration at which tipping of the block about *A* will commence assuming sufficient friction to prevent slipping.

$\left(\text{Ans. : } \mu g; \dfrac{bg}{h}\right)$

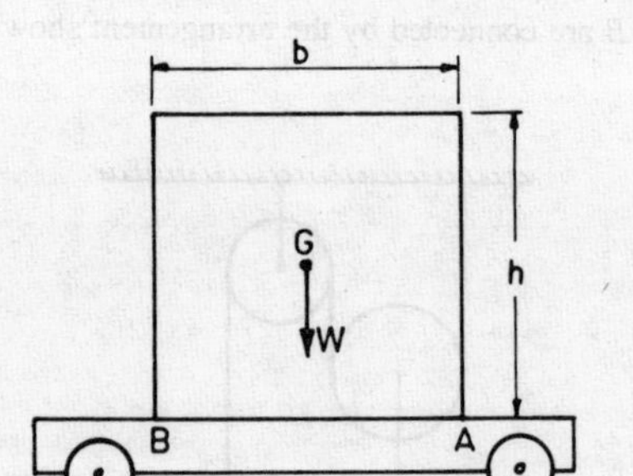

Fig. 7.24

7.25 A small trolley and its load weigh 5000 kgs. The trolley is partially counterweighted by a 250 kg weight. If the trolley is initially at rest, determine the force F required if the velocity of the trolley is to be 4 m/sec after it has moved 10 m up the incline. Neglect the frictional resistance.
(Ans. 6290.6).

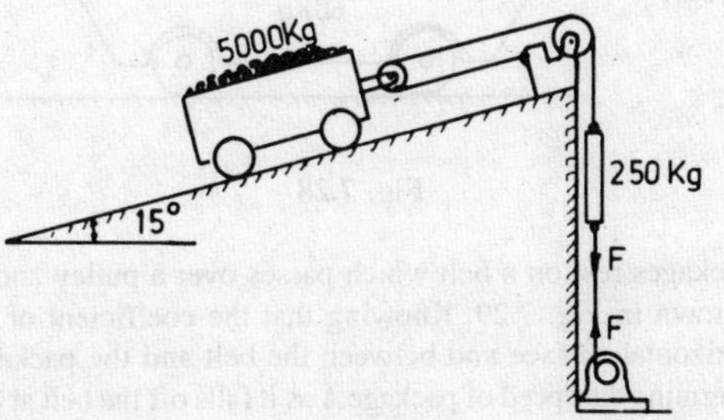

Fig.7.25

7.26 Two blocks A and B are connected as shown in Fig. 7.26. Neglecting the effect of friction, determine (a) the acceleration of block A, (b) the tension in the cable.
(Ans. 3 m/sec²; 100 N).

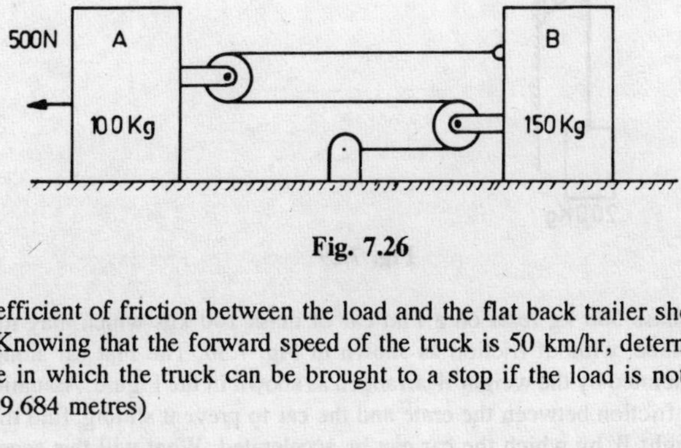

Fig. 7.26

7.27 The coefficient of friction between the load and the flat back trailer shown in Fig. 7.33 is 0.5. Knowing that the forward speed of the truck is 50 km/hr, determine the shortest distance in which the truck can be brought to a stop if the load is not to shift.
(Ans. 19.684 metres)

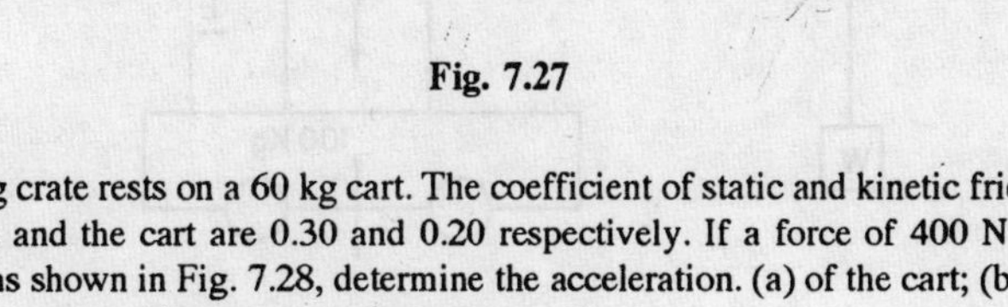

Fig. 7.27

7.28 A 100 kg crate rests on a 60 kg cart. The coefficient of static and kinetic friction between the crate and the cart are 0.30 and 0.20 respectively. If a force of 400 N is applied to the cart as shown in Fig. 7.28, determine the acceleration. (a) of the cart; (b) of the crate; (c) of the crate with respect to the cart.
(Ans. 3.40 m/sec²; 1.96 m/sec²; 1.44 m/sec²)

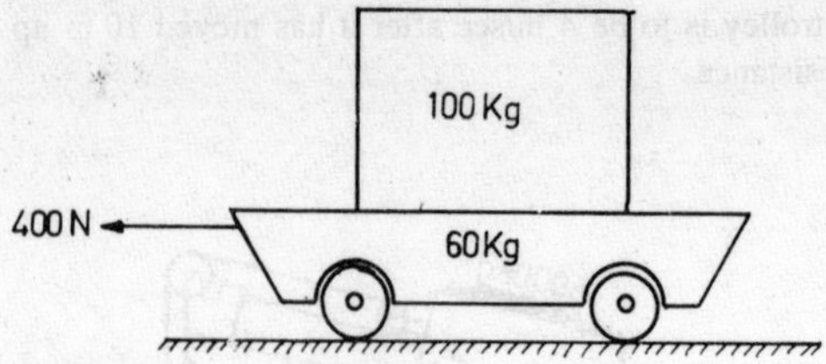

Fig. 7.28

7.29 Two 150 kg packages rest on a belt which passes over a pulley and is attached to a 200 kg weight as shown in Fig. 7.29. Knowing that the coefficient of friction between the belt and the horizontal surface and between the belt and the packages is 0.40 and 0.50 respectively, determine the speed of package *A* as it falls off the belt at C if it starts from rest. (Ans. 2.5044 m/sec).

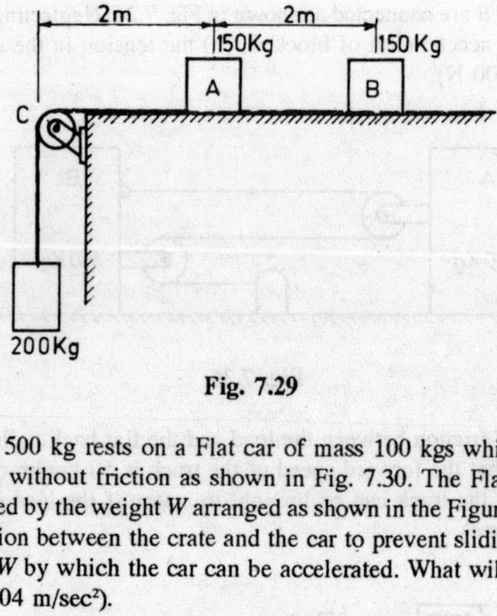

Fig. 7.29

7.30 A crate of mass 500 kg rests on a Flat car of mass 100 kgs which may roll along the horizontal plane, without friction as shown in Fig. 7.30. The Flat car along with crate is to be accelerated by the weight *W* arranged as shown in the Figure. Assuming that there is sufficient friction between the crate and the car to prevent sliding, find the maximum value of weight *W* by which the car can be accelerated. What will this acceleration be? (Ans. 125 kg; 2.04 m/sec²).

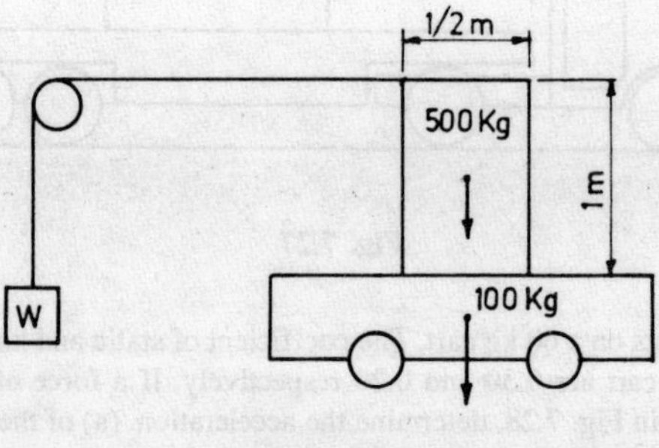

Fig. 7.30

7.31 A flexible but inextensible chain of length 'l' m is held on a smooth table with an initial overhang h as shown in Fig. 7.31. If the chain weighs W kg/m, calculate the velocity with which the chain will leave the table if released.

$$\left(\text{Ans. } V = \sqrt{\frac{g}{l}\,(l^2 - h^2)}\right)$$

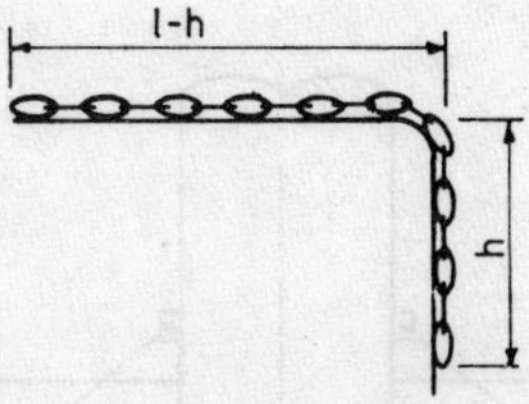

Fig. 7.31

7.32 The fork-lift truck shown in the Fig. 7.32 is used for shifting packages from shop floor to the loading yard. The mass of the fork-lift truck is 1800 kgs. The fork lift truck is moving forward at a speed of 15 kms/hr with a package of mass 2000 kgs on it, when the brakes are applied in all the four wheels. Determine the shortest distance in which the truck can be brought to a stop if the package is not to slide, and if the truck is not to tip forward. The coefficient of friction between the package and the fork lift is 0.4.
(Ans. 2.214 metres).

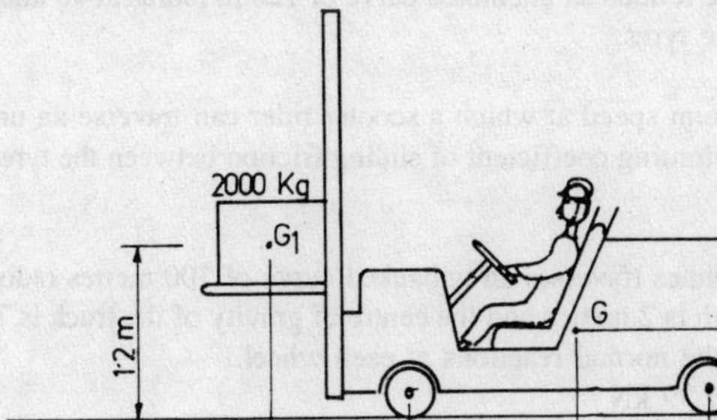

Fig. 7.32

Motion in A Circle

7.33 A body of mass 1 kg resting on a smooth table, is whirled in a horizontal circle at the end of a spring. The spring stiffness is 6 kN/m. The natural unstreched length of the spring is 200 mm. Calculate the radius of rotation of the weight and the stretch in the spring when the speed of rotation is 420 rpm.
(Ans. 295.27 mm; 95.27 mm).

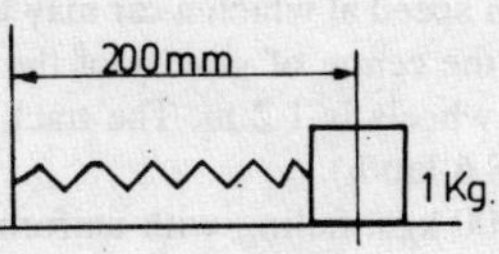

Fig. 7.33

7.34 A centrifugal clutch with its axis mounted vertically is shown in Fig. 7.34 in rest position. The rotating bobs are each of mass 200 gms and the spring stiffness 7 kN/m. The centre of mass of each bob is at 150 mm radius in the rest position. The gap between the bob and clutch face, in rest position is 10 mm. Calculate the radial force on the clutch face at 840 rpm.
(Ans. 177.8 N)

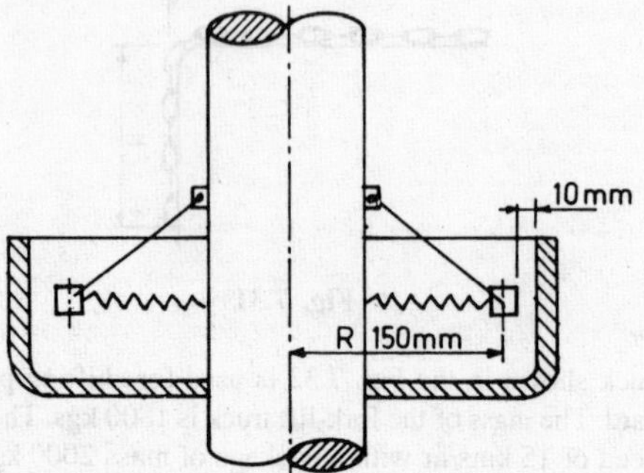

Fig. 7.34

7.35 A car of mass 1 tonne rounds an unbanked curve of 120 m radius at 40 kms/hr. Calculate the side thrust on the tyres.
(Ans. 1029 N).

7.36 Calculate the maximum speed at which a scooter rider can traverse an unbanked curve of 40 m radius. The limiting coefficient of sliding friction between the tyres and the road is 0.4.
(Ans. 45 kms/hr.).

7.37 A truck of mass 6 tonnes traverses an unbanked curve of 100 metres radius at 40 km/h. The wheel track width is 2 metres and the centre of gravity of the truck is 700 mm above the road. Calculate the normal reactions at each wheel.
(Ans. outer wheel 31.99 kN
inner wheel 26.807 kN).

Superelevation of Tracks

7.38 A car travels at 80 kms/h on a track at 30° to the horizontal. The limiting coeffieicnt of friction between the tyres and the road is 0.6. Calculate the minimum radius of curvature of the road if the car is not to skid.
(Ans. 42.813 metres).

7.39 Calculate the superelevation of the outside rail of a curved track if a train is to traverse the curve without side thrust on the rails at 60 km/hr. The radius of curve is 200 m and the track guage is 1400 mm. What is the lateral force on the track due to a 200 tonne vehicle at 100 kms/h with this angle of banking.
(Ans. 198 mm; 493.844 kN).

7.40 Calculate the maximum speed at which a car may traverse a banked track of 50 m radius without overturning if the centre of gravity of the car is 0.8 m above ground level and the track width of the wheels is 1.2 m. The track is banked at 20° to the horizontal.
(Ans. 27.4 m/sec or 98.6 km/h).

7.41 A parachute of mass 100 kgs. falling with uniform acceleration from rest, descends 75 metres in first ten seconds. Find the resultant pressure of air on the parachute.
(Ans. 830 N).

7.42 A mass of 2 kgs is hung from a spring balance in a lift. What is spring balance reading when lift is

(a) at rest;

(b) accelerating downward at 4 m/sec^2

(c) Accelerating upwards at 4 m/sec^2

(d) Moving downwards and retarding at 2 m/sec^2 ?

(Ans. 19.6 N; 11.6 N; 27.6 N; 23.6 N).

7.43 In a lift ascending under an acceleration, a spring balance indicates that the mass of a certain body is 15 kg. If the lift descends with twice the acceleration than it had while going upwards, the mass appears in the spring balance to be 12 kgs. Find the actual mass of the body and upward acceleration of the lift take g = 9.8 m/sec^2.
(Ans. 14 kg; 0.7 m/sec^2).

7.44 The tool force on a shapping machine during the cutting stroke is 250 N and the reciprocating parts are equivalent to a moving mass of 50 kgs. If the power is suddenly shut off what would be the further distance cut by the tool if the cutting speed at the time of power shut off was 1 m/sec? Assume the cutting forcc to be independent of the shutting speed.
(Ans. 10 cms).

7.45 A car of total mass 1,500 kgs travels down an incline 1 in 15. The wheels are 600 mm in diameter and tractive resistance is 500 N. The engine torque is 60 Nm and gear reduction ratio between engine and axle is 14 : 1. Find the acceleration of the car, neglecting rotational inertias, friction and transmission losses.
(Ans. 1.2533 m/sec^2)

Impulse and Momentum

7.46 A bullet of mass 50 gms leaves the barrel of a gun with a velocity of 400 m/sec. Find the impulse produced by the discharge of the bullet. If the force lasts for .002 secs, find the average impulsive force.
(Ans. 20 kg m/sec; 10 kN)

7.47 A bullet of mass 80 gm is fired horizontally with a velocity of 1000 m/sec to hit a wooden block of mass 10 kgs which is hanging freely at the end of a string 4 m long. Find the angle through which the bullet embedded the wooden block swings.
(Ans. 78° 40′)

7.48 A wagon of mass 200 tonnes moving at 20 km/hr collides with the back of a wagon, of mass 60 tonnes moving in the same direction at 5 km/hr. What is the velocity of the 200 tonne wagon after the impact and the impulse between them if the 60 tonne wagon moves off at 30 km/hr after the impact.
(Ans. 12.5 km/hr; 416666.6 kg m/sec).

7.49 A pile driving hammer of mass 800 kgs falls 3.6 metres from rest on to a pile of mass 160 kgs. There is no rebound and the pile is driven 200 mm into the ground. Calculate

(a) The common velocity after impact

(b) The average resistance offered by the ground

(Ans. 7.0 m/sec; 117.6 kN).

7.50 A rocket propelled missile ejects 150 kgs of exhaust gases per sec. at a speed of 1200 m/sec relative to the engine. Calculate the thrust of the engine if the forward speed of the missile is 400 m/sec.
(Ans. 180 kN).

.51 A jet plane discharges a jet at the rate of 50 kg/sec with a velocity of 1400 m/sec relative to the plane. If the forward speed of the plane is 960 km/hr, what is the thrust on the plane and the power developed at the jet?
(Ans. 70 kN; 18.66 MW).

8. WORK ENERGY AND POWER

After having mastered the content of this chapter, you will be able to:

- Explain the meaning of work.
- Calculate the work done by a constant and linerly varying force.
- Calculate the resultant work done by a number of forces acting on a body.
- Explain the meaning of energy and its relationship with work.
- Differentiate between potential, kinetic and total mechanical energy.
- Apply the principle of work and energy to the solution of problems of rigid bodies.
- Explain the principle of conservation of mechanical energy.
- Apply the principle of conservation of mechanical energy to problems of rigid bodies undergoing rectilinear, curvilinear and rotational motion.
- Differentiate between work and power.
- Calculate power required for doing certain work in given time.

CHAPTER 8

Work Energy and Power

This chapter deals with the concept of work and energy and establishes the relationship between the two, known as the principle of work and energy. Expressions for the work and energy principle applied to bodies undergoing translation and rotation have been derived in the ensuing articles. It also includes the concept of potential energy, total mechanical energy and the principle of conservation of energy. The chapter ends with the concept of power, and how it is related with the work done.

The application of all the above concepts and principles is illustrated through a number of solved problems.

8.1 Work

In day to day life, the word work means exertion, which may be either physical or mental. But in science and technology, the word "work" means differently. *Work is said to be done by a force, when firstly it moves the body on which it is acting, and secondly the distance has a component in the direction of the force.* Let us consider the case of a man lifting a block of weight W. If we look to the free body diagram of the block, there are two forces W and P acting on this block. P is the force applied by the man to lift the block and W is the weight of the block. As the block is raised through a height 'h', both the forces W and P will do work, because the distance moved by the block is in the direction of their line of action (Figure 8.1). In case the block is raised

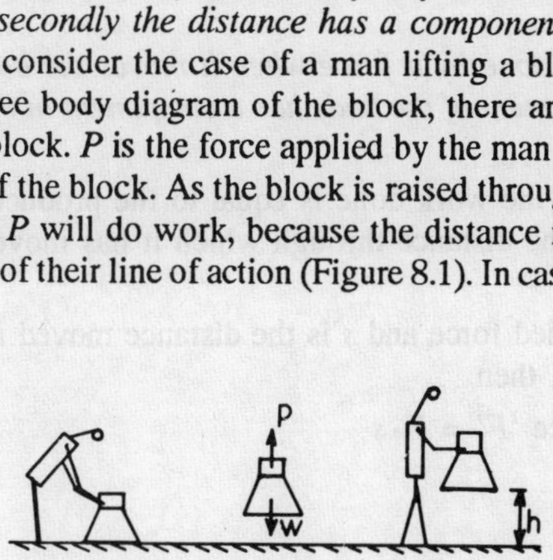

Fig. 8.1

with a constant velocity, P will be equal to W and the work done by both the forces will be equal in magnitude. Let us consider another case when the man is just holding the block in his hand (Fig. 8.2). In this case no work is done by the forces W and P because the block does not move at all.

In another case (Fig. 8.3) the man is holding the block at a constant level and at the same time moving on a level surface. No work will be done in

Fig. 8.2

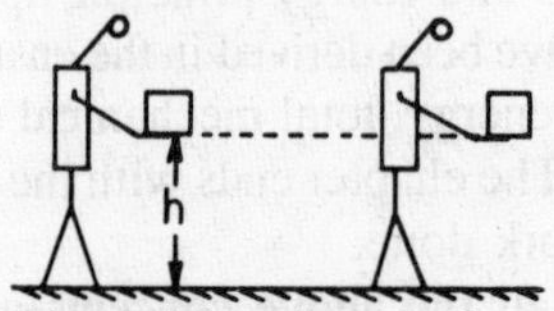

Fig. 8.3

this case also. This is so because the displacement of the block is normal to the line of action of the forces, and has no component in their direction.

Work is, therefore, done by a force when it moves the body on which it is acting and the movement of the body has a component in the direction of the force.

The amount of the work done is equal to the product of the force and the component of the distance through which it has moved the body in its direction.

If F is the applied force and s is the distance moved in the direction of the force (Fig. 8.4), then

Work done by Force 'F' $= F \cdot s$

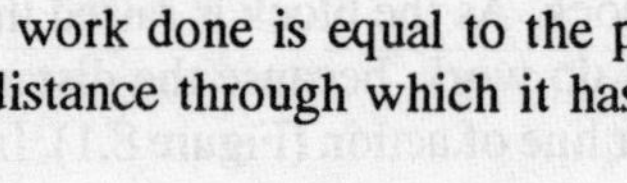

Fig. 8.4

Let us now consider a block of weight W being hauled over a level surface as shown in Figure 8.5. A force 'P' inclined at an angle 'θ' is applied to haul this block. Referring to Fig. 8.5 A, the forces acting on the block are

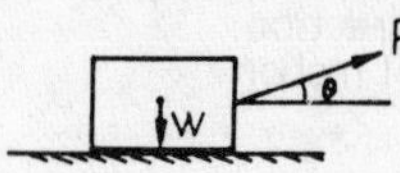

Fig. 8.5

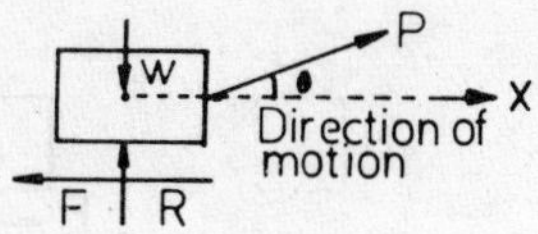

Fig. 8.5 A

P, the force of pull inclined at θ° to the direction of motion *x*

W, the weight of the block acting normally to the direction of motion,

R, the reaction offered by the surface. It is acting normal to the direction of motion,

F, the frictional resistance acting in a direction opposite to that of the direction of motion

In moving the block through a distance '*S*', the work done by the various forces acting on the block is as follows

Work done by P = *P* × displacement component in the direction of *P* (Fig. 8.6)

$$= P \times S \cos\theta$$
$$= P\,S \cos\theta$$

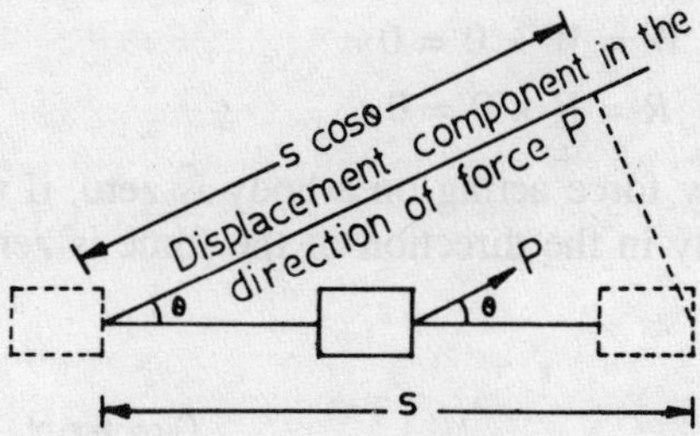

Fig. 8.6

The work done by the force *P* is called *positive work* or accelerating work, because the distance moved has a component in the same direction as that of the force. The work done by *P* can also be written as *P* cosθ × *s*, i.e. the work done is also equal to the force component in the direction of the distance moved multiplied by the displacement.

Work done by force F = *F* × displacement component in the direction of F. (Fig. 8.7)

$$= F \times (-S)$$
$$= -FS$$

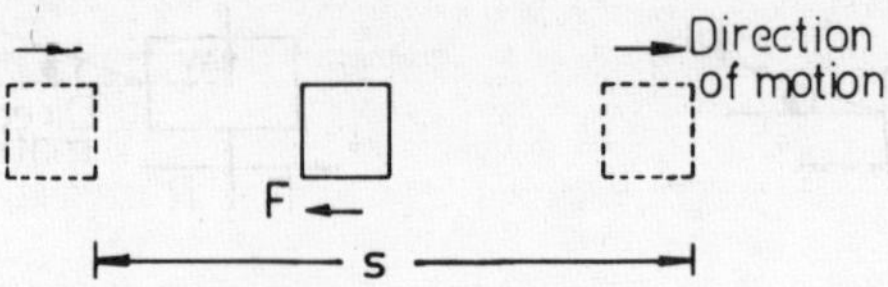

Fig. 8.7

F is the frictional force and acts on the body where it makes contact with the surface on which it moves or tends to move. The force acts in a direction opposite to that of the movement of the body and as such the distance '*s*' moved by it is assigned a negative sign. It opposes or retards the movement of the body and accordingly, the work done by it is called 'negative' or 'retarding' work.

It can therefore be said that work done by any force is negative when the displacement of the body on which it acts is in a direction opposite that of its own.

Work done by Forces W & R

As the block moves through distance 's', we know that the component of this displacement in the direction of the forces *W* and *R* is zero. As such, the work done by the forces *W* and *R* will be:

$$\text{Work done by } W = W \times 0 = 0$$

$$\text{Work done by } R = R \times 0 = 0$$

The work done by any force acting on a body is zero, if the displacement component of the body in the direction of the force is zero.

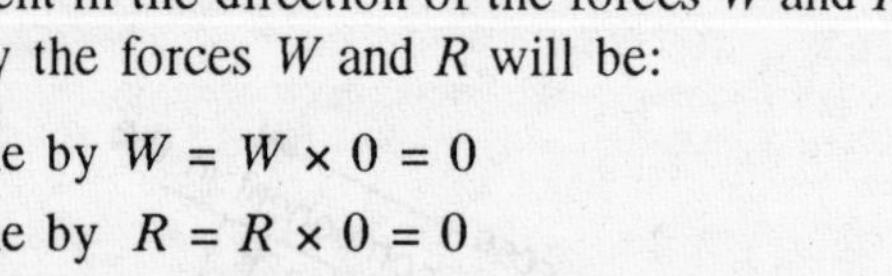

Fig. 8.8

8.1.1 *Units of Work*

Work is a scalar quantity i.e. it has a magnitude and sense, but no direction. We know that work is a product of two vector quantities i.e. force and displacement; therefore, its units will be the same as the units of the product of force and displacement. In SI units, the units of force and displacement are Newton(N) and metre(m) respectively, therefore

Unit of work = Newton × metre

= N m

One Newton-metre is called a Joule(J) and is equal to the work done by a constant force of 1 Newton acting on a body and moving it through a distance of 1 m in its direction.

In the MKS system, the units of force and displacement are kilogram force (kgf) and metre (m) respectively.

Therefore

Unit of work = kgf × m

= kgf m

Since 1 kgf = 9.81 N,

1 kgf m = 9.81 Nm

= 9.81 J

Example 8.1

What is the work done by a man in raising a bucket of water of mass 10 kgs from the ground floor to the first floor. The height of the balcony of the first floor from the ground floor is 3.5 metres.

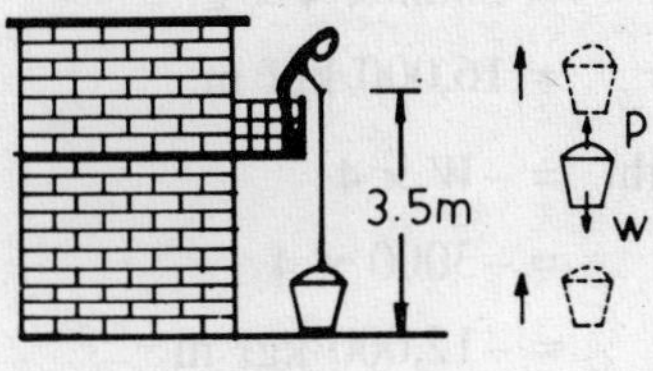

Fig. 8.9

Solution

Work done by the man in raising the bucket = Force applied (P) × Distance moved in the direction of P

When the bucket is raised at constant velocity

$P = W$

= 10 × 9.81

= 98.1 N

∴ Work done by the man = 98.1 × 3.5

= **343.3 J**

Example 8.2

Determine the work done by a force P of 2000 kgf in raising a block of weight 3000 kgf through a height of 4 m. The block is raised along an incline making an angle of 30° to the horizontal and the force P is applied parallel to the incline as shown in Figure 8.10. Also find out the work done by the weight of the block.

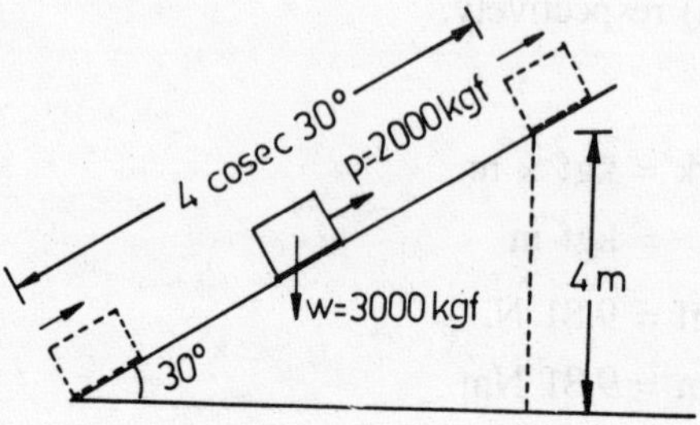

Fig. 8.10

Solution

Work done by the force $P = P \times$ Component of displacement in the direction of force P

$$= P \times 4 \text{ cosec } 30°$$

$$= 2000 \times 4 \times 2$$

$$= 16{,}000 \text{ kgf m.}$$

Work done by the weight $= -W \times 4$

$$= -3000 \times 4$$

$$= -12{,}000 \text{ kgf m}$$

The work done by the load W is negative, because the load moves up, in a direction opposite to that of gravitation.

Example 8.3

A box of mass 150 kg is being hauled along an inclined surface making an angle of 30° with the horizontal. A constant force P of 1200 N inclined at an angle 30° to the contact surface is applied to drag the box as shown in Figure 8.11. Coefficient of kinetic friction μ_k is given as 0.2. Calculate the work done by each force and the resultant work, when the box is moved along the incline for a distance of 2.5 metres. Take $g = 9.8 \text{ ms}^{-2}$.

Solution

Draw the free body diagram of the box. Let the direction of motion of the box be represented by the X-axis (Figure 8.11)

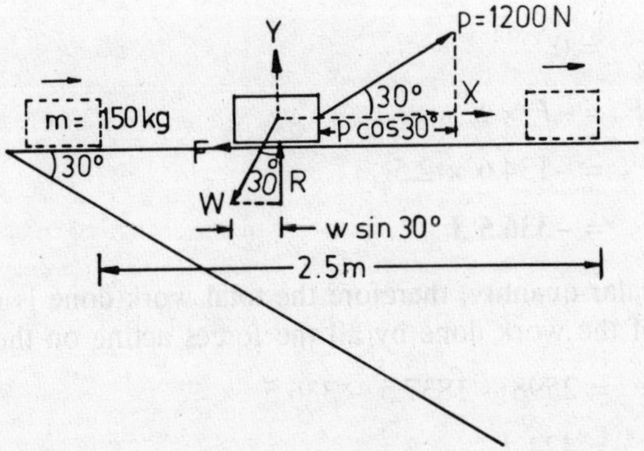

Fig. 8.11

The various forces acting on the box are

P, the force of pull
W, the weight of the box
R, the normal reaction offerred by the inclined surface
F, the frictional resistance.

Out of the above forces, forces *R* and *F* are unknown. As this is the case of rectilinear motion, therefore

$\Sigma F_y = 0$

$\therefore P \sin 30° + R - W \cos 30° = 0$

$R = W \cos 30° - P \sin 30°$

$= 150 \times 9.8 \cos 30 - 1200 \sin 30$

$= 673$ N

$F = \mu k \cdot R$

$= 0.2 \times 673 = 134.6$ N

Work done by *P* = Component of *P* in the direction of displacement × displacement

$= P \cos\theta \times S$

$= 1200 \cos 30 \times 2.5$

$= 1200 \times \frac{\sqrt{3}}{2} \times 2.5$

$= 2598$ J

Work done by *W* $= -W \sin\theta \times S$

$= -150 \times 9.8 \times 0.5 \times 2.5$

$= \mathbf{-1837.5}$ **J**

Work done by R $= R \times 0$

$= 0$

Work done by F $= -F \times S$

$= -134.6 \times 2.5$

$= \mathbf{-336.5\ J}$

As work is a scalar quantity, therefore the total work done is equal to the algebraic sum of the work done by all the forces acting on the body.

Total work done $= 2598 - 1837.5 - 336.5$

$= 424$ J

8.1.2 *Work Done by Constant Torque*

You must have seen a job being turned on a lathe. when the job rotates, a cutting tool remains fixed in a particular position and goes on cutting the metal on the periphery of the rotating job. For cutting the metal the tool overcomes a certain resistance. There is, thus, a force applied by the tool for cutting the metal. This force acts tangential to the rotating job as shown in Fig. 8.13. The tangential cutting force may be constant or a varying in magnitude.

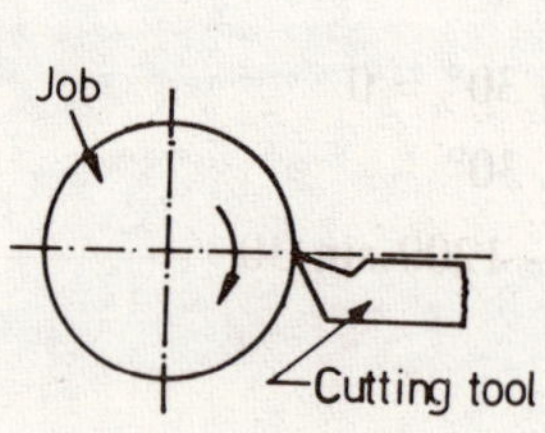

Fig. 8.12

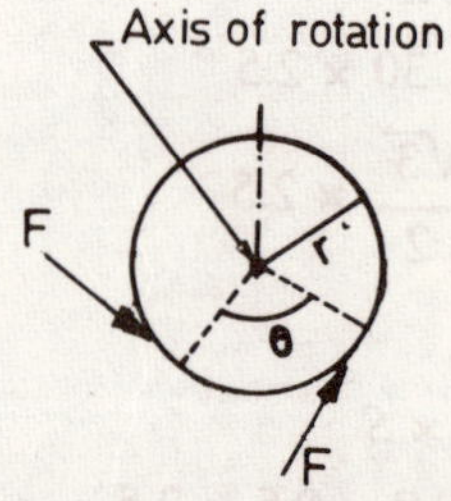

Fig. 8.13

If the tangential cutting force acting at any instant of time is F, and the radius of the job is r, then it is said that torque T is acting on the job. *Torque (T) is thus the moment of the tangential force about the axis of rotation of the job.* It will be constant, if the tangential cutting force remains constant.

The work done by the constant tangential force F, when the job moves through an angle θ will be worked out as under

Work done = Force × Distance moved in the direction of force.

$= F \times r\,\theta = T\,\theta$

where r = Normal distance between the tangential force and the axis of rotation

θ = Angle in radians through which the constant force F acts

T = Torque = $F \cdot r$

If T is in Newton metres, and θ in radians, the unit of work is in Newton metres.

Example 8.4

A bar is rough turned to a diameter of 110 mm and then finish turned to a diameter of 105 mm. Determine the work done through one revolution of the bar in both rough turning and finished turning, when the tangential cutting forces in rough and finished turning are 2000 N and 500 N respectively.

Solution

(a) Work done in rough turning

$= \text{Force} \times r \times \theta$

$= 2000 \times \dfrac{110}{2 \times 1000} \times 2\pi$

$= 691.43$ Nm

(b) Work done in finished turning

$= 500 \times \dfrac{105}{2 \times 1000} \times 2\,\pi$

$= 165$ Nm

8.1.3 *Work Done by Variable Forces and Torques*

There are cases, when a force acting on a body does not remain constant, but varies with the displacement undergone by the body. For example, when a spring is stretched, the force applied to the spring varies with its stretch.

Let a graph shown in Figure 8.14 represent the force F which varies when the distance 's' moved by the body increases. Force is assumed to be acting in the direction of the distance moved.

Considering force 'F' to be constant for a differential displacement dS, the work done for this displacement will be equal to $F \cdot dS$. This represents the area of the small shaded rectangle. From this, we can say that the total

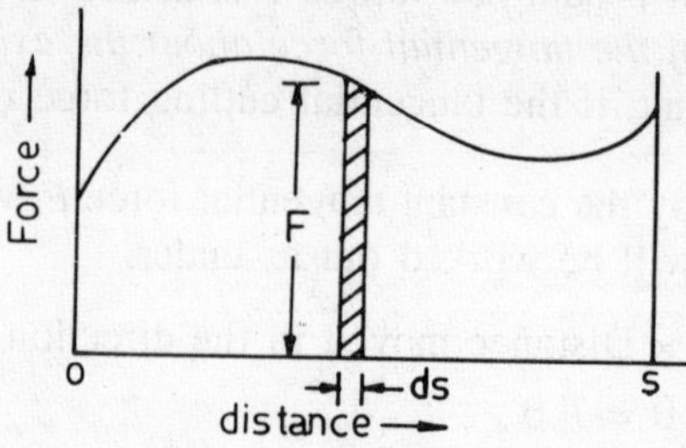

Fig. 8.14

work done by the variable force F over a displacement 'S' is equal to the area under the force displacement diagram. Mathematically, we can say that

The work done by a variable force F

in undergoing displacement 'S' $= \int_0^S F\,dS$

Let us now take the case of a spring in which the force applied F to stretch it is proportional to be stretch 'S'. Expressing mathematically,

$$F \propto S$$

or $$F = K \cdot S.$$

where 'K' is the proportionality constant of the spring. It is also called the stiffness of the spring or *spring modulus*, and is the force required for unit deformation of the spring. The relation between the force and the corresponding stretch of the spring is, therefore, linear and can be represented by a straight line as shown in Fig. 8.15.

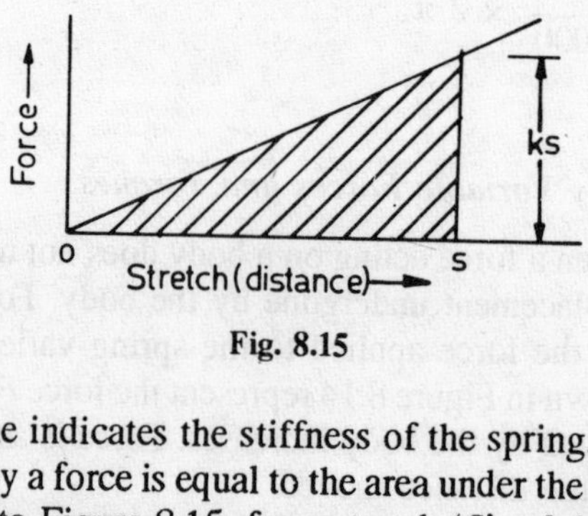

Fig. 8.15

The slope of the line indicates the stiffness of the spring. We already know that the work done by a force is equal to the area under the force displacement diagram. Referring to Figure 8.15, for a stretch 'S', of the spring, the force applied to the spring will be K.S.

∴ The work done in stretching the spring for a length of S =Area under the Force displacement diagram.

$$= \frac{1}{2} K.S.S.$$

$$= \frac{1}{2} KS^2$$

Example 8.5

Determine the work done in stretching a spring through a length of 50 mm. The stiffness of the spring is 5 N/mm.

Solution

$$\text{Work done} = \frac{1}{2} K\, S^2$$

$$= \frac{1}{2} \cdot 5.50^2$$

$$= 6250 \text{ N.mm or } 6.25 \text{ J}$$

Example 8.6

A spring, when unstretched, has a length of 150 mm. On the application of a load, it extends to a length of 200 mm. On the removal of the load, if the spring does a work of 6 Joules in returning to its unstretched position, determine the stiffness of the spring and the maximum load applied to the spring.

Solution

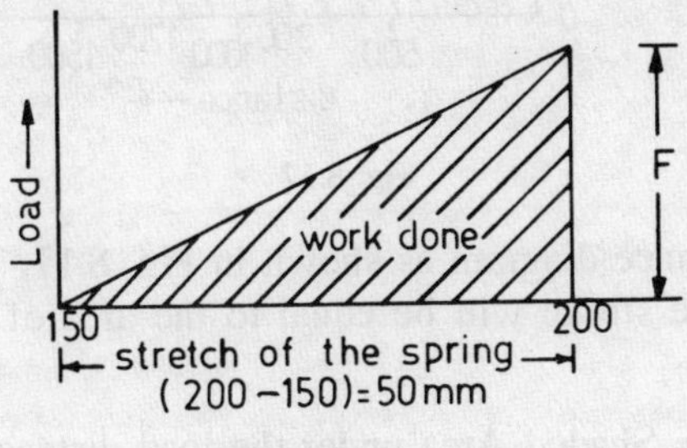

Fig. 8.16

Figure 8.16 shows the load stretch diagram of the spring. On the application of the maximum load, the length of the spring = 200 mm

unstretched length of the spring = 150 mm

∴ Stretch 's' of the spring = 200 − 150 = 50 mm.

Let the maximum load applied to the spring be F, then the work done by the spring on removal of the load will be equal to the area of the load-stretch diagram.

Work done $= \frac{1}{2} F s$, Work done = 6 J

$$6 = \frac{1}{2} \times F \times .050$$

$$\therefore \quad F = \frac{6 \times 2}{.050} = 240 \text{ N}$$

$$\text{Spring stiffness} = \frac{\text{Load}}{\text{Stretch}} = \frac{240}{50} = 4.8 \text{ N/mm}$$

Example 8.7

A broach has a total stroke of 1200 mm. While cutting, it has to overcome a variable resistance. The initial resistance is 1200 N. At the end of three quarters of the stroke the resistance rises uniformly to 7200 N. In the last quarter of the stroke the resistance falls uniformly from the maximum value of 7200 N to 3600 N. Compute the work done per stroke.

Solution

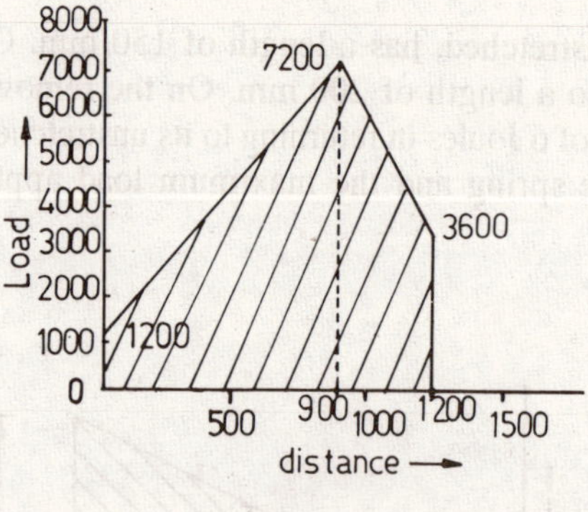

Fig. 8.17

Draw the load distance diagram as shown in Fig. 8.17. The work done by the broach in single stroke will be equal to the area of the Load-distance diagram.

Work done/stroke = Shaded Area under the load distance diagram.

$$= \frac{1}{2}(1200 + 7200) \times \frac{900}{1000}$$

$$+ \frac{1}{2}(7200 + 3600) \times \frac{300}{1000}$$

$$= 5400 \text{ Nm.}$$

Example 8.8

Fig. 8.18 shows the force distance graph of planning job for a stroke of length 300 mm. The curved portion of the graph satisfies the equation

Fs = 15000 N.mm. Determine the work done per stroke using middle ordinate rule and taking five equal intervals of 60 mm.

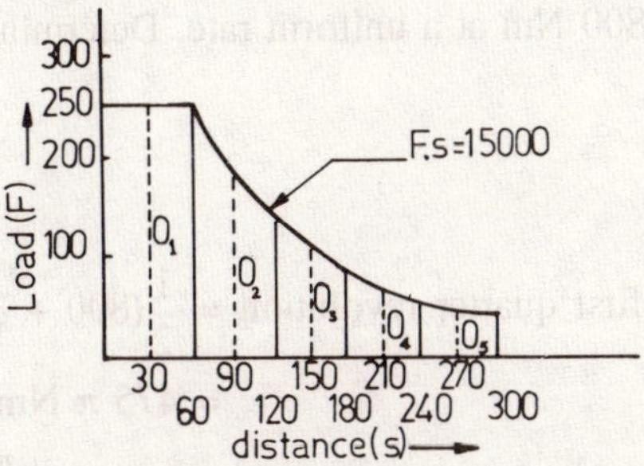

Fig. 8.18

Solution

The force-distance diagram has been divided into five equal strips. O_1, O_2, O_3, O_4 and O_5 are the middle ordinates of these strips and represent the forces (F) applied by the tool when it is at a distance of 30 mm, 90 mm, 150 mm, 210 mm, and 270 mm from the start.

From the diagram $O_1 = F_{30} = 250$ N.

In the curved section we know that, $Fs = 15000$

$$\therefore\ F = \frac{15,000}{s}$$

$$O_2 = F_{90} = \frac{15,000}{90} = 166.67 \text{ N}$$

$$O_3 = F_{150} = \frac{15,000}{150} = 100.00 \text{ N}$$

$$O_4 = F_{210} = \frac{15,000}{210} = 71.43 \text{ N}$$

$$O_5 = F_{270} = \frac{15,000}{270} = 55.55 \text{ N}$$

Work done = Area under the Force-distance diagram

$= \text{Width of strip } (O_1 + O_2 + O_3 + O_4 + O_5)$

$$= \frac{60}{1000}(250 + 166.67 + 100 + 71.43 + 55.55)$$

$$= \frac{60}{1000} \times 643.65$$

$= 38.62$ Nm

Example 8.9

A variable torque is applied in driving a shaft. In the first quarter revolution the torque varies uniformly from 800 Nm to 3000 Nm. In the next quarter, it remains constant at 3000 Nm. For the remaining half revolution, it falls from 3000 Nm to 800 Nm at a uniform rate. Determine the work done per revolution.

Solution

Work done = $T\,\theta$

Work done in the first quarter revolution $= \frac{1}{2}(800 + 3000) \times \frac{\pi}{4}$

$= 475\,\pi$ Nm

Work done in the next quarter revolution $= 3000 \times \frac{\pi}{4}$

$= 750\,\pi$ Nm

Work done in the remaining half revolution $= \frac{1}{2}(3000 + 800) \times \frac{\pi}{2}$

$= 950\,\pi$ Nm

Work done per revolution $= (475\,\pi + 750\,\pi + 950\,\pi)$ Nm

$= 2175\,\pi$ Nm

$= 6832.96$ Nm

8.2 Energy

Energy may be considered as the *capacity to do work.* There are many forms of energy such as heat energy, mechanical energy, electrical energy, chemical energy, etc. These different forms of energy are transformable from one form to another. We are here concerned only with the mechanical energy of a body. Mechanical energy of a body consists of two parts; the first is due to its position and is known as its potential energy, the second is due to its motion and is called kinetic energy.

8.2.1 *Kinetic Energy, Work Energy Principle*

As has been said earlier, the kinetic energy of a body is due to its motion, e.g. a car in motion or a bullet in motion will possess kinetic energy.

Let us consider a particle of mass 'm' to which a force 'F' of constant magnitude and direction is applied. Under the action of the force, the particle will move along a constant direction with a uniform acceleration 'a'. From Newton's second law of motion, we know that

$$F = ma \qquad \text{...... (i)}$$

Let the velocity of the particle at any instant of time be v_1. After moving a distance S, let the velocity of the particle be v_2. During this displacement of the particle, the work done by the force F is as under.

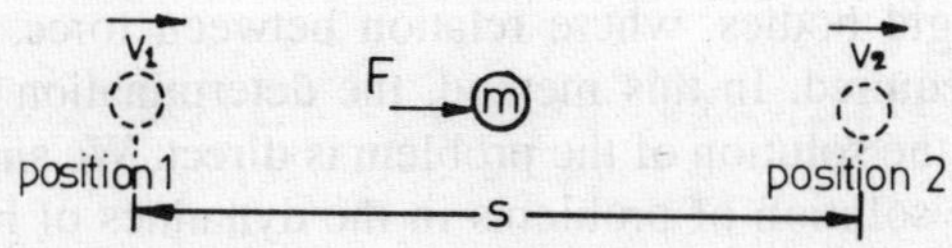

Fig. 8.19

Work done by the force $F = F \cdot S$ .... (ii)

Substituting for F from (i), we get

$$FS = maS \quad \text{.... (iii)}$$

We also know that

$$2\,a\,S = V_2^2 = V_1^2 \quad \text{.... (iv)}$$

$$\therefore\ a\,S = \frac{1}{2}\,(V_2^2 - V_1^2)$$

Substituting for $a\,S$ in (iii), we get

$$F\,S = \frac{1}{2}\ m\,(V_2^2 - V_1^2)$$

$$= \frac{1}{2}\ mV_2^2 - \frac{1}{2}\ mV_1^2 \quad \text{.... (v)}$$

The left hand side of the expression (v) indicates the work done by the constant force F acting on the particle through a distance S. In the right hand side of the expression, the term $\frac{1}{2}\ mV^2$ is defined as the kinetic energy of the particle. Like work, it is also a scalar quantity and from its expression, it is obvious, that it will be always positive irespective of the direction of the motion of the particle. Since the right hand side of the expression is equal to its left hand side, therefore both sides should have the same units. Thus the units of kinetic energy will be the same as that of work i.e. *Joule* (J).

If we now study equation (v), the right hand side, shows $\frac{1}{2}\ mV_1^2$ as the kinetic energy of the particle, when it is at position 1, and $\frac{1}{2}\ mV_2^2$ as its kinetic energy when it is at position 2. The right hand side of the equation, thus, indicates the change in the kinetic energy of the particle during its motion from position 1 to position 2. We can, therefore, conclude that the *work done by a force 'F' on a particle in moving it through a distance is equal to the change in the kinetic energy of the particle. This is known as the work energy principle.*

If instead of one force, there are several forces acting on the particle, then the change in the kinetic energy of the particle will be equal to the algebraic sum of the work done by all the forces acting on the particle.

The principle of work and energy is very useful in solving problems of dynamics of rigid bodies, where relation between force, velocity and displacement is required. In this method, the determination of acceleration is eliminated and the solution of the problem is direct. We shall now apply this principle to the solution of problems in the dynamics of bodies undergoing rectilinear, curvilinear and rotational motion.

8.3 Application of Work and Energy Principle to Rigid Bodies Undergoing Rectilinear Translation

It is known that when a rigid body is having a rectilinear translation, then all its particles undergo the same displacement and have equal velocity and acceleration. Such a motion of a body can, therefore, be represented by the motion of its centre of gravity. We also know that the effect of a number of forces acting on a body is equal to that of a single force which is called the resultant of these forces. Also the motion of the body takes place in a straight line and in a direction in which the resultant acts. Therefore, in a rectilinear motion, if the direction of the resultant is assumed as X-axis, then

$$\Sigma F_X = R$$

and $\Sigma F_Y = 0$

where ΣF_X = Algebraic sum of the components of all the forces in the X direction

ΣF_Y = Algebraic sum of the components of all the forces in the Y direction.

R = Resultant of the forces acting on the body.

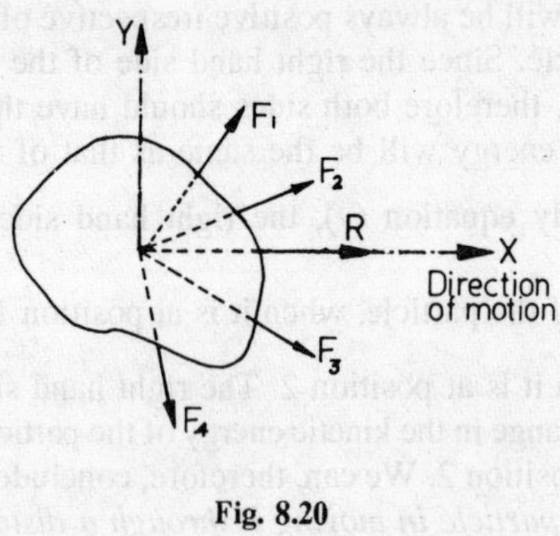

Fig. 8.20

Applying the principle of work and energy to rigid bodies in rectilinear translation we can write,

$$W_R = E_K$$

where, W_R = Algebraic sum of the work done by all the forces acting on the body.

E_K = Change in the kinetic energy of the body

$= \frac{1}{2} m (V_2^2 - V_1^2)$

where

m = mass of the body

V_1, V_2 = velocity of the body at position 1 and 2 respectively.

Example 8.10

A car having a mass of 1000 kgs is running at a velocity of 15 ms^{-1}. Determine its kinetic energy.

Solution

$$E_K = \frac{1}{2} m V^2$$

$$= \frac{1}{2} \times 1000 \times 15^2$$

$$= \mathbf{112{,}500\ J}$$

Example 8.11

A stone of mass '*m*' kg is dropped from the top of a building. Determine the velocity of the stone, before it strikes the ground. The top of the building is at a height '*h*' metres above the ground surface. Neglect the frictional resistance due to air.

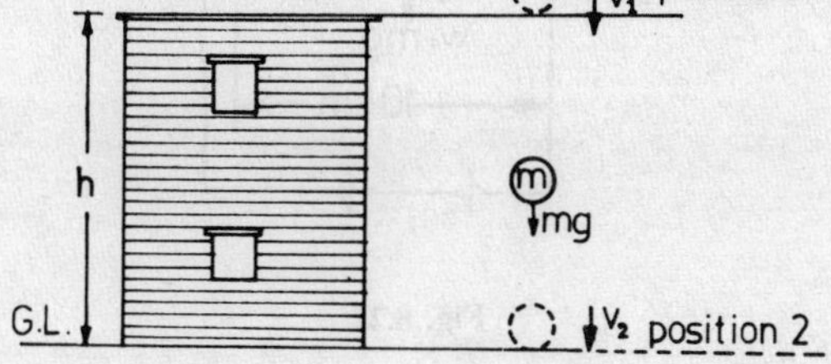

Fig. 8.21

Solution

Let the velocity of the stone at the top of the building and just before it strikes the ground be v_1 and v_2 respectively, then

$v_1 = 0$

(i) *Forces acting on the stone*

The only force acting on the stone is its weight '*W*', acting vertically downwards which will be equal to *mg*.

(ii) *Kinetic energy*

KE of the stone at top of building = 0

KE of the stone just before it strikes the ground = $\frac{1}{2}\ mV_2^2$

$\therefore$ Change in KE (E_K) = $\frac{1}{2}\ m\ V_2^2$

(iii) *Work done*

Work done by the stone = $mg \cdot h$

(iv) *Applying the principle of work and energy*

Resultant work done = Change in KE

$$\therefore \quad mgh = \frac{1}{2}\ m\ V_2^2$$

or

$$V_2 = \sqrt{2\ g\ h}\ ms^{-1}$$

Example 8.12

A bullet of mass 20 gm moving at a velocity of 150 ms^{-1} strikes a wooden log and penetrates through a distance of 10 cms. Calculate the average retarding force offered by the log in stopping the bullet.

Solution

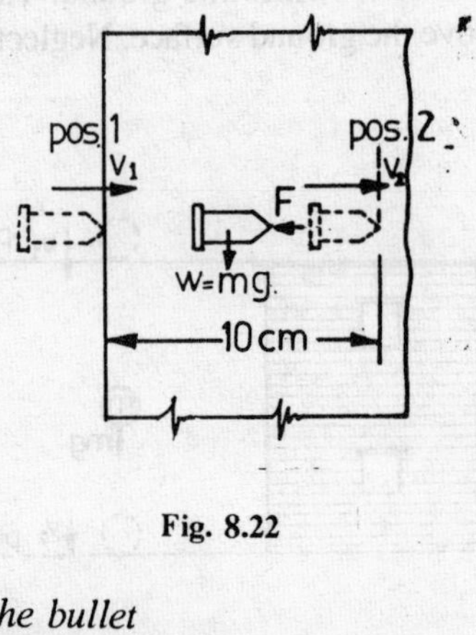

Fig. 8.22

(i) *Forces acting on the bullet*

1 The force of retardation F ←

2 The weight of the bullet W ↓

(ii) *Kinetic energy*

Let v_1 and v_2 be the velocity of the bullet at the time when it strikes the log and when it has penetrated a depth of 10 cms respectively, then

KE of the bullet at position 1 = $\frac{1}{2}\ mV_1^2$

$$= \frac{1}{2} \times \frac{20}{1000} \times 150^2$$

$$= 225 \text{ J}$$

KE of the bullet at position 2 = $0(\because V_2 = 0)$

$\therefore$ Change in the kinetic energy $= 0 - 225$

$$= -225 \text{ J}$$

(iii) *Work*

Resultant work done $= -F \times 0.10 + W \times 0$

$$= -0.10 \text{ F J}$$

(iv) *Applying the work energy principles*

Resultant work done = Change in KE.

$$-0.10\, F = -225$$

$\therefore$ $F =$ **2250 N**

Example 8.13

A car of mass 1500 kg is moving down a hill having a slope of 15° to the horizontal. At the time, when the car is moving at a speed of 10 ms^{-1}, the driver applies the brakes. Calculate the average force applied parallel to the hill slope that will stop the car in a distance of 30 metres.

Solution

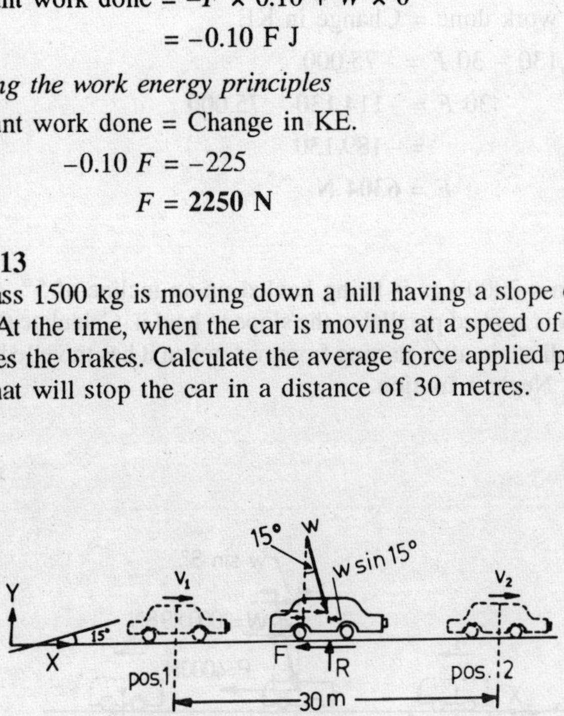

Fig. 8.23

(i) *Forces acting on the car*

W, the weight of the car = 1500 × 9.8 = 14,700 N

R, the reaction normal to road surface

F, Average retarding force parallel to the road surface.

(ii) *Kinetic energy*

KE of the car at 1 $= \frac{1}{2}\, m\, V_1^2$

$$= \frac{1}{2} \times 1500 \times 10^2 = 75{,}000 \text{ J}$$

KE of the car at 2 = 0 ($\because V_2 = 0$)

$\therefore$ Change in KE = –75,000 J

(iii) *Work*

Resultant work done = $W \sin 15° \times 30 + R \times 0 - F \times 30$

$= 14{,}700 \times 30 \times 0.2588 - 30\,F$

$= 114{,}130 - 30$ F

(iv) *Applying work energy principle*

Resultant work done = Change in KE

$114{,}130 - 30\,F = -75{,}000$

$-30\,F = -114{,}130 - 75{,}000$

$= -189{,}130$

$\therefore$ F = **6304 N**

Example 8.14

A trolley of mass 2000 kgs is being hauled up an incline of 5°. A constant force of 4000 N is applied parallel to the slope to haul it. Calculate the velocity of the trolley will attain after starting from rest, when it has travelled a distance of 200 metres. Neglect friction.

Solution

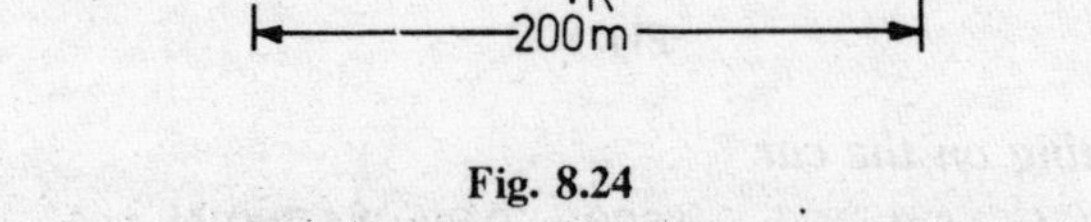

Fig. 8.24

(i) *Forces acting on the trolley*

W, the weight of the trolley = 2000 × 9.8 = 19,600 N

P, the force of pull = 4000 N

R, the reaction of the contact surface

(ii) *Kinetic Energy*

K E of the trolley at 1 = 0 ($\because V_1 = 0$)

$$\text{KE of the trolley at 2} = \frac{1}{2} \times 2000 \times V_2^2$$
$$= 1000\ V_2^2$$
$$\text{Change in KE} = 1000\ V_2^2$$

(iii) *Work*

$$\text{Resultant work done} = -W \sin 5° \times 200 + P \times 200 + R \times 0$$
$$= -2000 \times 9.8 \times 0.0872 \times 200 + 4000 \times 200$$
$$= 458{,}176 \text{ J}$$

(iv) *Applyiing the principle of work and energy*

$$\text{Change in KE} = \text{Resultant work done}$$
$$1000\ V_2^2 = 458{,}176$$
$$v_2^2 = \sqrt{\frac{458176}{1000}}$$
$$= \mathbf{21.4\ ms^{-1}}$$

Example 8.15

A block A of weight 150 kgf is being hauled up an incline of 20° by a block B of weight 300 kgf moving downwards over a pulley. If the system is released from rest, determine the velocity of block A after the block B has moved down by a distance of 10 metres. The coefficient of friction between the block A and the surface is 0.20. The pulley may be considered as weightless and frictionless.

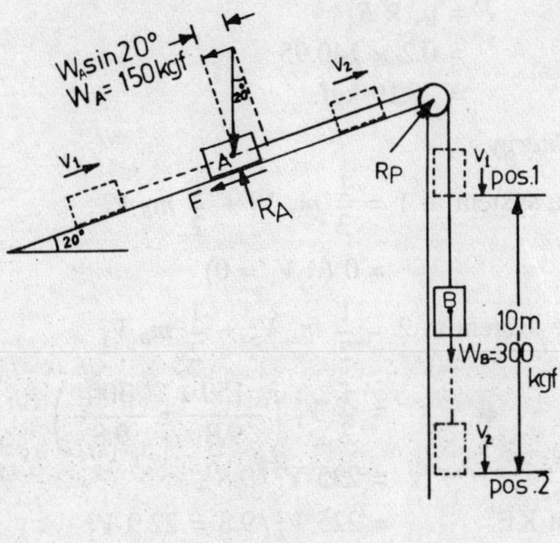

Fig. 8.25

Solution

(i) *Forces acting on the system are*

W_A, the weight of block A

R_A, the normal reaction offerred by the inclined surface
F, the frictional resistance at the contact surface of block A
R_P, the reaction at the pulley
W_B, the weight of block B

Out of these forces R_A, F and R_P are unknown. As R_A and R_P do not do any work, therefore we are interested only in finding the force F.

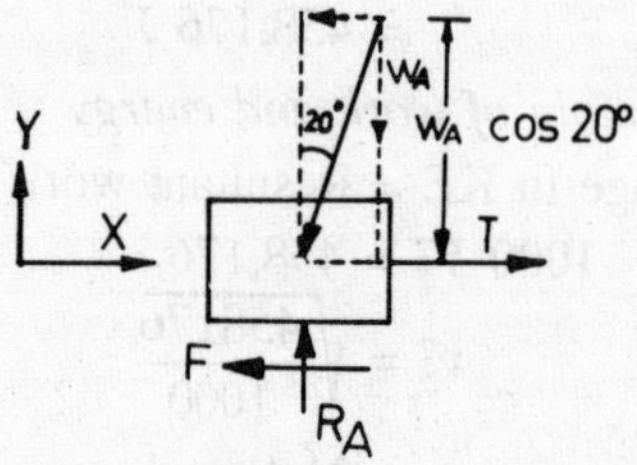

Fig. 8.26

From the free body diagram of the block A(Fig. 8.26)

$\therefore\ \Sigma F_Y = 0$ ($\because$ Block A is undergoing a rectilinear motion in the direction X)

$$R_A - W\cos 20° = 0$$

$$R_A = W\cos 20° = 150 \times 0.9397 = 140.95 \text{ kgf.}$$

$$F = \mu_k \times R_A = 0.2 \times 140.95 = 28.19 \text{ kgf.}$$

(ii) *Kinetic Energy*

$$\text{KE of the system at 1} = \frac{1}{2} m_A V_1^2 + \frac{1}{2} m_B V_1^2 = 0 \ (\because V_1 = 0)$$

$$\text{KE of the system at 2} = \frac{1}{2} m_A V_2^2 + \frac{1}{2} m_B V_2^2 = \frac{1}{2} V_2^2 \left(\frac{150}{9.8} + \frac{300}{9.8}\right) = 225\, V_2^2/9.8$$

$$\text{Change in KE} = 225\, V_2^2/9.8 = 22.9\, V_2^2$$

(iii) *Work*

$$\text{Resultant work done} = -W_A \sin 20° \times 10 - F \times 10 + R_A \times 0 + R_P \times 0 + W_B \times 10$$

$$= -150 \times 0.342 \times 10 - 28.19 \times 10 + 0 + 0$$

$$+ 300 \times 10$$
$$= -513 - 281.9 + 3000$$
$$= 2205.1 \text{ kgf m.}$$

(iv) Applying the principle of work and energy

Change in KE = Resultant work done

$22.9\ V_2^2 = 2205.1$

$$V_2 = \sqrt{\frac{2205.1}{22.9}} = 9.81 \text{ ms}^{-1}$$

Example 8.16

A block of mass '*m*' is held by a spring having a stiffness of '*k*' Newtons/metre. The block is pulled down through a distance of *d* metres from its static equilibrium position and then released. Determine its velocity when it passes through static equilibrium position.

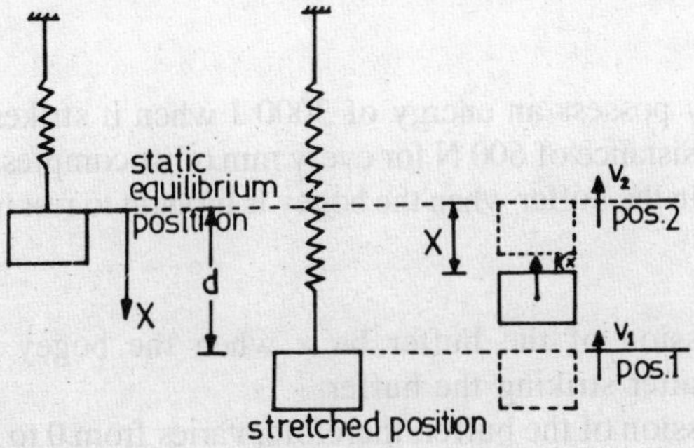

Fig. 8.27

(i) *Kinetic Energy*

Let the velocity of the block at the time when it is released from the stretched position be V_1

Also let the velocity of the block when it passes in its upward journey from the static equilibrium position be V_2, then

Kinetic energy of the block at position 1 = 0 ($\because V_1 = 0$)

$$\text{KE of the block at postion 2} = \frac{1}{2}\ m\ V_2^2$$

$$\text{Change in the KE} = \frac{1}{2}\ mV_2^2$$

(ii) *Work*

Resultant force acting on the block = Force in the spring over and above the force which balances the weight of the block

This force is variable and proportional to the stretch of the spring.

As the block moves between position 1 and 2, the stretch of the spring varies from d to 0. As the force in the spring is proportional to its stretch, its magnitude varies from kd at position 1, to 0 at position 2.

$$\therefore \quad \text{Work done} = \frac{1}{2}(kd + 0) \times d$$

$$= \frac{1}{2} kd^2$$

(iii) *Applying the principle of work and energy*

Change in KE = Resultant work done

$$\frac{1}{2} m V_2^2 = \frac{1}{2} kd^2$$

$$V_2 = \frac{kd^2}{m}$$

$$V_2 = d\sqrt{k/m}$$

Example 8.17

A moving bogey possess an energy of 3000 J when it strikes a buffer. The buffer offers a resistance of 600 N for every mm of its compression. Calculate the compression in the buffer when the bogey is brought to rest instantaneously.

Solution

Let the compression of the buffer be x when the bogey comes to rest instantaneously after striking the buffer.

The compression of the buffer, therefore, varies from 0 to x. It is 0, when the bogey comes in contact with the buffer and reaches x, when the bogey has come to an instant rest.

Fig. 8.28 shows the graph of the resistance offered by the buffer to its compression in mm.

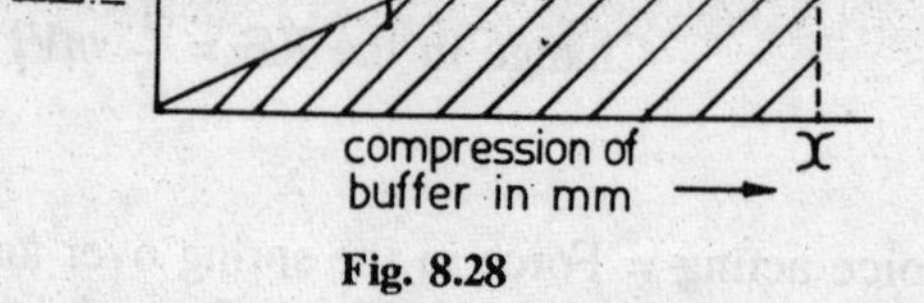

Fig. 8.28

Let the compression in the buffer when it comes to instant rest

$= F = 600\,x$ N

Then the work done in compressing the buffer through x mm

$$= \text{Area under the Load-distance diagram}$$

$$= \frac{1}{2} F \cdot x$$

$$= \frac{1}{2} \cdot 600\ x^2$$

$$= 300\ x^2$$

As the energy in the bogey is equal to the work done in compressing the buffer

$$\therefore\ 300\ x^2 = 3000 \times 10^3 \text{ N mm}$$

$$x^2 = \frac{3000 \times 10^3}{300}$$

$$x^2 = 10{,}000 \text{ mm}$$

$$x = 100 \text{ mm.}$$

Maximum compression in the buffer $= 600 \times 100 = 60{,}000$ N.

8.4 Application of Work Energy Principle to Rigid Bodies Undergoing Curvilinear Translation

We already know that in curvilinear translation, the motion of a body is along a curved path. Also all the particles of the body are assumed to undergo the same displacement, and thus have the same velocity and accleration at any instant of time. This definition requires that the body be considered as a point, and its motion be represented by its centre of gravity. This is possible in case the size of the body is very very small as compared to the path traversed by it.

Let us consider a particle of mass 'm' traversing a curved path as shown in Figure 8.29. Let a resultant force F be acting on it. This force F can be

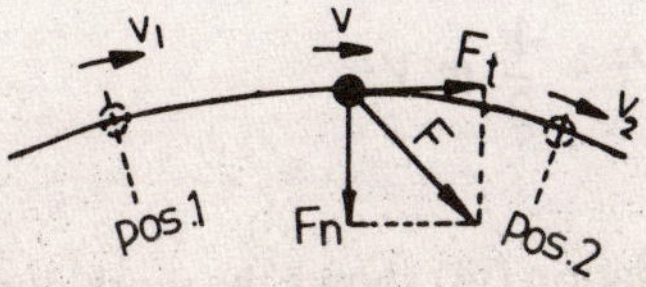

Fig. 8.29

resolved into two components, F_t and F_n, which are tangential and normal to the path. As per equation of motion of curvilinear translation,

$$F_t = m.\ a_t = m\ \frac{dv}{dt} \qquad \text{.... (i)}$$

$$F_n = m.\ a_n = m\ \frac{V^2}{r} \quad \text{.... (ii)}$$

where a_t, a_n = Tangential and normal components of acceleration of the particle

For a small displacement dS of the particle, the work done by the resultant force F acting on the particle will be

Work done by Force 'F' $= F_t \cdot dS$ .... (iii)

($\because$ the work done by the resultant force F is equal to the work done by its tangential component F_t and the work done by its normal component F_n, which is zero. This is so because the direction of F_n is normal to the displacement).

From (i) and (ii), we have

$$F_t \cdot dS = m\ a_t \cdot dS \quad \text{.... (iv)}$$

Also we have

$$a_t = \frac{dV}{dt}$$

and $V = \dfrac{dS}{dt}$

$\therefore\ dS = V\ dt$

and $$a_t \cdot dS = \frac{dv}{dt}\ V\ dt$$

$$= V \cdot dV$$

Substituting in (iv), we get

$$F_t \cdot dS = mv\ dV \quad \text{.... (v)}$$

In the movement of the particle from position 1 to position 2 we have

$$\int_{S_1}^{S_2} F_t \cdot ds = \int_{V_1}^{V_2} mv\ dv$$

or $$\int_{S_1}^{S_2} F_t \cdot ds = \frac{1}{2}\ mV_2^2 - \frac{1}{2}\ m\ V_1^2 \quad \text{.... (vi)}$$

The left hand side of equation (vi) shows the work done by the resultant force acting on the particle in moving from position 1 and 2. The right hand side of the equation shows the change in the kinetic energy of the particle. Equation (vi) is, thus, the work energy equation for curvilinear translation.

Example 8.18

A particle of mass 'm' slides down a smooth curved path. Determine its velocity at position 2, when it starts from rest at position 1.

Solution

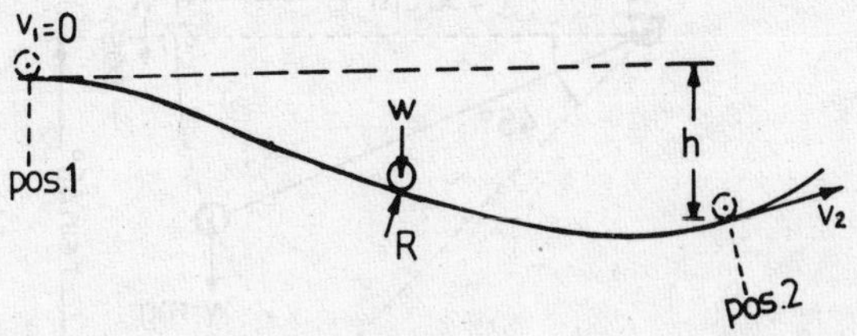

Fig. 8.30

(i) *Forces Acting on the Particle*

W, the weight of the particle and
R, the reaction normal to the contact surface

(ii) *Kinetic Energy*

KE of the particle at position 1 = 0

KE of the particle at position 2 = $\frac{1}{2}\ m\ V_2^2$

$$\text{Change in the KE} = \frac{1}{2}\ m\ V_2^2$$

(iii) *Work*

Resultant work done = $W\ h + R \times 0$ (∵ R is always normal to the displacement)

$$Wh = mgh\ (\because\ W = mg)$$

(iv) *Applying the Principle of Work and Energy*

Change in KE = Resultant work done

$$\frac{1}{2}\ m\ V_2^2 = mgh$$

$$\therefore \quad V_2 = \sqrt{2\ gh}$$

Example 8.19

A simple pendulum as shown in Figure 8.31 is released from rest at position 1. Determine the velocity attained by the bob, after it has traversed through an angle of 45°. The mass of the bob is 1 kg. Also find the tension in the cord at this position.

Solution

(i) *Forces Acting on the Bob*

W, the weight of the bob.

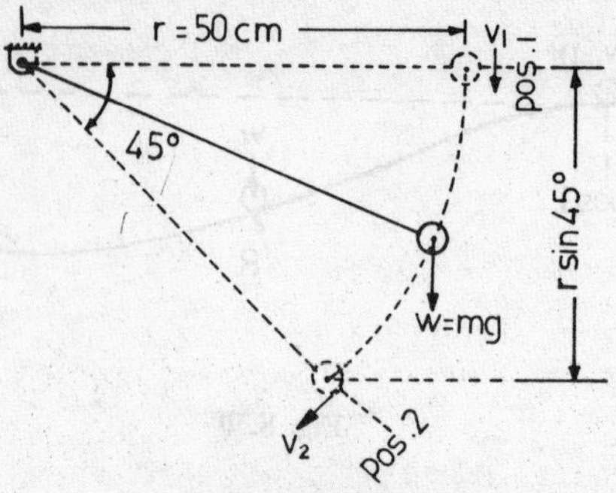

Fig. 8.31

(ii) *Kinetic Energy*

KE of the bob at position 1 = 0 ($\because$ V_1 = 0)

KE of the bob at position 2 = $\frac{1}{2}\ m\ V_2^2$

$$\text{Change in the KE} = \frac{1}{2}\ m\ V_2^2$$

(iii) *Work*

Resultant work done = $mg.\ r \sin 45°$

$$= 1 \times 9.8 \times 0.5 \times \frac{1}{\sqrt{2}}$$

$$= 3.46 \text{ J}$$

(iv) *Applying the Principle of Work and Energy*

Change in KE = Resultant work done

$$\frac{V_2^2}{2} = 3.46$$

$$V_2 = \sqrt{3.46 \times 2}$$

$$= 2.63\ m\ s^{-1}$$

The tension in the cord can be found by applying D'Alemberts' principle, which states that the impressed forces acting on a body are in dynamic equilibrium with the inertial forces of the body. The inertial force of a body is the product of its mass and acceleration, and always acts in a direction opposite to that of the acceleration of the body. Laws of equilibrium as stated in Chapter 3 are, therefore, applicable to a dynamic problem if the intertial forces are introduced.

In the present problem, the bob experiences acceleration, the normal and tangential components of which are shown in Figure 8.32. The normal and

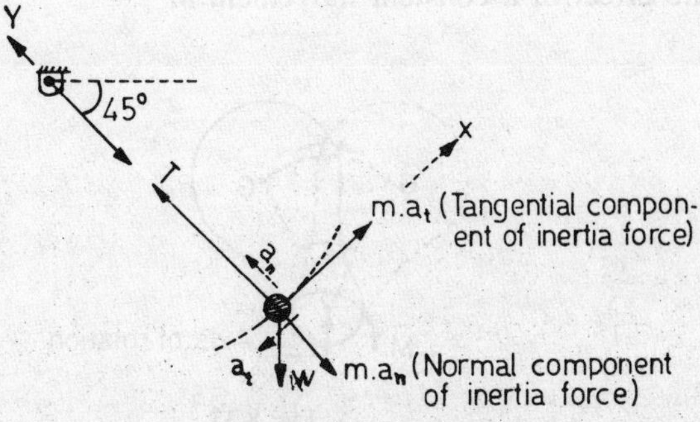

Fig. 8.32

tangential components of the intertial force of the bob are $m \cdot a_n$ and $m \cdot a_t$ respectively. In addition to these innertial forces, other forces acting on the bob are

T = the tension in the string

W = weight of the bob

Selection coordinate axes X and Y as shown in Figure 8.32, we can write

$$\Sigma F_Y = 0$$

$$\therefore\ T - m \cdot a_n - W \cos 45° = 0$$

$$T - m \cdot \frac{V_2^2}{r} - mg \cos 45° = 0$$

$$\text{or } T - 1 \times \frac{2.63^2}{0.5} - 1 \times 9.8 \times \frac{1}{\sqrt{2}} = 0$$

$$\therefore\ T = 1 \times \frac{2.63^2}{0.5} + \frac{9.8}{\sqrt{2}}$$

$$= \mathbf{20.76\ N}$$

8.5 Application of Work Energy Principle to Bodies Undergoing Rotation about a Fixed Axis

As studied earlier, a body is said to rotate about a fixed axis, if all the particles constituting the body move such that their distance from the fixed axis remains constant for all positions of the body. It means that the path described by each particle is a circle. For the particles of the body placed at an equal distance from the fixed axis, their displacement, velocity and acceleration will be equal. But for the particles placed at unequal distances from the fixed axis, their linear displacement, linear velocity and linear acceleration will be different.

Let a rigid body as shown in Figure 8.33 be rotating about a fixed axis under the effect of a constant movement M

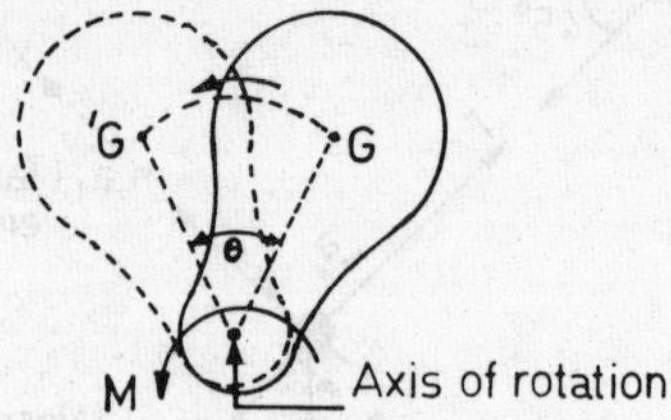

Fig. 8.33

The constant moment 'M' can be considered as equivalent to a force F applied at a distance 'r' from the axis of rotation, such that

$$M = F \cdot r \quad \text{.... (i)}$$

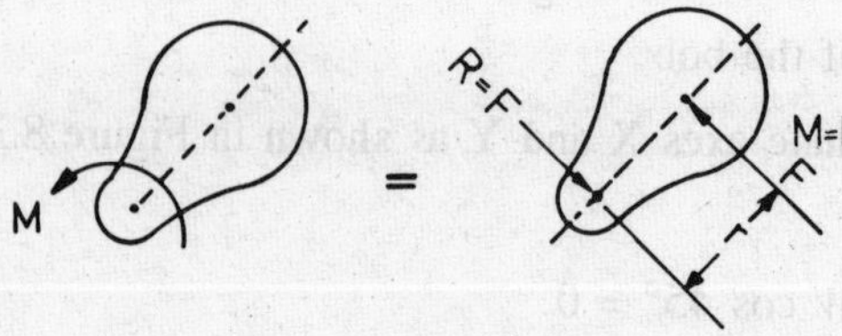

Fig. 8.34

We already know that the work done by a force F

$$= F \cdot S \quad \text{.... (ii)}$$

where S is the displacement component in the direction of the force F. When the body rotates through an angle θ from position 1 to position 2 (Fig. 8.35)

$$S = r \cdot \theta \quad \text{.... (iii)}$$

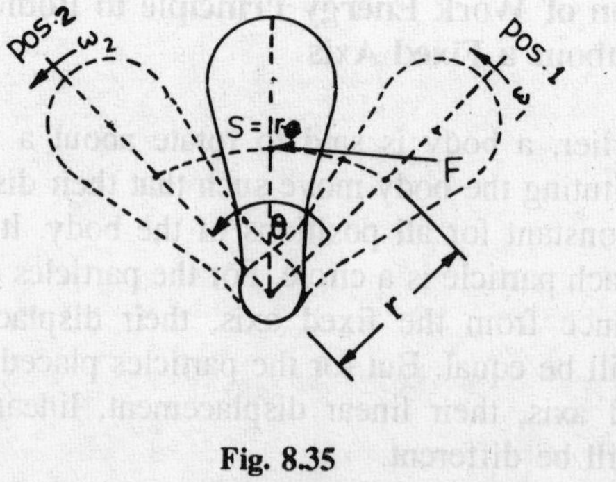

Fig. 8.35

$\therefore$ Substituting for s in (ii), we get

Work done $= F\,r\,\theta$

$$= M\,\theta \quad (\because M = F\,r) \qquad \text{.... (iv)}$$

Applying Newton's second law of motion to bodies under rotation, we have

$$M = I \propto \qquad \text{.... (v)}$$

where M = Resultant moment about the axis of rotation of all the external forces acting on the body.

I = Moment of inertia of the body about the axis of rotation.

$\propto$ = Angular acceleration of the body.

Multiplying both sides of equation (v) by θ, we get

$$M\,\theta = I \propto \theta \qquad \text{.... (vi)}$$

From the circular motion of the bodies we know, that

$$2 \propto \theta = \omega_2^2 - \omega_1^2$$

$$\therefore \qquad \propto \theta = \frac{1}{2}\,(\omega_2^2 - \omega_1^2)$$

Substituting for $\propto \theta$ in (vi), we get

$$M\,\theta = \frac{1}{2}\,I\,(\omega_2^2 - \omega_1^2) \qquad \text{...(vii)}$$

Equation (vii) is the work energy equation for bodies undergoing rotation. Left hand side of the equation is the resultant work done by all the forces acting on the body between positions 1 and 2. Right hand side of the equation indicates the change in the kinetic energy of the body, ω_1 and ω_2 being the angular velocities of the body at positions 1 and 2 respectively.

Example 8.20

A flywheel having a rotational moment of inertia of 20 kg–m² is rotating at a speed of 10 rad. s⁻¹ at an instant of time. Find the angular velocity of the fly wheel, after it has rotated through 6 π rad. A constant moment of 200 Nm is applied to the flywheel. Neglect friction at the bearing.

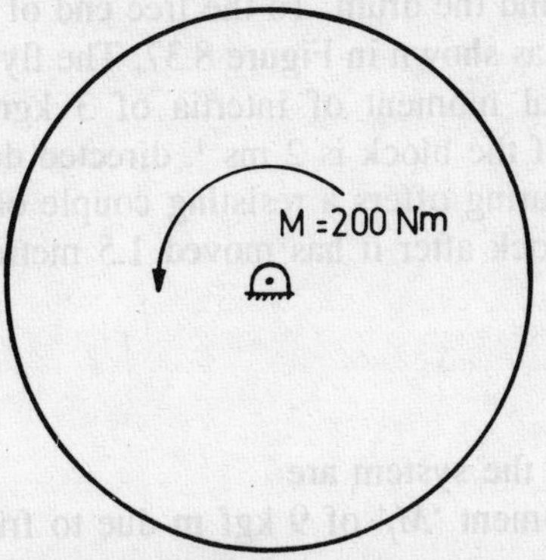

Fig. 8.36

Solution

(i) *Kinetic Energy*

$$\text{K E at position 1} = \frac{1}{2} I \omega_1^2$$

$$= \frac{1}{2} \times 20 \times 10^2 = 1000 \text{ Nm}$$

$$\text{K E at position 2} = \frac{1}{2} \times I \omega_2^2$$

$$= \frac{1}{2} \times 20 \times \omega_2^2$$

$$= 10 \omega_2^2$$

$$\text{Change in K E} = 10 \omega_2^2 - 1000$$

(ii) *Work*

$$\text{Resultant work done} = M \theta$$

$$= 200 \times 6 \pi$$

$$= 3769.9 \text{ J}$$

(iii) *Applying the Principle of Work and Energy*

$$\text{Change in KE} = \text{Resultant work done}$$

$$10 \omega_2^2 - 1000 = 3769.9$$

$$10 \omega_2^2 = 3769.9 + 1000$$

$$= 4769.9$$

$$\omega_2^2 = \frac{4769.9}{10}$$

$$\omega_2 = \sqrt{\frac{4769.9}{10}}$$

$$\omega_2 = \mathbf{21.84 \text{ Rad. s}^{-1}}$$

Example 8.21

A drum of radius 50 cms is rigidly attached to a flywheel. An inextensible cable is wrapped around the drum. To the free end of this cable, a block of 200 kgf is suspended as shown in Figure 8.37. The flywheel and drum have a combined centroidal moment of intertia of 3 kgm–m². At the instant shown, the velocity of the block is 2 ms^{-1}, directed downwards. Assuming that friction at the bearing offers a resisting couple of 9 kgf. m, determine the velocity of the block after it has moved 1.5 metres downward.

Solution

(i) *Forces*

Forces acting on the system are

(a) Resisting moment 'M_f' of 9 kgf m due to friction at the bearing.

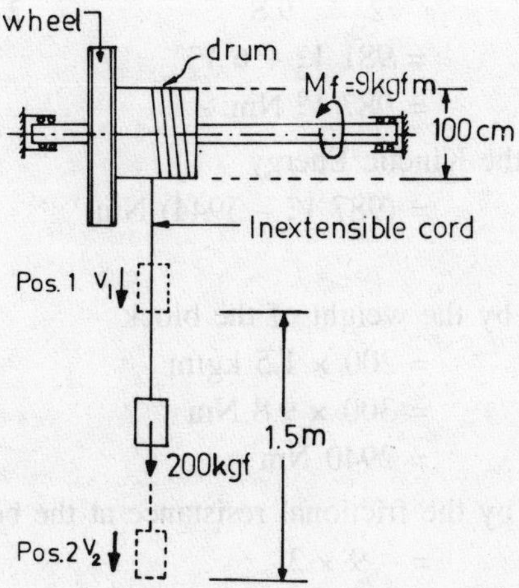

Fig. 8.37

(b) Weight '*W*' of the block = 200 kgf

Let the angle turned by the drum, when the block moves down a distance of 1.5 m be θ then

$$r\,\theta = 150 \text{ cms}$$

$$\theta = \frac{150}{50}$$

$$= 3 \text{ rad}$$

(ii) *Kinetic Energy*

Initial velocity of the block $v_1 = 2\ \text{ms}^{-1}$
Initial angular velocity (ω_1) of the drum

$$= \frac{2}{0.5} = 4 \text{ rad s}^{-1}$$

$$\text{K E at position } 1 = \frac{1}{2}\, m \text{ block } V_1^2 + \frac{1}{2}\, \text{I}\, \omega_1^2$$

$$= \frac{1}{2}\ \frac{200}{9.8}\ 2^2 \times 9.81 + \frac{1}{2} \times 3 \times 4^2$$

$$= 3944 \text{ Nm}$$

Let the velocity of the block at position 2 be v_2, then angular velocity of the flywheel and drum at this position will be

$$\omega_2 = \frac{V_2}{0.5} = 2\, V_2$$

$$\text{KE at position } 2 = \frac{1}{2}\ \text{m (block)}\ V_2^2 + \frac{1}{2}\ \text{I}\ \omega_2^2$$

$$= \frac{1}{2} \times \frac{200}{9.8} V_2^2 \times 9.81 + \frac{1}{2} \times 3 \times (2\ v_2)^2$$

$$= 981\ V_2^2 + 6\ V_2^2$$

$$= 987\ V_2^2 \text{ Nm.}$$

Change in the Kinetic Energy

$$= (987\ V_2^2 - 3944) \text{ Nm}$$

(iii) *Work*

Work done by the weight of the block

$$= 200 \times 1.5 \text{ kgfm}$$

$$= 300 \times 9.8 \text{ Nm}$$

$$= 2940 \text{ Nm}$$

Work done by the frictional resistance at the bearings

$$= -9 \times 3$$

$$= -27 \text{ kgf.m.}$$

$$= -27 \times 9.8 = 264.6 \text{ Nm.}$$

Resultant work done

$$= 2940 - 264.6 = 2675.4$$

(iv) *Applying the Principle of Work and Energy*

Change in KE = Resultant work done

$$986\ v_2^2 - 3944 = 2675.4$$

$$986\ V_2^2 = 2675.4 + 3944$$

$$= 6619.4$$

$$V_2 = \sqrt{\frac{6619.4}{986}}$$

$$= 2.59 \text{ ms}^{-1}$$

Example 8.22

A flywheel consists of a flat circular disc and has a mass of 200 kgs. The flywheel is brought to rest through a braking system as shown in Figure 8.38. Determine the magnitude of the force *P* as shown in Figure 8.38, in order

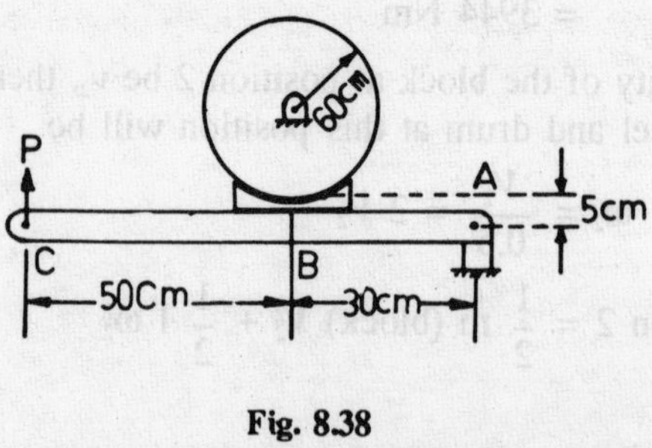

Fig. 8.38

to bring the flywheel to rest in 25 revolutions. The angular velocity of the wheel at the time of applying the brakes is
(a) 400 rpm clockwise
(b) 400 rpm anticlockwise.
The coefficient of friction between wheel and the brake is 0.2

Solution

Moment of intertia '*I*' of the wheel about its

axis of rotation $= \frac{1}{2}\, mr^2$

$= \frac{1}{2} \times 200 \times .6^2 = 36$ kg m²

(a) *When Angular Velocity of the Wheel is 400 rpm Clockwise*

(*i*) *Work*

Let the frictional force applied by the brake to bring the wheel to rest in 25 revolutions be *F* Newtons. Forces and moments acting on the flywheel are shown in its freebody diagram in Figure 8.39.

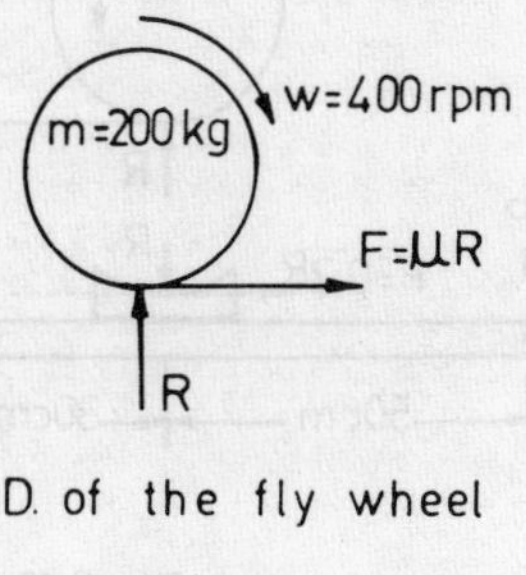

Fig. 8.39

Retarding moment M_r acting on the wheel due to the frictional resistance *F*

$$M_r = F \cdot r$$
$$= F \times 0.6$$
$$= -0.6 \text{ F Nm.}$$

Work done by the retarding moment M_r in 25 revolutions

$$= Mr\ \theta$$
$$= -0.6\ F \times 2\pi \times 25 = -94.24 \text{ F}$$

Work done by $R = 0$ ($\because$ displacement is normal to the line of action of reaction *R*)

Resultant work done = $-94.25\ F$

(ii) *Kinetic Energy*

$$\text{KE at position 1} = \frac{1}{2} I \omega_1^2$$

$$= \frac{1}{2} \times 36 \left(\frac{400}{60} \times 2\pi \right)^2 = 31600 \text{ J}$$

$$\text{KE at position 2} = 0 \quad (\because \omega_2 = 0)$$

$$\text{Change in K E} = 0 - 31600 \text{ J}$$

$$= -31600 \text{ J}$$

(iii) *Applying the Principle of Work and Energy*

Resultant work done = Change in KE

$$-94.25\, F = -31600 \text{ J}$$

$$\because \quad F = 335 \text{ N}$$

(iv) *To Find out the Magnitude of P*, let us draw the FBD of the brake system as shown in Figure 8.40. Taking moment of all the forces about *A*, we have for equilibrium $\Sigma M = 0$

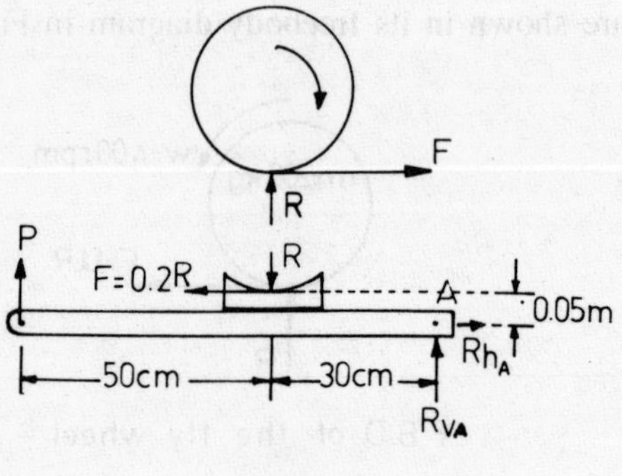

Fig. 8.40

$$P \times 0.8 - F \times 0.05 - R \times 0.3 = 0$$

We know that,

$$F = \mu \cdot R$$

$$= 0.2 \text{ R}$$

$$\therefore \quad R = 5\, F$$

Substituting for *R* in the above equation

$$0.8\, P - 0.05\, F - 5\, F \times 0.3 = 0$$

$$\text{or} \quad 0.8\, P - 0.05\, F - 1.5 \text{ F} = 0$$

$$\text{or} \quad 0.8\, P = 1.55\, F$$

$$\text{or} \quad P = \frac{1.55}{0.8} \times 335 = 649 \text{ N}$$

(b) *When the initial angular velocity is 400 rpm anticlockwise*, the only change will be that the direction of the frictional force F will reverse.

The free body diagram of the brake system is shown in Figure 8.41. Taking moments of all forces about A. For equilibrium, $\Sigma M = 0$

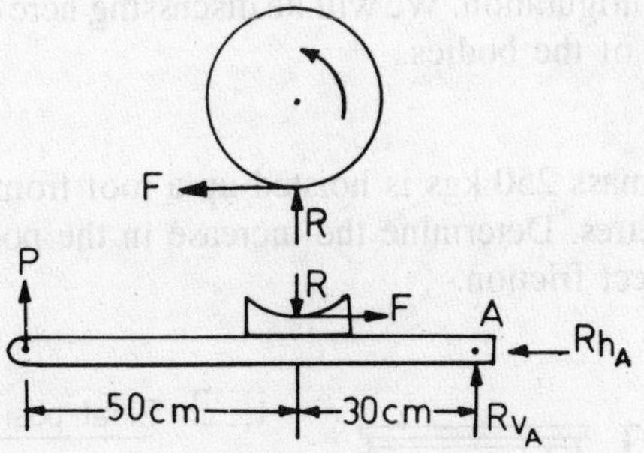

Fig. 8.41

$$P \times 0.8 + F \times 0.05 - R \times 0.3 = 0$$

Substituting for R, we get

$$0.8\,P + 0.05\,F - 5\,F \times 0.3 = 0$$

$$0.8\,P + 0.05\,F - 1.5\,F = 0$$

$$0.8\,P = 1.45\,F$$

$$P = \frac{1.45}{0.8} \times 335 = \mathbf{607\ N}$$

8.6 Potential Energy

Let us take the example of a block of mass '*m*' being lifted by a man through a height '*h*'. In doing this job, the man has done work against the gravitational pull acting on the block directed vertically downwards. The work

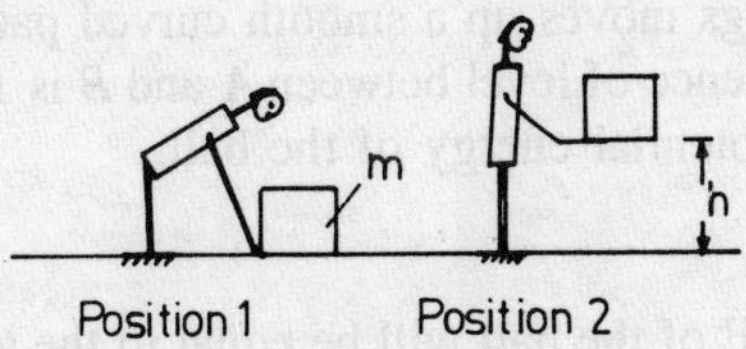

Fig. 8.42

done by the man will be *mgh*. From this new position, if the ball is allowed to fall back on the ground, it has a capacity of doing work equal to *mgh*. This capacity of doing work gained by the ball because of its position above the reference level (ground) is called its gravitational potential energy or simply potential energy. *Potential energy of a body is thus the energy possessed by it due to its position.* It is always expressed relative to some reference level, such as earth's surface etc. There are also other forms of potential energy, such as elastic, chemical, electrical and nuclear. These are due to the internal mechanism or configuration. We will be discussing here only the gravitational potential energy of the bodies.

Example 8.23

A stone slab of mass 250 kgs is hoisted up a roof from the ground through a height of 3 metres. Determine the increase in the potential energy of the stone slab. Neglect friction.

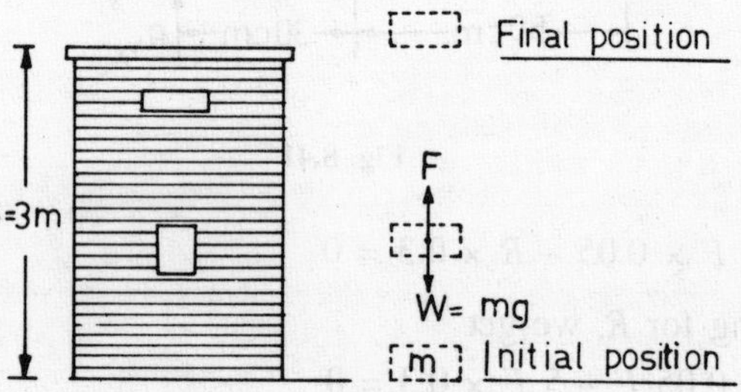

Fig. 8.43

Solution

The increase in the potential energy of the stone slab will be equal to the work done on raising it from the ground floor to the roof level.

PE = Work done = $F \cdot S$

$= mg.\ S$

$= 250 \times 9.8 \times 3$

$= 7{,}350$ J

Example 8.24

A ball of mass 100 kgs moves up a smooth curved path from position A to position B. The difference of level between A and B is 10 metres. Determine the increase in the potential energy of the ball.

Solution

The increase in the PE of the ball will be equal to the work done in moving the ball up from A to B.

$\therefore$ PE = Work done = mgh

$= 100 \times 9.8 \times 10$

= 9,800 J

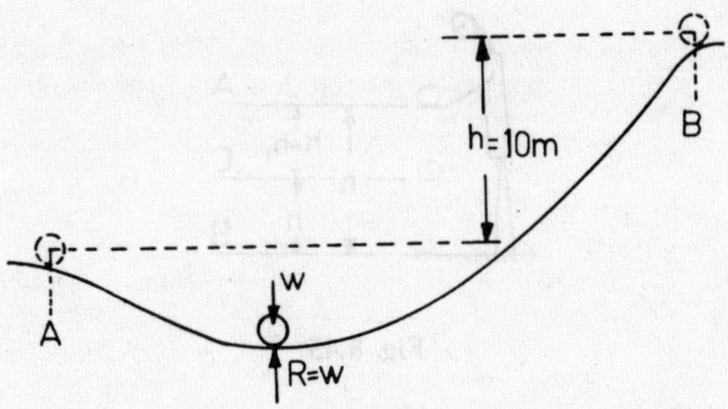

Fig. 8.44

8.7 Conservation of Energy

Let us consider a man holding a ball of mass '*m*' at *A* which is '*h*' metres above the ground surface (Fig. 8.45). Considering the ground surface as the reference level for measuring potential energy, the ball at *A* has a potential energy of *mgh*. The kinetic energy of the ball at A will be zero, because it is at rest and has no velocity. The total *mechanical energy* (sum of potential and kinetic energy) of the ball at *A* is, therefore, *mgh*. As the ball is released from *A*, it starts gaining velocity ignoring frictional resistance due to air, the velocity v_c of the ball at any position C when it is at a height h_1 from the ground surface, will be $\sqrt{2g\,(h - h_1)}$ (Refer Example 8.5). At this position,

PE of the ball $= mgh_1$

KE of the ball $= \frac{1}{2}\, m\, V_c^2$

$$= \frac{1}{2}\, m \left[\sqrt{2\, g\, (h - h_1)}\right]^2$$

$$= \frac{1}{2}\, m.\, 2\, g\, (h - h_1)$$

$$= mg\, (h - h_1)$$

Total energy $= \text{PE} + \text{KE}$

$$= mgh_1 + mg\, (h - h_1)$$

$$= mg\, (h_1 + h - h_1)$$

$$= mgh$$

When it just strikes the ground at B, it will have attained a velocity equal to $\sqrt{2\, gh}$. Thus at *B*, the kinetic energy of the ball is $\frac{1}{2}$ m $(\sqrt{2\, gh})^2 = mgh$, and its potential energy is zero. The total mechanical energy of the ball at

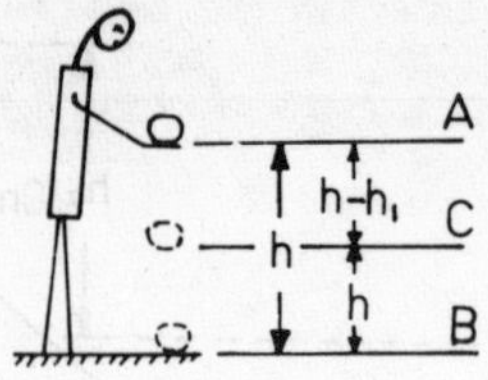

Fig. 8.45

B, just before it strikes the ground is also, therefore, equal to *mgh*. From above, we can conclude that

(1) the total mechanical energy of the ball at *A*, *B* or at any other position *C* in between them is the same, provided frictional resistance due to air is neglected.

(2) No energy is destroyed or created, but only its form is changed.

These conclusions lead us to the *Principle of conservation of energy which states that in any reaction within a system, or in any process, no energy is either created or destroyed. It only changes from one body to another or from one form to another*. Applying this principle to rigid bodies in motion, we can say that the total energy of a rigid body i.e. the sum of its potential and kinetic energy under the action of conservative forces remains constant. It is true that in many processes, some of the energy becomes unavailable, e.g. in the case of the ball falling freely from position *A* to *B* if resistance due to air is considered, then the velocity attained by the ball at any position *C* will not be the same as mentioned earlier, but will be less than this. It will mean a decrease in the kinetic energy of the ball, and thus, a decrease in the total available energy of the ball. Thus decrease in the total energy of the ball is due to the work done by the frictional resistance offered by the air, which is converted into heat energy and is wasted in heating the ball as well as the particles surrounding it in the air. The energy is, thus, not lost, but converted into a form in which it is not available for doing any useful work and is wasted. In such a case

Total energy of the ball at *A* = Total energy of the ball at B + Work done against friction.

Example 8.25

A car of mass 2000 kgs rolls down a curved path of length 200 metres as shown in Figure 8.46. If the car is allowed to roll freely from rest at *A*, determine its velocity when it passes *B*. Assume that a constant force of 500 N parallel to the road surface is retarding the motion of the car.

Solution

Taking B as the reference level for measuring potential energy

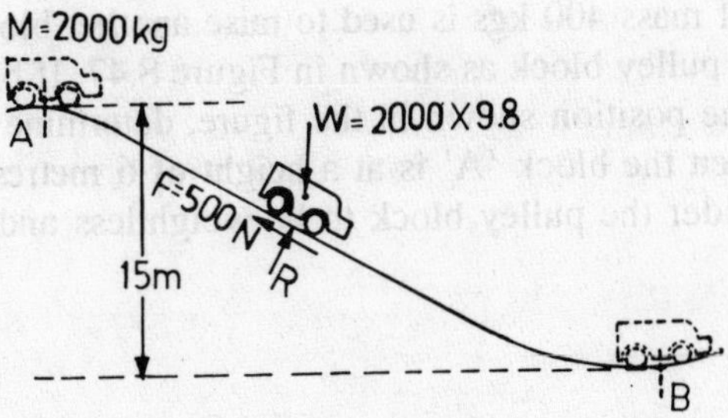

Fig. 8.46

(i) *At A*

PE of the car $= mgh$

$= 2000 \times 9.8 \times 15$

$= 294{,}000$ Nm

K E $= 0$ ($\because V_A = 0$)

Total energy at A = PE + KE

$= 294{,}000$ Nm

(ii) *At B*

PE of the car = 0

K E of the car $= \frac{1}{2} m V_B^2$

$= \frac{1}{2} \times 2000\ V_B^2$

$= 1000\ V_B^2$

(iii) *Work Done by the Retarding Force from A to B*

$= F \cdot S$

$= 500 \times 200$

$= 100{,}000$ Nm

(iv) *Applying the Principle of Conservation of Energy*

Total energy at A = Total energy at B + Work done against friction

$294{,}000 = 1000\ V_B^2 + 100{,}000$

$1000\ V_B^2 = 294{,}000 - 100{,}000$

$= 194{,}000$

$V_B^2 = \sqrt{194}$

$= 13.943$ m s^{-1}

Example 8.26

A block 'A' of mass 400 kgs is used to raise another block 'B' of mass 200 kgs through a pulley block as shown in Figure 8.47. If block 'A' is released from rest at the position shown in the figure, determine the velocity of the block 'B', when the block 'A' is at a height of 6 metres above the ground surface. Consider the pulley block to be weightless and neglect friction.

Solution

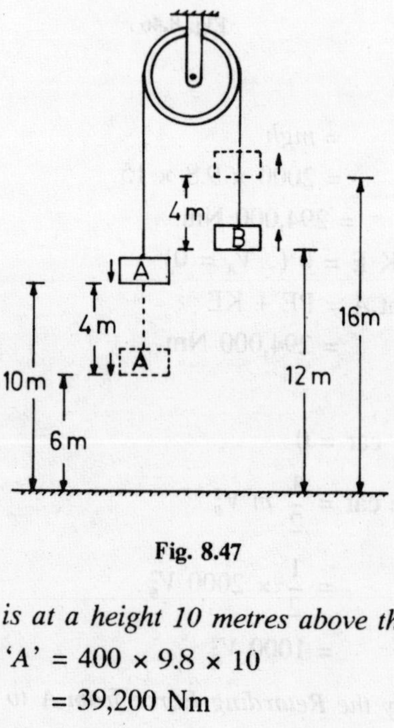

Fig. 8.47

(1) *When Block 'A' is at a height 10 metres above the ground level*

PE of the block 'A' = $400 \times 9.8 \times 10$

= 39,200 Nm

PE of the block B = $200 \times 9.8 \times 12$ = 23,520 Nm

K E of block 'A' = K E of block B = 0

= ($\because$ Velocity of block 'A' and 'B' is zero)

Total energy of the system = 39,200 + 23,520 = 62,720 Nm

(2) *When the Block 'A' is at a height 6 metres above the ground*

PE of block 'A' = $400 \times 9.8 \times 6$ = 23,520 Nm

PE of block 'B' = $200 \times 9.8 \times 16$ = 31,360 Nm

KE of block 'A' = $\frac{1}{2} m_A v^2 = \frac{1}{2} \times 400 \times V^2 = 200\ V^2$

KE of block B = $\frac{1}{2} \times 200\ V^2 = 100\ V^2$

Total energy of the system = $23{,}520 + 31{,}360 + 200\ V^2 + 100\ V^2$

= $54{,}880 + 300\ v^2$

(3) *Applying the Principle of Conservation of Energy*

$$54{,}880 + 300\, v^2 = 62{,}720$$

$$300\, v^2 = 62{,}720 - 54{,}880 = 7840$$

$$v^2 = \frac{7840}{300}$$

$$v = \sqrt{\frac{7840}{300}} = 5.1 \text{ m s}^{-1}.$$

8.8 Power

Let us take the case of hoisting a stone block from the ground floor to the roof of a building. It is understood that in doing this job, a hand operated winch will take much more time as compared to a power operated winch, though the amount of work done in both the cases is the same. Power operated winch is said to be more powerful as compared to the hand operated winch, because it can do the same work much faster. *Power is therefore defined as the time rate of doing work.* If some amount of work has been done in time '*t*', the average power '*p*' will be expressed as $P = \frac{\text{work done}}{\text{time taken}}$

8.8.1 *Units of Power*

In S.I. Units, the units of work is Nm or Joule. As power is the time rate of doing work, therefore the unit of power will be a Joule per second (J s^{-1}), which is called Watt(W).

In MKS units, the unit of power is a metric horse power(*CV*) which is equal to 4500 *kgf m* per minute or 75 *kgf m* per second.

Example 8.27

A man of mass 80 kgs can climb up a height of 10 metres by running over a ramp in 10 seconds. What is his muscular power?

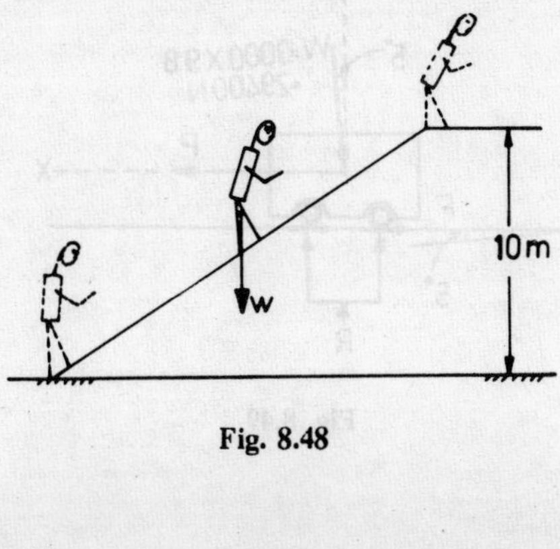

Fig. 8.48

Solution

(1) Work done by the man $= 80 \times 9.8 \times 10 = 7840$ Nm

(2) Time taken to do the work = 10 secs

$$\text{Average Power} = \frac{\text{Work done}}{\text{Time taken}}$$

$$= \frac{7840}{10} = 784 \; W.$$

Example 8.28

What power is needed to fill an overhead tank of capacity 60,000 litres in 2 hours, if the water is to be lifted through a height of 30 metres?

Solution

$$\text{Work done} = 60{,}000 \times 9.8 \times 30 = 1764,\,0000 \text{ J}$$

$$\text{Time} = 2 \times 60 \times 60 = 7200 \text{ seconds}$$

$$\text{Power required} = \frac{1764{,}0000}{7200} = 2450 \; W$$

Example 8.29

A truck of mass 3000 kgs is moving up an incline of 5° at a constant velocity of 45 km h^{-1}. Determine the power required. Rolling resistance offered parallel to the road surface is 5 N/kN.

Solution

(1) *Forces Acting on the Vehicle*
The various forces acting on the vehicle are

(a) *W*, the weight of the truck $= 3000 \times 9.8 = 29{,}400$ N

(b) *F*, the rolling resistance $= 294{,}00 \times \dfrac{5}{1000} = 147$ N

(c) *P*, force applied by the engine in driving the vehicle

(d) *R*, the reaction offered by the road surface.

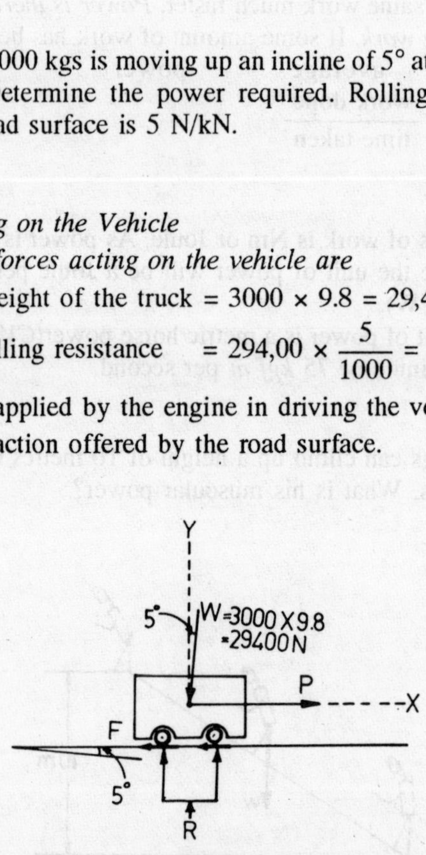

Fig. 8.49

As the truck is moving with a constant velocity, it does not have any acceleration, and thus the resultant force acting on the truck will be zero. In other words, the truck will be in equilibrium. Therefore,

$$\Sigma F_x = 0$$
$$\Sigma F_y = 0$$

We are interested in finding *P*, the motive force applied by the engine in order to find its power.

$$\because \Sigma F_x = 0$$
$$\therefore P - F - W \sin 5° = 0$$
$$P = F + W \sin 5°$$
$$= 147 + 29{,}400 \sin 5° = 147 + 29{,}400 \times 0.0872$$
$$= 147 + 2560 = 2707 \text{ N}$$

We know that,

$$\text{Average power} = \frac{\text{Work done}}{\text{Time taken}}$$

$$\text{Work done} = F \cdot \Delta S$$

where ΔS = displacement in the direction of the force *F*. If Δt is the time taken in undergoing the displacement, then

$$\text{Average Power} = F \cdot \frac{\Delta S}{\Delta t}$$

$$\frac{\Delta S}{\Delta t} = V_{av.} \text{ (The average velocity)}$$

$$\therefore \quad \text{Average Power} = F \cdot V_{av.}$$

In the present case,

The power expended by the engine of the truck will be

$$= P \times V_{av}$$

$$= 2707 \times \frac{45 \times 10^3}{3600} = 33{,}837.5 \text{ } W$$

Example 8.30

A constant motive force of 350 kgf is applied to haul a trolley on a level road. Determine the horse power expended if the trolley is moved through a distance of 500 metres in 3 minutes. Given that the motive force applied is in a direction

(a) Parallel to the road surface

(b) Inclined at 10° to the road surface.

Solution

(a) Work done by the motive force

$$= 350 \times 500 = 175{,}000 \text{ kgf. m.}$$

$$\text{Time taken } (t) = 3 \text{ mins or } 180 \text{ secs}$$

$$\therefore \text{ Average power expended} = \frac{175000}{180 \times 75} = 12.96 \text{ hp}$$

(b) Work done by the motive force = 350 cos 10° × 500

$$= 350 \times 0.9848 \times 500$$

$$= 172{,}340 \text{ kgf. m.}$$

$$\text{Time taken} = 180 \text{ secs}$$

$$\therefore \quad \text{Average power expended} = \frac{172{,}340}{180 \times 75} = 12.76 \text{ hp}$$

Example 8.31

At a time when a work piece of diameter 180 mm is being machined on a lathe, the turning moment(M_t) on the spindle of the lathe is 750 Nm. If the cutting speed (peripheral velocity) is 80 m/min., determine the power expended in the cutting process as well as the cutting force.

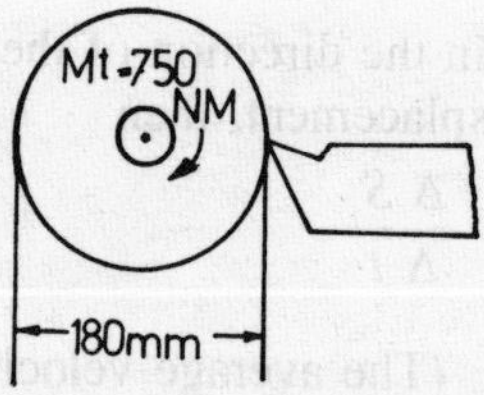

Fig. 8.50

Solution

$$M_t = F \cdot r$$

$$\therefore \quad F = \frac{M_t}{r}$$

$$= \frac{750}{0.09} = 8333 \text{ N}$$

$$\text{Power} = F \times V_{\text{peripheral}}$$

$$= 8333 \times \frac{80}{60} = 11{,}111.1 \text{ W}$$

Summary

Work. Work is said to be done by a force when it moves a body through a distance and the distance thus moved has a component parallel to the direction of the force. It is a scalar quantity. In SI system, its unit of measurement is a Joule, which is the work done by a force of 1 Newton when its point of application moves through a distance of 1 metre in the direction of the force. It is expressed as under:

Work done by force $F = F \times S$

where S = component of the distance moved in the direction of the force F.

Work is regarded as positive, when the displacement has the same sense as that of the force.

Work done by a variable force is equal to the area under the curve representing the force and the distance moved.

Energy. Energy is the capacity to do work.

Kinetic Energy. The energy possessed by a body due to its motion is called its kinetic energy. It is expressed as KE $= \frac{1}{2} m V^2$

where m, v are the mass and velocity of the body.

Work and Energy Principle. The principle states that the resultant work done by a force or a number of forces acting on a body is equal to the change in its kinetic energy.

Potential Energy. The energy possessed by a body due to its position is called its gravitational potential energy.

Total Mechanical Energy It is the sum of the kinetic energy and gravitational potential energy of a body.

Principle of Conservation of Energy. The principle states that in any reaction within a system, or in any process, no energy is either created or destroyed. It only changes from one body to another or from one form to another. The principle can be expressed as follows

Total energy of a body at any position A = Total energy of the body at any other position B + Work done against friction.

Power. Power is the rate of doing work. The units of power in SI system is Watt (W) which is equal to 1 Joule of work done in a period of 1 second. In MKS units the unit of power is a metric horse power, which is equal to a work of 75 kgf m done in a period of 1 second. Power is expressed as under:

$$\text{Power} = \frac{\text{Work done}}{\text{time taken}} = \frac{F.\ \Delta S}{\Delta t} = F \cdot V_{\text{average}}.$$

Review Questions

1 Name two situations, where the force moves, but does not do any work.
2 Is the work a scalar or a vector quantity?
3 State, whether any work is done by a man pushing against a car which does not move. Explain.

4 A man is swinging a ball attached to an inextensible string in a horizontal circle. Is any work being done by the pull in string? Explain.

5 What is the work done by the weight of a body, if it is moved horizontally? Explain.

6 A car is moving at a constant speed on a level road. In this condition, the net force acting on the car is zero. Is any work done on the car? Explain.

7 Two bodies having the same mass are lifted through a height of 6 metres, one in 2 minutes and the other in 5 seconds. In which case is the work done more? In case it is equal, explain why.

8 What is the work done by the resultant of two forces acting on a body, if one of them is acting in the direction of motion of the body and the other normal to the motion of the body. Explain.

9 Differentiate between torque and work.

10 Two identical vehicles are moving on a level surface such that one has a speed double than that of the other. How do their kinetic energies compare?

11 A ball is thrown up in the air. Does its kinetic energy increase as it goes up? Explain.

12 Is it possible for two bodies of different mass to possess the same kinetic energy? If so, state the condition.

13 If the velocity of a body is increased 'n' times, what will be the increase/decrease in its kinetic energy?

14 State the Principle of work and energy and explain with an example.

15 Is only the work done by the motive force, or the resultant work done by all the forces acting on the system considered in the application of work energy principle to a system?

16 A bullet moving at velocity V pierces a body by 10 cms. Had the bullet been moving at a velocity of 2 V, how far would it have pierced the body? Assume the average retarding force to be the same in both the cases.

17 State the difference between potential energy and kinetic energy.

18 If two bodies of same mass have been raised to heights h and 2 h compare their potential energy.

19 State the principle of conservation of energy.

20 What is a conservative force? Is frictional force a conservative force? How is it taken care of in the principle of conservation of energy?

21 What is power? How does it differ from work?

22 Is the displacement or the velocity of the force that effects the power expended?

23 What are the units of power in SI system?

24 You are asked to determine your power, how would you proceed to find it out?

25 If turning moment remains constant and angular velocity is increased, how will it effect power?

26 If power is constant and angular velocity is increased, how will it affect the turning moment?

27 If in the first case, 100 litres of water has to be raised through a height of 8 metres in 8 secs, and in the second case, 50 litres of water has to be raised through a height of 6 metres in 2 secs, in which case a more powerful motor will have to be used?

Problems

8.1 What work will be done in lifting a tin of paint weiging 5 kgf. from the ground floor to the second floor of a building? The height from the ground floor to the second floor is 8 metres.
(Ans. 40 kgf. m).

8.2 What is the work done by a constant horizontal push(P) of 200 N in moving a block through a distance of 10 metres along an incline of 5° as shown in Figure 8.2.
(Ans. 1992.4 Nm).

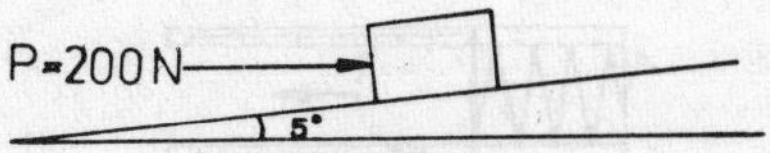

Fig. Prob 8.2

8.3 In ploughing land a tractor applies a motive force of 20 kN for the first 200 metres. For the next 50 metres, 75 metres and 100 metres, it applies motive force of magnitude 15 kN, 25 kN and 18 kN respectively. Determine the total work done by the tractor.
(Ans. 8425×10^3 J).

8.4 What is the work done in raising a block of mass 200 kgs through a height of 15 metres along a smooth curved surface as shown in Figure Prob 8.54.
(Ans. 29,400 Nm).

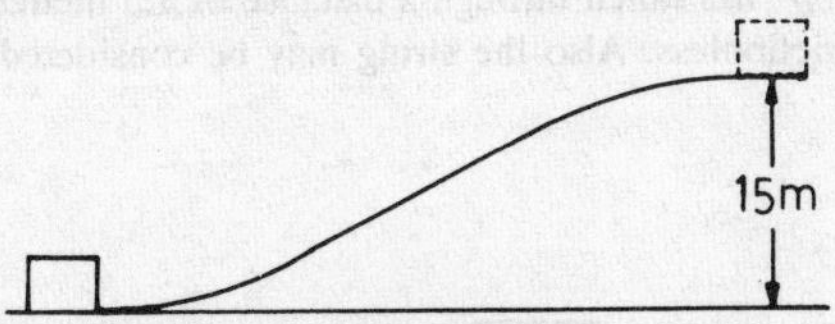

Fig. Prob 8.4

8.5 A locomotive of mass 80,000 kgs is moving up an incline of 6°. A constant tractive force of 150 kN is applied for a distance of 500 metres along the track. If the rolling resistance is 10 N/kN. Determine the resultant work done by all the forces acting on the locomotive.
(Ans. 30080×10^3 J).

8.6 A car of mass 1500 kg starts from rest and moves on a level path. What motive force should be applied to the car in order that it catches up a speed of 20 ms^{-1} after having travelled a distance of 600 metres. Neglect friction.
(Ans. 500 N).

8.7 A stone weighing 10 kgf is dropped from the top of the building. As it reaches the ground, it attains a velocity of 14 m s^{-1}. Determine the kinetic energy of the stone as it strikes the ground. Also determine the height of the roof top from the ground level
(Ans. 100 kgf. m.
10 metres).

8.8 A car having a mass of 1200 kgs coasts down an incline of 10°. At the time when the speed of the car is 40 km h^{-1} the driver applies the brakes. What forces must be applied to the car parallel to the road in order to stop it in a distance of 30 metres?
(Ans. 4511.2 N)

8.9 A package is thrown up along an incline of 30° with an initial velocity of 10 m s^{-1}. It goes upto a point Q and then slides back to the point P, from where it was thrown up. If the distance between P and Q is 6 metres, determine the coefficient of friction between the package and the incline surface. Also determine the velocity of the package when it returns to P.
(Ans. 0.444

$\sqrt{14} = 3.74$ m s^{-1}).

8.10 A device as shown in Figure 8.10 is used to throw a ball. The spring is initially compressed by an amount of 5 cm. Find the velocity of the ball at the time of leaving the tube. The

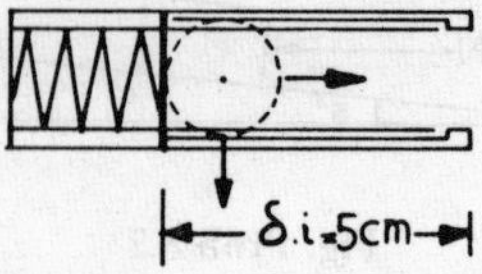

Fig. Prob. 8.10

stiffness 'k' of the spring is 5 kgf/cm. Neglect friction. Take the weight of ball as 50 kgf.

(Ans. 15.65 m s^{-1}).

8.11 Figure 8.11 shows two block '*P*' and '*Q*' having a mass of 400 kg and 100 kg respectively. The coefficient of friction between the block '*P*' and the horizontal surface is 0.2. If the system as shown in the figure is released from rest, determine the velocity of the block '*P*', after the block '*Q*' has fallen through a distance of 1.5 metres. Consider the pulley as weightless and frictionless. Also the string may be considered as inextensible.

(Ans. 1.084 m s^{-1})

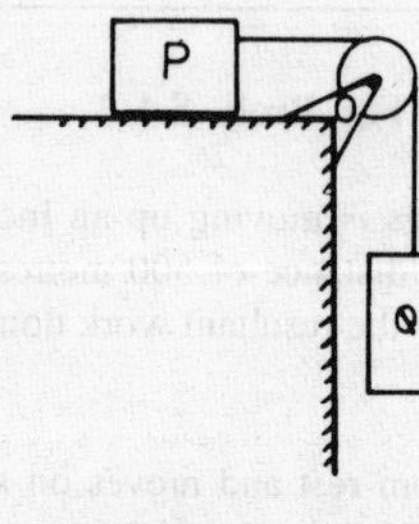

Fig. Prob. 8.11

8.12 A scooter is coasting at a speed of 6 m s^{-1}. It comes to a dip which is 4 metres deep and then rises to a height of 1 m from the bottom of the dip as shown in Figure 8.12. Find the force exerted by the road on the scooter at the dip *A*, if the radius of curvature there is 5 metres. Also calculate the minimum radius of curvature at *B*. The total weight of the scooter and the rider may be taken as 200 kgf. Neglect friction.

(Ans. 467 kgf, 9.57 metres).

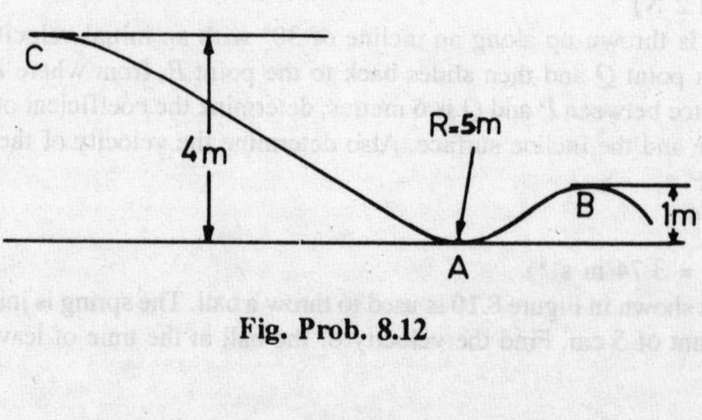

Fig. Prob. 8.12

8.13 A simple pendulum of length 45 cms is hanging from a moving vehicle. The vehicle is moving at a constant velocity of 1.5 m s^{-1}. Determine the maximum amplitude θ of the pendulum, if the vehicle is suddenly brought to rest.
(Ans. 42°)

8.14 Determine the velocity of the bob of a simple pendulum at $\theta = 45°$ from the horizontal, when it is released from its horizontal position as shown in Figure 8.14. The length of bob is 50 cms.
(Ans. 2.632 m s^{-1})

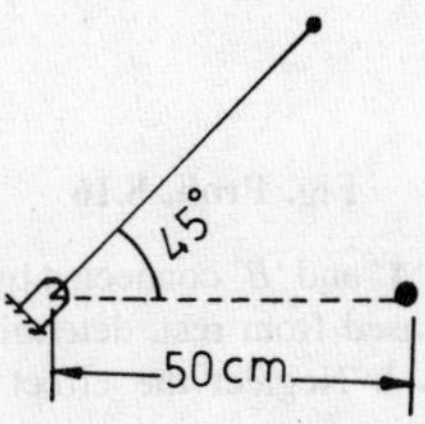

Fig. Prob. 8.14

8.15 A block '*P*' of mass 50 kgs drops 4 metres from rest before it strikes the floor. The pulley is mounted on an axle which is 15 cms in diameter (Figure 8.15). There is a constant axle friction of 200 N. Determine the number of turns made by the pulley after the block '*P*' strikes the ground. Also compute the tension in the cord just before the block '*P*' strikes the ground. The pulley has a centroidal moment of Inertia of 10 kg m^2.
(Ans. 11, 25 N)

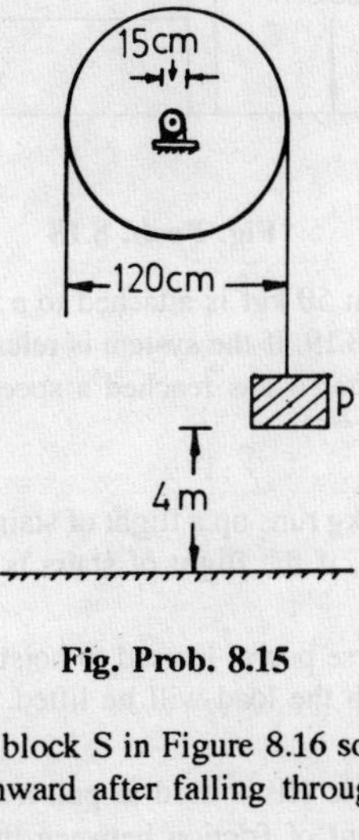

Fig. Prob. 8.15

8.16 Determine the weight of the block S in Figure 8.16 so that the block P is able to attain a velocity of 5 m s^{-1} downward after falling through a distance of 6 metres from rest.
(Ans. 7327 N)

8.17 1000 litres of water has been raised from the ground level and stored in a tank. If the average height to which the water has been raised is 9 metres, calculate the increase in its potential energy.
(Ans. 88,200 J)

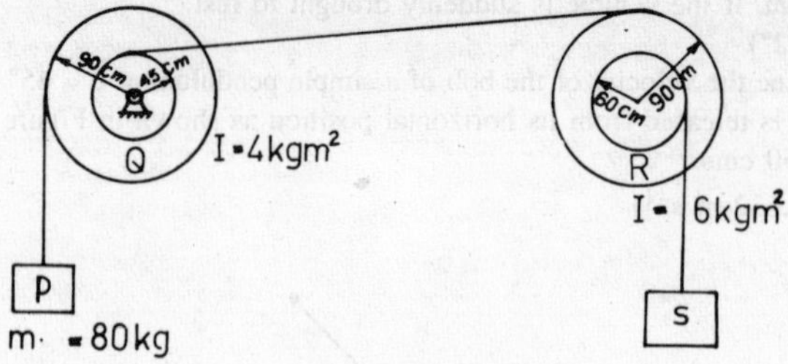

Fig. Prob. 8.16

8.18 Figure 8.18 shows two blocks '*A*' and '*B*' connected by an inextensible cord. If the system as shown in the figure is released from rest, determine the position of block '*A*' when its velocity reaches 1.5 m s^{-1}. Neglect the effect of friction. Use the principle of conservation of energy.
(Ans. 4.08 cm)

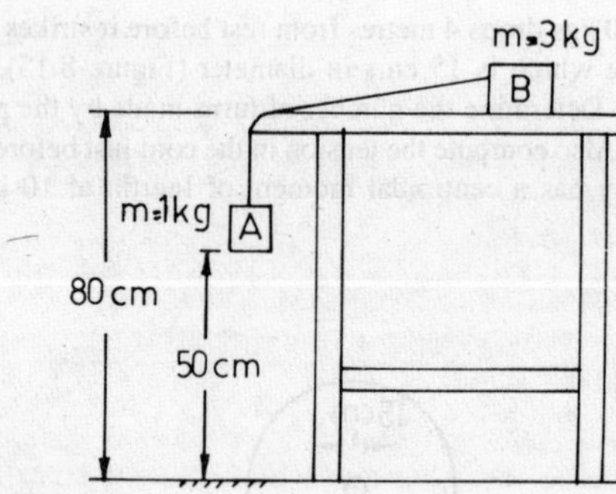

Fig. Prob. 8.18

8.19 A slider block '*A*' of weight 50 kgf is attached to a block '*B*' weighing 20 kgf through a cable as shown in Figure 8.19. If the system is released from rest, find out the distance travelled by the block *A* after it has reached a speed of 2 m s^{-1}. Use the principle of conservation of energy.
(Ans. 66.3 cm)

8.20 A man having a mass of 80 kg runs up a flight of stairs in 5 seconds. Determine the horse power required by the man, if the flight of stairs is 3.5 metres high.
(Ans. 548 W).

8.21 A machine of 20 metric horse power is used to hoist a load of 2000 kgf. Determine the maximum speed with which the load will be lifted.
(Ans. 0.75 m s^{-1})

8.22 A machine of power 5 metric HP is used to pull a block on a level path with a speed of 5 m s^{-1}. If the coefficient of friction between the block and the level path is 0.2, determine the weight of the block.
(Ans. 375 kgf).

8.23 A lift is used to carry 8 persons through a height of 30 metres in 2 minutes. If the average mass of each person is 95 kgs, determine the power required.
(Ans. 1274 W).

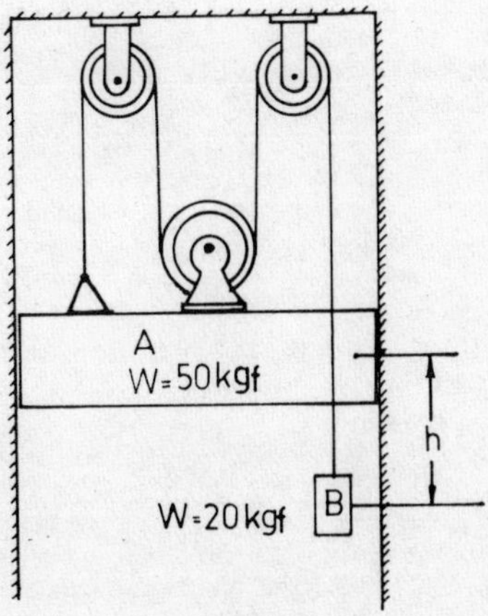

Fig. Prob. 8.19

8.24 In ascending a hill which is 2000 metres in height, a car with its occupants takes 1 hour. If the weight of the car and the occupants is 2000 kgf, determine the metric horse power required by the engine. Neglect friction.
(Ans. 14.8 CV).

8.25 A train having a mass of 100,000 kgs is being pulled up an incline of 5°. The train offers a constant frictional resistance of 5 N/kN. If the speed of the train is increased uniformly 10 m s^{-1} to 20 m s^{-1} in a distance of 500 metres, determine the power developed by the engine.
(Ans. 1805 kW).

9. MACHINES

After having mastered the content of this chapter, you will be able to:

- State with examples the meaning of a '*machine*'
- Define mechanical advantage, velocity ratio, and efficiency of a machine.
- Apply the principle of work and energy, principle of conservation of energy and principle of equilibrium to simple machines for determining their mechanical advantage, velocity ratio and efficiency.
- State the law of the machine
- Formulate equation of a machine from the given experimental data.

CHAPTER 9

Machines

In this chapter we shall be dealing with some simple machines and apply the principles of mechanics in understanding their working. This will help in the application of the same principles to other machines not dealt with in this chapter.

9.1 Machine

Look at a screw jack. It can lift a heavy load by the application of a small force at the tip of its tommy bar. And the direction in which the force is applied to the tommy bar is different from the direction in which the load is being raised. Take now the case of a metal cutting lathe. It is put into motion by an electric motor. The electric motor receives electric energy from a generating station and transforms it into mechanical energy which is used in the running of the lathe. Similarly, at the electric generating station, electricity is produced by a dynamo which, in turn, is run by a hydro or a thermal unit. In the above examples, we see that a unit, say screw jack or the Lathe machine, receives energy from some outside source in a certain form and then transforms it into another form for performing useful work.

A machine is, thus, a mechanical apparatus or contrivance which receives energy from an outside source and transforms it into a form suitable for doing some specific work, after certain losses. It modifies the force or motion in magnitude or direction, or both in magnitude and direction. It is designed to gain an advantage. The advantage gained may be in force, or the distance moved, or in the direction of motion. However, it must be borne in mind, that the machine does not provide any gain in the work done. According to the principle of conservation of energy, the work done by the machine, i.e. the output of the machine, can at best be equal to the work done on the machine by an outside source, i.e. the input to the machine. As a matter of fact, some work is always done in overcoming the friction in the machine; and therefore, the output of the machine in general is always less than the input.

The general law of the machine is

Work done on the machine (input) = work done by the machine (output) + work done to overcome friction.

Before we proceed to study the working, as well as the principles of working, of some simple machines, it would be appropriate to know certain basic concepts.

9.2 Basic Concepts

9.2.1 *Load*

Load is the useful resistance overcome by a machine, or the weight lifted by it. For example an inclined plane is used to raise a block of weight '*W*' through a certain height. The load is the weight of the block '*W*'.

9.2.2 *Effort*

Effort is the force applied by an outside agency in overcoming the resistance encountered by the machine or in lifting the weight by the machine. It is generally denoted by the letter '*P*'.

9.2.3 *Input*

Input is the work done on the machine, and is equal to the effort '*P*' multiplied by its displacement s_i

$\therefore$ Input $= P \times s_i$

9.2.4 *Output*

Output is the useful work done by the machine. It is equal to the load '*W*' on the machine multiplied by its displacement S_o

$$\text{Output} = W \times S_o$$

9.2.5 *Mechanical Advantage (MA)*

MA is the ratio of the load 'W' lifted or the useful resistance overcome, by the machine to the effort 'P' applied to the machine.

$$MA = \frac{W}{P}$$

For example, in the case of a screw jack, if an effort *P* is applied to the tommy bar is 20 N in raising a load of 700 N, then the mechanical advantage of the screw jack will be $\frac{700}{20} = 35$

9.2.6 *Displacement Ratio (DR)*

DR is the ratio between the distance moved by the effort P applied to the machine and the corresponding distance moved by the load W in the same interval of time.

$$DR = \frac{S_i}{S_o}$$

where S_i, S_o are the distances moved by the effort *P* and the load *W* in the same interval of time.

9.2.7 *Velocity Ratio*

Velocity ratio of the machine is the ratio between the velocity of the effort P applied to the machine and the velocity of the load W on the machine. It is also equal to the Displacement ratio of the machine.

For example in the case of a screw jack, if for one turn of the tommy bar, the load is raised by a distance h, then

$$DR = VR = \frac{2\pi r}{h}$$

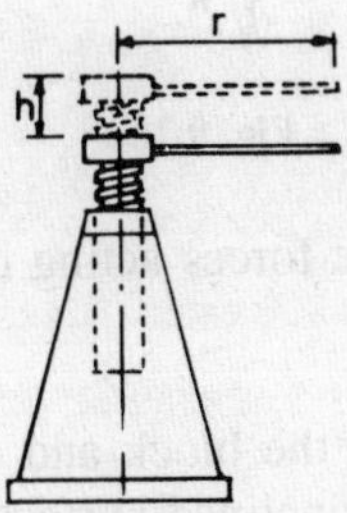

Fig. 9.1

where r is the perpendicular distance from the centre of jack to the line of action of the applied force P.

9.2.8 *Efficiency of the Machine*

It is the ratio between the output work and the input work, or the output power to the input power. The efficiency 'η' of the machine can, therefore, be expressed as

$$\eta = \frac{\text{Output work}}{\text{Input work}}$$

$$= \frac{W \times S_o}{P \times S_i}$$

$$= \frac{W}{P} \cdot \frac{S_o}{S_i} = \frac{W}{P} \cdot \frac{1}{\frac{S_i}{S_o}}$$

$$= MA/VR$$

9.3 Simple Machines

9.3.1 *Inclined Plane*

An inclined plane is a simple machine used for raising or lowering a load with greater ease. Let us, consider an inclined plane AB making an angle $\propto$

to the horizontal. Fig. 9.2a shows that a force P is applied at an angle ϕ to the incline for moving a block of weight 'W' up this inclined plane. X and Y are the reference axes parallel and normal to the incline.

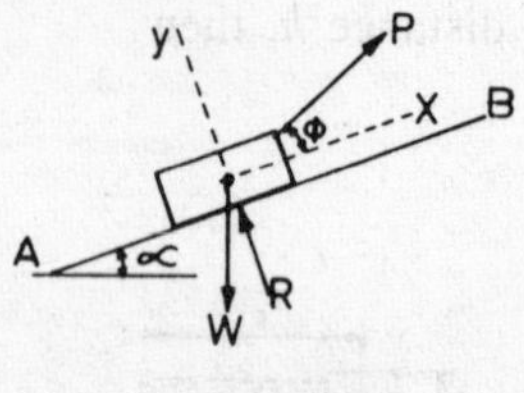

Fig. 9.2(a)

If friction is neglected, then the forces acting on the block are as shown in Fig. 9.2(a), i.e.

W, the weight of the block

P, the force of pull applied to the block and

R, the reaction offered by the inclined surface to the block.

In case, the block is at rest or is moving up the plane with a constant speed, then for equilibrium (Figure 9.2b)

$$\Sigma F_y = 0$$

$$\Sigma F_x = 0$$

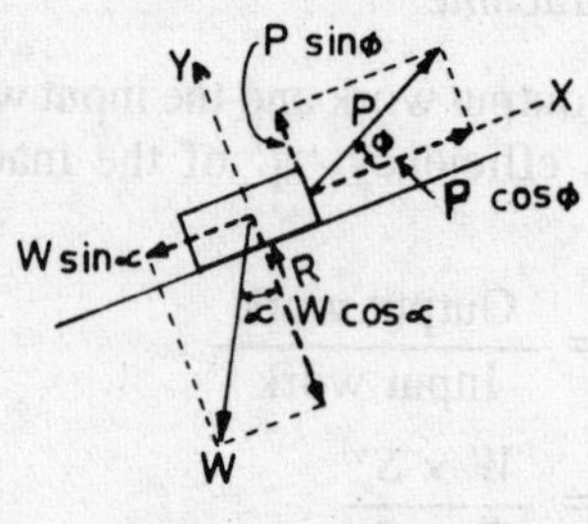

Fig. 9.2(b)

For $\Sigma F_y = 0$,

$$R - W \cos \propto + P \sin \phi = 0$$

$$\therefore \; R = W \cos \propto - P \sin \phi \quad \text{.... (i)}$$

For $\Sigma F_x = 0$

$$P \cos \phi - W \sin \propto = 0$$

$$\therefore \; P \cos \phi = W \sin \propto$$

or $$P = W \frac{\sin \propto}{\cos \phi} \quad \text{.... (ii)}$$

Therefore, the force '*P*' required to move the block of weight *W* up the incline is given by expression (ii). For a given inclined plane, the value of *P* is minimum, when the value of cos ϕ is maximum. Maximum value of cosine of an angle is 1, i.e. when $\phi = 0$ or the force '*P*' is applied in a direction parallel to the inclined plane. In this case

$$P = W \sin \propto \quad \text{.... (iii)}$$

Mechanical Advantage

$$MA = \frac{\text{Load raised}}{\text{Effort applied}}$$

$$= \frac{W}{P}$$

$$= \frac{W}{\frac{W \sin \propto}{\cos \phi}} \qquad P = W \frac{\sin \propto}{\cos \phi}, \text{ when friction between the contact surfaces is neglected.}$$

$$= \frac{\cos \phi}{\sin \propto} \quad \text{.... (iv)}$$

When the force 'P' is applied in a direction parallel to the inclined plane

$$\phi = 0 \quad \therefore \cos \phi = 1$$

$$\text{and } MA = \frac{1}{\sin \propto} \quad \text{.... (v)}$$

From (*v*) it can be concluded that as the inclination of the plane to the horizontal increases, *MA* will decrease i.e. more and more *P* will be required to raise the load *W*.

When the applied force P is horizontal

$$\phi = \propto$$

$$MA = \frac{\cos \propto}{\sin \propto}$$

$$= \cot \propto \quad \text{.... (vi)}$$

For $\propto$ less than 45°, cot $\propto$ is always greater than 1. At $\propto = 45°$, cot $\propto = 1$, and for $\propto > 45°$, cot $\propto < 1$. From this, it can be concluded that for $\propto$ less than 45°, *P* required will always be less than *W*; and for $\propto > 45°$, *P* required will be greater than *W*. All these conclusions are based on the assumption that friction between the contact surfaces is neglected.

Displacement and Velocity Radio

$$VR = \frac{\text{Distance moved by effort}}{\text{Distance moved by the load}}$$

$$= \frac{S \cos \phi}{S \sin \propto}$$

$$= \frac{\cos \phi}{\sin \propto} \quad \text{.... (vii)}$$

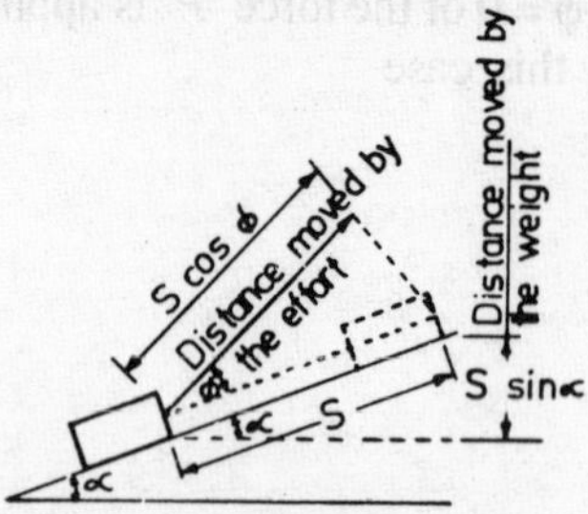

Fig. 9.2(c)

Comparing (iv) and (vii), we find that *MA* will be equal to the *VR* of a machine. This is so because, friction between the contact surfaces has been neglected.

Efficiency

$$\eta = \frac{\text{Output work}}{\text{Input work}}$$

Output work = $W.\ s \sin \alpha$

Input work = $P.\ s \cos \phi$

$$\therefore \quad \eta = \frac{W.\ s \sin \alpha}{P.\ s \cos \phi} \qquad \text{..... (viii)}$$

When friction is neglected,

$$P = \frac{W \sin \alpha}{\cos \phi}$$

Substituting for *P* in (viii), we get

$$\eta = \frac{W.\ s \times \sin \alpha}{W \dfrac{\sin \alpha}{\cos \phi}.\ s \cos \phi} \times 100 = 100 \text{ per cent} \qquad \text{.... (ix)}$$

From (viii) and (ix) it can be concluded as follows:

(i) If the machine is considered ideal, i.e. the effect of frictional forces and body forces is neglected, then the output will be equal to the input and the efficiency of the machine is 100 per cent.

(ii) As the input is equal to the output, the gain in force will mean more distance to be travelled by the effort *P* than that by the load *W* in the same duration of time. If friction between the contact surfaces is taken into consideration, the efficiency of the machine will be always less than 1. Let

μ be the coefficient of friction between the contact surfaces, then from Figure 9.2(d) for equilibrium,

$$\Sigma F_y = 0$$

$$\Sigma F_x = 0$$

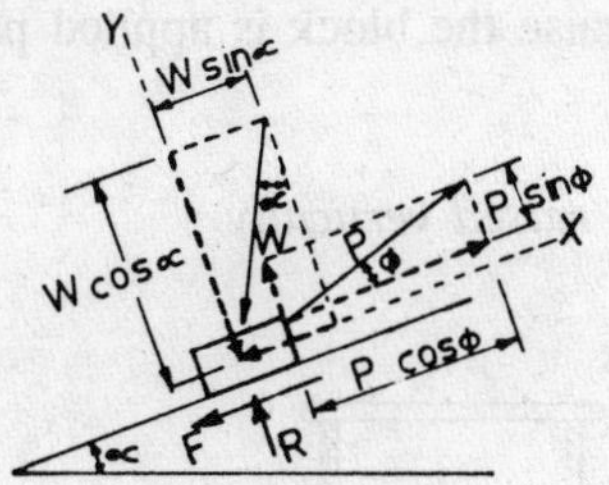

Fig. 9.2(d)

For $\Sigma F_y = 0$ $\quad R - W\cos\alpha + P\sin\phi = 0$

$$\therefore\ R = W\cos\alpha - P\sin\phi$$

As $F = \mu R$

$$F = \mu\,(W\cos\alpha - P\sin\phi)$$

For $F_x = 0$, $P\cos\phi - W\sin\alpha - F = 0$

$$\text{or } P\cos\phi - W\sin\alpha - \mu\,(W\cos\alpha - P\sin\phi) = 0$$

$$\text{or } P\,(\cos\phi + \mu\sin\phi) = W\,(\sin\alpha + \mu\cos\alpha)$$

$$\therefore \qquad P = \frac{W(\sin\alpha + \mu\cos\alpha)}{(\cos\phi + \mu\sin\phi)} \qquad \text{.... (x)}$$

$$MA = \frac{W}{P}$$

$$= \frac{\cos\phi + \sin\phi}{\sin\alpha + \mu\cos\alpha} \qquad \text{.... (xi)}$$

$$\text{and } \eta = \frac{MA}{VR} = \frac{(\cos\phi + \mu\sin\phi)}{\sin\alpha + \mu\cos\alpha} \Big/ \frac{\cos\phi}{\sin\alpha}$$

$$= \frac{\cos\phi\sin\alpha + \sin\phi\sin\alpha}{\sin\alpha\cos\phi + \cos\alpha\cos\phi}$$

Dividing the numerator and the denomenator by $\sin\alpha\cos\phi$, we get

$$= \frac{1 + \mu\tan\phi}{1 + \mu\cot\alpha} \qquad \text{.... (xii)}$$

Example 9.1

A block of mass 20 kg is to be raised from the ground floor to the first floor

of a building. The height of the first floor from the ground floor is 4 metres. Determine the effort required when the block is raised
(a) vertically
(b) along an inclined plane making an angle of 30° with the horizontal.
In case (b) also find the *MA*, *VR* and efficiency of the plane when (i) friction is neglected between the contact surfaces (ii) the coefficient of friction between the contact surfaces is 0.20
The effort *P* used to raise the block is applied parallel to the plane.

Solution
(a) *When the block is raised vertically*

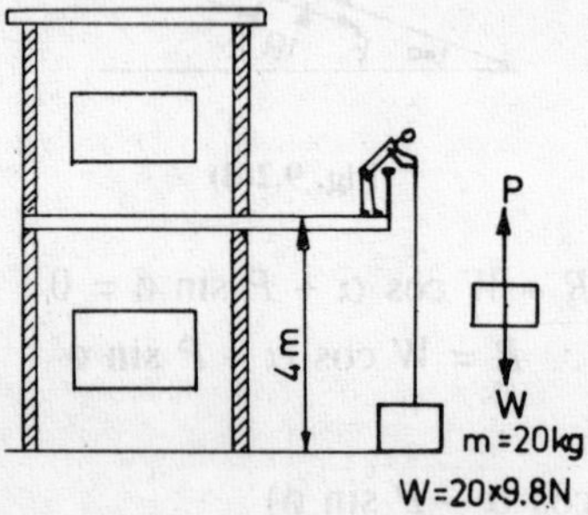

Fig. 9.3

Effort '*P*' required to raise = Weight of the block
the load *W*

$$= 20 \times 9.8 = 196 \text{ N}$$

(b) (i) *When the block is raised along the incline and friction is neglected*

Fig. 9.4(a)

For equilibrium,

$$\Sigma F_x = 0$$

$$P - W \sin \alpha = 0$$

$$\therefore \quad P = W \sin \alpha$$

$$= 196 \times \frac{1}{2}$$

$$= 98 \text{ N}$$

$$MA = \frac{W}{P}$$

$$= \frac{196}{98}$$

$$= 2$$

$$VR = \frac{\text{Distance travelled by the effort}}{\text{Distance travelled by the load}}$$

$$= \frac{4 \text{ cosec. } 30°}{4}$$

$$= \frac{4 \times 2}{4}$$

$$= 2$$

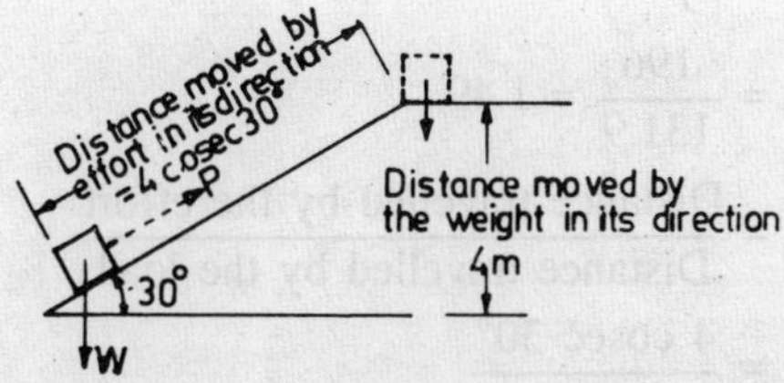

Fig. 9.4(b)

$$\text{Efficiency } (\eta) = \frac{MA}{VR} \times 100$$

$$= \frac{2}{2} \times 100 = 100 \text{ per cent}$$

(b) (ii) *When friction is taken into account*

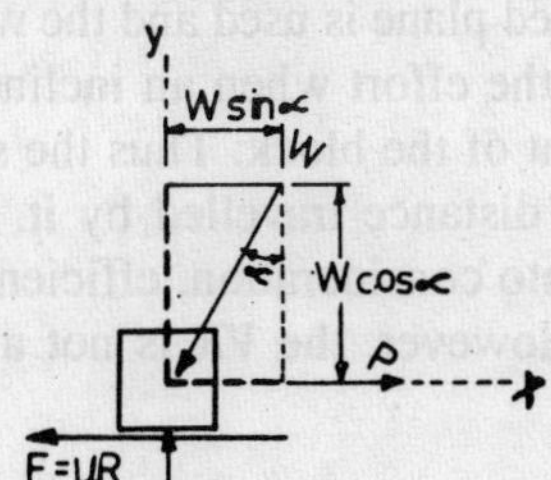

Fig. 9.4(c)

For equilibrium,

$$\Sigma F_y = 0$$

$$R - W \cos \alpha = 0$$

$$\therefore \quad R = W \cos \alpha$$

$$= 196 \cos 30°$$

$$= 196 \frac{\sqrt{3}}{2} = 98\sqrt{3}$$

$$F = \mu \cdot R$$

$$= 0.2 \times 98\sqrt{3}$$

$$= 33.9 \text{ N}$$

For $\quad \Sigma F_x = 0,$

$$P - W \sin \alpha - F = 0$$

or $\quad P = W \sin \alpha + F = 196 \times \frac{1}{2} + 33.9 = 131.9 \text{ N}$

$$MA = \frac{W}{P}$$

$$= \frac{196}{131.9} = 1.49$$

$$VR = \frac{\text{Distance travelled by the effort}}{\text{Distance travelled by the load}}$$

$$= \frac{4 \text{ cosec } 30°}{4}$$

$$= \frac{4 \times 2}{4} = 2$$

$$= \frac{MA}{VR} = \frac{1.49}{2} \times 100$$

$$= 74.5 \text{ per cent}$$

Following observations can be made from this example.

1 By using a smooth incline plane the effort applied is 98 N as compared to 196 N, when no inclined plane is used and the weight is lifted vertically.
2 The distance moved by the effort when an incline is used is double than that moved by the weight of the block. Thus the saving in the effort used is at the cost of greater distance travelled by it.
3 When friction is taken into consideration, efficiency as well as the *MA* of the machine decrease. However, the *VR* is not affected.

9.3.2 *Wedge*

A wedge is a form of an inclined plane, and its shape is that of a triangular prism. Surface *ABCD* is called the head of the wedge, whereas the side surfaces

BCFE and *ADFE* are called its checks (Fig. 9.5). Angle *AEB* is called the angle of the wedge, and is considerably less than the other two angles *ARE* and *BAE*

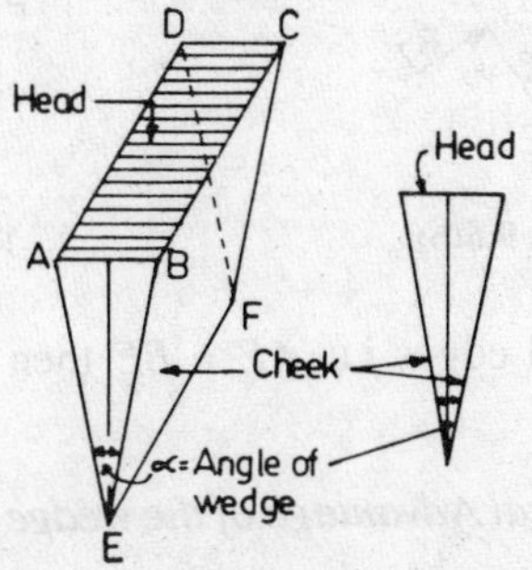

Fig. 9.5

Let us consider a wedge penetrating a block as shown in Fig. 9.6(a). Let the driving force applied to the wedge be P (Fig. 9.6a). The penetration of the wedge inside the block is resisted and the block offers reactions R_1 and R_2 normal to the cheeks of the wedge *AE* and *BE* at the points of contact *L* and *M* respectively.

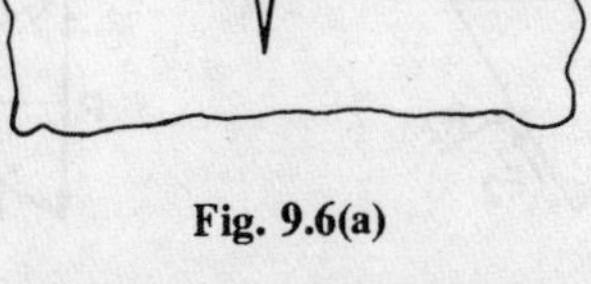

Fig. 9.6(a)

Neglecting friction and considering that the wedge is penetrating the block at a constant speed, for equilibrium the three forces, P, R_1 and R_2 acting on the wedge as shown in Fig. 9.6(a) must balance. The lines of action of these three forces must, therefore, intersect at the same point say O. Representing these forces P, R_1 and R_2 by vectors *ST*, *TU* and *US* respectively, they will form a closed triangle of forces *STU* as shown in Fig. 9.6(c)

$\angle TUS = \alpha$ (R_1 and R_2 are normal to the sides *AE* and *EB* and the angle between any two sides is equal to the angle between their normals)

$\therefore$ ΔAEB and ΔTUS are similar

$\therefore$ $P : R_1 : R_2 : AB : AE : EB$

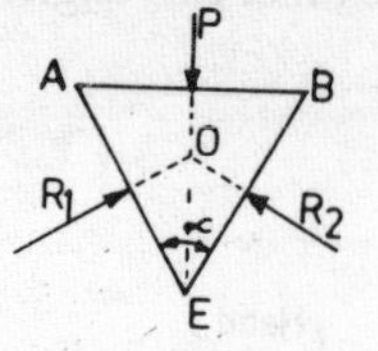

Fig. 9.6(b)

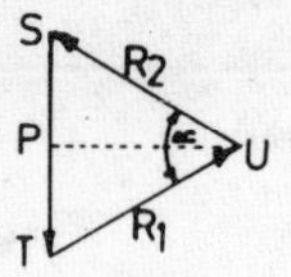

Fig. 9.6(c)

If the wedge has equal edges i.e. $AE = BE$ then $R_1 = R_2 = R$ and $\frac{R}{P} = \frac{AE}{AB}$

Therefore the Mechanical Advantage of the wedge $\left[\frac{\textit{Resistance overcome}}{\textit{Effort applied}} = \frac{R}{P}\right]$ *is equal to the ratio of its length of cheeks to the thickness of the head.* The smaller the angle 'α' of the wedge, and thinner the head, the greater will be the gain in the force.

Properties of the wedge are used to a great advantage in splitting of wood by cutting tools, in connections where pins of wedge shape are used in shuttering, frame work, and many other situations.

If friction is taken into account, then for the same effort 'P', the resistance overcome will be less (Fig. 9.7 and 9.8) and the Mechanical Advantage will thus decrease. The frictional resistance offered to the penetration of the wedge will depend upon the magnitude of reaction R, which is further dependent upon the angle of the wedge α. Lesser the angle α of the wedge, greater will be the value of R. *Thus the force of friction increases as the angle of the wedge decreases.*

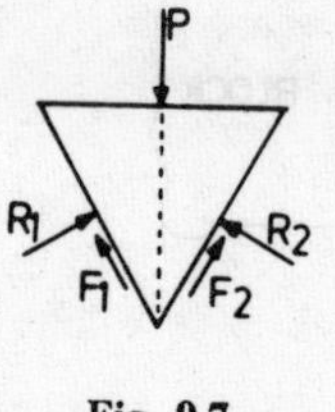

Fig. 9.7

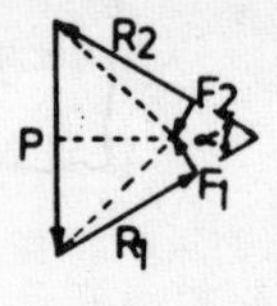

Fig. 9.8

For example, it is easier to split wood with a thick headed axe than that with the carpenters hatchet because in the first case, the input striking energy is more due to the greater mass of the axe; and secondly the thick headed axe can be taken out easily, if the wood is not completely split.

Example 9.2

A wedge ABC with cheeks of the same length as shown in Figure 9.9 is used for splitting wood. If the force 'P' applied to the wedge is 400 N, determine the reactions R_1 and R_2 on the cheeks of the wedge. Neglect friction. Also determine the mechanical advantage.

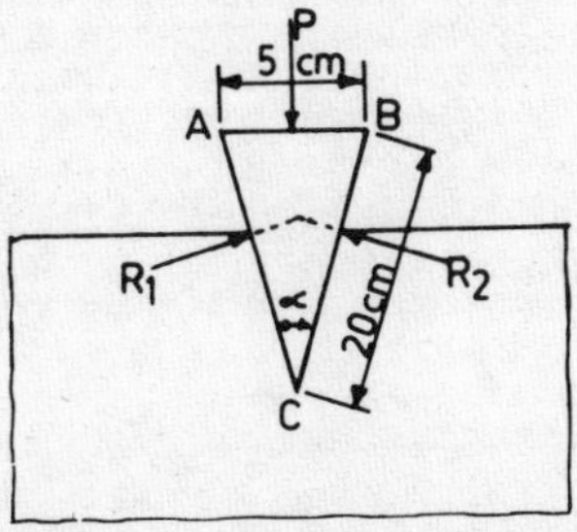

Fig. 9.9(a)

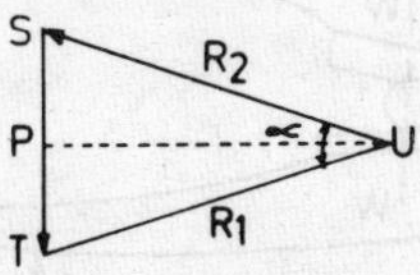

Fig. 9.9(b)

Solution

We know that,

$$\frac{R_1}{P} = \frac{AC}{AB}$$

$$\frac{R_2}{P} = \frac{BC}{AB}$$

As $AC = BC \therefore R_1 = R_2$

$$R_1 = R_2 = P\,\frac{AC}{AB}$$

$$= 400 \times \frac{20}{5}$$

$$= 1600 \text{ N}$$

$$MA = \frac{R}{P}$$

$$= \frac{1600}{400} = 4$$

9.3.3 *Lever*

You must have seen a betelnut cracker, a boatman's oar or a jumper. All these are examples of a lever. By applying a small force to the handles of a betelnut cracker, it is possible to crack a betelnut which is very hard and difficult to cut otherwise. Similarly, a jumper is used to lift or move a heavy stone or any such heavy article by the application of a comparatively small force at its end.

A lever a is device which is used independently, or as an integral part of a

machine or some mechanism, either to gain advantage in the effort applied (force input) or in the distance to be moved by the effort. A simple lever is a straight or a bent rod capable of rotating about a point called the fulcrum.

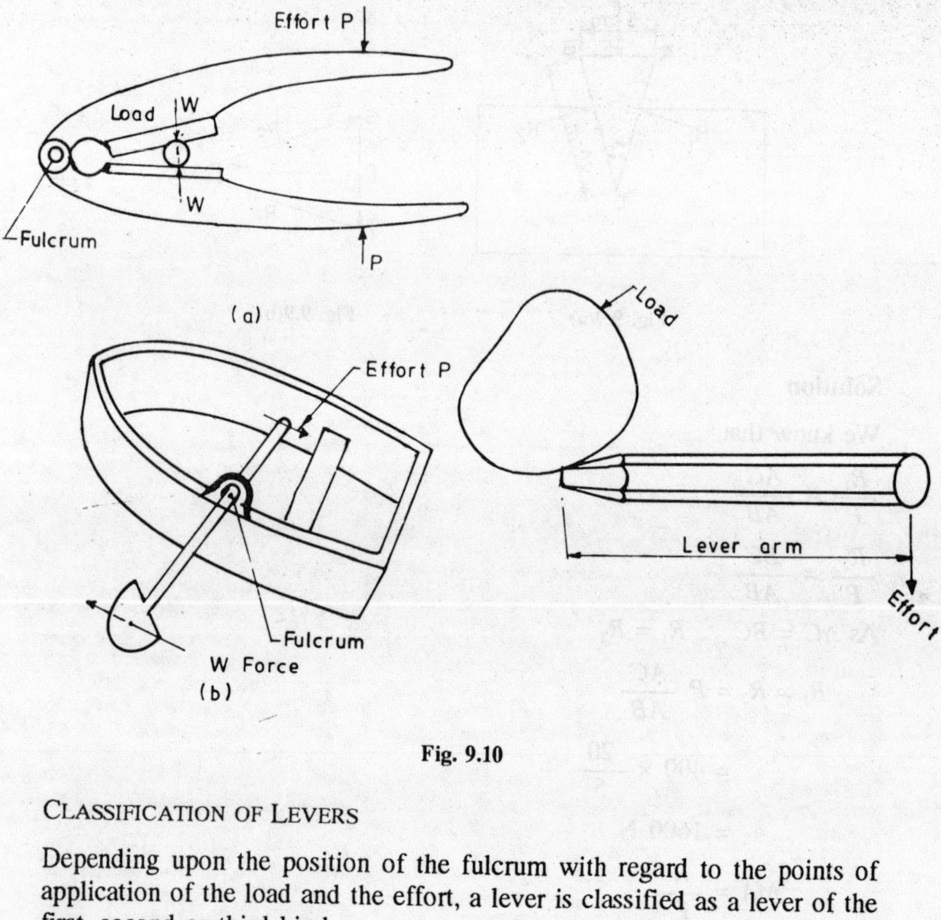

Fig. 9.10

CLASSIFICATION OF LEVERS

Depending upon the position of the fulcrum with regard to the points of application of the load and the effort, a lever is classified as a lever of the first, second or third kind.

Lever of the First Kind

A lever is said to be of the first kind, if the fulcrum is in between the load and the effort. It can either be straight or bent. Examples of this lever are a nail extractor, a see-saw, a scissor, a jumper, etc. Fig. 9.11 shows a diagramatic model of this lever.

O is the fulcrum of the lever. '*a*' and '*b*' are the perpendicular distances from the fulcrum *O* to the lines of action of the load '*W*' and the effort *P*. Distances *a* and *b* are also known as the lever arms of the load *W* and the effort *P* respectively.

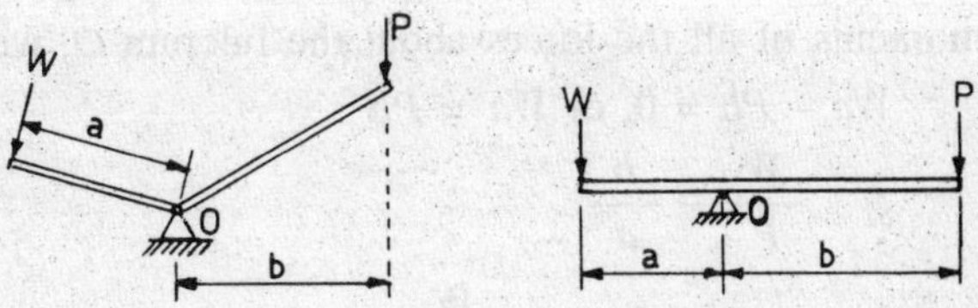

Fig. 9.11

Let us now determine the effort P required to raise a load W.
We know that for equilibrium $\Sigma M = 0$
Taking moments of the forces about the fulcrum O and neglecting the friction at the bearings (This will eliminate the moment of the reaction R), we have

$$Pb - Wa = 0, \text{ or}$$

$$Pb = Wa \qquad \text{.... (i)}$$

$$P = W\frac{a}{b} \qquad \text{.... (ii)}$$

$$\text{Mechanical advantage} = \frac{\text{Resistance overcome (Load Raised)}}{\text{Effort applied}}$$

$$= \frac{W}{P} = \frac{b}{a} \qquad \text{From (i)}$$

The ratio $\frac{b}{a}$ is called the *leverage.* Therefore, the mechanical advantage is equal to the leverage.

Lever of the Second Kind

A lever is said to be of the second kind if the point of application of the load is in between the fulcrum and the point of application of effort. Examples of this lever are a betelnut cracker, a hand driven rickshaw etc. Figure 9.12 shows lever of this kind. W is the load and P is the effort applied to raise it. a and b are the horizontal distances from the fulcrum O to the lines of action of the load W and effort P respectively.

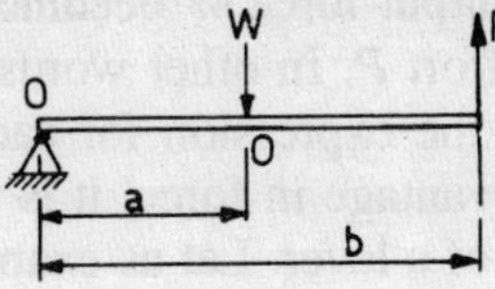

Fig. 9.12

For equilibrium, $\Sigma M = 0$

Taking moments of all the forces about the fulcrum O, we have

$$Wa - Pb = 0, \text{ or } Wa = Pb$$

$$\frac{W}{P} = \frac{b}{a}$$

Again the Mechanical advantage $\frac{W}{P}$ is equal to the leverage $\frac{b}{a}$. More is this leverage, more will be the mechanical advantage.

Lever of the Third Kind

In the third kind of lever the point of application of the effort is in between the fulcrum and the load. Examples of this lever are a Doctors pair of tongs, human arm etc. Figure 9.13 shows lever of the third kind. W is the load on the lever. P is the effort applied to raise the load. a and b are the lever arms of the load W and the effort P.

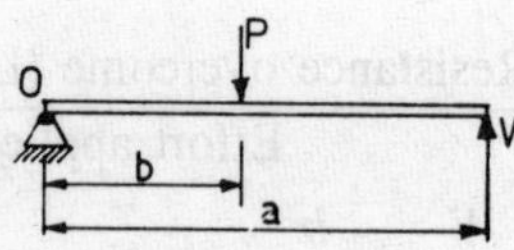

Fig. 9.13

For equilibrium,

$$\Sigma M = 0$$

Taking moments of all the forces about the fulcrum 0, we get

$$Wa - Pb = 0, \text{ or } Wa = Pb$$

$$\therefore \quad \frac{W}{P} = \frac{b}{a}$$

$\frac{W}{P}$ is the mechanical advantage of the lever, and is equal to $\frac{b}{a}$, the leverage of the lever as before.

In all the above types of levers, friction at the bearings has been neglected. If friction at the bearing of the fulcrum is taken into consideration, then the useful work done by the output force W becomes less than the input work or the work done by the effort P. In other words, the MA of the lever will be less than that given by the expression for each type of lever.

Apart from gaining advantage in force, it is possible to gain advantage in displacement by the use of a lever. Let us examine a lever AOB as shown in Figure 9.14. The lever is capable of rotating about its fulcrum O. If l_1 and l_2 are the lengths of its arms OA and OB, then for a rotation α of the rod about O,

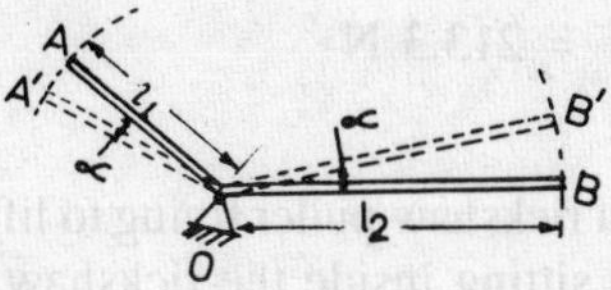

Fig. 9.14

Displacement of end $A = l_1 \alpha$

Displacement of end $B = l_2 \alpha$

$$\frac{\text{Displacement of } B}{\text{Displacement of } A} = \frac{l_2 \alpha}{l_1 \alpha} = \frac{l_2}{l_1}$$

the displacement of end B will, therefore, depend upon the ratio $\frac{l_2}{l_1}$. By giving a small displacement to A it is possible to achieve a large displacement of B by increasing the ratio $\frac{l_2}{l_1}$. This property of the lever is very extensively used in the construction of measuring instruments.

Example 9.3

A nail extractor as shown in Figure 9.15 has been used to extract a nail. If for pulling the nail, an effort P of 80 N is applied at the end of the handle of the nail extractor as shown in Figure 9.15, determine the maximum resistance offered by the nail.

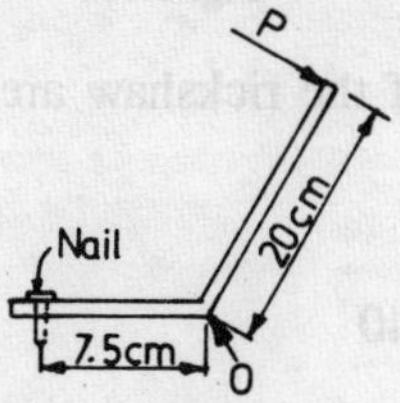

Fig. 9.15

Solution

The resistance offered by the nail has been denoted by W.

For equilibrium $\Sigma M = 0$

Taking moment of all the forces about the fulcrum O, thereby eliminating the moment of the reaction at O, we get

$$-W \times 7.5 + P \times 20 = 0$$

$$\text{or } 7.5\,W = 80 \times 20$$

$$W = \frac{1600}{7.5}$$

$$= 213.3 \text{ N}$$

Example 9.4

Figure 9.16 shows a rickshaw puller trying to lift the handle of a rickshaw. The weight of the man sitting inside the rickshaw is 70 kgf. Determine the vertical pull to be applied by the rickshaw puller in order to lift the handles of the rickshaw.

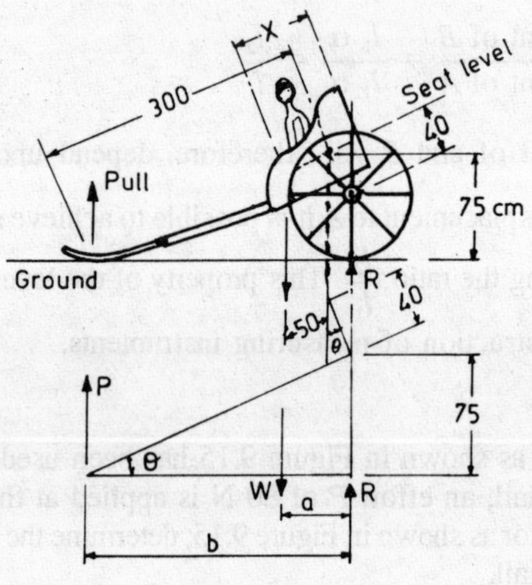

Fig. 9.16

The forces on the handle of the rickshaw are shown in Fig. 9.16.

$b^2 = 300^2 - 75^2$

$b = 290.5$ cms

$x = 50 + 75/290.5 \times 40$

$= 60.3$ cms

$$a = 60.3 \times \frac{290.5}{300}$$

$= 58.40$ cms

For equilibrium, $\Sigma M = 0$

Taking moments of all forces about O, we get

$Pb - Wa = 0$, or

$P \times 290.5 - 70 \times 58.40 = 0$

$$\therefore\ P = \frac{70 \times 58.40}{290.5} = 140.5 \text{ kgf}$$

Compound Lever

When it is desired to achieve more mechanical advantage, which otherwise is difficult to attain by the use of a simple lever, compound levers are used. A compound lever is a combination of simple levers linked with one another as shown in Figure 9.17.

AB is a simple lever connected to another simple lever *CD* with the help of a link *BC*.

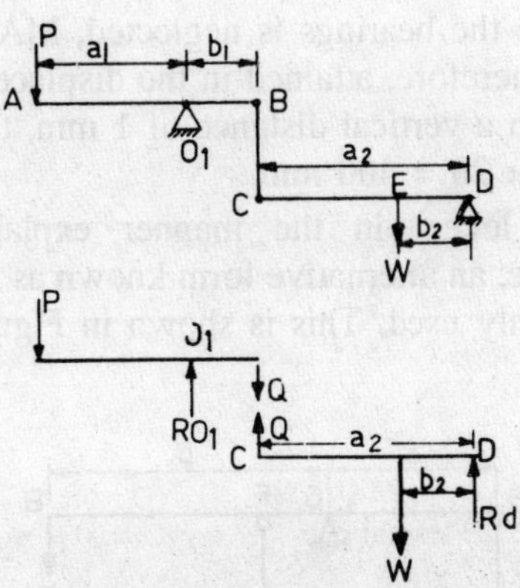

Fig. 9.17, 9.18

Let us now determine the effort *P* that must be applied at A to lift a load *W* at E. Draw the free body diagram of the levers AB and CD (Figure 9.18) Let Q be the force in the link BC,

Then for equilibrium, $\Sigma M = 0$.

Taking moments of the forces about O_1,

$$Pa_1 - Q.\, b_1 = 0$$

$$\therefore \quad Q = P.\, \frac{a_1}{b_1} \qquad \text{.... (1)}$$

From the free body diagram of the lever CD, taking moments of the forces about D, we have

$$Q.\, a_2 - Wb_2 = 0$$

$$\therefore \qquad Q = W\, \frac{b_2}{a_2} \qquad \text{.... (2)}$$

From (1) and (2)

$$P\, \frac{a_1}{b_1} = W\, \frac{b_2}{a_2}$$

$$P = W\, \frac{b_2}{a_2} \cdot \frac{b_1}{a_1}$$

$$\text{Mechanical advantage} = \frac{W}{P} = \frac{a_1}{b_1} \cdot \frac{a_2}{b_2}$$

Let $\dfrac{a_1}{b_1} = \dfrac{a_2}{b_2} = 20$

If only lever *AB* is employed, the mechanical advantage is 20. However, if two levers *AB* and *CD* are combined, the mechanical advantage increases to $20 \times 20 = 400$. To further increase the mechanical advantage more such levers can be combined.

As the friction in the bearings is neglected, MA will be equal to VR.

Similar gain is, therefore, attained in the displacement. e.g. if the point E is displaced through a vertical distance of 1 mm, then the end A will be displaced by $1 \times 20 \times 20 = 400$ mm.

Combining the levers in the manner explained above becomes cumbersome, therefore, an alternative form known as *differential or floating lever* is most commonly used. This is shown in Figure 9.19.

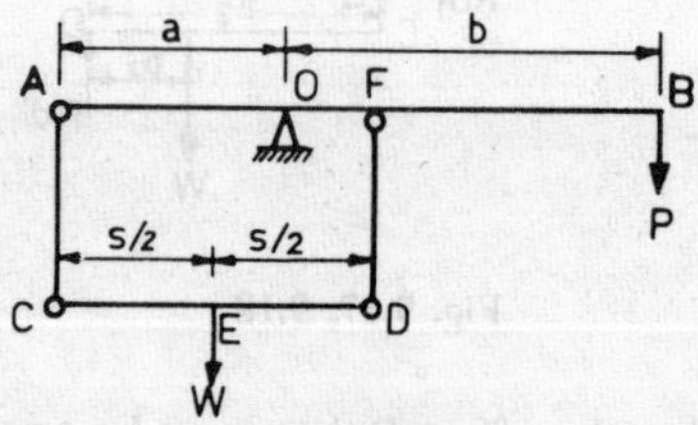

Fig. 9.19

It consists of a lever AB which is connected with a cross piece CD by means of two equal and parallel transverse links AC and FD. Load *W* is applied at the mid point of the cross piece CD. Let us now find out the mechanical advantage in this case.

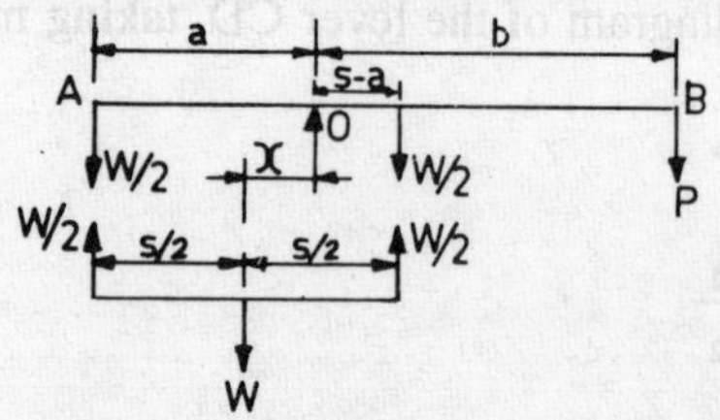

Fig. 9.20

As the load *W* is applied at the centre of the cross piece CD, the forces in each of the arms AC and FD will be equal to *W*/2.

Taking moments about O of all the forces acting on the freebody diagram of the lever AB, we get,

$$-W/2\ a + P\cdot b + W/2(s - a) = 0$$

$$\therefore \quad Pb = \frac{W}{2}(a - s + a)$$

$$= W(a - s/2)$$

If the distance between the fulcrum O and the line of action of W is assumed to be x

then $x = a - s/2$

and $Pb = W\cdot x$

$$\therefore \quad \frac{W}{P} = b/x$$

By reducing the value of x, the mechanical advantage can be increased to any extent.

MA will also be equal to the VR of the lever, if the friction at the bearings is neglected.

Example 9.5

Figure 9.21 shows a compound lever. It is desired to raise a load of 40 kN applied at point E of the lever. Determine the effort P to be applied at A that would just raise this load. Also determine the Mechanical advantage of the lever.

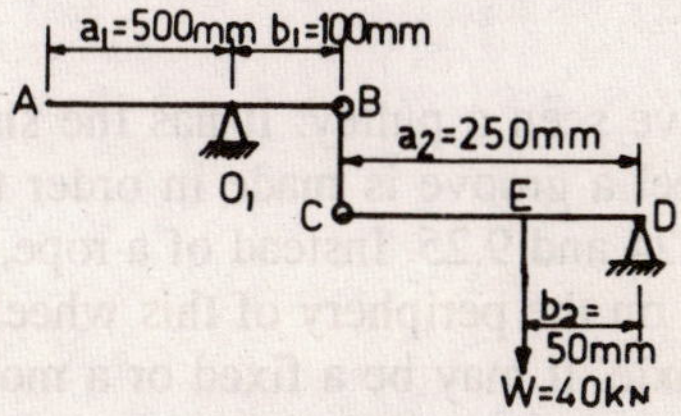

Fig. 9.21

Solution

Draw the freebody diagram of the lever AB (Figs. 9.22 and 9.23)

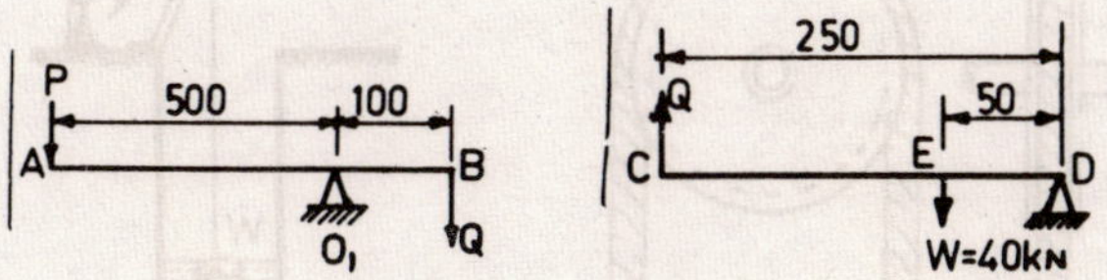

Fig. 9.22 and 9.23

For equilibrium $\Sigma M = 0$

Taking moments of the forces about O, we have

$-P \times 500 + Q \times 100 = 0$

$$\therefore \quad P = Q \times \frac{100}{500} = Q/5$$

Draw the freebody diagram of the lever CD (Figure 9.23).

Taking moments about D, we have

$Q \times 250 - 40 \times 50 = 0$

$$\therefore Q = \frac{40 \times 50}{250} = 8 \text{ kN}$$

$$P = \frac{Q}{5}$$

$$= \frac{8}{5} = 1.6 \text{ kN}$$

Alternatively, $P = W. \dfrac{b_2}{a_2} \cdot \dfrac{b_1}{a_1}$

$$= 40 \times \frac{50}{250} \times \frac{100}{500} = 1.6 \text{ kN.}$$

Mechanical advantage $= \dfrac{40}{1.6} = 25$

9.3.4 *Pulleys, Blocks and Tackles*

PULLEY

Most of you must have seen a pulley. It has the shape of a wheel. At the periphery of this wheel a groove is made in order to accommodate a rope as shown in Figure 9.24 and 9.25. Instead of a rope, if a chain is used, then sprocket teeth are cut on the periphery of this wheel. The pulley is capable of rotating about its axle. It may be a fixed or a movable pulley depending upon whether its axle remains stationary at a position or moves with the movement of the rope or chain.

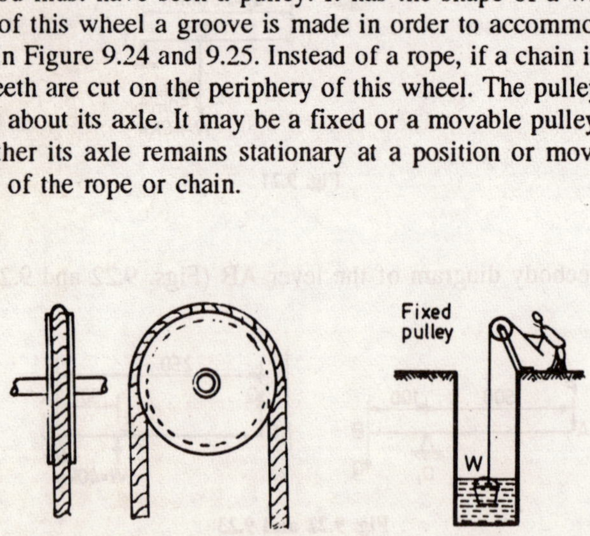

Fig. 9.24 and 9.25

Fixed Pulley

As mentioned earlier, a fixed pulley is that whose axle does not move from its position, when the rope or chain is moved over it. (Refer to Fig. 9.26)

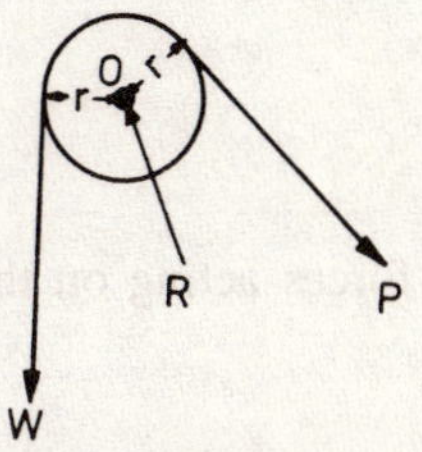

Fig. 9.26

Let P be the effort required to lift a load 'W', neglecting friction at the contact surfaces, we have for equilibrium,

$$\Sigma M = 0$$

Taking moments of all the forces acting on the pulley about O (Figure 9.27), we have

$$Pr - Wr = 0$$
$$Pr = Wr$$
$$P = W$$

From above we can say that no gain is obtained in the effort applied by the use of a fixed pulley. It only helps in affecting a change in the direction of application of the effort.

Movable Pulley

A movable pulley is that which does not remain stationary and changes its position when the work is being done. Such a pulley has been shown in Figure 9.27. In this case, one end of the rope is fixed to a support and the other end is passed round the movable pulley from below, and then over a fixed pulley as shown in the figure.

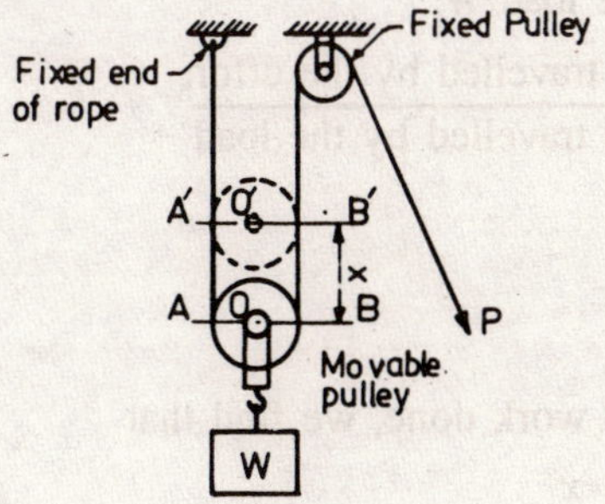

Fig. 9.27

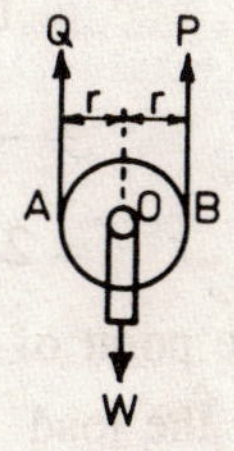

Fig. 9.28

The weight to be raised is hung from the hook attached to the movable pulley and the effort P is applied at the running end of the rope. Let us now try to find the effort P required to raise the load W, neglecting friction between the contact surfaces. The free body diagram of the movable pulley is shown in Fig. 9.28.

For equilibrium,

$$\Sigma M = 0$$

Taking moments of all the forces acting on the pulley about O, we get,

$$Pr - Qr = 0 \quad \text{.... (i)}$$

$$Pr = Qr \quad \text{.... (ii)}$$

$$P = Q$$

Now taking moments of all the forces about A. For equilibrium

$$\Sigma M = 0, \qquad Wr - P.\ 2r = 0 \quad \text{.... (iii)}$$

$$Wr = 2\ pr$$

$$W = 2\ P$$

$$\text{or} \qquad P = \frac{W}{2} \quad \text{.... (iv)}$$

Mechanical Advantage

$$MA = \frac{W}{P}$$

$$= \frac{W}{W/2}$$

$$= 2$$

Velocity Ratio

Referring to Figure 9.27, when the load W is raised to O′ through a height x, then the effort P will be displaced through a distance AA′ + BB′ = $2x$. The displacement of the effort P is, therefore, twice the displacement of the load. The gain in the effort applied has, therefore, been at the cost of P having travelled double the distance than the load W.

$$\text{Velocity Ratio } (VR) = \frac{\text{Distance travelled by the effort}}{\text{Distance travelled by the load}}$$

$$= \frac{2\,x}{x}$$

$$= 2$$

Looking it from the point of view of work done, we find that

Work done by the load $W = W \cdot x$ (v)

Work done by the effort $P = P \cdot 2\,x$ .... (vi)

When friction is neglected, we know from (iv) that $P = W/2$

Substituting in (vi), we get

Work done by effort $P = W/2.\ 2\ x = Wx$

which means that the work done by the effort P is equal to the work done by the load W, which is as it should be according to the principle of conservation of energy, when friction is neglected.

SYSTEMS OF PULLEYS

We already know that for gaining more mechanical advantage, a number of simple levers are combined into a system called a compound lever. Similarly, pulleys are also combined into a system of pulleys to gain more mechanical advantage. There are in all, three systems of pulleys, i.e. the first, second and third system.

First System of Pulleys

In this system, all the pulleys used are movable. Figure 9.29 shows such a system of 3 pulleys.

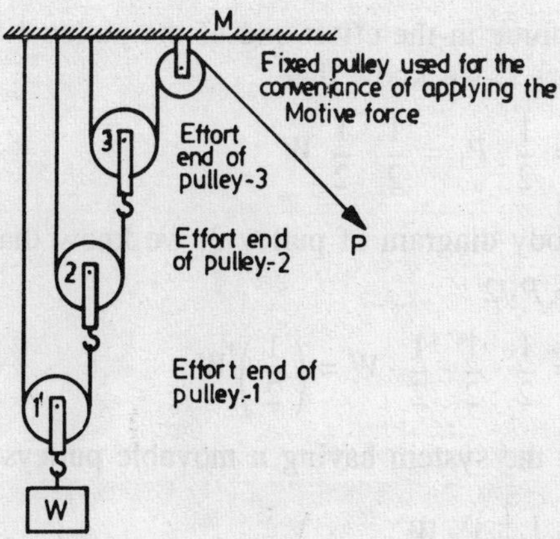

Fig. 9.29

A separate rope is passed round each movable pulley. One end of this rope is fastened to a horizontal beam by means of a hook and the other end of the rope, called the effort end is fastened to the axle of the next upper pulley.

The motive force or the effort P is applied to the effort end of the rope which passes round the uppermost movable pulley (in this case pulley No. 3). A fixed pulley 'M' is generally used in order to apply the motive force P in a convenient direction.

Neglecting the friction between the contact surfaces, let us find out the motive force P required to raise a load W.

We have from the freebody diagram (Figure 9.30) of the movable pulley 1,

$$P_1 = \frac{1}{2} W$$

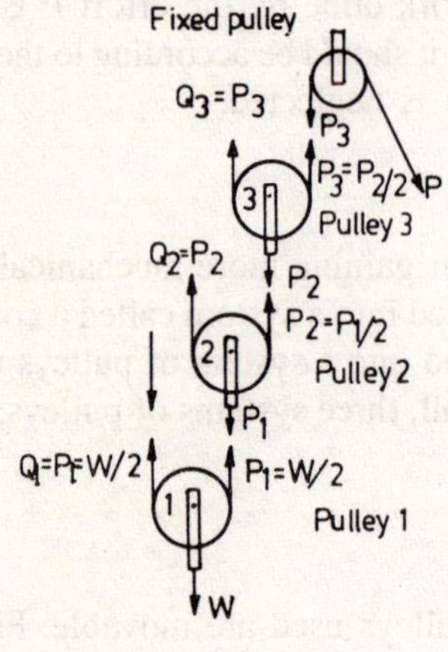

Fig. 9.30

where P_1 is the force in the effort end of the pulley 1. From the free body diagram of pulley 2, we know that

$$P_2 = \frac{1}{2} \cdot P_1 = \frac{1}{2} \cdot \frac{1}{2} W$$

From the free body diagram of pulley 3, we know that

$$P_3 = P_2/2$$

$$= \frac{1}{2} \cdot \frac{1}{2} \cdot \frac{1}{2} \cdot W = \left(\frac{1}{2}\right)^3 W$$

Generalizing for the system having n movable pulleys, we can write

$$P_n = \left(\frac{1}{2}\right)^n \cdot W$$

where n = No. of movable pulleys in the system. From the freebody diagram of the fixed pulley, we find that

$$P = P_n$$

$$= \left(\frac{1}{2}\right)^n W$$

Mechanical Advantage

The mechanical advantage (MA) of this system of pulleys is

$$\text{MA} = \frac{\text{Load raised}}{\text{Effort applied}} \text{ (Resistance overcome)}$$

$$= \frac{W}{\left(\frac{1}{2}\right)^n W} = 2^n$$

As no energy is lost in the friction according to the assumption made earlier, the input work must be equal to the output work

Input work = Work done by the effort P

$= P \times s$

Output work = $W \times h$

where s = displacement of P

h = height to which W is raised

$\therefore \quad P \times s = W \times h$

or $$\frac{s}{h} = \frac{W}{P} = \text{MA} = 2^n$$

where s/h is the displacement of the velocity ratio of the system, and is equal to the MA i.e. 2^n. For illustration, if the number of movable pulleys in such a system is 3, then (neglecting friction)

$$\text{MA} = \text{VR} = 2^n = 2^3 = 8$$

Second System of Pulleys

In this system of pulleys, both fixed and movable pulleys are used. Figure 9.31 shows such a system. Pulleys 1, 2, and 3 are the fixed pulleys arranged vertically one over the other and encased in a casing or strap. Pulleys 4, 5, and 6 are the movable pulleys, which are again arranged one over the other vertically and encased in a casing L. One end of the rope is fastened by means of a hook to the casing, M. The other end of rope is taken round the pulleys one by one, successively, and finally, passed over the uppermost fixed pulley. To the free end of the rope, which is also called the running end, effort *P* is applied as shown in Figure 9.31.

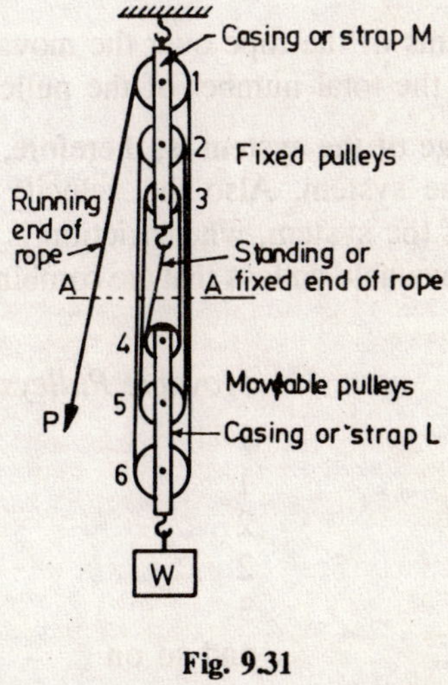

Fig. 9.31

Let us now find out the effort *P* required to raise a load *W* attached by means of a hook to the casing *L* of the movable pulleys.

Take a section at AA and draw the free body diagram of the movable pulleys (Fig. 9.32). As there is only one rope which passes over the various pulleys, therefore the tension everywhere in this rope will be equal. Thus the tension in each segment of the rope is equal and assumed to be *T*.

For equilibrium, $\Sigma V = 0$

$$\therefore \quad 6T = W$$

or
$$T = \frac{W}{6}$$

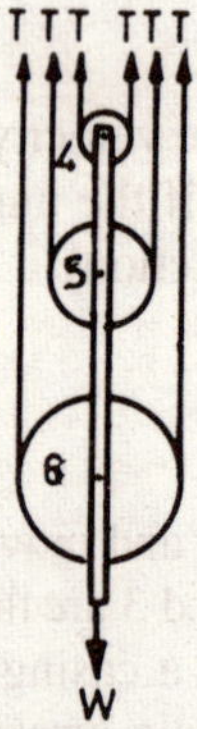

Fig. 9.32

The effort P is equal to the tension in the rope and is, therefore, equal to *W*/6. Generalizing for any such system of pulleys, we can say that

$$P = \frac{W}{n}$$

where n = No. of segments of the rope over the movable pulleys, which is also equal to the total number of the pulleys in the system.

Mechanical advantage of the system is, therefore, equal to n', the total number of pulleys in the system. Also the velocity ratio will equal the mechanical advantage of the system, when friction is neglected.

The number of fixed and movable pulleys that are combined to form the system can be as follows:

Fixed Pulleys	*Movable Pulleys*
1	1
2	1
2	2
3	2
3	3
	and so on

These combinations are shown in Figure 9.33.

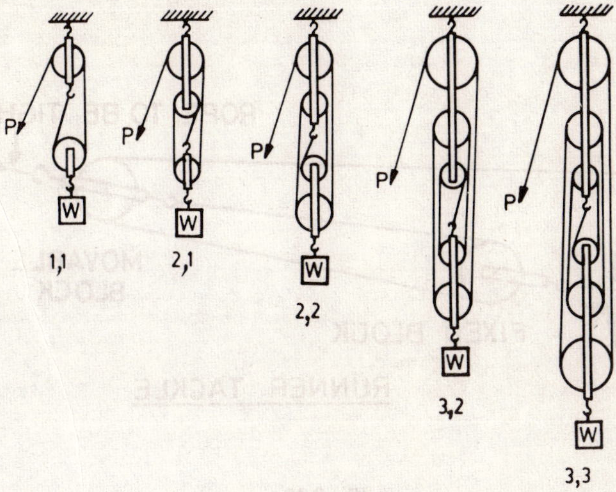

Fig. 9.33

Block

Instead of arranging the pulleys one upon another vertically in a casing, *the pulleys are usually arranged on the same horizontal axle and then encased. Such an arrangement is called a block*. A block may have one, two or three pulleys. A block having one pulley is called a single block. Similarly a block having 2 pulleys is called a double block, and that having three pulleys is called a triple block.

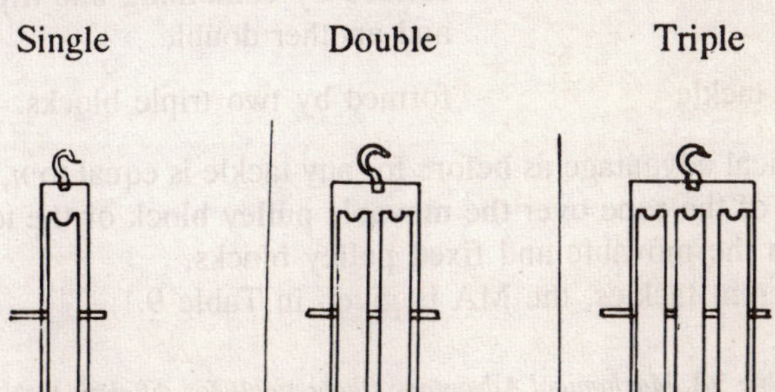

Fig. 9.34

Tackle

When blocks of pulleys (fixed and movable) are grouped together by winding a rope or a chain around them, it is called a tackle. A tackle may be used either for lifting purpose or for tightening the guy ropes of a supporting structure.

In case the tackle is used for the purpose of lifting, it is called a lifting tackle. When the tackle is used for tightening the guy ropes, it is called a running tackle. Both these tackles have been shown in Figure 9.35.

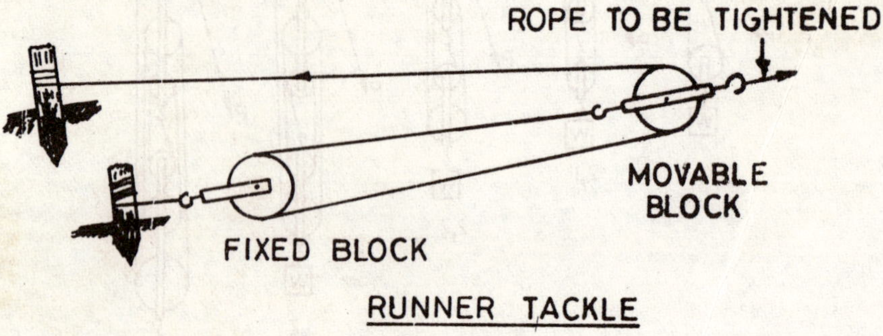

Fig. 9.35

Depending upon the pulley blocks used and their combinations, different tackles are termed as under:

Single-single tackle	—	formed by combining two single blocks
Double-single tackle	—	formed by combining one double and one single block
Double-double tackle	—	formed by combining two double blocks
Triple-double tackle	—	formed by combining one triple block and another double.
Triple-triple tackle	—	formed by two triple blocks.

The mechanical advantage as before for any tackle is equal to n, the number of segments of the rope over the movable pulley block or the total number of pulleys in the movable and fixed pulley blocks.
For the different tackles, the MA is given in Table 9.1.

Table 9.1 *Machanical Advantage (theoretical) for different tackles*

Tackle	*Mechanical Advantage*
1/1	2
2/1	3
2/2	4
3/2	5
3/3	6

However, in actual practice, there is friction at the contact surfaces of the rope with the pulleys and values of (MA given in the table above get decreased).

Third System of Pulleys

In this system of pulleys, one of the pulleys used is a fixed pulley and all the other are movable pulleys. This system of pulleys is rarely used in practice. Figure 9.36 shows this system of pulleys. Pulley 1 is a fixed pulley, whereas all the other pulleys, i.e. 2, 3, ... are movable pulleys.

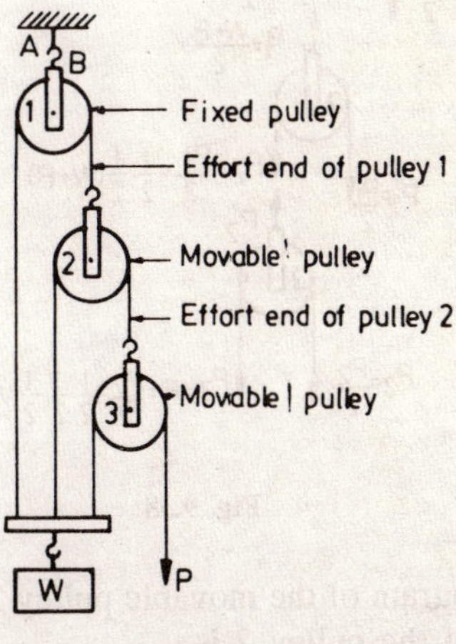

Fig. 9.36

Let us try to find the effort P required to raise a load W through this system of pulleys.

The whole system is supported by the string AB. Neglecting the weight of the ropes and the pulleys, the Tension in the string AB = $W + P$

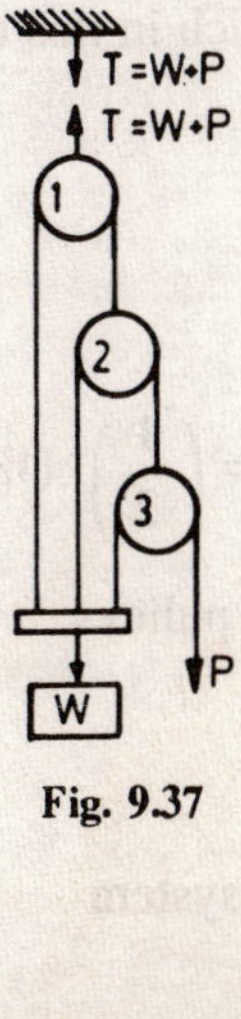

Fig. 9.37

From the free body diagram of the fixed pulley 1 (Figure 9.38). The tension P_1, in the effort end of the pulley 1 is

$$P_1 = \frac{W + P}{2}$$

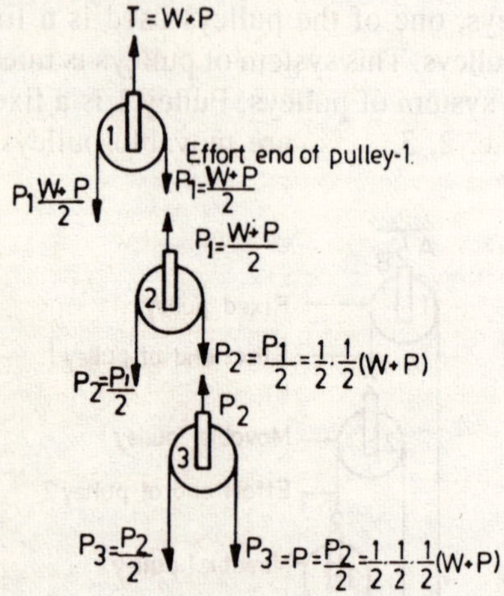

Fig. 9.38

From the freebody diagram of the movable pulley 2, (Fig. 9.38), the tension P_2 in the effort end of the pulley 2 is

$$P_2 = \frac{1}{2} \cdot P_1$$

$$= \frac{1}{2} \cdot \frac{1}{2} (W + P)$$

From the freebody diagram of the movable pulley 3, the tension P_3 in the effort end of the pulley 3, which in this case will be equal to P, is given by (Fig. 9.38).

$$P_3 = P_2/2$$

$$= \frac{1}{2} \cdot \frac{P_1}{2}$$

$$= \frac{1}{2} \cdot \frac{1}{2} \cdot \frac{1}{2} (W + P) = \left(\frac{1}{2}\right)^3 (W + P)$$

Generalizing for a system of 'n' pulleys

$$P_n = \left(\frac{1}{2}\right)^n (W + P)$$

where n = No. of pulleys in the system

As $P = P_n$

$$\therefore \quad P = \left(\frac{1}{2}\right)^n (W + P)$$

or $2^n P = (W + P)$

$\therefore$ $P(2^n - 1) = W$

$$P = \frac{W}{(2^n - 1)}$$

$$\text{MA} = \frac{W}{P} = \frac{W}{\dfrac{W}{(2^n - 1)}} = (2^n - 1)$$

If the number of pulleys in the system are 3, then the mechanical advantage will be $2^3 - 1 = 7$. Also the distance travelled by the effort P will be 7 times the distance by which the load W is raised.

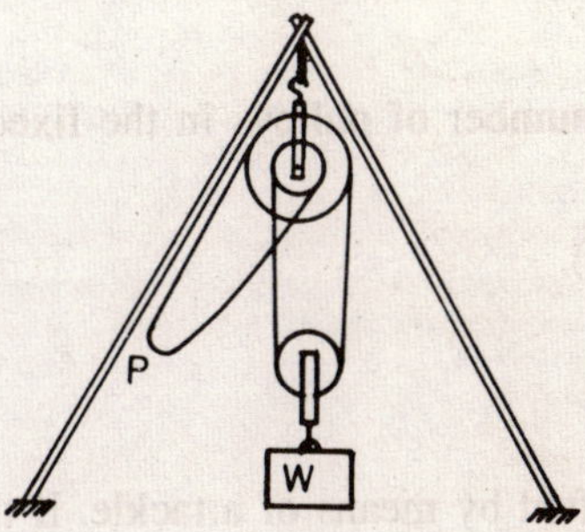

Fig. 9.39

Example 9.6

In raising a load of 2 kN, determine the effort to be applied, when,

(i) a first system of pulleys is used, and the number of pulleys in the system is 2.

(ii) a third system of pulleys is used and the number of pulleys in the system is 4.

(iii) a 3/2 tackle is used.

Neglect friction

Solution

(i) In the first system of pulleys

$$P = \frac{W}{2^n}$$

$$= \frac{2000}{2^2}$$

$$= \frac{2000}{4}$$

$$= \mathbf{500\ N}$$

(ii) In the third system of pulleys,

$$P = \frac{W}{(2^n - 1)}$$

$$= \frac{2000}{2^4 - 1}$$

$$= \frac{2000}{16 - 1}$$

$$= \mathbf{133.3\ N}$$

(iii) When 3/2 tackle is used

$$P = \frac{W}{n}$$

where n = Total number of pulleys in the fixed and movable blocks

$$\therefore \quad P = \frac{2000}{(3 + 2)}$$

$$= \mathbf{400\ N}$$

Example 9.7

A load of 3*t* is to be lifted by means of a tackle. If the permissible tension in the rope to be used for the tackle is 575 kgf, select the tackle that you would use for lifting this load.

Solution

As the effort to be applied for raising the load cannot exceed the maximum permissible tension in the rope, therefore,

Max. effort applied 'P' = 575 kgf.

We know that in a lifting tackle

$$P = \frac{W}{n}$$

$$\therefore\ n = \frac{W}{P}$$

$$= \frac{3000}{575}$$

$$= 6$$

$\therefore$ We use a 3/3 Tackle.

9.3.5 *Differential Pulley Block*

Instead of using the systems of pulleys described earlier, a differential pulley block can be used to gain the same mechanical advantage by using a far less number of pulleys. A differential pulley block consists of two blocks of pulleys, one fixed and the other movable. The fixed block consists of two pulleys of different radii R and r, arranged on the same horizontal axle. An endless chain is passed round the pulleys as shown in Figures 9.40, 9.41, 9.42. For raising the load, effort is applied to the segment of the chain passing over the pulley of the bigger radius R as shown in figure 9.40. For lowering the load, effort is applied to the segment of the rope passing over the pulley of the shorter radius r as shown in Figure 9.41.

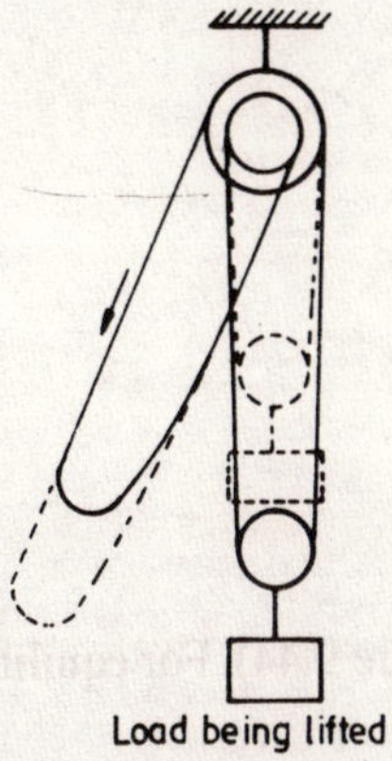

Fig. 9.40

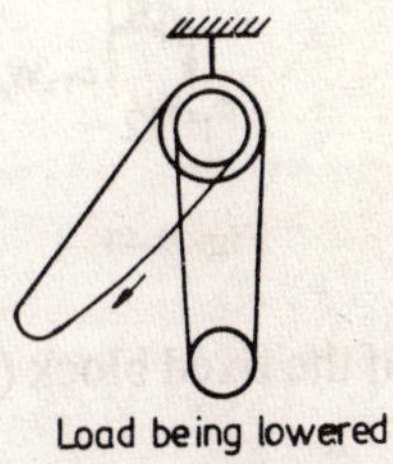

Fig. 9.41

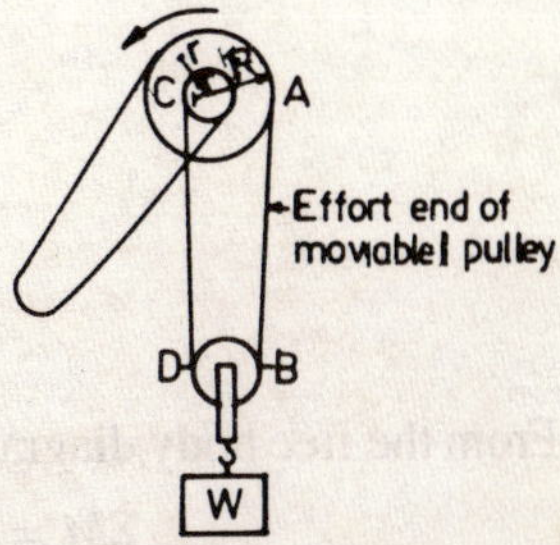

Fig. 9.42

Let us now determine the mechanical advantage, VR and efficiency of this machine.

Velocity Ratio

As the pulleys in the fixed block move through one revolution in an anticlockwise direction, segment AB of the chain moves up by a distance $2\pi R$ and the segment CD of the chain moves down by a distance $2\pi r$. Therefore, the net upward movement of the effort and of the movable pulley is $2\pi(R - r)$ and the upward displacement of the movable block or in other words, the weight W will be $\dfrac{2\pi R - 2\pi r}{2} = \pi(R - r)$

$$\text{VR} = \frac{\text{Distance moved by the effort}}{\text{Distance moved by the load}}$$

$$= \frac{2\pi R}{\pi(R - r)} = \frac{2R}{(R - r)} \qquad \text{.... (i)}$$

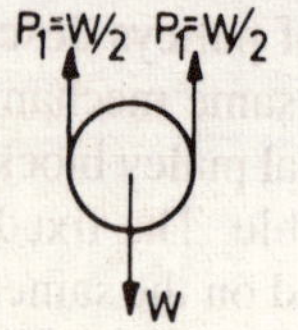

Fig. 9.43

Mechanical Advantage

From the free body diagram of the movable pulley (Fig. 9.44)

$$P_1 = W/2$$

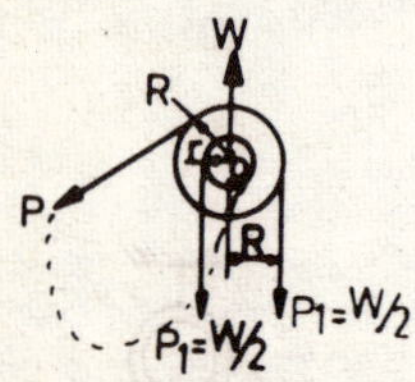

Fig. 9.44

From the free body diagram of the fixed block (Figure 9.44). For equilibrium,

$$\Sigma M = 0,$$

Taking moments of all the forces about O,

$$\frac{W}{2}R - \frac{W}{2}r - PR = 0$$

$$PR = \frac{W}{2}(R - r)$$

$$\therefore P = \frac{W(R - r)}{2R}$$

$$\text{MA} = \frac{W}{P} = \frac{2R}{R - r} \quad \text{.... (ii)}$$

The mechanical advantage thus gained can be increased to any extent by decreasing the value of $(R - r)$. Also we find that the MA is equal to the VR. This is so because the friction at the contact surfaces has been neglected and the machine is considered ideal. However, if the friction between the contact surfaces is taken into consideration, then the effort P required to raise the load W will be more than $\frac{W(R - r)}{2R}$.

Let in this case the effort required be P', then,

Work done by the effort (work input) $= P' \times 2\,\pi\,R$
in one revolution of the axle

Work done by the Load (work output) $= W \times \pi\,(R - r)$

$$\text{Efficiency '}\eta\text{'} = \frac{\text{Work output}}{\text{Work input}}$$

$$= \frac{W\,\pi\,(R - r)}{P' \times 2\,\pi\,R}$$

$$= \frac{W}{P'}\;\frac{R - r}{2\,R}$$

$$\text{Mechanical Advantage (MA)} = \frac{W}{P'}$$

$$= \eta \cdot \frac{2\,R}{(R - r)}$$

Example 9.8

A machine of mass 2000 kgs is to be lifted by using a differential pulley block. The radii of the bigger and smaller pulleys of the fixed block are 300 mm and 200 mm, respectively. If the efficiency of the pulley block is 80 per cent, determine the effort required to raise the load.

Solution

$$\text{MA of a differential pulley block} = \frac{2\,R}{(R - r)} \cdot \eta$$

$$= \frac{2 \times 300}{(300 - 200)} \times 0.8$$

$$= \frac{600 \times 0.8}{100}$$

$$= 4.8$$

$$\therefore \quad \frac{W}{P} = 4.8$$

$$P = \frac{W}{4.8}$$

$$= \frac{2000 \times 9.8}{4.8}$$

$$= \mathbf{4083.3\ N}$$

9.3.6 *Wheel and Axle (Simple Windlass)*

This simple machine is very commonly used for hoisting purposes, and is part of many a machine. It consists of a drum '*D*' and a wheel *E* fixed to

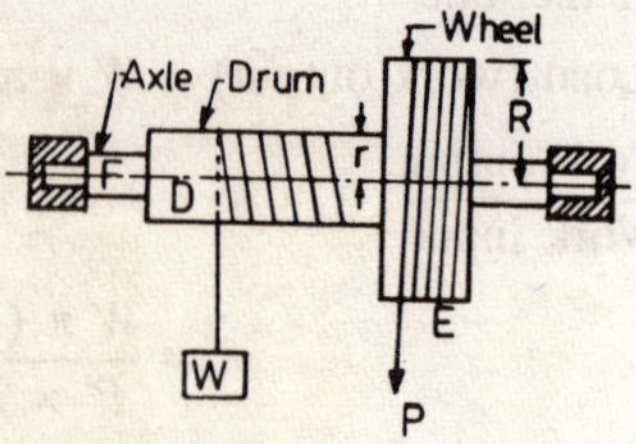

Fig. 9.45

an axle '*F*' as shown in Figure 9.45. Two ropes are used in this case. One is used for raising or lowering the load, and is wound over the drum. The other is used for applying the effort and is wound round the wheel. If the rope on the drum is wound in a clockwise direction, then that on the wheel is wound in the anticlockwise direction, so that when the rope on the wheel is being unwound, the rope on the drum gets wound up and the load is raised. Instead of using a wheel, a handle is generally linked with the axle of the drum for applying the motive force as shown in Figure 9.46. Sometimes, radiating arms are attached to the wheel for applying the effort in order to avoid the use of the rope on the wheel (Figure 9.47).
Velocity ratio, Mechanical advantage and efficiency of this machine have been determined as under.

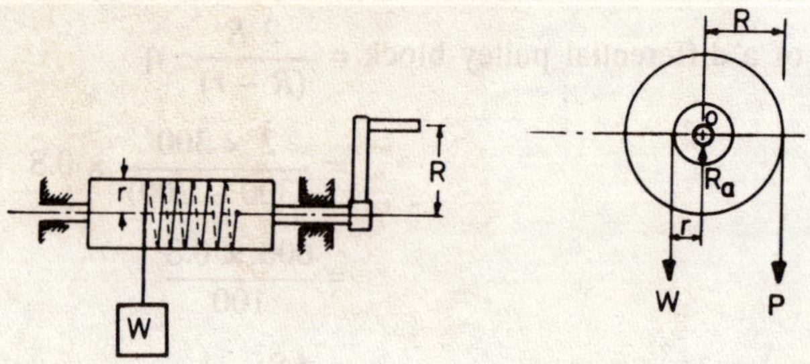

Fig. 9.46

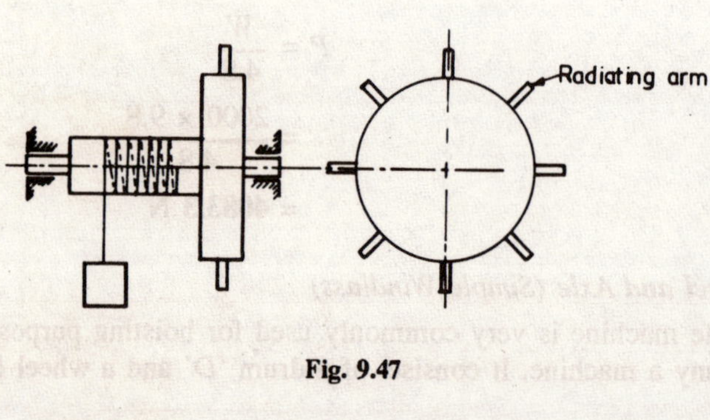

Fig. 9.47

Velocity Ratio

For one complete rotation of the axle,

Distance moved by the effort 'P' = $2\pi R$

Distance moved by the load = $2\pi r$

where R and r = radii of the wheel and the drum respectively.

$$\text{Velocity ratio (VR)} = \frac{\text{Distance moved by the effort}}{\text{Distance moved by the load}}$$

$$= \frac{2\pi R}{2\pi r}$$

$$= \frac{R}{r}$$

Mechanical Advantage (MA)

Neglecting friction at the contact surfaces, the forces acting on the system are shown in Figure 9.46.

For equilibrium, $\Sigma M = 0$

Taking moments of all forces about O, we get,

$$PR - Wr = 0$$

$$PR = Wr$$

$$\therefore\ P = W\frac{r}{R}$$

$$MA = \frac{W}{P} = \frac{R}{r}$$

Note: Actually R = Radius of the wheel + $\frac{1}{2}$ thickness of the rope

r = radius of the drum + $\frac{1}{2}$ thickness of the rope

As the thickness of the rope is very small, as compared to the radius of the drum and wheel, it is neglected for all practical purposes.

MA is equal to the velocity ratio. This is so, because friction is neglected and as a result of this efficiency of the machine will be 100 per cent.

If frictional resistance is taken into consideration as is true in any practical case, the effort required to raise the load will not be $W\frac{r}{R}$, but it will be greater than this. If 'P', is the effort required to raise the load, when friction is considered, then for one complete revolution of the axle,

Work done by the effort $P' = P' \times 2\pi R$
(Input work)

Work done by the load $W = W \times 2\pi r$

$$\text{Efficiency of the machine '}\eta\text{'} = \frac{\text{Output work}}{\text{Input work}}$$

$$\eta = \frac{2\,r.\,W}{2\,R.\,P'}$$

$$= \frac{W}{P'} \cdot \frac{r}{R}$$

$$\text{From this, } P' = W \cdot \frac{r}{R} \cdot \frac{1}{\eta}$$

9.3.7 *Differential Wheel and Axle or Differential Windlass*

It consists of a stepped drum or in other words a drum having two portions of different diameters D and d as shown in Figure 9.48. The drum is mounted on a spindle, which is supported at its ends on the bearings. A single rope is wound round the drum in such a manner that when the spindle rotates, the rope gets wound up on one portion of the drum and unwounds itself on the other portion. This permits the load to move up or down. For example, in Figure 9.48, when the spindle moves in clockwise direction (looking from the left end), the rope gets unwound on the portion of the drum with smaller diameter, and at the same time is wound up on the portion of the drum with larger diameter. This will raise the load. To accomplish this, one end of the rope is fastened at a point, say C on the portion of the drum of smaller diameter. The rope is then given a number of turns round this portion of the drum in a clockwise direction. It is then passed round a movable pulley block, and finally taken around a portion of the drum of larger diameter and given a number of turns on it in a clockwise direction.

For applying the motive force or the effort, two arrangements are generally used.

In the first arrangement, a wheel is mounted on the spindle as was done in the case of a simple windlass. On this wheel a few turns of a rope are given. Motive force is applied to the running end of this rope.

In the second arrangement, a crank is mounted on the spindle as shown in Figure 9.48. The effort is applied through the handle of this crank.

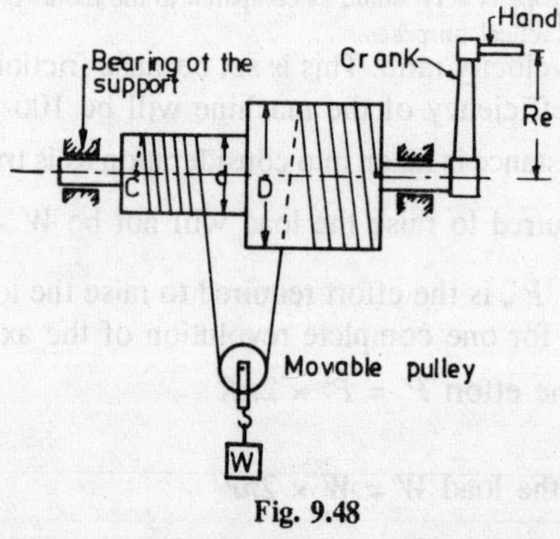

Fig. 9.48

Velocity Ratio

For one complete revolution of the wheel or the crank, the effort 'P' will undergo a displacement equal to $2\pi R_e$, where R_e is the radius of the circle described by the effort applied to the crank. In case a wheel is used for applying the effort, R_e is approximately taken equal to the radius of the wheel.

Also for one complete revolution of the wheel or the crank, say in a clockwise direction (refer Fig. 9.49), segment QR of the rope goes up by a distance equal to the circumference of the portion of the drum of bigger diameter, i.e. πD, and segment ST of the rope falls down by a distance equal to the circumference of the portion of the drum of smaller diameter, i.e.

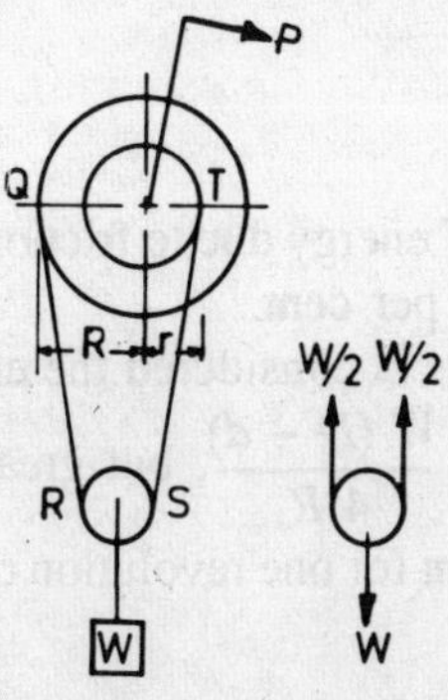

Fig. 9.49

Fig. 9.50

πd. The net distance by which the rope moves up is, therefore, equal to $\pi (D - d)$. Assuming the segments QR and ST to be vertical (though these cannot be vertical because they are converging) the distance by which the load is raised will be equal to $\dfrac{\pi D - \pi d}{2} = \dfrac{\pi}{2}(D - d)$

$$\therefore \text{ Velocity ratio} = \frac{\text{Distance moved by the effort}}{\text{Distance moved by the load}}$$

$$= \frac{2\pi R_e}{\frac{\pi}{2}(D - d)} = \frac{4 R_e}{(D - d)}$$

Mechanical Advantage

From the free body diagram of the movable pulley (Fig. 9.50), if it is assumed that the segment QR and ST are vertical, then for equilibrium, $\Sigma V = 0$ and the tension in each of the segments

QR and ST will be $W/2$

From the free body diagram of the drum shown in Figure 9.50 we see that (Neglecting the friction at the bearings), for equilibrium

$$\Sigma M = 0$$

Taking moments of all forces about O, we get

$$P \times R_e - W/2.\ \frac{D}{2} + \frac{W}{2}.\ \frac{d}{2} = 0$$

or
$$P.\ R_e = \frac{W}{2}\left(\frac{D}{2} - \frac{d}{2}\right)$$

$$\therefore \qquad P = \frac{W}{2}\ \frac{(D - d)}{4\ R_e}$$

$$\text{Mechanical Advantage} = \frac{W}{P}$$

$$= \frac{4\ R_e}{(D - d)}$$

which is the same as the VR because loss of energy due to friction is neglected and the efficiency of the machine is 100 per cent.

However, if the friction at the bearings is considered the effort required to lift the same load will not be equal to $\frac{W(D - d)}{4\ R_e}$, but greater than this.

Let the effort required in this case be P', then for one revolution of the spindle, the work done by the effort P'

(work input) $= P' \times 2\pi R_e$

the work done by the Load W

(work output) $= W \frac{\pi}{2} (D - d)$

$$\text{Efficiency of the machine} = \frac{\text{Output work}}{\text{Input work}}$$

$$= W \cdot \frac{\pi}{2} (D - d)/P' \cdot 2\pi R_e$$

$$= \frac{W}{P'}\ \frac{(D - d)}{4 \cdot R_e}$$

$$\text{Also } P' = W \frac{(D - d)}{4 \cdot R_e}\ \frac{1}{\eta}$$

$$\text{MA (actual)} = \frac{W}{P} = \frac{4\ R_e}{(D - d)} \cdot \eta$$

Example 9.10

A differential wheel and axle is used to lift a box of mass 4000 kgs. The larger

and smaller radii of the beam are 40 cms and 30 cms respectively. The length of the crank arm is 45 cms. If the efficiency of the machine is 75 per cent, determine the motive force required to lift the box, the velocity ratio and the mechanical advantage gained.

Solution

(1) Velocity Ratio $= \dfrac{4 \cdot R_e}{(D - d)}$

$= \dfrac{4 \times 45}{(40 - 30)}$

$= 18$

(2) Effort required P' $= \dfrac{W\,(D - d)}{4 \cdot R_e \cdot \eta}$

$= 4 \times 10^3 \times 9.8 \;\dfrac{(40 - 30)}{4 \times 45 \times 0.75}$

$= 2900$ N

(3) Mechanical Advantage $= \dfrac{\text{Load lifted}}{\text{Effort applied}}$

$= \dfrac{4 \times 10^3 \times 9.8}{2900}$

$= \mathbf{13.52}$

9.3.8 *Pulley Drives and Gear Drives*

If two pulleys 1 and 2 mounted on separate axles are connected by means of a belt, then the rotation of one pulley will cause rotation of the other pulley. If the belt is an open one, then the rotation of both the pulleys will be in the same direction. This is shown in Figure 9.51. If a crossed belt is used, then the pulleys will rotate in an opposite direction as shown in Figure 9.52.

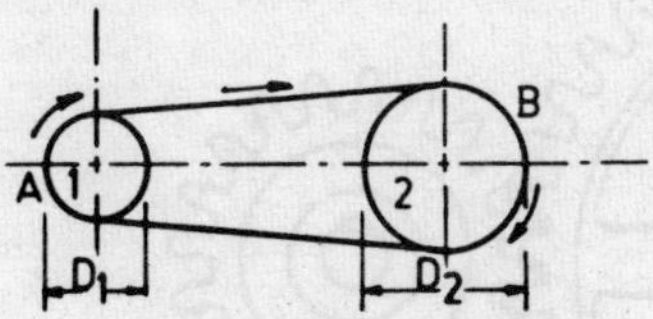

OPEN BELT PULLEY DRIVE

Fig. 9.51

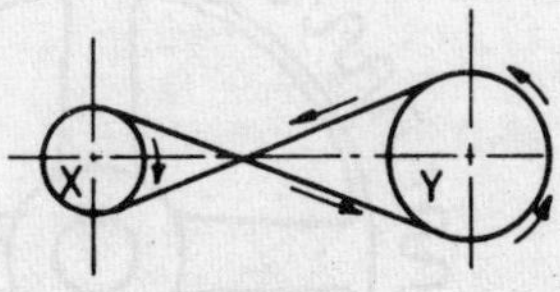

CROSSED BELT DRIVE

Fig. 9.52

Whatever may be the direction of rotation, the speed of two points say A and B anywhere on the belt will remain the same.

Linear speed of pulley '1' $= \pi D_1 N_1$

Linear speed of pulley '2' = $\pi D_2 N_2$

where N_1 and N_2 are the revolutions per second made by pulleys 1 and 2 respectively.

Equating the linear speed of both the pulleys, we get,

$$\pi D_1 N_1 = \pi D_2 N_2$$

or
$$\frac{N_1}{N_2} = \frac{D_2}{D_1}$$

By the use of a belt drive on pulleys, it is possible to make a gain in the number of revolutions moved by a pulley. Also rotational speed of pulleys in this case is inversely proportional to their diameters.

Gear Drives

Figure 9.53 shows two gears 1 and 2 meshed against one another. A gear is like a wheel with teeth cut on its periphery.

When the teeth of two or more gears mesh with one another, the linear speed of the gears will remain the same. The rotational speed of the gears will be inversely proportional to their number of teeth.

i.e.
$$\frac{N_1}{N_2} = \frac{T_2}{T_1}$$

where N_1 and N_2 are the revolutions made by gears 1 and 2 per unit of time. T_1 and T_2 are the number of teeth on gears 1 and 2 respectively.

When two spur gears mesh with one another, the direction of motion of these gears will be opposite to one another as shown in Figure 9.53. In order to make the gears rotate in the same direction, an idler gear can be used as shown in Figure 9.54. This gear does not affect the rotational speed of gears 1 and 2. It only helps in making the direction of motion of the gears 1 and 2 the same.

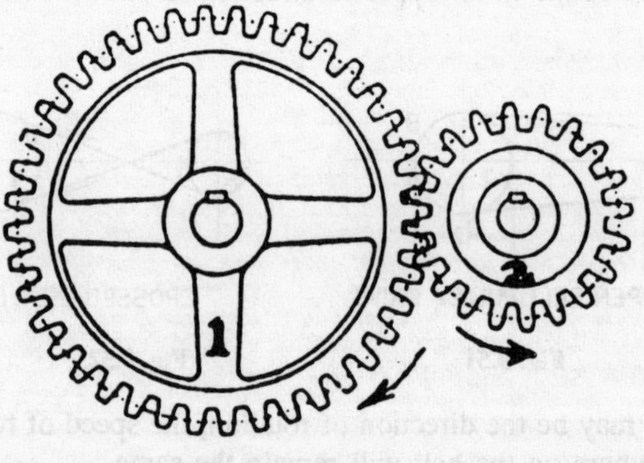

Fig. 9.53

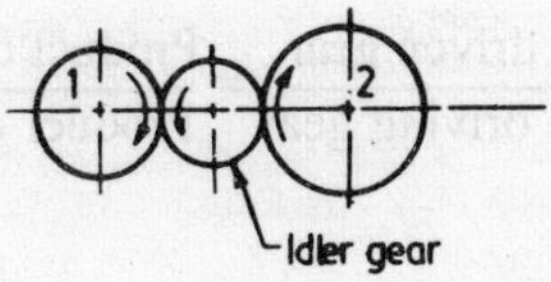

Fig. 9.54

Compound Gear Train

A compound gear train is the one in which all the gears in the train do not mesh with one another. In a simple gear train shown in Figure 9.54, all the gears mesh with one another. A compound gear train is shown in Figure 9.55. Assuming that power is supplied to gear A, then gear A is called the driver gear and the gear B is the driven gear. Gear C is fixed to the same axle as gear B, and becomes the driver gear, and gear D becomes the driven gear.

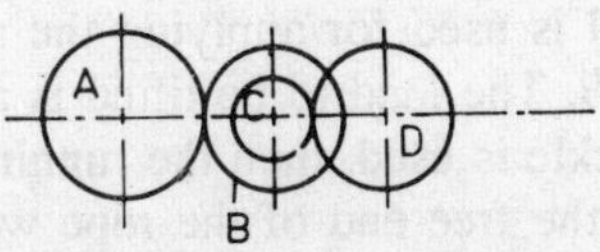

Fig. 9.55

Assuming gears A, B, C, D have number of teeth equal to T_A, T_B, T_C and T_D, and rotate at N_A, N_B, N_C and N_D revolutions per second respectively, then we have,

$$\frac{N_A}{N_B} = \frac{T_B}{T_A}$$

$$\text{or } N_B = N_A \times \frac{T_A}{T_B}$$

$$\text{Also } \frac{N_C}{N_D} = \frac{T_D}{T_C}$$

$$\text{or } N_D = N_C \times \frac{T_C}{T_D}$$

But $N_C = N_B$, ($\because$ Gears B and C are mounted on the same axle)

$$\therefore N_D = N_B \times \frac{T_C}{T_D}$$

Substituting for N_B, we have

$$N_D = N_A \times \frac{T_A}{T_B} \times \frac{T_C}{T_D}$$

$$\text{or}\quad \frac{N_D}{N_A} = \frac{T_A}{T_B} \times \frac{T_C}{T_D}$$

$$\text{or}\quad \frac{\text{Circular speed of final driven gear}}{\text{Circular speed of initial driving gear}} = \frac{\text{Product of Teeth on driving gears}}{\text{Product of teeth on driven gears}}$$

9.3.9 *Winch Crab*

Winch Crab is a machine most commonly used for lifting loads in conjunction with the tackles. The running end of the rope from the tackle is attached to the crab for the purpose of applying the motive force.

A winch crab can be a single purchase or a double purchase.

Single Purchase Winch Crab

A single purchase winch crab consists of a toothed wheel *M* (Refer fig. 9.56) mounted on a spindle S_1, which in turn is supported on its end bearings. A barrel *B* is also mounted on this spindle. To the toothed wheel *M* is geared a toothed pinion *T* which is mounted on another spindle S_2 parallel to the spindle S_1 and supported on its end bearings. A crank '*C*' is attached to one end of the spindle S_2 and is used for applying the motive force. A rope is wound round the barrel *B*. The load to be lifted is attached to the free end of this rope. In case a tackle is used, then the running end of the rope from the tackle is fastened to the free end of the rope wound round the barrel.

Velocity Ratio

For complete one revolution of the crank '*C*' or the spindle S_2, distance moved by the effort $= 2\,\pi\, r_e$ (1)

where, r_e = Lever arm of the crank '*C*'

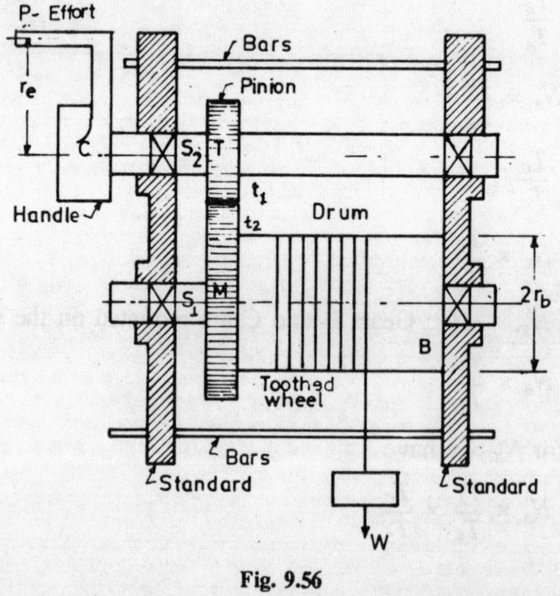

Fig. 9.56

If the number of teeth in the pinion T is t_1, and that in the toothed wheel M is t_2, then for one revolution of the pinion, the number of revolutions undergone by the wheel M

$$= 1 \times \frac{t_1}{t_2}$$

As the wheel M and the barrel B are fixed on the same spindle, therefore the number of revolutions undergone by the barrel will be equal to that undergone by the wheel, i.e. $\frac{t_1}{t_2}$

$$\therefore \text{ Displacement of load } W = 2\,\pi\, r_b \times \frac{t_1}{t_2} \qquad \text{.... (2)}$$

where, r_b = radius of the barrel 'B'

$$\text{Velocity ratio} = \frac{\text{Displacement of effort}}{\text{Displacement of load}}$$

$$= \frac{2\,\pi\, r_e}{\frac{t_1}{t_2} \times 2\,\pi\, r_b}$$

$$= \frac{r_e}{r_b} \cdot \frac{t_2}{t_1} \qquad \text{.... (3)}$$

Mechanical Advantage

The work done by the effort will be equal to the work done by the load, provided friction is neglected.

For one revolution of the spindle S_2

Work done by the effort $P = P \times 2\pi r_e$

Work done by the load $W = W \times \frac{t_1}{t_2} \times 2\pi r_b$

Equating, we get

$$P \times 2\pi r_e = W \times \frac{t_1}{t_2}\, 2\pi r_b$$

$$P = W \times \frac{t_1}{t_2} \cdot \frac{r_b}{r_e} \qquad \text{.... (4)}$$

$$\text{MA} = \frac{W}{p} = \frac{t_2}{t_1} \cdot \frac{r_e}{r_b} \qquad \text{.... (5)}$$

If friction is taken into account, then the effort required will be more than that given by equation (4). Let in this case, the effort be P'.

Work done by the effort $P' = P' \times 2\pi r_e$
(Input work)

Work done by the load $W = W \times \frac{t_1}{t_2} \cdot 2\,\pi\, r_b$
(Output work)

$$\text{Efficiency } (\eta) = (W \cdot \frac{t_1}{t_2} 2\pi r_b)/(P' \times 2\pi r_e)$$

$$= \frac{W}{P'} \frac{t_1}{t_2} \frac{r_b}{r_e} \quad .... (6)$$

$$P' = W \cdot \frac{t_1}{t_2} \frac{r_b}{r_e} \cdot \frac{1}{\eta} \quad .... (7)$$

Double Purcahse Winch Crab

In a double purchase winch crab, the arrangement is the same as that of the single purchase winch crab except that one toothed wheel X and a pinion Y fixed to a spindle S_3 are added. The wheel X and the pinion Y are geared to the pinion T and wheel M respectively. The whole arrangement is shown in figure 9.57.

Velocity Ratio

For one complete revolution of the crank, which is used for applying an effort P in a direction normal to the plane containing the handle and the axis of the spindle S_2,

No. of revolutions made by the pinion T for one revolution of crank C = Revolution made by the crank C = 1 ($\because$ Pinion T and the crank 'C' are mounted on the same spindle S_2)

No. of revolution made by wheel X $= 1 \times \frac{t_1}{t_2}$

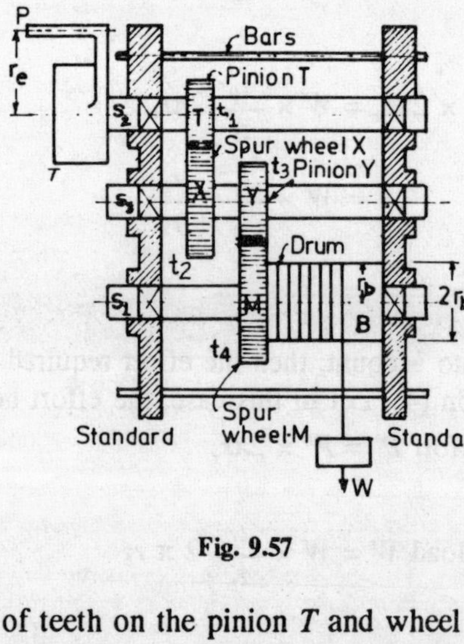

Fig. 9.57

where t_1, t_2 = No. of teeth on the pinion T and wheel X respectively.

No. of revolutions made by pinion Y = Revolutions made by the wheel X ($\because$ pinion Y and wheel X are mounted on the same spindle S_3)

$$= 1 \times t_1/t_2$$

No. of revolutions made by the wheel M = No. of revolutions made by pinion Y $\times \dfrac{t_3}{t_4}$

$$= 1 \times \frac{t_1}{t_2} \times \frac{t_3}{t_4}$$

where, t_3, t_4 = No. of teeth on the pinion Y and wheel M, respectively.

No. of revolutions made by the barrel B = No. of revolutions made by wheel M ($\because$ barrel B and wheel M are mounted on the same spindle S_1)

$$= 1 \times \frac{t_1}{t_2} \times \frac{t_3}{t_4}$$

$\therefore$ For one complete revolution of the crank 'C'

Distance moved by the load $W = 2\,\pi\, r_b \times 1 \times \dfrac{t_1}{t_2} \times \dfrac{t_3}{t_4}$

Distance moved by the effort $P = 2\,\pi\, r_e$

where, r_e = perpendicular distance from the axis of the spindle S_2 to the line of action of the effort P

r_b = radius of the barrel B

$$\text{Velocity ratio} = \frac{2\,\pi\, r_e}{2\,\pi\, r_b \cdot \dfrac{t_1}{t_2} \times \dfrac{t_3}{t_4}}$$

$$= \frac{r_e}{r_b} \cdot \frac{t_2}{t_1} \cdot \frac{t_4}{t_3}$$

Mechanical Advantage

In case loss of energy due to friction is neglected, then

$$\text{MA} = \text{VR}$$

$$= \frac{r_e}{r_b} \cdot \frac{t_2 \cdot t_4}{t_1 \cdot t_3}$$

If the loss of energy due to friction is not neglected, but taken into consideration, then

$$\text{MA} = \frac{r_e}{r_b} \cdot \frac{t_2 \cdot t_4}{t_1 \cdot t_3} \times \eta \text{ (where } \eta \text{ is the efficiency of the system)}$$

Example 9.11

A double purcahse winch crab is used to hoist a load in conjunction with a 3/2 tackle. If the effort applied P' to the crank of the crab is 200 N, determine the maximum load W that can be lifted. Data regarding the crab is given as under. Efficiency of the crab and the tackle is 40 per cent and 60 per cent respectively. Also find the VR of the whole system.

Length of the crank arm = 600 mm
No. of teeth in the first pinion = 15
No. of teeth in the second pinion = 20
No. of teeth in the first spur wheel = 45
No. of teeth in the second spur wheel = 40
Radius of the drum = 400 mm

Solution

In the case of a double winch crab, we know that

$$\text{MA} = \frac{r_e}{r_b} \cdot \frac{t_2 \cdot t_4}{t_1 \cdot t_3} \times \eta_{\text{crab}}$$

$$= \frac{600}{400} \cdot \frac{45 \times 40}{15 \times 20} \times 0.4$$

$$= 3.6$$

∴ the load that can be raised by the crab or the effort P_1 that can be applied to the running end of the rope from the tackle is

Effort $P_1 = P' \times 3.6$
$= 200 \times 3.6$
$= 720$ N

MA of the tackle = $n \times \eta$ tackle where η is the total number of pulleys in the movable and immovable blocks

$$\text{MA}_{\text{tackle}} = (3 + 2) \times 0.6 = 3$$

∴ Maximum Load that can be lifted by means of the tackle

$= P_1 \times 3$
$= 720 \times 3$
= 2160 N

Overall MA of the system = 3.6 × 3 = 10.8

Overall 'η' of the system $= \eta_{\text{crab}} \times \eta_{\text{tackle}}$
$= 0.6 \times 0.4$
$= 0.24$ or 24 per cent

$$\text{VR} = \frac{\text{MA}}{\eta} = \frac{10.8}{0.24} = 45$$

9.3.10 *Screw Jack*

This simple machine is very commonly used for raising very heavy loads through small distances, e.g. raising of the front or rear portion of a vehicle for the purpose of changing the wheel or tyres etc. It consists of a base, the shape of which is that of a hollow frustrum *A*. A hole is made in this hollow case which is threaded in order to receive a strong square threaded screw *S*. On the top of the screw a collar *D* is fitted in such a manner that the screw is free to rotate in it. On the top of the collar *D* the load '*W*' to be raised is placed. Effort *P* is applied at the end of a tommy bar, which can be inserted into any of the holes provided at the top of the screw. As the effort is applied at the end of the tommy bar to turn the screw in the hole, the top of the screw will either go up, or down, depending upon the direction in which the tommy bar is turned.

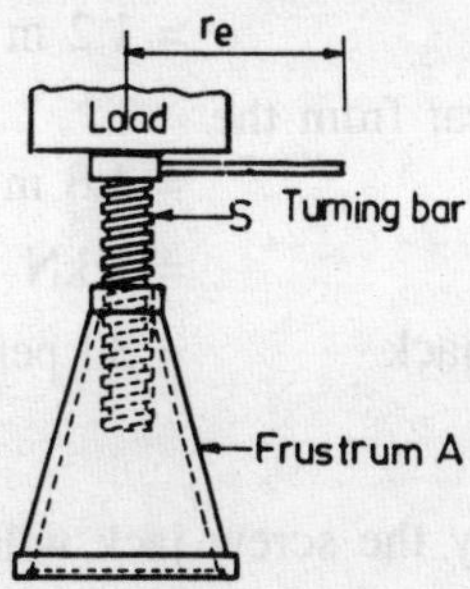

Fig. 9.58

If r_e is the radius at which effort *P* is applied, and '*p*' is the pitch of the screw, then for one revolution of the screw,

Distance moved by the effort $P = 2\pi r_e$

Distance moved by the load $W = p$

$$\therefore \text{VR} = \frac{2\pi r_e}{p}$$

If friction is neglected, then the MA will be the same as the VR. However, if friction is considered, then,

$$\text{MA} = \frac{2\pi r_e}{p} \cdot \eta$$

where η = efficiency of the machine.

Example 9.12

The front wheel of a vehicle is punctured. A screw jack is used to raise the front portion of the vehicle. Determine the effort that must be applied at the end of the tommy bar normal to plane containing the bar and the screw with the following given data.

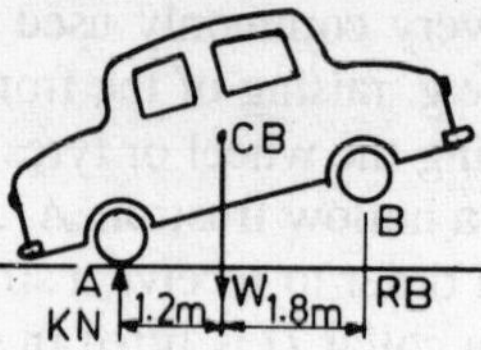

Fig. 9.59

Distance from the axis of the screw to the line of action of the effort	= 30 cm
Pitch '*P*'	= 6 mm
Distance of CG of the car from the axle of hind wheels	= 1.2 m
Distance of CG of the car from the axle of the front wheel	= 1.8 m
Weight of the vehicle	= 1 kN
Efficiency of the screw jack	= 30 per cent

Solution

The load being raised by the screw jack will be equal to the reaction R_B offered by it.

Taking moments of the forces about A,

For equilibrium, $M = 0$

$1000 \times 1.2 - R_B\ 3 = 0$

Load on the screw jack (R_B) $= \dfrac{1200}{3}$

$= 400$ N

$$MA = \frac{2 \pi r_e}{p}$$

$$= \frac{2 \pi \times 300}{6} \times 0.3$$

$$= 30$$

We know that MA $= \dfrac{W}{P'}$

$$\therefore \quad P' = \frac{W}{MA}$$

$$= \frac{400}{30} = \mathbf{4.24\ N}$$

9.4 Equations for a Machine Expressing Relationship between the Motive Force and the Resistance Overcome

In the preceding articles, we have found out expressions for the velocity ratio and mechanical advantage of various simple machines. In arriving at these expressions, it was always assumed that the machine is ideal, i.e. there is no friction between the contact surfaces of the moving parts or the parts which tend to move. Secondly, the effect of body forces and ropes not being taut completely were neglected.

Now when all the above factors present in an actual machine are considered, we find that the expression for the velocity ratio remains the same. This is so, because VR is the ratio between the displacement of the effort, and that of the load or the resistance overcome, and is not affected by any of the above factors. When we think of the Mechanical advantage of a machine then the expression, already worked out for each machine does not hold good. Really speaking, it is not possible to find the Mechanical advantage of any machine analytically. So the approach that is followed in determining the mechanical advantage of any machine is by finding out the relationship between the motive force (effort) and the resistance overcome (load) by a machine experimentally. Let us take, for example, the case of a screw jack. In order to find out the relationship between the effort and the load, an experiment can be done on this by replacing the tommy bar of the screw jack with a wheel. A cord is wound round this wheel and passed over a pulley for applying the motive force (Figure 9.60).

The effort *P*, required for raising load *W* of varying magnitudes is determined. Figure 9.61 shows a plot between the effort *P* required for different values of *W*. In most of the machines, it is generally possible to represent the relationship between the effort *P* and the load *W*.

The equation of a straight line given below relates the effort *P* and the load *W*. This equation is sometimes called the *law of the machine*, and is very useful for finding the effort required for raising load of any given magnitude.

$$P = m\,W + C \tag{i}$$

where m = slope of the line

C = intercept of the line on the P axis

From any two sets of values of P and W from the graph, it is possible to find the values of the constants m and c; e.g. let us say that for $W_1 = 1000$, $P_1 = 45$ and for $W_2 = 400$, $P_2 = 21$.

Substituting these values in equation (i),

we get $45 = m\,1000 + C$

$21 = m\;400 + C$

Solving these equations, the values of m and C can be found as $\frac{1}{25}$ and 5 respectively.

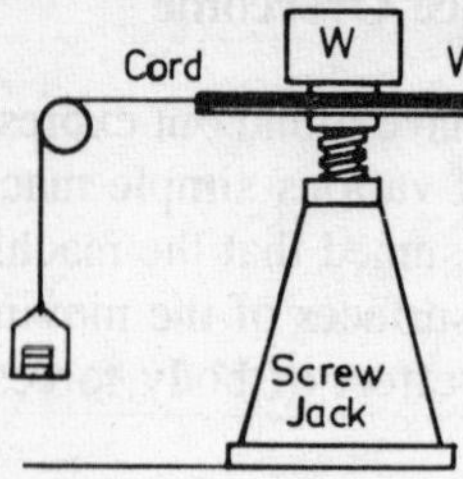

Fig. 9.60

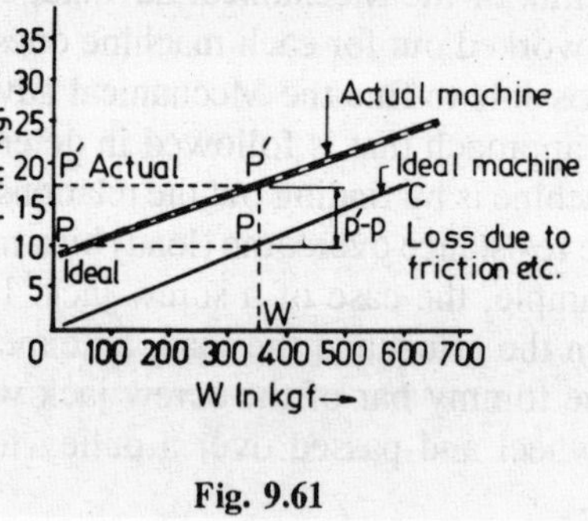

Fig. 9.61

∴ The equation of the machine then becomes

$$P = \frac{1}{25} W + 5.$$

Considering the machine to be ideal, relationship between *P* and *W* can be represented by the straight line OC. For any load *W*, the effort required will be '*P*' as shown in Figure 9.61. However, in an actual machine, the effort required would be *P'* as shown in the figure. The difference *P'* – *P* is the excess effort required to overcome friction and other body forces.

Let us now look to the relationship between the load lifted and the efficiency of the machine. We know that,

$$\text{Efficiency} = \frac{\text{Work output}}{\text{Work input}}$$

$$= \frac{W \times \text{Displacement of load } W}{P \times \text{Displacement of effort } P} \qquad \text{(ii)}$$

$$= \frac{\text{MA}}{\text{VR}}$$

When the machine is ideal, we know that MA is equal to VR, and the efficiency, therefore, is constant and does not vary with the variation in the load W.

In the case of an actual machine,

We know from (i) that $P = mW + C$

Substituting the value of P in (ii),

$$= \frac{W}{mW + C} \cdot \frac{1}{VR} \quad \text{(iii)}$$

where VR the velocity ratio.

$$\text{Simplifying} = \frac{1}{VR\left(m + \frac{C}{W}\right)} \quad \text{(iv)}$$

By increasing W, the value of $\frac{C}{W}$ will decrease, which will, in turn, increase the efficiency. In an actual machine, therefore, the efficiency is not a constant, but increases with the increase in the load. The maximum or the limiting value of efficiency of a machine can be $\frac{1}{m\,(VR)}$ when $\frac{C}{W}$ approaches zero. The relationship between load W and efficiency is shown in Figure 9.62.

The mechanical advantage of the machine

$$= \frac{\text{Load}}{\text{Effort}} = \frac{W}{P}$$

$$= \frac{W}{mW + C} \quad (\because P = mW + C)$$

$$= \frac{1}{m + \frac{C}{W}}$$

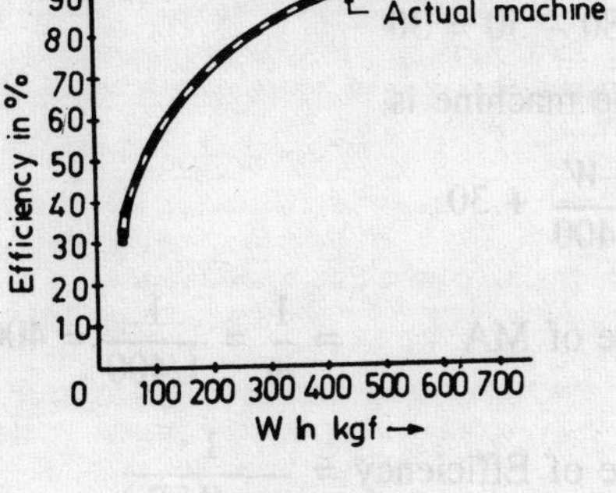

Fig. 9.62

As the value of W increases, the MA of the machine also increases. The maximum value of MA for a machine can be $\frac{1}{m}$ when $\frac{C}{w}$ approaches zero. This is also called the limiting value of MA of a machine.

Example 9.13

In an experimental test conducted on a hoisting machine, it was observed that an effort of 80 N was applied to lift a load of 20 kN, whereas an effort of 60 N was required to lift a load of 12 kN. The velocity ratio of the machine was 500. Determine the following:

(i) Law of the machine
(ii) The limiting mechanical advantage
(iii) The limiting efficiency
(iv) The effort required for a load of 18 kN, as also the mechanical advantage and efficiency at this moment.

Solution

(i) The Law of machine is

$$P = mW + C$$

Substituting the values from the test results, we get,

$$80 = m \times 20{,}000 + C \quad (1)$$

$$60 = m \times 12{,}000 + C \quad (2)$$

Solving (1) and (2), we get

$$m = \frac{1}{400}$$

Substituting the value of m in (1), we get

$$80 = \frac{1}{400} \times 20{,}000 + C$$

$$80 = 50 + C$$

$$\therefore \quad C = 80 - 50 = 30$$

The law of the machine is

$$P = \frac{W}{400} + 30$$

(ii) Limiting value of MA $= \frac{1}{m} = \frac{1}{1/400} = 400$

(iii) Limiting value of Efficiency $= \frac{1}{m \times (\text{V.R.})}$

$$= \frac{1}{\frac{1}{400} \times 500} = \frac{400}{500}$$

$$= 80 \text{ per cent}$$

(iv) We know that

$$P = \frac{W}{400} + 30$$

For $W = 18{,}000$ N

$$P = \frac{18{,}000}{400} + 30$$

$$= 45 + 30$$

$$= 75 \text{ N}$$

$$\text{MA} = \frac{W}{P}$$

$$= \frac{18{,}000}{75}$$

$$= 240$$

$$\text{Efficiency} = \frac{\text{MA}}{\text{VR}}$$

$$= \frac{240}{500}$$

$$= 48 \text{ per cent.}$$

Summary

Machine It is a device which modifies force or motion in magnitude or direction, or both in magnitude and direction. It receives energy from some outside source and transforms it into another form or does some useful work.

Load It is the useful resistance overcome or the weight '*W*' lifted by a machine.

Effort It is the force applied to the machine and is also called the input force '*P*'.

Displacement and Velocity Ratio It is the ratio between the distance moved by the effort (S_i) and the load (S_o)

$$\text{VR} = \text{DR} = \frac{S_i}{S_o}$$

Mechanical Advantage It is the ratio between the load W and the effort P, and is expressed as

$$MA = \frac{W}{P}$$

For an ideal machine, i.e. when friction between the contact surfaces is neglected, and no energy is wasted, the mechanical advantage is equal to the velocity ratio.

Input Work It is the work done on the machine and is equal to $P \times S_i$.

Output Work It is the useful work done by the machine and equal to $W \times S_o$.

Efficiency It is the ratio between the output and input work.

$$= \frac{\text{Output work}}{\textit{Input work}} = \frac{W \times S_o}{P \times S_i} = \frac{MA}{VR}$$

In case the friction between the contact surfaces is neglected, efficiency of the machine is 100 per cent. However, in actual machines, friction is always present and the efficiency is never 100 per cent. It is always less than this.

General Law of the Machine Input work = Output work + Work done in overcoming friction.

Equation Relating Effort and Load on an Actual Machine Actual Mechanical advantage of a machine is mostly found experimentally. A straight line relation exists between the effort P and load W which is expressed as under:

$$P = mW + C$$

where m = slope of the line

C = intercept on the P–axis.

Review Questions

1 What is a machine?
2 Differentiate between an ideal and an actual machine.
3 Why is the output work always less than the input work in an actual machine?
4 What is velocity ratio of a machine? How is it different from mechanical advantage.
5 State reasons for the Mechanical Advantage to be equal to the velocity ratio in the case of an ideal machine.
6 Does the friction between the contact surfaces affect the velocity ratio or mechanical advantage? Explain.
7 In a smooth inclined plane, what should be the direction of application of the effort, so that its magnitude is a minimum for raising any load?
8 In what way the inclination of an inclined plane with the horizontal affects its mechanical advantage?
9 How is it possible to achieve a gain in force by the use of a machine? Explain.
10 How does the angle of a wedge affects its mechanical advantage?
11 Name any three situations, where the use of a wedge is made?
12 Will the resistance to penetration increase or decrease with the decrease of the wedge angle?

13 Will the mechanical advantage of a lever decrease or increase when the effort *P* is applied at an angle to the lever rather than in a direction normal to the lever?

14 What is the mechanical advantage of a compound lever, if one lever gives a MA of 2 and the other 3?

15 Differentiate between a block and a tackle.

16 What do you mean by a double or a triple block?

17 What do you mean by a 3/2 lifting tackle?

18 To which block (movable or fixed) will the standing end of the rope be tied in a 2/2 lifting tackle?

19 If friction is neglected, what is the mechanical advantage of a 2/1 tackle?

20 What are the advantages of a differential pulley block and a differential windlass?

Problems

9.1 Determine effort *P* required to move a block of mass 300 kgs up an inclined plane. The inclined plane makes an angle of 30° with the horizontal. The effort *P* is applied parallel to the inclined plane and the coefficient of friction is 0.1.
(Ans. 1724 N)

9.2 Two blocks of mass 50 kgs and 70 kgs are resting on inclined planes making an angle of α_1 and α_2 with the horizontal as shown in Figure Prob. 9.2. The blocks are joined together by means of a chord which passes over a smooth pulley. If α_1 = 30° and the blocks are in equilibrium, determine α_2.

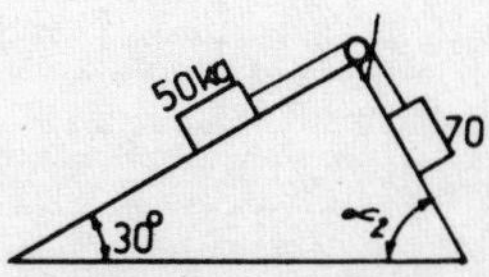

Fig. Prob. 9.2

(Ans. 20.9°)

9.3 Refer to Fig. 9.3 and determine the force *P* required to balance a weight *W* of 8 kN.
(If AB = 200 mm, BD = 800 mm, BC = 200 mm
EG = 199 mm, GF = 201 mm)

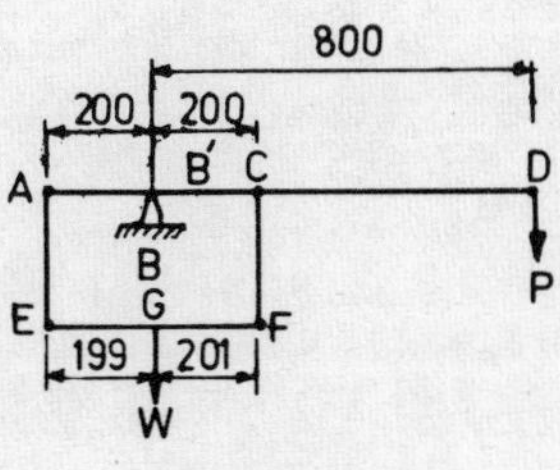

Fig. Prob. 9.3

(Ans. 10 N)

9.4 Determine the effort *P* required to raise a block of mass 500 kgs with the help of 3/2 tackle. The efficiency of the tackle is 40 per cent.
(Ans. 2.45 kN)

9.5 Determine the mechanical advantage of a differential pulley block, when the diameter of the sheaves of the upper block are
D = 300 mm, d = 280 mm
(Ans. 30)

9.6 Calculate the effort required to raise a load of 2 kN with the help of a differential pulley block. The efficiency of the machine is 35 per cent. The diameter of the sheaves of the upper block are

D = 200 mm L = 180 mm

(Ans. 285.7 N)

9.7 A differential wheel and axle is used to lift a load of 2.5 kN. The diameters of the larger and smaller portions of the drum are 200 and 150 mm respectively. The length of the handle is 400 mm. Determine the velocity ratio and the mechanical advantage if an effort of 80 N is required to lift the load. Also determine the efficiency of the machine.

VR = 32

MA = 51.25

η = 97.6 per cent

9.8 In a double purchase winch crab, the number of teeth in the pinions are 20 and 30, whereas the number of teeth of the spur wheels are 40 and 50 respectively. If the diameter of the barrel is 200 mm, determine the effort required at the end of the handle to lift a load of 1.5 kN. The effort handle is 35 cm along and the efficiency of the machine is 35 per cent.

(Ans. 368 N).

9.9 The threads of a screw jack have a pitch of 10 mm. Determine the effort *P* required to raise a load of 21 kN. The efficiency of the screw is 40 per cent. The effort is applied to the end of the tommy bar at a distance of 300 mm from the centre of the screw.

(Ans. 26.5 N).

APPENDICES

APPENDIX-1

Introduction to SI Units

1.1 Introduction

SI is the abbreviation for 'Systemé International D' Unites' in French and also in many other languages. It is the purest form of metric system, and an extension and refinement of the traditional metric system. The SI derives all the quantities needed in all technologies from only six basic and arbitrarily defined units. This contrasts with the traditional metric system now being used in which additional quantities, such as calorie and horsepower, are arbitrarily, and indeed differently, defined in different countries. In SI the main departure from the traditional metric system is the use of 'newton' (N) as the unit of force. In other respects, however, the use of SI units, instead of technical metric units, has little effect on every day life. The metre and the kilogram still remain the units of linear measure and mass; and the litre, internationally accepted for most purposes as a special name for cubic decimetre, is commonly used as a unit of volume. The SI system possesses features that make it logically superior to any other system, and also more convenient as it is coherent, rational and comprehensive.

A system of units is coherent, if the product or quotient of any two unit quantities in the system is the unit of the resultant quantiy–for example, in any coherent system, unit area results when unit length is multiplied by unit length; unit velocity when unit length is divided by unit time; and unit force when the unit mass is multiplied by unit acceleration. Whatever be the system of units, and whether it be coherent or noncoherent, the magnitudes of some physical quantities must be arbitrarily selected and declared have unit value. These magnitudes form a set of standards and are called 'basic units'. All other units are derived units related to the basic units by definitions.

1.2 Definitions of Basic SI Units

The six basic units of SI with their symbols are listed in Table 1 and defined below:—

(i) *Unit of Length—Metre (m)*
The metre is the length equal to 1650763.73 wavelengths in vacuum of the radiation corresponding to the transition between the levels $2P_{10}$ and $5d_5$ of krypton-86 atom.

(ii) *Unit of Mass—Kilogram (kg)*
The kilogram is the unit of mass and is equal to the mass of the international prototype of the kilogram.

(iii) *Unit of time—Second (s)*

The second is the duration of 9 192 631 770 periods of the radiation corresponding to the transition between the two hyperfine levels of the ground state of the caesium-133 atom.

(iv) *Unit of electric current—Ampere (A)*

The ampere is that constant current which, if maintained in two straight parallel conductors of infinite length, of negligible circular crossection, and placed one metre apart in vacuum, would produce between these conductors a force equal to 2×10^{-7} newton per metre length.

(v) *Unit of thermodynamic temperature—Kelvin (K)*

The kelvin, unit of thermodynamic temperature, is the fraction 1/273.16 of the thermodynamic temperature of the tripple point of water.

Note 1 The kelvin may also be used for expressing a temperature interval.

Note 2 The degree Celsius (°C) is a unit of the International Practical Temperature Scale on which the thermodynamic temperature of the zero point is 273.16 K.

Note 3 The degree Celsius is equal to kelvin (1°C = 1 K).

Note 4 The degree Celsius may also be used for expressing a temperature interval.

(vi) *Unit of luminous intensity—Candela (cd)*

The candela is the luminous intensity, in the perpendicular direction, of a surface of 1/600 000 square metre of a black body at the temperature of freezing platinum under a pressure of 101 325 newtons per square metre.

Table 1 Basic SI Units

S. No.	*Physical quantity*	*Name of SI Unit*	*Symbol*
1	Length	Metre	m
2	Mass	Kilogram	kg
3	Time	Second	s
4	Electric current	Ampere	A
5	Thermodynamic temperature	Kelvin	K
6	Luminous intensity	Candela	cd

1.3 Derived Units

The expressions for the derived units in SI are stated in terms of basic units. Some derived units having special names are given in Table 2. Some other examples of derived units are given in Table 3.

Table 2 Some Derived SI Units Having Special Names

S. No.	*Physical quantity*	*Name of SI Unit*	*Symbol*
1	Frequency	Hertz	Hz = 1 c/s
2	Force	Newton	N = kg.m/s^2
3	Work, energy, quantity of heat	Joule	J = Nm
4	Power	Watt	W = J/s
5	Electric charge	Coulomb	C = As
6	Electric potential	Volt	V = W/A
7	Electric capacitance	Farad	F = As/V
8	Electric resistance	Ohm	Ω = V/A
9	Magnetic flux	Weber	Wb = Vs
10	Magnetic flux density	Tesla	T = Wb/m^2
11	Inductance	Henry	H = Vs/A
12	Luminous flux	Lumen	lm = cd sr
13	Illumination	Lux	lx = lm/m^2

Note: It may sometimes be advantageous to express derived units in terms of other derived units having special names, that is, SI unit of electric dipole moment (Asm) is usually expressed as (Cm).

Table 3 Some Derived SI Units with Complex Names

S. No.	*Physical quantity*	*Name of SI Unit*	*Symbol*
1	Area	Square metre	m^2
2	Volume	Cubic metre	m^3
3	Density (mass density)	Kilogram per cubic metre	kg/m^3
4	Velocity	Metre per second	m/s
5	Angular velocity	Radian per second	rad/s
6	Acceleration	Metre per second squared	m/s^2
7	Angular acceleration	Radian per second squared	rad/s^2
8	Pressure (stress)	Newton per square metre	N/m^2
9	Dynamic viscosity	Newton second per square metre	Ns/m^2
10	Kinematic viscosity	Square metre per second	m^2/s
11	Electric field strength	Volt per metre	V/m
12	Magnetic field strength	Ampere per metre	A/m
13	Luminance	Candela per square metre	cd/m^2
14	Wavenumber	1 per metre	m^{-1}
15	Entropy	Joule per kelvin	J/K
16	Specific heat	Joule per kilogram kelvin	J/kg. K
17	Thermal conductivity	Watt per metre kelvin	W/mK
18	Electric intensity	Watt per steradian	W/sr
19	Activity (of radioactive source)	1 per second	s^{-2}
20	Permittivity	Farad per metre	F/m
21	Magnetomotive force	Ampere	A
22	Permeability	Henry per metre	H/M

1.4 Decimal Multiples and Submultiples of Units

Decimal multiples and submultiples of SI units are formed by means of Prefixes as given below:

Factor by which the Unit is Multiplied	*Prefix*	*Symbol*
1 000 000 000 000 = 10^{12}	tera	T
1 000 000 000 = 10^{9}	giga	G
1 000 000 = 10^{6}	mega	M
1 000 = 10^{3}	kilo	k
100 = 10^{2}	hecto	h
10 = 10^{1}	deca	da
0.1 = 10^{-1}	deci	d
0.01 = 10^{-2}	centi	c
0.001 = 10^{-3}	milli	m
0.000 001 = 10^{-6}	micro	μ
0.000 000 000 = 10^{-9}	nano	n
0.000 000 000 000 = 10^{-12}	pico	p
0.000 000 000 000 000 = 10^{-15}	*femto	f
0.000 000 000 000 000 000 = 10^{-18}	*atto	a

Note The prefixes hecto, deca, deci and centi should be avoided as far as possible. SI units must be expressed in "Preferred standard form" in which the multiplier is 10^{3n} where n is a positive or negative integer.

1.5 Rules for Writing Symbols of Units

1.5.1 *International Symbols for Units*—In cases where international symbols for units exist, these are to be used. They are printed in Roman upright type, remain unaltered in the plural, are written without final full-stop (period) and placed after the complete numerical value in the expression for a quantity.

1.5.2 It is a general rule that they are printed in lower case letters except when the name of the unit is derived from a proper name.

Examples : m metre
g gram
s second
A Ampere
Wb weber.

1.5.3 *Combination of Symbols for Units*—When a combined unit is formed by multiplication of two or more units, this may be indicated in one of following ways:

$$\text{Nm, N m, N.m, N} \times \text{m}$$

Note: When using a prefix a symbol which coincides with the symbol of unit, special care should be taken to avoid confusion. The unit newton metre for torque should be written, for example, Nm or m.N, to avoid confusion with mN, the millinewton.

1.5.4 When a combined unit is formed by dividing one unit by another this may be indicated in one of the following ways:

$\frac{m}{s}$, m/s or by writing the product of m and s^{-1} that is ms^{-1}.

In no case, should more than one solidus sign (/) on the same line be included in such a combination unless parentheses be inserted to avoid all ambiguity. In complicated cases negative powers or parentheses should be used.

1.5.5 *Abbreviations for Names of Units*—In cases where no internationally adopted symbol exists for a unit, and where abbreviations for the names of units exists, these abbreviations are printed in Roman upright type. They are in some cases followed by a full stop.

1.5.6 *Prefixes Indicating Decimal Multiples and Sub-multiples of Units*—Symbols for prefixes are printed in Roman upright type without space between the prefix and the symbol for the unit. Compound prefixes are avoided, for example, we write nm (nanometre) instead of mμm.

APPENDIX-2

Second Moment of Areas (Moment of Inertia) of Plane Sections

Moment of Inertia of a body about any axis is measured as the Summation of Products of elementary mass (δm_1) and square of its, distance from the axis (Fig. A1), i.e.

$$M.I. = \Sigma\ (\delta m_1\ d_1^2)$$

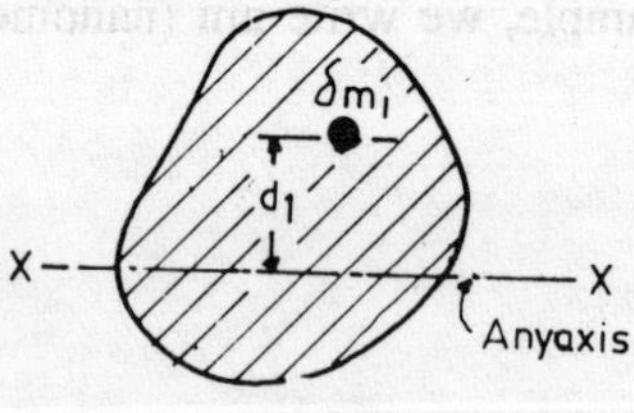

Fig. A1

In case of plane surface, the mass of body is directly proportional to the avea of plane surface (of homogeneous and uniform thickness), and hence, the moment of Inertia in this case is referred to as *second moment of area.* Thus, second moment of area of any plane surface about any axis is measured as summation of Products of elementary areas (δ_A) and square of their centroidal distances from the axis (Fig. A2) i.e.

$$I_{xx} = \Sigma\ (\delta_A\ d^2)$$

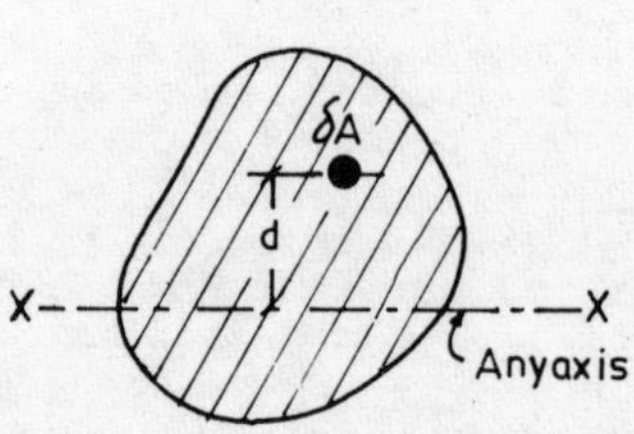

Fig. A2

2.1 Parallel Axis Theorem

Second moment of area of any plane surface of area 'A' about any axis

MM parallel to centroidal axis $M_c M_c$ is the sum of second moment of area about $M_c M_c$ and product of area and square of distance '*d*' between the axies, i.e.

$$I_{MM} = I_{MC} + AD^2,$$

where d = distance between the axis *MM* and centroidal axis $M_c M_c$; A = area of the surface. Similarly,

$$I_{xx} = I_{xc} + AC_y^2$$

$$I_{yy} = I_{xc} + AC_x^2$$

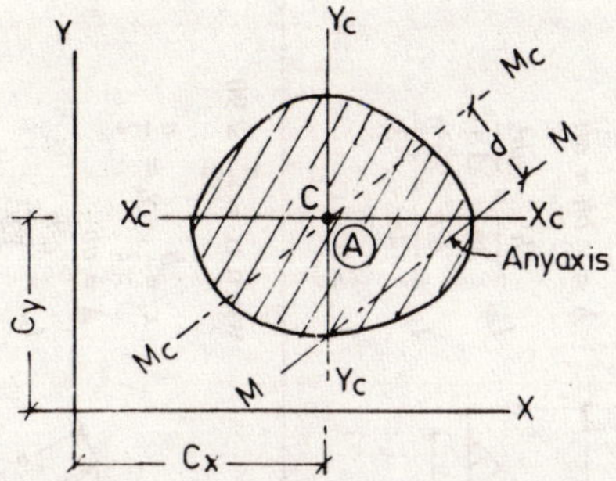

Fig. A3

2.2 Polar Axis Theorm

Second moment of area (polar moment of inertia) about its centroidal Polar axis at right angles to two principal centroidal plane axes of any plane area is equal to sum of second moment of area about its two principal plane centroidal axes i.e.

$$I_p = I_{xc} + I_{yc}$$

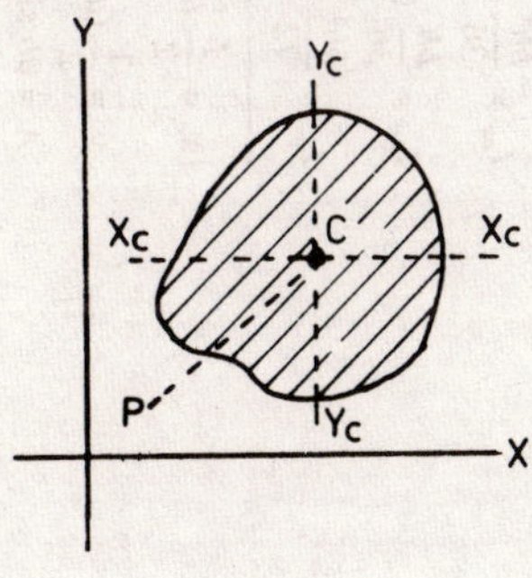

Fig. A4

Table Appendix II : *Second Moment of Areas of Plane Sections*

Position of Centroid, Area and Second Moment of Area	*Shape*	*Specific Cases Centroid, Area, and Second Moment of area*	*Specific Shape*
(1)	(2)	(3)	(4)
$x_c = \frac{1}{3}(a+b),\ y_c = \frac{1}{3}h$ $A = \frac{1}{2}bh$ $I_{xxc} = \frac{bh^3}{36}$ $I_{yyc} = \frac{bh}{36}(b^2 - ab + a^2)$ $I_{xyc} = \frac{bh^2}{72}(2a - b)$	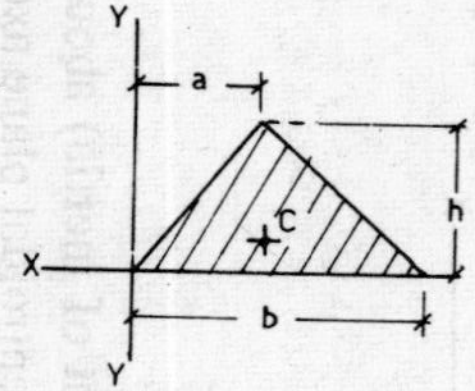 Triangle	when $\beta = 2a$ $x_c = a,\ y_c = \frac{h}{3}$ $A = \frac{1}{2}bh = ah$ $I_{xxc} = \frac{bh^3}{36} = \frac{ah^3}{18}$ $I_{yyc} = \frac{b^3h}{48} = \frac{a^3h}{3}$ $I_{xyc} = 0$	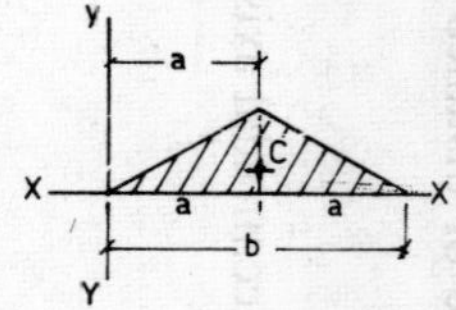 Symmetrical Triangle
$x_c = \frac{1}{2}(b + a\cos\alpha)$ $y_c = \frac{1}{2}(a\sin\alpha)$ $A = ab\sin\alpha$ $I_{xxc} = \frac{a^3b}{12}\sin^3\alpha$ $I_{yyc} = \frac{ab}{12}\sin\alpha\,(b^2 + a^2\cos^2\alpha)$ $I_{xyc} = \frac{a^3b}{12}(\sin^2\alpha\cos\alpha)$	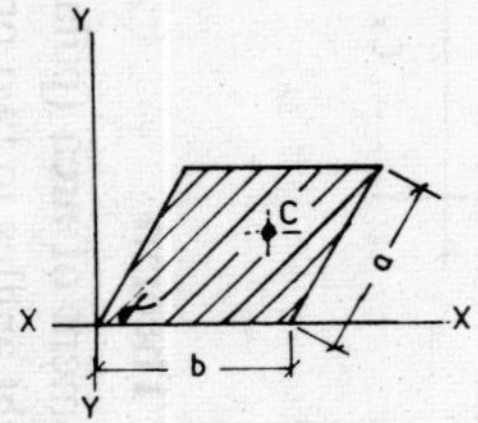 Parallelogram	when $\alpha = \frac{\pi}{2} = 90°$ $x_c = \frac{b}{2},\ y_c = \frac{a}{2}$ $A = ab$ $I_{xxc} = \frac{a^3b}{12}$ $I_{yyc} = \frac{ab^3}{12}$ $I_{xyc} = 0$	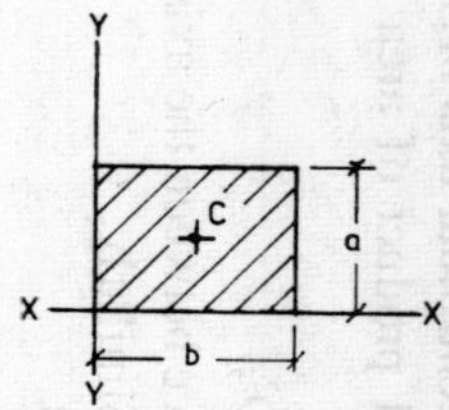 Rectangle

(1)	(2)	(3)	(4)
$x_c = \frac{2r}{3} \sin\alpha,\ y_c = 0$ $A = r^2 \alpha$ $I_{xxc} = \frac{r^4}{4} (\alpha - \frac{\sin 2\alpha)}{2}$ $I_{yyc} = \frac{r^4}{4} (\alpha + \frac{\sin 2\alpha)}{2}$ $I_{xyc} = 0$	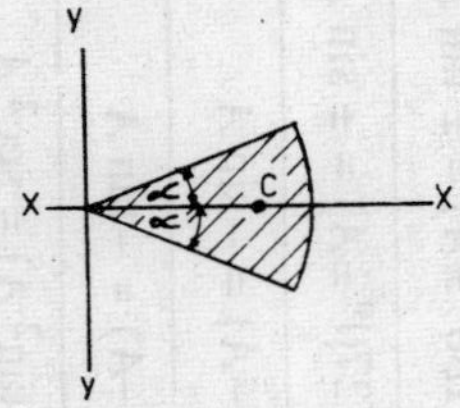 Sector of a circle	when $\alpha = \pi = 180$ $x_c = 0,\ y_c = 0$ $A = \pi r^2 = \frac{\pi (d^2)}{4}$ $I_{xxc} = \frac{\pi r^4}{4} = \frac{\pi (d^4)}{64}$ $I_{yyc} = \frac{\pi r^4}{4} = \frac{\pi (d^4)}{64}$ $I_{xyc} = 0$	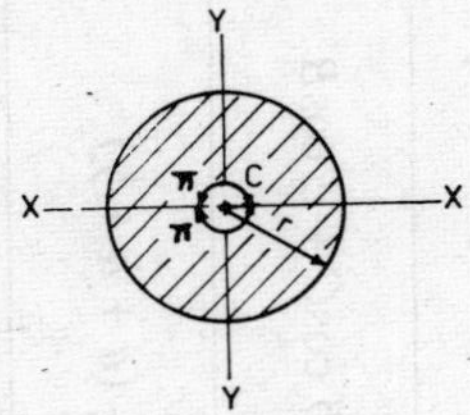 Circle
$x_c = \frac{2r \sin^3\alpha}{3\left(\alpha - \frac{\sin 2\alpha}{2}\right)}$ $y_c = 0$ $A = r^2 \left(\alpha - \frac{\sin 2\alpha}{2}\right)$ $I_{xxc} = \frac{r^4}{4}\left(\alpha - \frac{\sin 2\alpha}{2}\right)\left[1 - \frac{2(\sin^3\alpha \cos\alpha)}{3\left(\alpha - \frac{\sin 2\alpha}{2}\right)}\right]$ $I_{yyc} = \frac{r^4}{4}\left(\alpha - \frac{\sin 2\alpha}{2}\right)\left[1 + \frac{2(\sin^3\alpha \cos\alpha)}{\left(\alpha + \frac{\sin 2\alpha}{2}\right)}\right]$	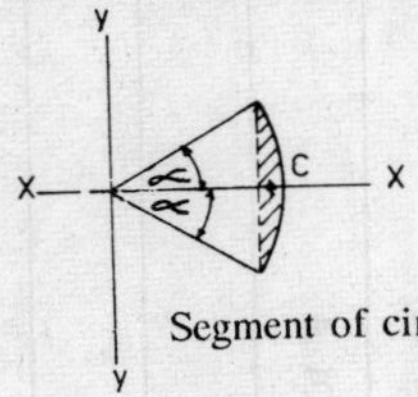 Segment of circle	when $\alpha = \pi/2 = 90°$ $x_c = \frac{4r}{3\pi},\ y_c = 0$ $A = \frac{\pi r^2}{2}$ $I_{xxc} = \frac{\pi r^4}{8}$ $I_{yyc} = \frac{\pi r^4}{8}$ $I_{xyc} = 0$	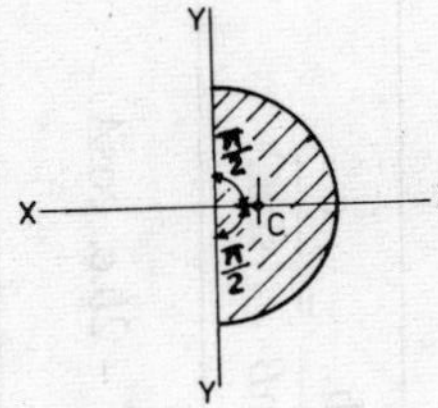 — Semicircle

Appendix III : *Properties of Triangles*

a, b, c are lengths of sides opposite angle A, B, C of triangle

$\frac{a}{\sin A} = \frac{b}{\sin B} = \frac{c}{\sin C}$	$a = b\cos C + c\cos B$
$a^2 = b^2 + \chi^2 - 2b.c.\cos A.$	$s = \frac{1}{2}(a + b + \chi)$
$\sin A = \frac{2}{bc}\sqrt{s(s-a)(s-b)(s-c)}$	$\text{Area} = \frac{1}{2}a.\ b.\ \sin c = \sqrt{s(s-a)(s-b)(s-c)}$

Appendix IV : *Plane Trigonometry (Angle 'A'* is less than 90°)

$\sin(90° \pm A) = +\cos A$	$\cos(90° \pm A) = \mp \sin A$
$\tan(90° \pm A) = \mp \cot A$	$\sin(180° \pm A) = \mp \sin A$
$\cos(180° \pm A) = -\cos A$	$\tan(180° \pm A) = \pm \tan A$
$\sin(270° \pm A) = -\cos A$	$\cos(270° \pm A) = \pm \sin A$
$\tan(270° \pm A) = \mp \cot A$	$\sin(-A) = -\sin A$
$\cos(-A) = +\cos A$	$\tan(-A) = -\tan A$
$\sin^2 A + \cos^2 A = 1$	$(1 + \tan^2 A) = \sec^2 A$
$1 + \cot^2 A = \operatorname{cosec}^2 A$	$\sin(A \pm B) = \sin A\cos B \pm \cos A\sin B$
$\cos(A \pm B) = \cos A\cos B \pm \sin A\sin B$	$\tan(A \pm B) = \frac{\tan A \pm \tan B}{1 \pm \tan A\tan B}$
$\sin^2 A - \sin^2 B = \sin(A + B)\sin(A - B)$	$\sin 2A = 2\sin A\cos A$
$\cos 2A = \cos^2 A - \sin^2 A = 1 - 2\sin^2 A = 2\cos^2 A - 1$	$\tan 2A = \frac{2\tan A}{1 - \tan^2 A}$
$\sin\frac{A}{2} = \sqrt{\{\frac{1}{2}(1 - \cos A)\}}$	$\cos\frac{A}{2} = \sqrt{\{\frac{1}{2}(1 + \cos A)\}}$
$\tan\frac{A}{2} = \frac{\sin A}{1 + \cos A}$	$\sin A + \sin B = 2\sin\frac{A+B}{2}\cos\frac{A-B}{2}$

$\sin A - \sin B = 2 \cos \frac{A+B}{2} \sin \frac{A-B}{2}$ $\cos A + \cos B = 2 \cos \frac{A+B}{2} \cos \frac{A-B}{2}$

$\cos A - \cos B = 2 \sin \frac{A+B}{2} \sin \frac{B-A}{2}$ 1 radian $= \frac{180°}{\pi} = 57°17'44''$

For very small angles θ in radians

$\sin \theta = \theta \frac{\theta^3}{\lfloor 3} + \frac{\theta^5}{\lfloor 5} ......$ $\cos \theta = 1 \frac{\theta^2}{\lfloor 2} + \frac{\theta^4}{\lfloor 4} ......$

Appendix V : *Important Differential Coefficients*

$\frac{d}{dx}(x^n) = n\, x^{n-1}$ $\frac{d}{dx}(a.\, u) = a\frac{du}{dx}$

$\frac{d}{dx}(a) = 0$ $\frac{d}{dx}(\log_e x) = \frac{1}{x} \log_a^e$

$\frac{d}{dx}(\log_a^u) = (\frac{1}{u} \log_a^e) \frac{du}{dx}$ $\frac{d}{dx}(a^x) = a^x \log_e^a$

$\frac{d}{dx}(a^v) = (a^v \log_e^a) \frac{dv}{dx}$ $\frac{d}{dx}(e^x) = e^x$

$\frac{d}{dx}(e^{ax}) = ae^{ax}$ $\frac{d}{dx}(\sin x) = \cos x$

$\frac{d}{dx}(\cos x) = -\sin x$ $\frac{d}{dx}(\tan x) = \sec^2 x$

$\frac{d}{dx}(\sec x) = \sec x.\tan x$ $\frac{d}{dx}(\text{cosec}\, x) = -\text{cosec}\, x.\cot x.$

$\frac{d}{dx}(\cot x) = -\text{cosec}^2 x$ $\frac{d}{dx}(y) = \frac{1}{\frac{dx}{dy}}$

$\frac{d}{dx}(y) = \frac{dy}{du} \cdot \frac{du}{dv} \cdot \frac{dv}{dx}$ $\frac{d}{dx}(u + v - w) \frac{du}{dx} + \frac{dv}{dx} + \frac{dw}{dx}$

$\frac{d}{dx}(u.v) = u\frac{dv}{dx} + v\frac{du}{dx}$ $\frac{d}{dx}\left(\frac{u}{v}\right) = \left(\frac{v.\frac{du}{dx} - u\frac{dv}{dx}}{v^2}\right)$

$\frac{d}{dx}(\cos u) = -\sin u \cdot \frac{du}{dx}$	$\frac{d}{dx}(u.\ v.\ w) = \frac{v.w dv}{dx} + \frac{uw dv}{dx} + \frac{uv.dw}{dx} +$
$\frac{d}{dx}(\sin u) = -\cos u \cdot \frac{du}{dx}$	$\frac{d}{dx}(\sin^{-1}u) = \frac{1}{\sqrt{1-u^2}} \cdot \frac{du}{dx}$
$\frac{d}{dx}(\cos^{-1}u) = \frac{1}{\sqrt{1-u^2}} \cdot \frac{du}{dx}$	$\frac{d}{dx}(\tan^{-1}u) = \frac{1}{1+u^2} \cdot \frac{du}{dx}$
$\frac{d}{dx}(\sec^{-1}u) = \frac{1}{u\sqrt{u^2-1}} \cdot \frac{du}{dx}$	$\frac{d}{dx}(\text{cosec}^{-1}u) = \frac{1}{u\sqrt{u^2-1}} \cdot \frac{du}{dx}$
$\frac{d}{dx}(\cot^{-1}u) = -\frac{1}{1+u^2} \frac{du}{dx}$	

INDEX